de Gruyter Series in Logic and Its Applications 3

Sy D. Friedman

# Fine Structure and Class Forcing

Walter de Gruyter
Berlin · New York 2000

*Author*

Sy D. Friedman
Institut für formale Logik
University of Vienna
Währingerstr. 25
1090 Vienna
Austria

Sy D. Friedman
Department of Mathematics
Massachusetts Institute of Technology
77 Massachusetts Avenue
Cambridge, MA 02139-4307
USA



*Mathematics Subject Classification 2000:* 03-02; 03E45, 03E15, 03E35, 03E55

*Keywords:* Fine structure, Forcing, Admissibility, Large cardinals, Descriptive set theory

♾ Printed on acid-free paper which falls within the guidelines of the ANSI to ensure permanence and durability

*Library of Congress – Cataloging-in-Publication Data*

Friedman, Sy D., 1953-
Fine structure and class forcing / Sy D. Friedman.
p. cm. – (De Gruyter series in logic and its applications ; 3)
Includes bibliographical references and index.
ISBN 3-11-016777-8 (alk. paper)
1. Forcing (Model theory). I. Title. II. Series.
QA9.7.F75 2000
511.3′22–dc21 00-023002
CIP

*Die Deutsche Bibliothek – Cataloging-in-Publication Data*

Friedman, Sy D.:
Fine structure and class forcing / Sy D. Friedman. – Berlin ; New York : de Gruyter, 2000
(De Gruyter series in logic and its applications ; 3)
ISBN 3-11-016777-8

ISSN 1438–1893



Printed in Germany.
Typesetting using the Author's TEX files: I. Zimmermann, Freiburg – Printing and binding: WB-Druck GmbH & Co., Rieden/Allgäu – Cover design: Rainer Engel, Berlin.

*For Lida and Scooty*

# Preface

This book provides a detailed analysis of the first two canonical approximations to the set-theoretic universe, given by the inner models $L$ and $L[0^\#]$. It is intended for the student who has completed one year of study in axiomatic set theory. Thus we assume familiarity with the basic principles of the axiom system ZFC and with the basic properties of Gödel's constructible universe $L$, but we assume background neither in forcing nor in fine structure theory.

$L$ has always played a central role in set theory. It is in the $L$-context that the detailed study of definability, derived from the field of higher recursion theory (Sacks [90]), can be thoroughly applied. What results is the striking fine structure theory of $L$ (Jensen [72]), which demonstrates that at least in one interpretation of set theory, a nearly complete analysis of the structure of the universe $V$ of all sets is possible.

The theory of *forcing* has taught us however that $L$ need not be a very good approximation to $V$. In the traditional forcing method we begin with a model $M$ of set theory, and choose both a partial-ordering $P \in M$ and a set $G \subseteq P$ which is "$P$-generic over $M$". Then the model $M[G]$ obtained by adjoining $G$ to $M$ is a new and larger model of the ZFC axioms. In this way, forcing suggests that $V$ should be larger than $L$, as for some partial-orderings $P$ in $L$, $P$-generic sets $G$ are likely to exist.

But for *which* $P$ in $L$ should a $P$-generic set exist? A possible criterion is for $P$ to be definable in $L$ without parameters. However, we face a more serious challenge when we turn to *class forcing*, where the partial-ordering $P$ is no longer required to be an element of $M$, but a class of $M$. Then we have:

**Fact 1.** There exist class forcings $P_0$, $P_1$ which are $L$-definable without parameters such that whenever $G_0$, $G_1$ are $P_0$, $P_1$-generic over $L$, respectively:

(a) ZFC holds in $\langle L[G_0], G_0\rangle$ and in $\langle L[G_1], G_1\rangle$.

(b) ZFC (indeed Replacement) fails in $\langle L[G_0, G_1], G_0, G_1\rangle$.

Thus we cannot have generics for all ZFC preserving class forcings which are $L$-definable without parameters. Finding a suitable generic existence criterion for $L$-definable forcings has become more difficult.

Fortunately, a natural such criterion is provided by Silver's theory of $0^\#$ (Silver [71]): We say that $L$ is *rigid* if there is no elementary embedding from $\langle L, \in\rangle$ to itself, other than the identity.

**Fact 2.** $L$ is rigid in class-generic extensions of $L$. If $L$ is not rigid then there is a smallest inner model in which $L$ is not rigid, and this inner model is $L[0^\#]$ where $0^\#$ is a real.

An $L$-definable forcing $P$ is *relevant* if there is a class which is $P$-generic over $L$ and which is definable in the inner model $L[0^\#]$. If $P_0$ and $P_1$ are relevant forcings then clearly generics for $P_0$ and for $P_1$ can coexist, as they both exist definably over $L[0^\#]$. Relevance thereby provides us with the desired generic existence criterion for forcings over $L$.

An important consequence, and indeed the major theme of this book, is that forcing now becomes not only a tool for obtaining consistency results, but also a technique for proving absolute results, in the theory ZFC $+0^\#$ exists. Thus we use the theory of relevant forcing to construct objects which actually exist (in the inner model $L[0^\#]$), rather than which *may* exist in a generic extension of the universe.

In this book we provide a thorough analysis of $L[0^\#]$ using the class forcing technique. In addition, we take a close look at the fine structure of $L$, which not only exposes its deepest combinatorial properties, but also supplies the tools necessary to construct some of the forcing partial-orderings used in our analysis of $L[0^\#]$.

Our emphasis in Chapter 1 (fine structure) is on the combinatorial properties $\diamond$, $\square$, Scale and Morass. The reader whose primary interest is with forcing rather than fine structure may choose to use this chapter solely as a reference for later chapters. Chapters 2, 3 and 4 provide the basic theory of class forcing, including a discussion of generic existence and Jensen coding. They are prerequisite to the rest of the book. What follows in the remaining four chapters, which are relatively independent of each other, are applications of the class forcing technique to resolve a number of problems concerning genericity, admissibility, descriptive set theory and set-theoretic definability. Included are solutions to the Genericity, $\Pi^1_2$-Singleton and Admissibility Spectrum problems, which are introduced in Section 4.1. We end with a list of open problems.

*Acknowledgments*. I should like to thank the Rockefeller Foundation for its support during my stay as a Resident Scholar at their center in Bellagio, during which time I wrote the first two chapters of this book. The National Science Foundation has provided me with considerable support through its research grants. I am grateful to the MIT Mathematics Department for providing me with such a congenial place to work, and especially to Bonnie Friedman and Jan Wetzel for their excellent preparation of the TEX files. In addition I thank the students of my MIT class forcing course, Elizabeth Brown, Peter Koellner, Boris Piwinger and Florian Rudolph, for their helpful comments. Lastly, I wish to acknowledge two intellectual debts; one to Gerald Sacks, who revealed to me the beauty and subtlety of the concept of definability, and the other to Ronald Jensen, the creator of both the fine structure theory and the coding method. It is my strong belief that set theory will be most powerfully understood through the use of these techniques, as they are applied to larger and larger approximations to the set-theoretic universe.

Vienna and Cambridge, MA, May 2000 *Sy D. Friedman*

# Contents

*Chapter 1*

# The $\Sigma^*$ Approach to the Fine Structure of $L$

In this chapter we reformulate the fine structure theory from Jensen [72]. We then use this reformulation to prove the $\square$ and Fine Scale Principles, and to construct Morasses. The material in this chapter will only be used in Sections 4.3, 7.2 and 8.3 of this book; for this reason, the reader may wish to begin with Chapter 2, and refer to the present chapter only when needed.

## 1.1 The $J$-Hierarchy

The most elegant hierarchy for Gödel's $L$ is obtained through iterated first-order definability. For any set $x$ let $\mathbf{Def}(x)$ denote $\{y \mid y \subseteq x, y \text{ is definable over } \langle x, \in\rangle \text{ by a first-order formula with parameters}\}$. Then $L$ is obtained as the union of all $L_\alpha$, where $L_0 = \emptyset$, $L_\lambda = \bigcup\{L_\alpha \mid \alpha < \lambda\}$, for limit $\lambda$ and:

$$L_{\alpha+1} = \mathbf{Def}(L_\alpha).$$

Unfortunately $L_{\alpha+1}$ is not closed under pairing and for this reason, Jensen [72] defined a modified hierarchy to get around this problem. We now present a description of the $J$-Hierarchy which, as above, is based on the idea of iterated definability.

Recall the Lévy hierarchy of formulas: A formula is $\Sigma_0$ ($= \Delta_0 = \Pi_0$) if it is built from atomic formulas through the use of logical connectives and bounded quantifiers $\forall x \in y$, $\exists x \in y$. A formula is $\Sigma_{n+1}$ if it is of the form $\exists \vec{x} \varphi$ where $\varphi$ is $\Pi_n$. Dually, a formula is $\Pi_{n+1}$ if it is of the form $\forall \vec{x} \varphi$ where $\varphi$ is $\Sigma_n$. Every formula is logically equivalent to a $\Sigma_n$ formula for some $n$, as it can be put into prenex normal form.

We want to define the $J$-hierarchy so that $J_{\alpha+1} \cap P(J_\alpha) = \mathbf{Def}(J_\alpha)$, $J_{\alpha+1}$ is closed under pairing and, in addition, $J_{\alpha+1}$ satisfies $\Sigma_0$-Comprehension. The latter is the statement that for any $x$ we can form $\{y \in x \mid \varphi(y)\}$, where $\varphi$ is a $\Sigma_0$ formula with arbitrary parameters. This is important for the construction of universal $\Sigma_n$ predicates, a notion that we define next.

A binary relation $W_n(e, x)$ on a transitive set $T$ is a *universal $\Sigma_n$ predicate* for $T$ if it is $\Sigma_n$-definable over $\langle T, \in\rangle$ without parameters and wherever $Y \subseteq T$ is $\Sigma_n$-definable over $\langle T, \in\rangle$ with parameters, there exists $e \in T$ such that:

$$Y = \{x \in T \mid W_n(e, x)\}.$$

Thus the sets $\{x \in T \mid W_n(e, x)\}$ are exactly the sets $\Sigma_n$-definable over $\langle T, \in\rangle$ with parameters, as $e$ varies over $T$.

**Lemma 1.1.** *Suppose that $T$ is a transitive set closed under pairing, satisfying $\Sigma_0$-Comprehension + "Every set has a transitive closure." Then there exists a universal $\Sigma_n$ predicate for $T$.*

*Proof.* It is enough to treat the case $n = 1$, as for example to get $W_2$ from $W_1$ we can just define $W_2(e, x) \iff \exists\, y \sim W_1(e, \langle x, y\rangle)$.

Let $\langle \varphi_i \mid i \in \omega\rangle$ be a recursive list of formulas with one free variable (with subformulas occuring earlier) and define $\mathrm{Sat}(z, i, x)$ to mean: $z$ is transitive, $x \in z$ and $\langle z, \in\rangle \models \varphi_i(x)$. Sat can be expressed by the following formula:

> $\mathrm{Sat}(z, i, x) \iff z$ is transitive, $x \in z$ and $\exists Y \subseteq (i+1) \times z$ such that $\{\forall j \le i$ [If $\varphi_j(x)$ is atomic then $\langle j, x\rangle \in Y \iff \varphi_j(x)$ is true; if $\varphi_j(x)$ is $\exists y\, \varphi_{j'}(\langle x, y\rangle)$ then $\langle j, x\rangle \in Y \iff \exists y \in z(\langle j', \langle x, y\rangle\rangle \in Y)$; if $\varphi_j(x)$ is $\sim \varphi_{j'}(x)$ then $\langle j, x\rangle \in Y \iff \langle j', x\rangle \notin Y$; if $\varphi_j(x)$ is $\varphi_{j_1}(x) \wedge \varphi_{j_2}(x)$ then $\langle j, x\rangle \in Y \iff (\langle j_1, x\rangle \in Y$ and $\langle j_2, x\rangle \in Y)$] and $\langle i, x\rangle \in Y\}$.

Using the fact that $T$ satisfies pairing and $\Sigma_0$ Comprehension, the above definition shows that when restricted to $T$, Sat is $\Sigma_1$-definable over $\langle T, \in\rangle$. Finally we set:

$$W_1(e, x) \iff \text{For some } \langle i, p\rangle \text{ and some transitive } z,\ e = \langle i, p\rangle \text{ and } \mathrm{Sat}(z, i, \langle x, p\rangle).$$

Using pairing, the existence of transitive closures and the persistence of $\Sigma_1$ formulas over transitive sets, one has that $W_1$ is universal. □

We're ready to define the $J$-hierarchy. By induction on $\alpha$ we define $J_\alpha$ to satisfy the hypotheses of Lemma 1.1. Let $W_n^\alpha(e, x)$ denote the canonical universal $\Sigma_n$ predicate coming from the proof of Lemma 1.1. For $\alpha = 0$ we have $J_0 = \emptyset$ and for $\alpha = 1$ we have $J_1 = L_\omega$. For $\alpha$ limit, $J_\alpha = \cup\{J_\beta \mid \beta < \alpha\}$. Note that for limit $\alpha$ the hypotheses of Lemma 1.1 are met by $J_\alpha$, given that they are met by each $J_\beta$, $\beta < \alpha$.

Suppose that $J_\alpha$, $W_n^\alpha(e, x)$ are defined for some $\alpha > 0$ and we wish to define $J_{\alpha+1}$. An *$n$-code* is a pair $(n, e)$ where $n \in \omega$ and $e \in J_\alpha$. By induction on $n$ define:

$$X(0, e) = e,$$
$$X(n+1, e) = \{X(n, f) \mid W_{n+1}^\alpha(e, f)\}.$$

Then $J_{\alpha,n} = \{X(n, e) \mid e \in J_\alpha\}$ and $J_{\alpha+1} = \cup\{J_{\alpha,n} \mid n \in \omega\}$.

**Lemma 1.2.** (a) $n \le m \implies J_{\alpha,n} \subseteq J_{\alpha,m}$.

(b) *$J_{\alpha,n}$ is transitive.*

(c) $\mathrm{ORD}(J_{\alpha,n}) = \omega\alpha + n$.

(d) $J_{\alpha+1} \models$ *Pairing* $+\Sigma_0$-*Comprehension.*

(e) $J_{\alpha+1} \cap P(J_\alpha) = \mathbf{Def}(J_\alpha)$.

*Proof.* (a) We define a $\Sigma_1(J_\alpha)$ function $F$ such that for all $(n, e) \in \omega \times J_\alpha$ we have $X(n, e) = X(n+1, F(n, e))$. For $n = 0$, let $F(0, e) = f$ where $\{g \mid W_1^\alpha(f, g)\} = e$; then $X(0, e) = e = \{g \mid W_1^\alpha(f, g)\} = X(1, f)$. Suppose that $F(n, e)$ has been defined for all $e$. Then let $F(n+1, e) = f$ where $\{g \mid W_{n+2}^\alpha(f, g)\} = \{F(n, h) \mid W_{n+1}^\alpha(e, h)\}$; $f$ exists as the latter set is $\Sigma_{n+1}(J_\alpha)$ with parameter $e$. Finally we get $X(n+2, f) = \{X(n+1, F(n, h)) \mid W_{n+1}^\alpha(e, h)\} = \{X(n, h) \mid W_{n+1}^\alpha(e, h)\}$ by induction, and the latter set is $X(n+1, e)$.

(b) $J_{\alpha,0} = J_\alpha$ is transitive by induction on $\alpha$, and if $x \in J_{\alpha,n+1}$ then $x \subseteq J_{\alpha,n}$ and hence $x \subseteq J_{\alpha,n+1}$ by (a).

(c) Clearly $\mathrm{ORD}(J_{\alpha,n}) \leq \omega\alpha + n$ since $x \in J_{\alpha,n+1} \Longrightarrow x \subseteq J_{\alpha,n}$. By induction on $n$, define $e_{n+1}$ such that $X(n+1, e_{n+1}) = \omega\alpha + n$: For $n = 0$ we can take $e_1$ so that $\omega\alpha = \{f \mid W_1^\alpha(e_1, f)\}$. If $e_{n+1}$ is defined take $e_{n+2}$ so that $\{f \mid W_{n+2}^\alpha(e_{n+2}, f)\} = \{F(n, g) \mid W_{n+1}^\alpha(e_{n+1}, g)\} \cup \{e_{n+1}\}$, where $F$ is from the proof of (a). Then $X(n+2, e_{n+2}) = X(n+1, e_{n+1}) \cup \{X(n+1, e_{n+1})\} = \omega\alpha + n + 1$.

(d) $J_{\alpha+1}$ is closed under pairing because all 2-element subsets of $J_{\alpha,n}$ belong to $J_{\alpha,n+1}$. For $\Sigma_0$-Comprehension it suffices to show that if $X \subseteq J_{\alpha,n}$ is definable over $\langle J_{\alpha,n}, \in\rangle$ then $X$ belongs to $J_{\alpha,m}$ for some $m$. But $\{e \mid X(n, e) \in X\}$ is a definable subset of $J_\alpha$. Choose $m$ so that this set is $\Sigma_m$-definable over $J_\alpha$ and using $F$ from the proof of (a), produce a $\Sigma_1(J_\alpha)$ $G$ such that for each $e$, $X(n, e) = X(m, G(e))$. Then $\{G(e) \mid X(n, e) \in X\}$ is $\Sigma_m$-definable over $J_\alpha$ and $X = \{X(m, G(e)) \mid X(n, e) \in X\}$ belongs to $J_{\alpha,m+1}$.

(e) We get $\mathbf{Def}(J_\alpha) \subseteq J_{\alpha+1}$ by (d). Conversely, if $X(n, e) \subseteq J_\alpha$ then $\{f \mid f \in X(n, e)\} = X(n, e)$ is a definable subset of $J_\alpha$, using the definition of $X(n, e)$. □

Of course now we may define $W_n^{\alpha+1}(e, x)$, using (d) of Lemma 1.2, thereby completing the definition of the $J$-hierarchy. In the future we will sometimes refer to the refined hierarchy $\langle \widetilde{J}_\alpha \mid \alpha \in \mathrm{ORD}\rangle$ defined by: $\widetilde{J}_{\omega\alpha+n} = J_{\alpha,n}$. Conveniently, $\mathrm{ORD}(\widetilde{J}_\alpha) = \alpha$ for all $\alpha$.

**Lemma 1.3.** (a) *$\langle \widetilde{J}_\alpha \mid \alpha < \lambda\rangle$ is $\Sigma_1(\widetilde{J}_\lambda)$ for limit $\lambda$, via a definition independent of $\lambda$.*

(b) *There is a $\Sigma_1(J_\alpha)$ well-ordering $<_\alpha$ of $J_\alpha$, via a definition independent of $\alpha$.*

(c) (Condensation) *If $\langle X, \in\rangle$ is $\Sigma_1$-elementary in $\langle J_\alpha, \in\rangle$ then $\langle X, \in\rangle \simeq \langle J_{\bar{\alpha}}, \in\rangle$ for some $\bar{\alpha}$.*

*Proof.* (a) We have $x = \widetilde{J}_\alpha \Longleftrightarrow$ There exists $\langle x_\beta \mid \beta < \gamma\rangle$ such that $\alpha < \gamma$ and $x = x_\alpha$, where: $x_n = L_n$ for finite $n < \gamma$, $x_\lambda = \cup\{x_\beta \mid \beta < \lambda\}$ for limit $\lambda < \gamma$ and for $\lambda + n < \gamma$, $\lambda$ limit, $x_{\lambda+n}$ is obtained from $x_\lambda$ as in the definition of $J_{\alpha,n}$ from $J_\alpha$, $\alpha > 0$. This definition works inside any $\widetilde{J}_\lambda$, $\lambda$ limit.

(b) Define well-orderings $<_\alpha$ of $J_\alpha$ as follows: $<_0 = \emptyset$, $<_1 =$ some fixed $L_\omega$-definable well-ordering of $L_\omega$; for limit $\lambda$, $x <_\lambda y \Longleftrightarrow x <_\alpha y$ for some $\alpha < \lambda$; $x <_{\alpha+1} y \Longleftrightarrow x <_\alpha y$ or for some $n$, $y \in J_{\alpha,n+1} - J_{\alpha,n}$ and either $x \in J_{\alpha,n}$ or $x \in J_{\alpha,n+1} - J_{\alpha,n}$ and ($<_\alpha$-least $e$ such that $X(n+1, e) = x$) $<_\alpha$ ($<_\alpha$-least $e$ such that $X(n+1, e) = y$). Then $<_\alpha$ is $\Sigma_1(J_\alpha)$, via a definition independent of $\alpha$.

(c) Let $\langle X, \in\rangle \simeq \langle \bar{X}, \in\rangle$ be the transitive collapse of $X$. Then $\langle \bar{X}, \in\rangle \models \Sigma_0$-Comprehension $+ \forall x \exists \beta (x \in \widetilde{J}_\beta) + \forall \beta \exists y (y = \widetilde{J}_\beta)$. But $\Sigma_0$-Comprehension gives $(\widetilde{J}_\beta)^{\bar{X}} = \widetilde{J}_\beta$ for $\beta \in \bar{X}$ so $\bar{X} = \widetilde{J}_{\omega\bar{\alpha}} = J_{\bar{\alpha}}$ where $\omega\bar{\alpha} = \mathrm{ORD}(\bar{X})$. □

## $\Sigma_1$-Skolem Functions

Condensation, as stated in Lemma 1.3(c), is a powerful tool for proving things about $L$. However, we must first provide a method for generating $\Sigma_1$-elementary submodels. Fix an ordinal $\alpha > 0$.

**Definition.** Suppose $X \subseteq J_\alpha$. The $\Sigma_1$-Hull of $X$ is the smallest $\Sigma_1$-elementary submodel of $J_\alpha$ containing $X$ as a subset. A $\Sigma_1$-Skolem function is a partial function $h : \omega \times J_\alpha \longrightarrow J_\alpha$ with $\Sigma_1$ graph such that for any $X \subseteq J_\alpha$, $\Sigma_1$-Hull of $X = \{h(n, x) \mid n \in \omega, x$ a finite sequence from $X\}$.

**Lemma 1.4.** *For any $X \subseteq J_\alpha$, the $\Sigma_1$-Hull of $X$ exists. Moreover there is a $\Sigma_1$-Skolem function for $J_\alpha$, with a $\Sigma_1$-definition independent of $\alpha$.*

*Proof.* Let $\varphi_0, \varphi_1, \ldots$ be a recursive list of formulas of 2 free variables and define $h^*(n, x) = <_\alpha$-least pair $(y, t)$ such that $x, y \in t$, $t$ is transitive and $\langle t, \in\rangle \models \varphi_n(x, y)$; if no such pair $(y, t)$ exists then $h^*(n, x)$ is undefined. Then $h(n, x) = y$ when $h^*(n, x) = (y, t)$. Any $\Sigma_1$-elementary submodel of $J_\alpha$ must be closed under $h$, and clearly for any $X \subseteq J_\alpha$, $\{h(n, x) \mid x$ a finite sequence from $X\}$ is a $\Sigma_1$-elementary submodel of $J_\alpha$. □

The key to Fine Structure Theory is to find a suitable generalization of Lemma 1.4 to higher levels of definability. We will take this up in the next section.

We close this section with an illustration of how $\Sigma_1$-hulls can be used to prove a version of Jensen's $\diamondsuit$-principle in $L$ (see Jensen [72]). Our version will include some technical conditions which are of use in our proof of Jensen's Coding Theorem (Section 4.3). Assume $V = L$ and let $\alpha$ be an infinite cardinal.

**Definition.** $C \subseteq \alpha^+$ is *closed unbounded* (CUB) if $\cup C = \alpha^+$ and $\cup(C \cap \beta) \in C$ whenever $\min C < \beta < \alpha^+$. $S \subseteq \alpha^+$ is *stationary* if $S \cap C \neq \emptyset$ for each CUB $C \subseteq \alpha^+$.

For $\mu < \alpha^+$, $\beta'(\mu)$ denotes the largest $\beta$ such that either $\beta = \mu$, or: $\mu < \beta$ and $J_\beta \models \mu$ is a cardinal greater than $\alpha$.

**Lemma 1.5.** *There exists $\langle D_\mu \mid \mu < \alpha^+\rangle$ such that $D_\mu \subseteq J_\mu$ and:*

(a) *If $D \subseteq J_{\alpha^+}$ then $\{\mu < \alpha^+ \mid D \cap J_\mu = D_\mu\}$ is stationary.*

(b) *$D_\mu$ is uniformly definable as an element of $J_{\beta'(\mu)}$, for $\mu < \alpha^+$.*

(c) *If $J_{\beta'(\mu)} \models \alpha^{++}$ exists or $\mu = \beta'(\mu)$ then $D_\mu = \emptyset$.*

*Proof.* Let $D_\mu = \emptyset$ if $J_{\beta'(\mu)} \vDash \alpha^{++}$ exists or if $\mu = \beta'(\mu)$. Otherwise let $\langle D_\mu, C_\mu \rangle$ be least in $J_{\beta'(\mu)}$ such that $C_\mu$ is CUB in $\mu$, $D_\mu \subseteq J_\mu$ and $\bar{\mu} \in C_\mu \Longrightarrow D_\mu \cap J_{\bar{\mu}} \neq D_{\bar{\mu}}$; if $\langle D_\mu, C_\mu \rangle$ does not exist then let $D_\mu = \emptyset$. We need only prove (a).

Suppose (a) fails and let $\langle D, C \rangle$ be least in $J_{\alpha^{++}}$ such that $D \subseteq J_{\alpha^+}$, $C$ is CUB in $\alpha^+$ and $\mu \in C \Longrightarrow D \cap J_\mu \neq D_\mu$. Let $\sigma < \alpha^{++}$ be least such that $\omega\sigma = \sigma$, $J_\sigma \vDash \alpha^+$ is the largest cardinal and $\langle D, C \rangle \in J_\sigma$. Let $H = \Sigma_1$-hull of $\alpha \cup \{\alpha^+\}$ in $J_\sigma$ and $\mu = H \cap \alpha^+$. Then $\langle H, \in \rangle \simeq \langle J_{\beta'}, \in \rangle$ for some $\beta'$ and since $J_{\beta'} \vDash \mu = \alpha^+$ we have $\beta' \leq \beta'(\mu)$. But now we have $\langle D_\mu, C_\mu \rangle = \langle D \cap J_\mu, C \cap \mu \rangle$ and since $\mu = \bigcup(C \cap \mu) \in C$, this is a contradiction. □

## 1.2 Fine Structure Theory

Our main goal is to develop a version of Lemma 1.4 for higher levels of definability. Specifically, we want to define the notion of $\Sigma_n^*$-formula so as to obtain:

(a) There is a universal $\Sigma_n^*$ predicate for $J_\alpha$ for each $n$.

(b) For any $X \subseteq J_\alpha$, the $\Sigma_n^*$-hull of $X$ in $J_\alpha$ exists for each $n$.

(c) There is a $\Sigma_n^*$-Skolem function for $J_\alpha$ for each $n$.

(d) Every formula is $\Sigma_n^*$ for some $n$.

What happens if we just take $\Sigma_n^* = \Sigma_n$? Then (a) holds by Lemma 1.1 and (d) is immediate. Next we demonstrate (b).

**Proposition 1.6.** *For any $X \subseteq J_\alpha$ and $n \in \omega$ there is a least $\Sigma_n$-elementary submodel of $J_\alpha$ containing $X$ as a subset.*

*Proof.* Let $M = \{y \in J_\alpha \mid$ For some $\Sigma_n$ formula $\varphi$ with parameters from $X$, $y$ is the $<_\alpha$-least solution to $\varphi$ in $J_\alpha\}$. Then $M$ is $\Sigma_n$-elementary in $J_\alpha$ since if $y_i$ is the $<_\alpha$-least solution to $\varphi_i$, $1 \leq i \leq n$ then $\langle y_1, \ldots, y_n \rangle$ is the $<_\alpha$-least solution to $\varphi_1((z)_1) \wedge \cdots \wedge \varphi_n((z)_n)$, where $(z)_i = i^{\text{th}}$ component of $z$. Suppose $X \subseteq N$, $N$ is $\Sigma_n$-elementary in $J_\alpha$ and $\varphi$ is a $\Sigma_n$ formula with parameters from $X$ with a solution in $J_\alpha$. Then $\varphi$ has a solution $y_0$ in $N$ and if $y_0$ is not the least solution then $N$ also has a solution $y_1 <_\alpha y_0$. Continuing in this way we see that in fact $N$ does contain the $<_\alpha$-least solution to $\varphi$ and hence we get $M \subseteq N$. □

What fails is property (c). The following argument is a refinement, due to Jensen, of the author's original argument using a subtle cardinal.

**Proposition 1.7.** *For some $\alpha$ there is no $\Sigma_2$ Skolem function for $J_\alpha$.*

*Proof.* Let $\kappa$ denote $\omega_1$. For each limit $\alpha < \omega_1$, $\alpha$ is the least $\beta$ such that $\widetilde{J}_{\kappa+\alpha} \models \kappa + \beta$ does not exist. If $\widetilde{J}_{\kappa+\alpha}$ has a $\Sigma_2$ Skolem function then $\alpha$ must be the unique solution in $\widetilde{J}_{\kappa+\alpha}$ to a $\Sigma_2$ formula $\exists x \forall y \varphi_\alpha$, $\varphi_\alpha$ $\Sigma_0$ with parameter $\kappa$. Suppose that each $\widetilde{J}_{\kappa+\alpha}$ has a $\Sigma_2$ Skolem function and by Fodor's Theorem choose $\varphi$ and $\alpha_0 < \kappa$ such that for a stationary set of $\alpha$, $\varphi_\alpha = \varphi$ and $\widetilde{J}_{\kappa+\alpha} \models \exists x \in \widetilde{J}_{\kappa+\alpha_0} \forall y \varphi$ holds at $\alpha$. But then choose any $\alpha < \beta$ in this stationary set with $\alpha_0 < \alpha$ and we have $\widetilde{J}_{\kappa+\alpha} \models \exists x \forall y \varphi$ holds at both $\alpha$ and $\beta$. Contradiction. □

Jensen [72] shows that for any $\alpha$ and any $n$ there is a partial $\Sigma_n$ function *with* parameters that can serve as a $\Sigma_n$-Skolem function for $\Sigma_n$-hulls without parameters. However this does not achieve our goal as the definition of the necessary parameters does not reflect to arbitrary $\Sigma_n$-elementary submodels that contain them.

Instead we take an approach based on the idea that in a certain sense $\Sigma_{n+1}$ can be viewed as $\Sigma_1$ relativized to $\Sigma_n$, for an arbitrary $J_\alpha$. Though this is only true for the usual Lévy hierarchy when awkward parameters are introduced, we define $\Sigma_n^*$ in such a way that this is true using only "standard" parameters, whose definitions relativize without difficulty to $\Sigma_n^*$-hulls.

## The $\Sigma_n^*$-Hierarchy

In order to define the notion of $\Sigma_n^*$ formula we must also define the auxiliary notions $n^{\text{th}}$ *reduct* and $n^{\text{th}}$ *standard parameter*, all by induction on $n$.

Let $M$ denote some fixed $J_\alpha$, $\alpha > 0$. We order finite sets of ordinals by the maximum difference order: $x < y$ iff $\beta \in y$, where $\beta$ is the largest element of $(y - x) \cup (x - y)$.

A $\Sigma_1^*$ formula is just a $\Sigma_1$ formula. The $\Sigma_1^*$ projectum of $M$, denoted $\rho_1^M$, is the least $\rho$ such that there is a subset of $\omega\rho$ which is $\Sigma_1^*$ over $M$ with parameters but is not an element of $M$. The 1st standard parameter of $M$, denoted $p_1^M$, is the least finite set of ordinals $p$ such that $A \cap \omega\rho_1^M \notin M$ for some $A$ which is $\Sigma_1^*$ with parameter $p$. We use $H_1^M$ to denote $J_{\rho_1^M}$ and for any $x \in M$, $A_1(x) = \{\langle y, m\rangle \mid \text{the } m^{\text{th}}\ \Sigma_1^* \text{ formula is true at } \langle y, x, p_1^M\rangle, y \in H_1^M\}$. The 1st reduct of $M$ relative to $x$, denoted $M_1(x)$, is the structure $\langle H_1^M, A_1(x)\rangle$.

For $n \geq 1$: a $\Sigma_{n+1}^*$ formula is one of the form $\varphi(x) \iff M_n(x) \models \psi$, where $\psi$ is $\Sigma_1$. The $\Sigma_{n+1}^*$ projectum of $M$, denoted $\rho_{n+1}^M$, is the least $\rho$ such that there is a subset of $\omega\rho$ which is $\Sigma_{n+1}^*$ with parameters but not an element of $M$. The $(n+1)^{\text{st}}$ standard parameter of $M$, denoted $p_{n+1}^M$, is $p_n^M \cup p$ where $p$ is the least finite set of ordinals such that $A \cap \omega\rho_{n+1}^M \notin M$ for some $A$ which is $\Sigma_{n+1}^*$ with parameter $p_n^M \cup p$. We use $H_{n+1}^M$ to denote $J_{\rho_{n+1}^M}$ and for any $x \in M$, $A_{n+1}(x) = \{\langle y, m\rangle \mid \text{the } m^{\text{th}}\ \Sigma_{n+1}^* \text{ formula is true at } \langle y, x, p_{n+1}^M\rangle, y \in H_{n+1}^M\}$. The $(n+1)^{\text{st}}$ reduct of $M$ relative to $x$, denoted $M_{n+1}(x)$, is the structure $\langle H_{n+1}^M, A_{n+1}(x)\rangle$.

This completes the definition of the $\Sigma_n^*$-hierarchy. Thus a $\Sigma_{n+1}^*$ formula is a formula expressing a $\Sigma_1$ property on $n^{\text{th}}$ reducts, uniformly.

**Lemma 1.8.** (a) *If $\varphi, \psi$ are $\Sigma_n^*$ formulas then $\varphi \vee \psi, \varphi \wedge \psi$ are equivalent to $\Sigma_n^*$ formulas.*

(b) *If $\varphi$ is a $\Sigma_n^*$ formula then both $\varphi$ and $\sim \varphi$ are equivalent to $\Sigma_{n+1}^*$ formulas.*

(c) *There is a universal $\Sigma_n^*$ formula, i.e., a $\Sigma_n^*$ formula $\varphi(e, x)$ such that if $\psi(x)$ is $\Sigma_n^*$ then for some $e \in \omega$, $\psi(x) \iff \varphi(e, x)$ for all $x$.*

(d) *The reduct $M_n(x) = \langle H_n^M, A_n(x)\rangle$ is amenable, i.e., if $y \in H_n^M$ then $y \cap A_n(x) \in H_n^M$.*

*Proof.* (a) is clear because a $\Sigma_{n+1}^*$ formula is of the form $\varphi(x) \iff M_n(x) \vDash \psi$, $\psi$ $\Sigma_1$ and $\Sigma_1$ is closed under $\vee, \wedge$.

(b) If $\varphi(x)$ is $\Sigma_n^*$ then so is $\varphi'(y, x, z) \iff \varphi(x)$ and choose $k$ so that $\varphi'$ is the $k^{\text{th}}$ $\Sigma_n^*$ formula. Then $\varphi(x) \iff \langle \emptyset, k\rangle \in A_n(x)$ so $\varphi$ is equivalent to a $\Sigma_{n+1}^*$ formula. Similarly for $\sim \varphi$ since $\sim \varphi(x) \iff \langle \emptyset, k\rangle \notin A_n(x)$.

(c) If $\psi$ is a universal $\Sigma_1$ formula then $\varphi(k, x) \iff \langle H_n^M, A_n(x)\rangle \vDash \psi(k, \emptyset) \iff \big\langle H_n^M, A_n(\langle k, x\rangle)\big\rangle \vDash \psi^*$ is a universal $\Sigma_{n+1}^*$ formula (where $\psi^*$ is $\Sigma_1$ and chosen to satisfy the last $\iff$).

(d) By (c) we have that $A_n(x)$ is $\Sigma_n^*$ (with parameter $p_n^M$) and hence $A_n(x) \cap y \in M$ for each $y \in H_n^M$. We must show that $A_n(x) \cap y$ in fact belongs to $H_n^M = J_{\omega\rho_n^M}$. But either $H_n^M = M$ or $\omega\rho_n^M$ is a cardinal of $M$. Using Proposition 1.6 and condensation, we show generally that if $\kappa$ is an $M$-cardinal then every bounded subset of $\kappa$ in $M$ actually belongs to $J_\kappa$: If $x \subseteq \gamma < \kappa$ and $x$ is $\Sigma_n$-definable with parameter $p$ over $\bar{M}$, a proper initial segment of $M$, then let $H = \Sigma_n$-Skolem hull of $\gamma \cup \{p\}$ in $\bar{M}$. Then $H \simeq J_\beta$ where $\beta$ is less than $\kappa$, since in $M$ the cardinality of $H$ is at most $\gamma$ (we may assume $\omega \le \gamma < \kappa$). But $x$ is definable over $J_\beta$, so $x \in J_{\beta+1} \subseteq J_\kappa$. □

As promised, we have the following analogue of Lemma 1.4, in the $\Sigma^*$ context.

**Lemma 1.9.** *For any $X \subseteq J_\alpha$, the $\Sigma_n^*$-hull of $X$ exists. Moreover there is a $\Sigma_n^*$-Skolem function for $J_\alpha$, via a $\Sigma_n^*$-definition independent of $\alpha$.*

*Proof.* By induction on $n$. The base case $n = 1$ is Lemma 1.4. Suppose that the result holds for $n \ge 1$ and we shall establish it for $n + 1$. Let $h_n(k, x)$ be a $\Sigma_n^*$-Skolem function for $J_\alpha$.

Lemma 1.1 holds uniformly for amenable structures so we may define a partial $\Sigma_{n+1}^*$ function $h(k, x)$ such that for each $x$, $H(x) = \{h(k, x) \mid k \in \omega, h(k, x) \text{ defined}\}$ is a $\Sigma_1$-elementary submodel of $M_n(x) = \langle H_n^M, A_n(x)\rangle$. Define

$$h_{n+1}(k, x) = h_n((k)_0, \langle h((k)_1, x), p_n^M\rangle)$$

where $k = \langle (k)_0, (k)_1\rangle$ is a pairing function on $\omega$. Now $h_{n+1}(k, x) = y \iff \exists z \in H_n^M (y = h_n((k)_0, \langle z, p_n^M\rangle) \wedge z = h((k)_1, x))$. As graph$(h_n)$ is $\Sigma_n^*$ and graph$(h)$ is $\Sigma_{n+1}^*$, this yields a $\Sigma_{n+1}^*$ definition of graph$(h_{n+1})$. If $H$ is a $\Sigma_{n+1}^*$-elementary submodel of $M$ then $H$ is closed under $h_n$ by induction, $H$ is closed under $h$ by $\Sigma^*_{n+1}$-elementarity and $H$ must contain $p^M_n$ since "$x = p^M_n$" is a $\Sigma^*_{n+1}$ formula. It follows that $H$ is closed under $h_{n+1}$.

It remains to show that $H = \{h_{n+1}(k, x) \mid k \in \omega\}$ is $\Sigma^*_{n+1}$-elementary in $M$. (It then follows that for any $X \subseteq M$, $\{h_{n+1}(k, x) \mid x$ a finite sequence from $X\}$ is $\Sigma^*_{n+1}$-elementary in $M$.) As $H$ is $\Sigma_1$-elementary in $M$ we know that $H$ satisfies extensionality so we may take the transitive collapse $\pi : \bar{M} \simeq H \subseteq M$. It will suffice to show that $\pi^{-1}[H \cap M_n(\pi(x))] = \bar{M}_n(\bar{x})$ for each $\bar{x} \in \bar{M}$, for then the closure of $H$ under $h$ guarantees $\Sigma^*_{n+1}$-elementarity. Now $M_n(\pi(\bar{x})) = \langle H^M_n, A_n(\pi(\bar{x}))\rangle$ where $H^M_n = J_{\omega\rho^M_n}$ and $A_n(\pi(\bar{x})) = \{\langle y, m\rangle \mid$ the $m^{\text{th}}$ $\Sigma^*_n$ formula is true at $\langle y, \pi(\bar{x}), p^M_n\rangle, y \in H^M_n\}$. Since by induction we have $\Sigma^*_n$-elementarity, it is enough to show:

$$\pi^{-1}[\rho^M_n] = \rho^{\bar{M}}_n, \pi^{-1}(p^M_n) = p^{\bar{M}}_n.$$

Let $\bar{\rho} = \pi^{-1}[\rho^M_n]$. Suppose $\bar{A} \subseteq J_{\bar{\rho}}$ is $\Sigma^*_n$-definable in $\bar{M}$ with parameter $\bar{q}$. For $\bar{\gamma} < \bar{\rho}$ we have $\bar{A} \cap J_{\bar{\gamma}} \in \bar{M}$ by $\Sigma_1$-elementarity of $\pi$ from $\bar{M}_n(\bar{q})$ to $M_n(\pi(\bar{q}))$. Note that if $\bar{p} = \pi^{-1}(p^M_n)$ then every element of $\bar{M}$ is of the form $h_n(k, \langle\bar{x}, \bar{p}\rangle), \bar{x} \in J_{\bar{\rho}}$ so the set $\{\langle k, \bar{x}\rangle \mid k \in \omega, \bar{x} \in J_{\bar{\rho}}, h_n(k, \langle\bar{x}, \bar{p}\rangle)$ defined, $\langle k, \bar{x}\rangle \notin h_n(k, \langle\bar{x}, \bar{p}\rangle)\}$ is $\Sigma^*_n$-definable in $\bar{M}$ with parameters and does not belong to $\bar{M}$. So $\bar{\rho} = \rho^{\bar{M}}_n$ and $\bar{p} \geq p^{\bar{M}}_n$.

Finally we show that $\bar{p} \leq p^{\bar{M}}_n$. Let $\bar{H} = \Sigma^*_n$-hull of $\{\bar{q} \mid \bar{q} < \bar{p}\}$. We may assume that $\bar{p} \neq \emptyset$ and therefore $\bar{\rho} \subseteq \bar{H}$. Now if $\bar{H} \simeq \bar{M}$ then we get $\bar{H} = \bar{M}$ and hence $\bar{p} \in \bar{H}$. But then by $\Sigma^*_n$-elementarity, $p^M_n \in \Sigma^*_n$-hull of $\{q \mid q < p^M_n\}$, which contradicts the definition of $p^M_n$. So $\bar{H} \simeq$ proper initial segment of $\bar{M}$ and therefore $\bar{A} \cap J_{\bar{\rho}} \in \bar{M}$ whenever $\bar{A}$ is $\Sigma^*_n$-definable in $\bar{M}$ from a parameter $\bar{q} < \bar{p}$. So $\bar{p} \leq p^{\bar{M}}_n$. □

Our next lemma helps to clarify the meaning of the standard parameters, as well as the relationship between $\Sigma^*_n$ and $\Sigma_n$.

**Lemma 1.10.** *Let $H = \Sigma^*_n$-Skolem hull of $\rho^M_n \cup \{p^M_n\}$ in $M$. Then $H = M$.*

*Proof.* Let $\pi : H \simeq \bar{M}$. Then $\bar{M} = M$ as $A \cap H^M_n$ is definable over $\bar{M}$ whenever $A$ is $\Sigma^*_n$-definable in $M$ with parameter $p^M_n$. So $M = \Sigma^*_n$-Skolem hull of $\rho^M_n \cup \{\pi(p^M_n)\}$. But we must have $\pi(p^M_n) = p^M_n$, as $\pi(p^M_n) < p^M_n$ contradicts the definition of $p^M_n$. □

**Corollary 1.11.** *For each $n$, $\Sigma_n \subseteq \Sigma^*_n$ and for $m < n$, $\Sigma^*_n$ is closed under existential quantification over $H^M_m$.*

*Proof.* We show that for each $n$, $\Sigma^*_n$ is closed under existential quantification over $M$. Then the first statement of the corollary follows and the second statement also follows, using the fact that "$x \in H^M_m$" is a $\Sigma^*_{m+1}$ formula. Now suppose that $\varphi$ is $\exists x\psi(x)$, $\psi$ $\Sigma^*_n$. We can write this as $\exists \bar{x} \in H^M_n \exists k \psi(h_n(k, \langle\bar{x}, p^M_n\rangle))$, which gives a $\Sigma^*_n$ formula equivalent to $\varphi$. □

**Remark.** With some effort, it can be shown that conversely, each $\Sigma^*_n$ formula is equivalent to a $\Sigma_n$ formula with parameters (see Jensen [72]). But we will have no use for this fact.

It will be useful to have approximations to the $\Sigma_n^*$-hulls and $\Sigma_n^*$-Skolem functions. For $n = 1$ and $\sigma < \omega\alpha = \mathrm{ORD}(M)$ we let $h_1^\sigma(k, x)$ be defined by restricting the $\Sigma_1$ definition of $h_1$ to $\widetilde{J}_\sigma$: if $h_1(k, x) = y \iff \exists z\varphi(x, y, z), \varphi\ \Sigma_0$ then $h_1^\sigma(k, x) = y \iff \exists z \in \widetilde{J}_\sigma\ \varphi(x, y, z)$. For any $n \geq 1$ and $\sigma < \omega\rho_n^M$ we define $h_{n+1}^\sigma(k, x) = h_n((k)_0, \langle h^\sigma((k)_1, x), p_n^M\rangle)$, where $h^\sigma$ is defined by restricting the $\Sigma_{n+1}^*$ definition of $h$ (from the proof of Lemma 1.9) to $\widetilde{J}_\sigma$: if $h(k, x) = y \iff M_n(k, x, y) \models \exists z\varphi, \varphi\ \Sigma_0$ then $h^\sigma(k, x) = y \iff M_n(k, x, y) \models \exists z \in \widetilde{J}_\sigma\ \varphi$. Also let $\Sigma_n^* \restriction \sigma$-hull $(X)$ denote $\{h_n^\sigma(k, x) \mid x$ is a finite sequence from $X\}$.

**Lemma 1.12.** *For any $X \subseteq J_\alpha$, $n \in \omega$, limit $\sigma < \omega\rho_{n+1}^M$, $\Sigma_{n+1}^* \restriction \sigma$-hull $(X)$ is $\Sigma_n^*$-elementary in $M$.*

*Proof.* This is clear if $n = 0$ and for $n > 0$ it suffices to show that the hull in question is closed under $h_n$. This follows from the facts that

$$\{h^\sigma(k, x) \mid x \text{ a finite sequence from } X\}$$

is closed under pairing and that $\{h_n(k, \langle y, p\rangle) \mid y$ a finite sequence from $Y\}$ is closed under $h_n$ for any $Y \subseteq M,\ p \in M$. □

The following fact about hull approximation is very useful.

**Lemma 1.13.** *Suppose $X \subseteq J_\alpha$, $\omega < \beta \leq \rho_{n-1}^M$, $\beta$ is a regular $M$-cardinal and $\beta \in \Sigma_n^*$-hull $(X)$ in $M$. Let $\bar\beta = \cup(\Sigma_n^*$-hull $(X) \cap \beta)$ and $\sigma = \cup(\Sigma_n^*$-hull $(X) \cap \rho_{n-1}^M)$. Then $\bar\beta = \beta \cap \Sigma_n^* \restriction \sigma$-hull $(X \cup \bar\beta)$ and for any $\bar{\bar\beta} < \bar\beta$, $x$ a finite sequence from $X$, $\beta \cap \Sigma_{n-1}^*$-hull $(\{x\} \cup \bar{\bar\beta})$ is bounded strictly below $\bar\beta$.*

*Proof.* Suppose $\gamma \in \beta \cap \Sigma_n^* \restriction \sigma$-hull $(X \cup \bar\beta)$. Then there exists a finite sequence $x$ from $X$ such that $\gamma \in \Sigma_n^* \restriction \bar\sigma$-hull $(\{x\} \cup \bar{\bar\beta})$ where $\bar\sigma, \bar{\bar\beta} \in \Sigma_n^*$-hull $(\{x\})$, $\bar\sigma < \sigma$, $\bar{\bar\beta} < \bar\beta$. But $\Sigma_n^* \restriction \bar\sigma$-hull $(\{x\} \cup \bar{\bar\beta}) \cap \beta$ belongs to $\Sigma_n^*$-hull $(X)$ and hence so does its supremum $\delta$. As $\beta$ is regular in $M$, $\delta < \beta$ and therefore $\delta < \bar\beta$. Since $\gamma < \delta$ we get $\gamma < \bar\beta$, as desired. The second conclusion of the lemma also follows, by Lemma 1.12. □

## The □ Principle

An important application of fine structure theory is to Jensen's □ Principle, which we now establish using the $\Sigma^*$ approach. We state □ in a strong form, which will be useful later, in our proof of Jensen's Coding Theorem.

□. Assume $V = L$. Then there is $\langle C_\mu \mid \mu$ a singular limit ordinal $\rangle$ such that

(a) $C_\mu$ is closed unbounded in $\mu$.

(b) ordertype $(C_\mu) < \mu$.

(c) $\bar{\mu} \in \lim C_\mu \Longrightarrow \bar{\mu}$ is singular and $C_{\bar{\mu}} = C_\mu \cap \bar{\mu}$.

(d) $\langle \widetilde{J}_\mu, C_\mu \rangle$ is amenable and if $\langle \widetilde{J}_{\bar{\mu}}, \bar{C} \rangle \longrightarrow \langle \widetilde{J}_\mu, C_\mu \rangle$ is $\Sigma_1$-elementary then $\bar{C} = C_{\bar{\mu}}$.

We refer the reader to Jensen [72] for background on and applications of the $\Box$ Principle.

Let $\mu$ be a singular limit ordinal. We wish to define $C_\mu$. Let $\beta(\mu) \geq \mu$ be the least limit ordinal $\beta$ such that $\mu$ is not regular with respect to $\widetilde{J}_\beta$-definable functions and let $n(\mu)$ be least such that there is a $\Sigma^*_{n(\mu)}(\widetilde{J}_{\beta(\mu)})$ partial function (with parameters) from an ordinal less than $\mu$ cofinally into $\mu$. Note that $\omega\rho^{\beta(\mu)}_{n(\mu)} \leq \mu$ (where $\rho^\beta_n$ denotes $\rho^{J_\beta}_n$) as otherwise such a partial function would belong to $\widetilde{J}_{\beta(\mu)}$, contradicting the leastness of $\beta(\mu)$. Also $\mu \leq \omega\rho^{\beta(\mu)}_{n(\mu)-1}$, else by Lemma 1.10 we have contradicted the leastness of $n(\mu)$.

For $X \subseteq \widetilde{J}_{\beta(\mu)}$ let $H(X)$ denote $\Sigma^*_{n(\mu)}$-hull $(X)$ in $\widetilde{J}_{\beta(\mu)}$. For some least parameter $q(\mu) \in \widetilde{J}_{\beta(\mu)}$, $H(\mu \cup \{q(\mu)\}) = \widetilde{J}_{\beta(\mu)}$. (Actually, $q(\mu) = p^{\beta(\mu)}_{n(\mu)} - \mu - p^{\beta(\mu)}_{n(\mu)-1}$.) Also let $\alpha(\mu) = \cup\{\alpha < \mu \mid \alpha = H(\alpha \cup \{q(\mu)\}) \cap \mu\}$. Then $\alpha(\mu) < \mu$ and (unless $\alpha(\mu) = \cup\emptyset = 0$), $\alpha(\mu) = H(\alpha(\mu) \cup \{q(\mu)\}) \cap \mu$. The former is because for large enough $\alpha < \mu$, $H(\alpha \cup \{q(\mu)\})$ contains both the domain and defining parameter for a $\Sigma^*_{n(\mu)}$ partial function from an ordinal less than $\mu$ cofinally into $\mu$.

If $\mu < \beta(\mu)$ let $p(\mu) = \langle q(\mu), \mu \rangle$ and if $\mu = \beta(\mu)$ let $p(\mu) = \emptyset$.

We are ready to define $C_\mu$.

Let $C^0_\mu = \{\bar{\mu} < \mu \mid \text{For some } \alpha \geq \alpha(\mu),\ \bar{\mu} = \cup(H(\alpha \cup \{p(\mu)\}) \cap \mu)\}$. Then $C^0_\mu$ is a closed subset of $\mu$. If $C^0_\mu$ is unbounded in $\mu$ then let $C_\mu = C^0_\mu$. If $C^0_\mu$ is bounded but nonempty then let $\mu_0 = \cup C^0_\mu$ and define $C^1_\mu = \{\bar{\mu} < \mu \mid \text{For some } \alpha,\ \bar{\mu} = \cup(H(\alpha \cup \{p(\mu), \mu_0\}) \cap \mu)\}$. If $C^1_\mu$ is unbounded then let $C_\mu = C^1_\mu$. If $C^1_\mu$ is bounded but nonempty then let $\mu_1 = \cup C^1_\mu$ and define $C^2_\mu = \{\bar{\mu} < \mu \mid \text{For some } \alpha,\ \bar{\mu} = \cup(H(\alpha \cup \{p(\mu), \mu_0, \mu_1\}) \cap \mu)\}$. Continue in this way, defining $C^k_\mu$ for $k \in \omega$ until $C^k_\mu$ is unbounded or empty for some least $k = k(\mu)$. To see that $k(\mu)$ exists, let $\alpha_k$ be least such that $\cup(H(\alpha_k + 1 \cup \{p(\mu), \mu_0, \ldots, \mu_{k-1}\}) \cap \mu$ contains an ordinal $\geq \mu_k$. We have $\alpha_k \in H(\{p(\mu), \mu_0, \ldots, \mu_k\})$. So $H(\alpha_k \cup \{p(\mu), \mu_0, \ldots, \mu_k\}) \cap \mu$ contains $H(\alpha_k + 1 \cup \{p(\mu), \mu_0, \ldots, \mu_{k-1}\}) \cap \mu$, which contains an ordinal $\geq \mu_k$ and therefore by definition of $\mu_k$ is unbounded in $\mu$. Hence $\alpha_{k+1} < \alpha_k$ and therefore $k(\mu)$ exists, else we get an infinite descending sequence of ordinals.

If $C^{k(\mu)}_\mu$ is unbounded in $\mu$ then let $C_\mu = C^{k(\mu)}_\mu$. If $C^{k(\mu)}_\mu = \emptyset$ then

$$H(\{p(\mu), \mu_0, \ldots, \mu_{k(\mu)-1}\}) \cap \mu$$

is unbounded in $\mu$. And $H(\{p(\mu), \mu_0, \ldots, \mu_{k(\mu)-1}\}) \cap \omega\rho^{\beta(\mu)}_{n(\mu)-1}$ is unbounded in $\omega\rho^{\beta(\mu)}_{n(\mu)-1}$, else this set belongs to $\widetilde{J}_{\beta(\mu)}$ and so $\mu$ is singular inside $\widetilde{J}_{\beta(\mu)}$, in contradiction to the leastness of $\beta(\mu)$. Let $\rho(\mu) = \omega\rho^{\beta(\mu)}_{n(\mu)-1}$, $p = \{p(\mu), \mu_0, \ldots, \mu_{k(\mu)-1}\}$ and let $h_n(k, x)$ be the $\Sigma^*_n$-Skolem function for $\widetilde{J}_{\beta(\mu)}$. Also let $\bar{\sigma}_m = \max(\{h_n(k, p) \mid k < m\} \cap \mu)$, $\sigma_m = \max(\{h_n(k, p) \mid k < m\} \cap \rho(\mu))$. We take $C_\mu = \{\delta_0, \delta_1, \ldots\}$ where $\delta_m$ is the ordertype of the transitive collapse of $\Sigma^*_{n(\mu)} \restriction \sigma_m$-hull $(\bar{\sigma}_m \cup \{p\})$. Note that $\delta_m < \mu$ as $\mu$ is regular inside $\widetilde{J}_{\beta(\mu)}$ and $\mathrm{card}(\delta_m) \leq \bar{\sigma}_m$ in $\widetilde{J}_{\beta(\mu)}$.

This completes the definition of $C_\mu$. Clearly $C_\mu$ is closed unbounded in $\mu$. The argument that $\alpha(\mu) < \mu$ also implies that ordertype $C_\mu < \mu$. So we need only show (c), (d) from the statement of $\square$.

**Lemma 1.14.** $\bar{\mu} \in C^k_\mu \Longrightarrow C^k_{\bar{\mu}} = C^k_\mu \cap \bar{\mu}$.

*Proof.* First suppose that $k = 0$. Given $\bar{\mu} \in C^0_\mu$ choose $\alpha < \bar{\mu}$ such that $\bar{\mu} = \bigcup(H(\alpha \cup \{p(\mu)\}) \cap \mu)$, where $H(X) = \Sigma^*_{n(\mu)}$-hull $(X)$ in $\widetilde{J}_{\beta(\mu)}$. Also let $\rho = \bigcup(H(\alpha \cup \{p(\mu)\}) \cap \omega\rho^{\beta(\mu)}_{n(\mu)-1})$. Let $H = \Sigma^*_{n(\mu)} \restriction \rho$-hull $(\bar{\mu} \cup \{p(\mu)\})$ and $\pi : \widetilde{J}_{\bar{\beta}} \simeq H \subseteq \widetilde{J}_{\beta(\mu)}$. By Lemma 1.13, $H \cap \mu = \bar{\mu}$ and therefore when $\mu < \beta(\mu)$, $\pi(\bar{\mu}) = \mu$. By Lemma 1.12 $\pi : \widetilde{J}_{\bar{\beta}} \longrightarrow \widetilde{J}_{\beta(\mu)}$ is $\Sigma^*_{n(\mu)-1}$-elementary, so we get that $\beta(\bar{\mu}) = \bar{\beta}$, $n(\bar{\mu}) \leq n(\mu)$. By the second conclusion of Lemma 1.13 we get $n(\bar{\mu}) > n(\mu) - 1$, so $n(\bar{\mu}) = n(\mu)$. Thus to conclude that $C^0_{\bar{\mu}} = C^0_\mu \cap \bar{\mu}$ we need only check that $\pi(q(\bar{\mu})) = q(\mu)$ when $\mu < \beta(\mu)$, and $\alpha(\bar{\mu}) = \alpha(\mu)$.

For the former, first note that $\mu \in H(\{\mu'\} \cup \{q(\mu)\})$ for some $\mu' < \mu$, $\mu' \in H(\{p(\mu)\})$, since $\mu, q(\mu) \in H(\{p(\mu)\}) \cap H(\mu \cup \{q(\mu)\})$. So in fact $\mu$ belongs to $\Sigma^*_{n(\mu)} \restriction \rho$-hull $(\bar{\mu} \cup \{q(\mu)\})$ and hence the latter is just $H$. Now let $\bar{q} = \pi^{-1}(q(\mu))$. We see that $\Sigma^*_{n(\bar{\mu})}$-hull $(\bar{\mu} \cup \{\bar{q}\}) = \widetilde{J}_{\beta(\bar{\mu})}$ and hence $\bar{q} \geq q(\bar{\mu})$. But $\bar{q} \in \Sigma^*_{n(\bar{\mu})}$-hull $(\bar{\mu} \cup \{q(\bar{\mu})\})$ in $\widetilde{J}_{\beta(\bar{\mu})}$ hence $q(\bar{\mu}) \geq \bar{q}$, else by applying $\pi$ we have contradicted the definition of $q(\mu)$. So $\pi(q(\bar{\mu})) = q(\mu)$. Now since $\alpha(\mu) < \bar{\mu}$ we get $\alpha(\mu) \leq \alpha(\bar{\mu})$. Conversely, $\alpha(\bar{\mu}) < \alpha$ where $\bar{\mu} = \bigcup(H(\alpha \cup \{p(\mu)\}) \cap \mu)$ so $H(\alpha(\bar{\mu}) \cup \{p(\mu)\}) \cap \mu = \alpha(\bar{\mu})$ and we get $\alpha(\bar{\mu}) \leq \alpha(\mu)$. So $\alpha(\bar{\mu}) = \alpha(\mu)$.

Now suppose $k = 1$. The above argument shows that $\bar{\mu} \in C^1_\mu \Longrightarrow C^0_{\bar{\mu}} = C^0_\mu \cap \bar{\mu}$ and hence, since $\mu_0 < \bar{\mu}$, we get $\bar{\mu}_0 = \mu_0$. Then the above argument shows that $C^1_{\bar{\mu}} = C^1_\mu \cap \bar{\mu}$. The general case $k \geq 0$ follows similarly. $\square$

To verify (c) in the statement of $\square$: if $\bar{\mu} \in \mathrm{Lim}\, C_\mu$ then we must have $C_\mu = C^k_\mu$ for some $k$ and so $C^k_{\bar{\mu}} = C_\mu \cap \bar{\mu}$ is unbounded in $\bar{\mu}$. Hence $C_{\bar{\mu}} = C^k_{\bar{\mu}} = C_\mu \cap \bar{\mu}$ as desired. Now we verify (d).

**Lemma 1.15.** (a) *If $A \subseteq \widetilde{J}_\mu$ and $A \in \widetilde{J}_{\beta(\mu)}$ then $A$ is $\Delta_1\langle \widetilde{J}_\mu, C_\mu\rangle$.*

(b) *Suppose $\pi : \langle \widetilde{J}_{\bar{\mu}}, \bar{C}\rangle \longrightarrow \langle \widetilde{J}_\mu, C_\mu\rangle$ is $\Sigma_1$-elementary. Then $\bar{C} = C_{\bar{\mu}}$ and $\pi$ extends uniquely to a $\Sigma^*_{n(\mu)}$-elementary $\tilde{\pi} : \widetilde{J}_{\beta(\bar{\mu})} \longrightarrow \widetilde{J}_{\beta(\mu)}$ such that $p(\mu) \in$ Range$(\tilde{\pi})$.*

*Proof.* First suppose that $C_\mu = C^k_\mu$ for some $k$. For $\mu' \in C_\mu$ form $H(\mu')$ as $H$ was formed in the proof of Lemma 1.14 for $\bar{\mu}$. Then $\pi(\mu') : \widetilde{J}_{\beta(\mu')} \longrightarrow \widetilde{J}_{\beta(\mu)}$, with range

$H(\mu')$, is $\Sigma^*_{n(\mu)-1}$-elementary and $\widetilde{J}_{\beta(\mu)} = \cup\{H(\mu') \mid \mu' \in C_\mu\}$. Also $\pi(\mu')$ is the identity on $\mu'$ and sends $p(\mu')$ to $p(\mu)$.

(a) If $A \subseteq \widetilde{J}_\mu$, $A \in \widetilde{J}_{\beta(\mu)}$ then $A \cap \widetilde{J}_{\mu'}$ is $\Sigma^*_{n(\mu)}$-definable as an element of $H(\mu')$ from some fixed parameter $x \in \widetilde{J}_\mu$, uniformly for sufficiently large $\mu' \in C_\mu$. So $A$ is $\Delta_1\langle \widetilde{J}_\mu, C_\mu\rangle$. This proves (a).

(b) Let $X = \text{Range}(\pi)$ and $\widetilde{X} = \Sigma^*_{n(\mu)}$-hull $(X \cup \{p(\mu)\})$ in $\widetilde{J}_{\beta(\mu)}$. If $y \in \widetilde{X} \cap \widetilde{J}_\mu$ then for some $\mu' \in C_\mu$, $y \in \Sigma^*_{n(\mu)}$-hull $((X \cap \widetilde{J}_{\mu'}) \cup \{p(\mu')\})$ in $\widetilde{J}_{\beta(\mu')}$, and $\mu'$ can be chosen in $\Sigma_1$-hull $(X)$ in $\langle \widetilde{J}_\mu, C_\mu\rangle$. It follows that $y \in (\Sigma_1\text{-hull } (X) \text{ in } \langle \widetilde{J}_\mu, C_\mu\rangle) = X$. So $\widetilde{X} \cap \widetilde{J}_\mu = X$ and if $\tilde{\pi} : \widetilde{J}_{\bar{\beta}} \simeq \widetilde{X} \subseteq \widetilde{J}_{\beta(\mu)}$ then $\tilde{\pi}$ is a $\Sigma^*_{n(\mu)}$-elementary embedding extending $\pi$ with $p(\mu)$ in its range. Let $\mu^* = \cup(X \cap \mu)$.

As $\widetilde{X} \simeq \Sigma^*_{n(\mu^*)}$-hull $(X \cup \{p(\mu^*)\})$ in $\widetilde{J}_{\beta(\mu^*)}$ we get that $\bar{\mu}$ is regular with respect to partial $\Sigma^*_{n(\mu)-1}(\widetilde{J}_{\bar{\beta}})$ functions and singular with respect to $\Sigma^*_{n(\mu)}(\widetilde{J}_{\bar{\beta}})$ partial functions. So we get $\beta(\bar{\mu}) = \bar{\beta}$, $n(\bar{\mu}) = n(\mu)$. Then the $\Sigma^*_{n(\mu)}$-elementarity of $\tilde{\pi}$ and the fact that $p(\mu) \in \text{Range}(\tilde{\pi})$ guarantee that $\overline{C} = C_{\bar{\mu}}$. The uniqueness of $\tilde{\pi}$ comes from the fact that $\widetilde{J}_{\beta(\bar{\mu})} = \Sigma^*_{n(\bar{\mu})}$-hull $(\bar{\mu} \cup \{p(\bar{\mu})\})$ and $\tilde{\pi} \restriction \bar{\mu}$ is determined by $\pi$. This proves (b).

In case $C^k_\mu = \emptyset$ for some $k$ then $C_\mu$ was defined as a special $\omega$-sequence cofinal in $\mu$. That definition was made precisely to enable the preceding arguments to also apply in this case. (Also note that in this case $\mu^* = \mu$.) $\square$

## Relativization

$\square$ and $\diamondsuit$ hold relative to *reshaped strings*, a fact which is useful in the proof of Jensen's Coding Theorem (Section 4.3). We state these versions here.

Assume that $A \subseteq ORD$, $L_\alpha[A] = H_\alpha$ for each infinite cardinal $\alpha$. For each such $\alpha$ define $S_\alpha$ to consist of all $s : [\alpha, |s|) \longrightarrow 2$, $\alpha \leq |s| < \alpha^+$ such that for all $\eta \leq |s|$, $L[A \cap \alpha, s \restriction \eta] \vDash \text{Card}(\eta) \leq \alpha$. These are the "reshaped strings" at $\alpha$.

We must also define *coding structures*. For $s \in S_\alpha$ define $\mu^{<s}$, $\mu^s$ inductively by: $\mu^{<\emptyset_\alpha} = \alpha$ (where $\emptyset_\alpha \in S_\alpha$, $|\emptyset_\alpha| = \alpha$ is the empty string), $\mu^{<s} = \cup\{\mu^t \mid t$ a proper intial segment of $s\}$ for $s \neq \emptyset_\alpha$, and $\mu^s =$ least $\mu > \mu^{<s}$ such that $\mu'\mu = \mu$ for $0 < \mu' < \mu$ and $L_\mu[A \cap \alpha, s] \vDash \text{Card}(|s|) \leq \alpha$. Also let $\hat{\mu}^s =$ largest $\mu > \mu^{<s}$ such that $\mu'\mu = \mu$ for $\mu' < \mu$, $L_\mu[A \cap \alpha, s] \vDash |s|$ is a cardinal, if exists; if there is no such $\mu$ then $\hat{\mu}^s = \mu^{<s}$. Then $\mathcal{A}^s = L_{\mu^s}[A \cap \alpha, s]$ and $\mathcal{A}^{<s} = \langle L_{\mu^{<s}}[A \cap \alpha, \hat{s}], A \cap \alpha, \hat{s}\rangle$, $\hat{\mathcal{A}}^s = \langle L_{\hat{\mu}^s}[A \cap \alpha, \hat{s}], A \cap \alpha, \hat{s}\rangle$ where $\hat{s} = \{\mu^{<s\restriction\eta} \mid s(\eta) = 1\}$. Note that the requirement "$0 < \mu' < \mu \Longrightarrow \mu'\mu = \mu$" for $\mu = \mu^s, \hat{\mu}^s$ implies that "$L$" can be replaced by "$\widetilde{J}$" in the definitions of $\mathcal{A}^s$, $\mathcal{A}^{<s}$, $\hat{\mathcal{A}}^s$; this is convenient for the proofs of Relativized $\square$, $\diamondsuit$ below.

And we must discuss *collapsibility*. A *J-model* is an amenable structure $\langle \mathcal{A}, C\rangle$ of the form $\langle \widetilde{J}_\mu[B], B, C\rangle$. We define $\mathcal{A}^+$ to be $\langle \widetilde{J}_{\mu^*}[B], B\rangle$ where $\mu^* \geq \mu$ is the least limit ordinal such that $\widetilde{J}_{\mu^*+\omega}[B] \vDash \mu$ is not a cardinal (if it exists). Now $\langle \mathcal{A}, C\rangle$ is *collapsible* if $\mathcal{A}^+$ exists and whenever $\pi : \langle \bar{\mathcal{A}}, \bar{C} \rangle \longrightarrow \langle \mathcal{A}, C \rangle$ is $\Sigma_1$ elementary then $\bar{\mathcal{A}}^+$ exists, $\bar{C}$ is definable over $\bar{\mathcal{A}}^+$ and $\pi$ lifts to a $\Sigma_1$ elementary $\pi^+ : \bar{\mathcal{A}}^+ \longrightarrow \mathcal{A}^+$.

**Relativized $\Box$.** *Suppose $\alpha$ is an uncountable limit cardinal. Then there exists $\langle C^s \mid s \in S_\alpha \rangle$ such that:*

(a) $s \neq \emptyset_\alpha \Longrightarrow C^s$ *is* CUB *in* $\mu^{<s}$, ordertype $C^s \leq \alpha$, $C^s \in \mathcal{A}^s$.

(b) $\mu \in \operatorname{Lim} C^s \Longrightarrow \mu = \mu^{<s \restriction \eta}$ *for some* $\eta \leq |s|$ *and* $C^{s \restriction \eta} = C^s \cap \mu$.

(c) $\langle \mathcal{A}^{<s}, C^s \rangle$ *is collapsible.*

(d) $s \neq \emptyset_\alpha$, $D \subseteq \mathcal{A}^{<s}$, $D \in (\mathcal{A}^{<s})^+ \Longrightarrow D$ *is* $\Delta_1 \langle \mathcal{A}^{<s}, C^s \rangle$.

**Relativized $\diamondsuit$.** *Suppose $\alpha$ is an uncountable limit cardinal. Then there exists $\langle D^s \mid s \in S_\alpha \rangle$ such that:*

(a) $D^s \subseteq \mathcal{A}^{<s}$, $\langle D^t \mid t$ *an initial segment of* $s \rangle \in \mathcal{A}^s$.

(b) *If* $D \subseteq \mathcal{A}^{<s}$, $D \in \hat{\mathcal{A}}^s \neq \mathcal{A}^{<s}$ *then* $\{\eta < |s| \mid D^{s \restriction \eta} = D \cap \mathcal{A}^{<s \restriction \eta}\}$ *is stationary in* $\hat{\mathcal{A}}^s$.

(c) *If* $\mu^{<s \restriction \eta} \in \operatorname{Lim} C^s$, $\eta < |s|$ *then* $D^{s \restriction \eta} = \emptyset$. *If* $\hat{\mathcal{A}}^s \vDash |s|^{++}$ *exists then* $D^s = \emptyset$. *And if* $\pi : \langle \mathcal{A}^{<\bar{s}}, \bar{C} \rangle \longrightarrow \langle \mathcal{A}^{<s}, C^s \rangle$ *is* $\Sigma_1$*-elementary,* $\pi(\bar{\alpha}) = \alpha$ *where* $\bar{s} \in S_{\bar{\alpha}}$ *then* $D^{\bar{s}} = \pi^{-1}[D^s]$.

*Proof.* For Relativized $\Box$, define $C^s$ using $\langle \widetilde{J}_{\beta(s)}[A \cap \alpha, \hat{s}], A \cap \alpha, \hat{s} \rangle$ as we defined $C_\mu$ using $\widetilde{J}_{\beta(\mu)}$, where $\beta(s) \geq \mu^{<s}$ is least so that $\mu^{<s}$ is not regular with respect to functions definable over this structure. Note that this structure belongs to $\mathcal{A}^s$. As before, we get property (a) and (b) follows from (the analogue of) Lemma 1.14. Properties (c), (d) follow from (the analogue of) Lemma 1.15.

For Relativized $\diamondsuit$, define $D^s$ using $\langle \widetilde{J}_{\beta(s)}[A \cap \alpha, \hat{s}], A \cap \alpha, \hat{s} \rangle$ as we defined $D_\mu$ in Lemma 1.5 using $\widetilde{J}_{\beta'(\mu)}$. Property (a) follows from (the analogue to) (b) of Lemma 1.5 and (b) follows from the same argument used to establish (a) of Lemma 1.5. Also, that argument in fact shows that (a) of Lemma 1.5 holds in the stronger form: if $D \subseteq J_{\alpha^+}$ then $\{\mu < \alpha^+ \mid D \cap J_\mu = D_\mu \text{ and } C^0_\mu = \emptyset\}$ is stationary; note that $C^0_\mu = \emptyset \Longrightarrow \mu \notin \operatorname{Lim} C_{\mu'}$ for $\mu' > \mu$. So by (the analogue to) this proof we may assume that the first statement of Relativized $\Box$ (c) holds. The second statement of (c) follows from (the analogue to) Lemma 1.5(c) and the final statement follows from (the analogues to) Lemma 1.15, Lemma 1.5(b). $\Box$

## The Fine Structure Principle

We summarize here those aspects of the $\Sigma^*$ theory that are used when establishing combinatorial principles in $L$. For any set $X$ let $\operatorname{Seq}(X)$ denote the set of all finite sequences from $X$ and recall the ordering $<$ on finite sets of ordinals: $p < q$ iff $\alpha \in q$ where $\alpha = \max((p - q) \cup (q - p))$. Also for any limit ordinal $\lambda$ let $M_\lambda$ denote $\tilde{J}_\lambda$ ($= J_\alpha$, where $\omega\alpha = \lambda$).

**(FSP).** There exists a sequence of recursive sets of formulas $\Sigma_1 = \Sigma_1^* \subseteq \Sigma_2^* \subseteq \cdots$ and partial functions $h_n^\lambda : \omega \times M_\lambda \longrightarrow M_\lambda$ for $\lambda$ limit, $n \in \omega$ such that

(1) $\bigcup\{\Sigma_n^* \mid n \in \omega\}$ = All first-order formulas, $\Pi_n^* = \{\sim \varphi \mid \varphi \in \Sigma_n^*\} \subseteq \Sigma_{n+1}^*$ and $\Sigma_n^*$ is closed under $\exists, \wedge, \vee$.

(2) $h_n^\lambda$ is $\Sigma_n^*$-definable and if $\varphi(x)$ is $\Sigma_n^*$ then for some $k$, $M_\lambda \models \varphi(x) \iff h_n(k, x)$ is defined.

(3) For any $X \subseteq M_\lambda$, $H_n^\lambda(X) = \{h_n^\lambda(k, x) \mid x \in \mathrm{Seq}(X), k \in \omega\}$ is the least $\Sigma_n^*$-elementary submodel of $M_\lambda$ containing $X$ as a subset.

(4) Let $\rho_n^\lambda$ = least ordinal $\rho$ such that for some $p \in Seq(\lambda)$ and $A \subseteq \lambda$, $A$ $\Sigma_n^*(M_\lambda)$ in parameter $p$, $A \cap \omega\rho \notin M_\lambda$. And let $p_n^\lambda$ = least such $p$. Then $p_n^\lambda \in H_{n+1}^\lambda(\emptyset)$ and $M_\lambda = H_n^\lambda(\omega\rho_n^\lambda \cup \{p_n^\lambda\})$. Also the formula "$x \in M_{\omega\rho_n^\lambda}$" is $\Sigma_{n+1}^*$.

(5) If $\pi : M_{\bar{\lambda}} \longrightarrow M_\lambda$ is $\Sigma_{n+1}^*$-elementary then $\pi^{-1}[\rho_n^\lambda] = \rho_n^{\bar{\lambda}}$ and $\pi^{-1}(p_n^\lambda) = p_n^{\bar{\lambda}}$.

(6) Approximations: $h_n^\lambda = \bigcup\{h_n^{\lambda,\sigma} \mid \sigma < \omega\rho_{n-1}^\lambda\}$ where $\sigma < \sigma' \Longrightarrow h_n^{\lambda,\sigma} \subseteq h_n^{\lambda,\sigma'}$, $\{\langle\sigma, k, x, y\rangle \mid h_n^{\lambda,\sigma}(k, x) = y\}$ is $\Sigma_n^* \cap \Pi_n^*$ and for each limit $\sigma$, $H_n^{\lambda,\sigma}(X) = \{h_n^{\lambda,\sigma}(k, x) \mid k \in \omega, x \in \mathrm{Seq}(X)\}$ is $\Sigma_{n-1}^*$-elementary in $M_\lambda$. (When $n = 1$ we take $\omega\rho_{n-1}^\lambda$ to be $\lambda$).

It is not difficult to verify that the proof of □ that we gave can be carried out directly from the FSP. In the next section we use the properties described in the FSP to construct morasses.

## 1.3 Morasses

A strong form of the gap-1 morass principle is used in the theory of strong coding (Section 7.2). We now establish a global form of this principle, which we call Morass with □.

In □ we found a uniform way of writing a singular ordinal as the union of a short sequence of smaller ordinals. In Morass we find a uniform way of writing an ordinal of regular cardinality as the direct limit of ordinals of smaller cardinality. These two principles interact in Morass with □.

Rather than begin with a statement of our principle, we first use the fine structure theory (as summarized by the Fine Structure Principle) to describe the actual object which will interest us. In this way it is easier to see the motivation behind a list of its combinatorial properties, expressed in Morass with $\Box$.

A limit ordinal $\alpha$ is *cardinal-correct* if whenever $\widetilde{J}_\alpha \models \kappa$ is a cardinal, then $\kappa$ really is a cardinal. Let $S^0 = \{\alpha > \omega \mid \alpha \text{ is cardinal-correct}\}$. Then $S^0$ is CUB in every uncountable cardinal. For $\alpha \in S^0$ let $S_\alpha = \{\nu \mid \alpha < \nu < \alpha^+, \nu \text{ is a limit ordinal}, \widetilde{J}_\nu \models \alpha \text{ is regular and } \alpha \text{ is the largest cardinal}\}$. Then $S_\alpha$ is a closed subset of $(\alpha, \alpha^+)$ and if $\alpha$ is not a cardinal, $\alpha < \beta$ in $S^0$ then $\cup S_\alpha < \beta$. We write $\nu_0 <_0 \nu_1$ iff $\nu_0 < \nu_1$ and for some $\alpha \in S^0$, $\nu_0$ and $\nu_1$ both belong to $S_\alpha$. When $\nu \in S_\alpha$ we write $\alpha(\nu) = \alpha$. (This is a different use of the notation $\alpha(\nu)$ than was made in the proof of $\Box$.) Let $S^1 = \cup\{S_\alpha \mid \alpha \in S^0\}$.

Now we come to the main definition. For $\nu \in S^1$, $\beta(\nu)$ is the least limit ordinal $\beta \geq \nu$ such that $\rho_n^\beta \leq \alpha(\nu)$ for some $n$, and $n(\nu)$ is the least such $n$. And $q(\nu)$ is the least $q \in \mathrm{Seq}(\beta(\nu))$ such that $\widetilde{J}_{\beta(\nu)} = H_{n(\nu)}^{\beta(\nu)}(\alpha(\nu) \cup \{q\})$. (Actually $q(\nu) = p_{n(\nu)}^{\beta(\nu)} - \alpha(\nu)$.) We write $\bar{\nu} <_1 \nu$ iff there is $\pi : \widetilde{J}_{\beta(\bar{\nu})} \longrightarrow \widetilde{J}_{\beta(\nu)}$ such that $\pi$ is $\Sigma^*_{n(\bar{\nu})}$-elementary, $n(\bar{\nu}) = n(\nu), \pi(\alpha(\bar{\nu})) = \alpha(\nu), \pi(q(\bar{\nu})) = q(\nu)$ and $\pi \restriction \alpha(\bar{\nu}) = \text{identity}$; in addition we impose the $Q$-condition:

(Q) Whenever $\varphi(x)$ is $\Sigma^*_{n(\nu)}$ with parameter $\bar{p} \in \widetilde{J}_{\beta(\bar{\nu})}$ then $\{\nu_0 < \nu \mid \widetilde{J}_{\beta(\nu)} \models \varphi(\nu_0, \pi(p))\}$ is bounded in $\nu$ iff $\{\bar{\nu}_0 < \bar{\nu} \mid \widetilde{J}_{\beta(\bar{\nu})} \models \varphi(\bar{\nu}_0, \bar{p})\}$ is bounded in $\bar{\nu}$.

If $\bar{\nu} <_1 \nu$ then $\pi$ as above is unique and we write $\pi_{\bar{\nu}\nu} = \pi \restriction \bar{\nu} : \bar{\nu} \longrightarrow \nu, \tilde{\pi}_{\bar{\nu}\nu} = \pi$.

The above structure, together with the $\Box$ sequence $\langle C_\alpha \mid \alpha \text{ singular limit}\rangle$ from the preceding section, constitutes our realization of Morass with $\Box$. Before stating this principle we make a few observations regarding the relation $<_1$. Using the fact that $\tilde{\pi}_{\bar{\nu}\nu}$ is $\Sigma^*_{n(\nu)}$-elementary and sends $(\alpha(\bar{\nu}), q(\bar{\nu}))$ to $(\alpha(\nu), q(\nu))$ it follows not only that $\tilde{\pi}_{\bar{\nu}\nu} = \pi$ is unique but also that $<_1$ is a tree and $\nu$ $<_0$-minimal, limit $\Longrightarrow \bar{\nu}$ $<_0$-minimal, limit. Also $\pi^{-1}[S_{\alpha(\nu)}] = S_{\alpha(\bar{\nu})} \cap \bar{\nu}$ and $\pi(\bar{\nu}_0^+) = \pi(\bar{\nu}_0)^+$ when $\pi(\bar{\nu}_0)^+ = (<_0 - \text{successor to } \pi(\bar{\nu}_0)) < \nu$. Also if $\bar{\nu}_0 <_0 \bar{\nu}$, $\nu_0 = \pi(\bar{\nu}_0)$ then $\pi \restriction \widetilde{J}_{\beta(\bar{\nu}_0)}$ is elementary from $\widetilde{J}_{\beta(\bar{\nu}_0)}$ into $\widetilde{J}_{\beta(\nu_0)}$ so $\bar{\nu}_0 <_1 \nu_0$ and $\pi_{\bar{\nu}_0\nu_0} = \pi_{\bar{\nu}\nu} \restriction \bar{\nu}_0$. Finally, $\bar{\bar{\nu}} <_1 \bar{\nu} <_1 \nu \Longrightarrow \pi_{\bar{\bar{\nu}}\nu} = \pi_{\bar{\nu}\nu} \circ \pi_{\bar{\bar{\nu}}\bar{\nu}}$ and $\{\alpha(\bar{\nu}) \mid \bar{\nu} <_1 \nu\}$ is always closed in $\alpha(\nu)$ and is unbounded if $\nu$ is not $<_0$-maximal. If $\{\alpha(\bar{\nu}) \mid \bar{\nu} <_1 \nu\}$ is unbounded then $\nu = \cup\{\mathrm{Range}(\pi_{\bar{\nu}\nu}) \mid \bar{\nu} <_1 \nu\}$.

There are four more properties of $\pi$ which take a bit of argument. First we claim that if $\nu < \beta(\nu)$ then $\nu \in \mathrm{Range}(\pi)$: If $n(\nu) > 1$ then this is clear because $\widetilde{J}_{\beta(\nu)} \models \nu = \alpha(\nu)^+$ and the property of being a cardinal is $\Sigma^*_2$. If $n(\nu) = 1$ then we claim that $q(\nu) - \nu$ is nonempty and hence if $\gamma \in q(\nu) - \nu$ we get $\nu = \alpha(\nu)^+$ of $\widetilde{J}_\gamma$ belongs to $H_1^{\beta(\nu)}(\{\alpha(\nu), q(\nu)\}) \subseteq \mathrm{Range}(\pi)$. The reason that $q(\nu) - \nu$ is nonempty is that as in the proof of Lemma 1.8(d), we can show that $\widetilde{J}_\nu$ is $\Sigma_1$-elementary in $\widetilde{J}_{\beta(\nu)}$ and hence $q(\nu) \subseteq \nu$ would contradict $H_1^{\beta(\nu)}(\alpha(\nu) \cup \{q(\nu)\}) = \widetilde{J}_{\beta(\nu)}$.

Second we claim that if $\bar{\nu}$ is $<_0$-limit, $\lambda = \cup\mathrm{Range}(\pi) < \nu$ then $\bar{\nu} <_1 \lambda$ and $\pi_{\bar{\nu}\lambda} = \pi_{\bar{\nu}\nu}$ : As in the proof of $\Box$ we form $H = H_{n(\nu)}^{\beta(\nu),\sigma}(\lambda \cup \{q(\nu)\})$ where $\sigma = \cup(\mathrm{Range}(\pi) \cap \rho_{n(\nu)-1}^{\beta(\nu)})$. Then as in the proof of Lemma 1.13, $H \simeq \widetilde{J}_{\beta(\lambda)}$ and $q(\nu)$ is sent to $q(\lambda)$ under this isomorphism. By composing with $\pi$, we get a $\Sigma^*_{n(\nu)}$-elementary embedding from $\widetilde{J}_{\beta(\bar{\nu})}$ into $\widetilde{J}_{\beta(\lambda)}$ sending $(\alpha(\bar{\nu}), q(\bar{\nu}))$ to $(\alpha(\lambda), q(\lambda))$. As the range of this embedding contains a cofinal subset of $\lambda$, the $Q$-condition is satisfied and $\bar{\nu} <_1 \lambda$, $\pi_{\bar{\nu}\lambda} = \pi_{\bar{\nu}\nu}$.

Third we claim that if $\bar{\nu} <_1 \nu$, $\pi_{\bar{\nu}\nu}$ is cofinal and for each $\bar{\nu}_0 <_0 \bar{\nu}$, $\alpha = \alpha(\nu_0')$ for some $\nu_0' <_1 \pi_{\bar{\nu}\nu}(\bar{\nu}_0)$ then $\alpha = \alpha(\nu')$ for some $\nu' <_1 \nu$. For, $H = H^{\beta(\nu)}_{n(\nu)}(\alpha \cup \{q(\nu)\}) = \bigcup\{H^{\beta(\nu),\sigma}_{n(\nu)}(\alpha \cup \{q(\nu)\}) \mid \sigma \in \mathrm{Range}(\tilde{\pi}_{\bar{\nu}\nu}), \sigma < \omega\rho^{\beta(\nu)}_{n(\nu)-1}\}$ and hence $H \cap \alpha(\nu) = \alpha$ since for each $\sigma$ as above, $\alpha = \alpha(\nu_\sigma')$ for some $\nu_\sigma' <_1 \nu_\sigma = \bigcup(\nu \cap H^{\beta(\nu),\sigma}_{n(\nu)}(\alpha \cup \{q(\nu)\}))$. Since $H \cap \nu$ is cofinal in $\nu$ (as we can assume that $\alpha \geq \alpha(\bar{\nu})$) we get $\alpha = \alpha(\nu')$, $\nu' =$ ordertype $(H \cap \nu)$.

Fourth we claim that if $\nu$ is a $<_0$-successor then so is $\bar{\nu}$: This is clear if $\nu < \beta(\nu)$ or $n(\nu) > 1$ as being the $<_0$-predecessor to $\nu$ is $\Pi_1(\widetilde{J}_\nu)$. If $(\beta(\nu), n(\nu)) = (\nu, 1)$ then we must use the $Q$-condition on $\pi$ to guarantee that $S_{\alpha(\bar{\nu})} \cap \bar{\nu}$ is bounded in $\bar{\nu}$.

The previous is our first use of the $Q$-condition on $\pi$. In strong coding (Section 7.2) we will use it to argue that if $\bar{\nu} <_1 \nu$ and $\nu$ is admissible then so is $\bar{\nu}$.

We now state Morass with □. We have shown that the structure defined above satisfies (a)–(f) in the list of properties below.

**Morass with □.** *There exist $\langle C_\alpha \mid \alpha$ singular limit$\rangle$, $\langle S_\alpha \mid \alpha \in S^0\rangle$, a binary relation $<_1$ on $S^1 = \bigcup\{S_\alpha \mid \alpha \in S^0\}$ and $\langle \pi_{\bar{\nu}\nu} \mid \bar{\nu} <_1 \nu\rangle$ such that*

(a) *For $\alpha$ a singular limit, $C_\alpha$ is* CUB *in $\alpha$, ordertype $C_\alpha < \alpha$, and for $\beta$ in* Lim $C_\alpha$, *$\beta$ is singular with $C_\beta = C_\alpha \cap \beta$.*

(b) *$S^0 \cap \kappa$ is* CUB *in $\kappa$ for every uncountable cardinal $\kappa$.*

(c) *For $\alpha \in S^0$, $S_\alpha$ is a closed subset of $(\alpha, \alpha^+)$. And:*

(c1) *$\alpha$ regular $\Longrightarrow S_\alpha = S^0 \cap (\alpha, \alpha^+)$.*

(c2) *$\alpha$ a singular cardinal $\Longrightarrow S_\alpha$ is a proper initial segment of $S^0 \cap (\alpha, \alpha^+)$.*

(c3) *$\alpha$ not a cardinal, $\alpha < \beta$ in $S^0 \Longrightarrow \bigcup S_\alpha < \beta$.*

**Notation.** (c) implies that when $\nu$ belongs to $S^1$ then there is a unique $\alpha$ such that $\nu$ belongs to $S_\alpha$; denote this by $\alpha(\nu)$. Write $\nu <_0 \nu'$ if $\nu < \nu'$ and $\alpha(\nu) = \alpha(\nu')$. For $\alpha \in S^0$, $\alpha$ not regular let $\nu(\alpha)$ denote max $(S_\alpha) < \alpha^+$. If $\nu \in S^1$, $\nu$ not $<_0$-maximal then $\nu^+$ denotes its $<_0$-successor.

(d) *$<_1$ is a tree and if $\bar{\nu} <_1 \nu$ then $\alpha(\bar{\nu}) < \alpha(\nu)$ and $\bar{\nu}$ is $<_0$-minimal, successor, limit iff $\nu$ is $<_0$-minimal, successor, limit.*

(e) *If $\bar{\nu} <_1 \nu$ then $\pi = \pi_{\bar{\nu}\nu} : \bar{\nu} \longrightarrow \nu$ is order-preserving, $\pi^{-1}[S_{\alpha(\nu)}] = S_{\alpha(\bar{\nu})} \cap \bar{\nu}$ and $\pi(\bar{\nu}_0^+) = \pi(\bar{\nu}_0)^+$ when $\pi(\bar{\nu}_0)^+ < \nu$. If $\bar{\nu}_0 <_0 \bar{\nu}$, $\nu_0 = \pi(\bar{\nu}_0)$ then $\bar{\nu}_0 <_1 \nu_0$ and $\pi_{\bar{\nu}_0\nu_0} = \pi \restriction \bar{\nu}_0$. If $\bar{\nu}$ is $<_0$-limit, $\lambda = \bigcup$* Range*$(\pi)$ then $\bar{\nu} <_1 \lambda$, $\pi_{\bar{\nu}\lambda} = \pi$; and if $\nu = \bigcup$* Range*$(\pi)$, $\alpha \in S^0$ and for each $\bar{\nu}_0 < \bar{\nu}$, $\alpha = \alpha(\nu_0')$ for some $\nu_0' <_1 \pi_{\bar{\nu}\nu}(\bar{\nu}_0)$, then $\alpha = \alpha(\nu')$ for some $\nu' <_1 \nu$.*

(f) $\bar{\bar{\nu}} <_1 \bar{\nu} <_1 \nu \Longrightarrow \pi_{\bar{\nu}\nu}\pi_{\bar{\bar{\nu}}\bar{\nu}} = \pi_{\bar{\bar{\nu}}\nu}$. *For* $\nu \in S^1$, $\{\alpha(\bar{\nu}) \mid \bar{\nu} <_1 \nu\}$ *is closed in* $\alpha(\nu)$ *and is unbounded unless* $\nu$ *is* $<_0$*-maximal. If* $\{\alpha(\bar{\nu}) \mid \bar{\nu} <_1 \nu\}$ *is unbounded in* $\alpha(\nu)$ *then* $\nu = \bigcup\{\text{Range}(\pi_{\bar{\nu}\nu}) \mid \bar{\nu} <_1 \nu\}$.

*Now let* $C'_\alpha$ *denote the limit points of* $C_\alpha$ *less than* $\alpha$*, for* $\alpha$ *singular limit.*

(g) *Suppose* $\nu$ *is* $<_1$*-limit and* $\alpha(\nu)$ *is singular. Then for* $\alpha$ *in a final segment of* $C'_{\alpha(\nu)}$ *there exists* $\nu_\alpha <_1 \nu$, $\nu_\alpha \in S_\alpha$, *and if* $\nu$ *is a* $<_0$*-limit,* $\lambda_\alpha = \bigcup\text{Range}(\pi_{\nu_\alpha\nu})$ *then* $\lambda_\alpha \in C'_\nu \cup \{\nu\}$ *and* $\alpha < \beta \Longrightarrow \lambda_\alpha \in \text{Range}(\pi_{\nu_\beta\nu}) \cup \{\nu\}$.

(h) *Suppose* $\nu$ *is* $<_1$*-minimal,* $<_0$*-limit. Then for* $\alpha$ *in a final segment of* $C'_{\alpha(\nu)}$, $\nu(\alpha)$ *is* $<_1$*-minimal,* $<_0$*-limit and there are* $\nu_\alpha <_0 \nu$ *such that* $\nu(\alpha) <_1 \nu_\alpha \in C'_\nu$, $\alpha < \beta \Longrightarrow \nu_\alpha \in \text{Range}(\pi_{\nu(\beta)\nu_\beta})$ *and* $\beta \in \text{Lim}\, C'_{\alpha(\nu)} \Longrightarrow \nu_\beta = \bigcup\{\nu_\alpha \mid \alpha \in C'_\beta\}$.

(i) *Suppose* $\nu$ *is a* $<_1$*-successor,* $<_0$*-limit. Let* $\bar{\nu} <_1^* \nu$ *express the property that* $\bar{\nu}$ *is the* $<_1$*-predecessor to* $\nu$*. Then for a final segment of* $\alpha \in C'_{\alpha(\nu)}$, $\nu(\alpha)$ *is* $<_1$*-successor,* $<_0$*-limit and there exist* $\nu_\alpha <_0 \nu$ *as in* (h) *such that in addition,* $\nu = \bigcup\text{Range}(\pi_{\bar{\nu}\nu}) \Longrightarrow \nu_\alpha = \pi_{\bar{\nu}\nu}(\bar{\nu}_\alpha)$ *where* $\bar{\nu}_\alpha <_1^* \nu(\alpha)$*, and* $\nu > \bigcup\text{Range}(\pi_{\bar{\nu}\nu}) = \lambda \Longrightarrow \lambda \in \text{Range}(\pi_{\nu(\alpha)\nu_\alpha})$, $\bar{\nu} <_1^* \nu(\alpha)$.

*Proof.* We have already established properties (a)–(f).

(g) Suppose $\alpha \in C'_{\alpha(\nu)}$. Then for some $k$, $\alpha \in \text{Lim}\, C^k_{\alpha(\nu)}$ and therefore for some $\gamma$:

$$\alpha = \bigcup\big(\alpha(\nu) \cap H^{\beta(\alpha(\nu))}_{n(\alpha(\nu))}(\gamma \cup \{p(\alpha(\nu)), \alpha(\nu)_0, \ldots, \alpha(\nu)_{k-1}\})\big)$$

where $\beta(\alpha(\nu))$, $n(\alpha(\nu))$, $p(\alpha(\nu)) = \langle q(\alpha(\nu)), \alpha(\nu)\rangle$ and $\alpha(\nu)_i$ are defined as in the proof of $\Box$. The fact that $\nu$ is a $<_1$-limit implies that $(\beta(\nu), n(\nu)) < (\beta(\alpha(\nu)), n(\alpha(\nu)))$. (I.e., either $\beta(\nu) < \beta(\alpha(\nu))$ or $\beta(\nu) = \beta(\alpha(\nu))$, $n(\nu) < n(\alpha(\nu))$. Note that as $\widetilde{J}_\nu \vDash$ There is a largest cardinal, $\beta(\nu)$ and $n(\nu)$ have the same meaning in this section as they did in the proof of $\Box$.) Thus for sufficiently large $\alpha \in C'_{\alpha(\nu)}$ we have by Lemma 1.13 that $\alpha = \alpha(\nu) \cap H^{\beta(\nu)}_{n(\nu)}(\alpha \cup \{q(\nu)\})$, where $q(\nu)$ is defined in this section. (We need only choose $\alpha$ large enough so that $H^{\beta(\nu)}_{n(\nu)}(\alpha \cup \{q(\nu)\}) \subseteq H^{\beta(\alpha(\nu))}_{n(\alpha(\nu))-1}(\alpha \cup \{p(\alpha(\nu))\})$.) To verify the $Q$-condition we must argue as follows. Either $\alpha$ can be chosen large enough so that $H^{\beta(\nu)}_{n(\nu)}(\alpha \cup \{q(\nu)\}) \cap \nu$ is cofinal in $\nu$, in which case the $Q$-condition is automatic, or we claim that the $Q$-condition implies that $H^{\beta(\nu)}_{n(\nu)}(\alpha \cup \{q(\nu)\})$ is $\Sigma^*_{n(\nu)+1}$-elementary in $\widetilde{J}_{\beta(\nu)}$. In the latter case the assumption that $\nu$ is a $<_1$-limit yields that in fact $(\beta(\nu), n(\nu)+1) < (\beta(\alpha(\nu)), n(\alpha(\nu)))$ and hence we get $H^{\beta(\nu)}_{n(\nu)}(\alpha \cup \{q(\nu)\}) = H^{\beta(\nu)}_{n(\nu)+1}(\alpha \cup \{q(\nu)\})$ obeys the $Q$-condition.

To prove the above claim suppose $\varphi(x)$ is $\Sigma^*_{n(\nu)+1}$ and note that $\varphi(x) \Longleftrightarrow \exists\gamma < \alpha(\nu)\exists k \in \omega(x \in y = h^{\beta(\nu)}_{n(\nu)}(k, \langle\gamma, q(\nu)\rangle), y \vDash \varphi(x), y\ \Sigma^*_{n(\nu)}\text{-elementary in } \widetilde{J}_{\beta(\nu)})$. To each $\sigma < \omega\rho^{\beta(\nu)}_{n(\nu)-1} = \rho$ associate the least $(\gamma(\sigma), k(\sigma))$ such that the above holds with $h^{\beta(\nu)}_{n(\nu)}$ replaced by $h^{\beta(\nu),\sigma}_{n(\nu)}$ and "$y\ \Sigma^*_{n(\nu)}$-elementary in $\widetilde{J}_{\beta(\nu)}$" replaced by "$y = H^{\beta(\nu),\sigma}_{n(\nu)}(y)$." Then $\varphi(x) \Longleftrightarrow A = \{\sigma \mid$ For some $\gamma, k, \sigma$ is least so that $(\gamma(\sigma), k(\sigma)) =$

$(\gamma, k)\}$ is bounded in $\rho \iff B = \{H^{\beta(\nu),\sigma}_{n(\nu)}(\alpha(\nu) \cup \{q(\nu)\}) \cap \nu \mid \sigma \in A\}$ is bounded in $\nu$. So $\nu$ is equivalent to the boundedness of a $\Sigma^*_{n(\nu)}$ subset of $\nu$, hence the $Q$-condition yields $\Sigma^*_{n(\nu)+1}$-elementarity.

Thus we have $\nu_\alpha <_1 \nu$ where $\nu_\alpha =$ ordertype $(\nu \cap H^{\beta(\nu)}_{n(\nu)}(\alpha \cup \{q(\nu)\})$. Again since $(\beta(\nu), n(\nu)) < (\beta(\alpha(\nu)), n(\alpha(\nu)))$, if $\alpha < \beta$ in $C'_{\alpha(\nu)}$ and $H^{\beta(\nu)}_{n(\nu)}(\alpha \cup \{q(\nu)\})$ is bounded in $\nu$ then its supremum below $\nu$ belongs to $H^{\beta(\nu)}_{n(\nu)}(\beta \cup \{q(\nu)\}) \supseteq H^{\beta(\nu)}_{n(\nu)+1}(\{\alpha, q(\nu)\}) \cap \nu$. So it only remains to show that $\lambda_\alpha = \bigcup(\nu \cap H^{\beta(\nu)}_{n(\nu)}(\alpha \cup \{q(\nu)\})) \in C'_\nu \cup \{\nu\}$. Note that if $p(\nu)$, $\nu_0$ are defined as in the proof of $\square$ then $q(\nu)$ as defined in this section belongs to $H^{\beta(\nu)}_{n(\nu)}(\{p(\nu), \nu_0\})$: $q(\nu)$ is just $p^{\beta(\nu)}_{n(\nu)} - p^{\beta(\nu)}_{n(\nu)-1} - \alpha(\nu)$, and so by definition of $p(\nu)$ we have $q(\nu) - \nu$ in $H^{\beta(\nu)}_{n(\nu)}(\{p(\nu)\})$; but if $q(\nu) \cap \nu$ is nonempty then it consists of a single ordinal $\delta$ and $\delta$ is largest so that $H^{\beta(\nu)}_{n(\nu)}(\delta \cup \{p^{\beta(\nu)}_n - \nu\}) \cap \nu = \delta$. This is precisely the ordinal used to provide a lower bound on $C^0_\nu$ in the proof of $\square$. As $C^0_\nu = \{\delta\}$ in this case we get $\nu_0 = \delta$. So if $\lambda_\alpha < \nu$ for sufficiently large $\alpha$ then $C_\nu$ is a final segment of $\{\bigcup(H^{\beta(\nu)}_{n(\nu)}(\alpha \cup \{q(\nu)\}) \cap \nu) \mid \alpha < \alpha(\nu)\}$. And the fact that $(\beta(\nu), n(\nu)) < (\beta(\alpha(\nu)), n(\alpha(\nu)))$ implies that $\lambda_\alpha \in C'_\nu$ for sufficiently large $\alpha$. Of course the alternative is that $\lambda_\alpha = \nu$ for sufficiently large $\alpha \in C'_{\alpha(\nu)}$ and so (g) is proved.

(h) There are two cases: either $(\beta(\nu), n(\nu)) = (\beta(\alpha(\nu)), n(\alpha(\nu)))$ or $(\beta(\nu), n(\nu) + 1) = (\beta(\alpha(\nu)), n(\alpha(\nu)))$. In the latter case we must have $H^{\beta(\nu)}_{n(\nu)}(\alpha \cup \{q(\nu)\}) \cap \nu$ bounded in $\nu$ for each $\alpha < \alpha(\nu)$, else we could contradict the $<_1$-minimality of $\nu$ by forming $H^{\beta(\nu)}_{n(\nu)}(\alpha_0 \cup \{q(\nu)\})$ where $\alpha_0 = \bigcup(\alpha(\nu) \cap H^{\beta(\nu)}_{n(\nu)}(\alpha \cup \{q(\nu)\}))$, $H^{\beta(\nu)}_{n(\nu)}(\alpha \cup \{q(\nu)\}) \cap \nu$ unbounded in $\nu$, $\alpha < \alpha(\nu)$.

First we treat the former case. Suppose $\alpha \in C'_{\alpha(\nu)}$. Then for some $k$, $\alpha \in C^k_{\alpha(\nu)}$ and so for some $\gamma$:

$$\alpha = \bigcup\big(\alpha(\nu) \cap H^{\beta(\nu)}_{n(\nu)}(\gamma \cup \{p(\alpha(\nu)), \alpha(\nu)_0, \ldots, \alpha(\nu)_{k-1}\})\big)$$

where $p(\alpha(\nu)) = \langle q(\alpha(\nu)), \alpha(\nu)\rangle$, $\alpha(\nu)_i$ are defined as in the proof of $\square$. Note that $q(\alpha(\nu))$ is precisely the $q(\nu)$ as defined in this section.

Write $p$ for $\{p(\alpha(\nu)), \alpha(\nu)_0, \ldots, \alpha(\nu)_{k-1}\}$ and $\rho$ for $\omega\rho^{\beta(\nu)}_{n(\nu)-1}$.

Now let $\nu_\alpha = \bigcup(\nu \cap H^{\beta(\nu)}_{n(\nu)}(\gamma \cup \{p\}))$, $\sigma_\alpha = \bigcup(\rho \cap H^{\beta(\nu)}_{n(\nu)}(\gamma \cup \{p\}))$. Then as $\alpha > \gamma(\alpha(\nu))$ we get $\nu_\alpha < \nu$, $\sigma_\alpha < \rho$ and by Lemma 1.13, $\alpha = \alpha(\nu) \cap H^{\beta(\nu),\sigma_\alpha}_{n(\nu)}(\alpha \cup \{p\})$. We get an embedding $\pi : \widetilde{J}_\beta \simeq H^{\beta(\nu),\sigma_\alpha}_{n(\nu)}(\alpha \cup \{p\})$ and $\pi^{-1}[\nu_\alpha] = \nu(\alpha)$, $\beta = \beta(\nu(\alpha))$, $n(\nu) = n(\nu(\alpha))$. In fact $\nu(\alpha) <_1 \nu_\alpha$ as $\pi$ maps $\nu(\alpha)$ cofinally into $\nu_\alpha$. It is not clear that $\nu(\alpha)$ is $<_1$-minimal as it is possible that there exists $\bar{\nu} <_1 \nu(\alpha)$, $\alpha(\bar{\nu}) \le \gamma(\alpha) = \gamma(\alpha(\nu))$, where $\gamma(\mu)$ is defined as was $\alpha(\mu)$ in the proof of $\square$. (We have changed notation to avoid confusion. ($\alpha(\bar{\nu}) > \gamma(\alpha)$ is ruled out because of the definition of $\gamma(\alpha)$.) However as $\nu$ is $<_1$-minimal the $Q$-condition must fail between $H^{\beta(\nu)}_{n(\nu)}(\gamma(\alpha) \cup \{q(\nu)\})$ and $\widetilde{J}_{\beta(\nu)}$ so we may choose $\alpha$ large enough in $C'_{\alpha(\nu)}$ so that this failure is captured by $H^{\beta(\nu),\sigma_\alpha}_{n(\nu)}(\alpha \cup \{p\})$, and therefore $\nu(\alpha)$ is $<_1$-minimal.

To see that $\nu_\alpha \in C'_\nu$ for sufficiently large $\alpha$, the same analysis as in the proof of (g) shows that if $q(\nu) \cap \nu \neq \emptyset$ then $C^1_\nu = \{\nu' < \nu \mid \text{For some } \gamma, \nu' = \cup(\nu \cap H^{\beta(\nu)}_{n(\nu)}(\gamma \cup \{q(\nu)\})\}$, and if $q(\nu) \cap \nu = \emptyset$ then $C^0_\nu$ is a final segment of this set, beyond an ordinal $\leq \gamma(\alpha(\nu))$. Thus it follows that either $C^1_\nu$ or $C^0_\nu$ agrees with $\{\nu_\alpha \mid \alpha \in C^0_{\alpha(\nu)}\}$ for $\alpha \geq \gamma(\alpha(\nu))$. If $C^0_{\alpha(\nu)}$ is unbounded in $\alpha(\nu)$ then we are done because then $C^1_\nu$ or $C^0_\nu$ as above is unbounded in $\nu$. If not then we need only note that $\alpha(\nu)_0 \in H^{\beta(\nu)}_{n(\nu)}(\{q(\nu), \nu^*\})$, $\nu^* \in H^{\beta(\nu)}_{n(\nu)}(\{q(\nu), \alpha(\nu)_0\})$ where $\nu^* = \cup C^1_\nu$ or $\cup C^0_\nu$ as above. Thus $\{\nu_\alpha \mid \alpha \in C^1_{\alpha(\nu)}\}$ agrees with $C^2_\nu$ or $C^1_\nu$ for $\alpha \geq \gamma(\alpha(\nu))$ and continuing in this way we get $\nu_\alpha \in C'_\nu$ for sufficiently large $\alpha$.

The last part of (h) is clear from the definition of the $\nu_\alpha$'s and the fact that $\alpha < \beta \Longrightarrow \nu_\alpha < \nu_\beta$.

Now we consider the case $(\beta(\nu), n(\nu) + 1) = (\beta(\alpha(\nu)), n(\alpha(\nu)))$ and recall that we have that $H^{\beta(\nu)}_{n(\nu)}(a \cup \{q(\nu)\}) \cap \nu$ is bounded in $\nu$ for each $\alpha < \alpha(\nu)$. Thus as in the proof of (g), for $\alpha \in C'_{\alpha(\nu)}$ we have $\alpha = \alpha(\nu) \cap H^{\beta(\nu)}_{n(\nu)}(\alpha \cup \{q(\nu)\})$ and $\hat{\nu}(\alpha) <_1 \nu_\alpha$ where $\nu_\alpha = \cup(\nu \cap H^{\beta(\nu)}_{n(\nu)}(\alpha \cup \{q(\nu)\}))$, $\hat{\nu}(\alpha) = \text{ordertype } (\nu \cap H^{\beta(\nu)}_{n(\nu)}(\alpha \cup \{q(\nu)\}))$. Also $\bar{\nu} <_1 \hat{\nu}(\alpha)$ implies as in the proof of (g) that $\tilde{\pi}_{\bar{\nu}\hat{\nu}(\alpha)}$ is $\Sigma^*_{n(\nu)+1}$-elementary, hence $\alpha(\bar{\nu}) \leq \gamma(\alpha(\nu))$; but as in the first part of the present proof, this is ruled out, for $\alpha$ sufficiently large. So for such $\alpha \in C'_{\alpha(\nu)}$ we have $\hat{\nu}(\alpha) = \nu(\alpha) <_1 \nu(\alpha)$ and $\nu(\alpha)$ is $<_1$- minimal. The proof that $\nu_\alpha \in C'_\nu$ for sufficiently large $\alpha$ is as in the proof of (g) and the last part of (h) is clear from the definition of $\nu_\alpha$ and the fact that $(\beta(\alpha(\nu)), n(\alpha(\nu)) > (\beta(\nu), n(\nu))$.

(i) As in (h) there are two cases: either $(\beta(\nu), n(\nu)) = (\beta(\alpha(\nu)), n(\alpha(\nu)))$, or $(\beta(\nu), n(\nu) + 1) = (\beta(\alpha(\nu)), n(\alpha(\nu)))$ and $\alpha < \alpha(\nu) \Longrightarrow H^{\beta(\nu)}_{n(\nu)}(\alpha \cup \{q(\nu)\}) \cap \nu$ is bounded in $\nu$.

We begin with the first case. As in (h), write $\alpha = \cup(\alpha(\nu) \cap H^{\beta(\nu)}_{n(\nu)}(\gamma \cup \{p\}))$ where $\gamma \geq \gamma(\alpha(\nu))$ if $C_{\alpha(\nu)} = C^0_{\alpha(\nu)}$ and $p = \{p(\alpha(\nu)), \alpha(\nu)_0, \ldots, \alpha(\nu)_{k-1}\}$. Also let $\sigma_\alpha = \cup(\rho \cap H^{\beta(\nu)}_{n(\nu)}(\gamma \cup \{p\}))$ where $\rho = \omega\rho^{\beta(\nu)}_{n(\nu)-1}$ and $\nu_\alpha = \cup(\nu \cap H^{\beta(\nu)}_{n(\nu)}(\gamma \cup \{p\}))$. Then $\nu(\alpha) <_1 \nu_\alpha$ and the $\nu_\alpha$'s obey the property expressed in (h). And there exists $\bar{\nu}_\alpha \leq_0 \bar{\nu} <^*_1 \nu$ such that $\bar{\nu}_\alpha <_1 \nu(\alpha)$ as we can take $\bar{\nu}_\alpha = \text{ordertype } (\nu \cap H^{\beta(\nu),\sigma_\alpha}_{n(\nu)}(\alpha(\bar{\nu}) \cup \{q(\nu)\}))$. As in (h) we can arrange that $\bar{\nu}_\alpha <^*_1 \nu(\alpha)$ for sufficiently large $\alpha$ by capturing a witness to the failure of the $Q$-condition between $H^{\beta(\nu)}_{n(\nu)}(\gamma(\alpha(\nu)) \cup \{q(\nu)\})$ and $\widetilde{J}_{\beta(\nu)}$. Note that in fact $k > 0$ and we must have $\gamma < \alpha(\bar{\nu})$ for sufficiently large $\alpha = \cup(\alpha(\nu) \cap H^{\beta(\nu)}_{n(\nu)}(\gamma \cup \{p\}))$ so we get $\nu_\alpha \in \text{Range}(\pi_{\bar{\nu}\nu})$ for $\nu_\alpha < \lambda = \cup \text{Range}(\pi_{\bar{\nu}\nu})$. Similarly, $\lambda \in \text{Range}(\pi_{\nu(\alpha)\nu_\alpha})$ when $\nu_\alpha > \lambda$ and we get $\bar{\nu}_\alpha = \bar{\nu}$.

Note that in the second case, $\pi_{\bar{\nu}\nu}$ is not cofinal. The argument now is very similar to the second case of the proof of (h), arranging $\bar{\nu} <^*_1 \nu(\alpha)$ as in the first case of the present proof. □

Our version of Morass with □ originates in Friedman [87] and is related to the concept of Morass with Linear Limits; see Donder [85].

The next principle arises in the proof of Jensen's Coding Theorem in the general case (Section 4.3) and bears some resemblance to the Squared Scales of Donder–Jensen–Stanley [85].

Again we first describe the object, obtained through the use of fine structure,, which satisfies this principle, before stating the principle itself. Let $T = \{\nu \mid \nu$ is a limit ordinal, $\widetilde{J}_\nu \models$ there is a largest cardinal $\alpha(\nu)$ and the cardinality of $\nu$ equals $\alpha(\nu)\}$. We do not require that $\alpha(\nu) = \mathrm{card}(\nu)$ is regular. Let $\beta(\nu) \leq \nu$ be the least limit ordinal such that for some $n$, $\rho_n^{\beta(\nu)} = \alpha(\nu)$ and let $n(\nu)$ be the least such $n$, $p(\nu) = \langle p_{n(\nu)}^{\beta(\nu)}, \alpha(\nu)\rangle$. Also $\hat{\beta}(\nu) = \beta(\nu) + \omega = T$-successor to $\nu$.

Now for any $k$ in $\omega$ and infinite cardinal $\alpha < \alpha(\nu)$ let $H(\nu, k, \alpha) = H_{n(\nu)+k}^{\beta(\nu)}(\alpha \cup \{p(\nu)\})$ and $\overline{H}(\nu, k, \alpha)$ its transitive collapse. Then $f(\nu, k, \alpha) = \alpha^+$ of $\overline{H}(\nu, k, \alpha)$. Note that $f(\nu, k, \alpha) \in T$ and $\alpha(f(\nu, k, \alpha)) = \alpha$.

For $\nu \in T$ we let $\widetilde{C}_\nu \subseteq \nu$ come from the $\Box$ Principle; then $\widetilde{C}_\nu$ is CUB in $\nu$, ordertype $(\widetilde{C}_\nu) \leq \alpha(\nu)$ and $\bar{\nu} \in \mathrm{Lim}\, \widetilde{C}_\nu \Longrightarrow \widetilde{C}_{\bar{\nu}} = \widetilde{C}_\nu \cap \bar{\nu}$. We let $C_\nu = \widetilde{C}_\nu$ if ordertype $(\widetilde{C}_\nu) < \alpha(\nu)$ and otherwise $C_\nu = \{\bar{\nu} < \nu \mid \text{For some } \alpha < \alpha(\nu), \bar{\nu} = \bigcup(\nu \cap H_{n(\nu)}^{\beta(\nu)}(\alpha \cup \{p(\nu)\}))\}$.

For $\nu \in T$, $k$ in $\omega$, $\alpha(\nu)$ an uncountable limit cardinal we define a CUB $D_{\nu,k} \leq \alpha(\nu)$ as follows. If $D = \{\alpha < \alpha(\nu) \mid \alpha = \alpha(\nu) \cap H_{n(\nu)+k+1}^{\beta(\nu)}(\alpha \cup \{p(\nu)\})\}$ is unbounded in $\alpha(\nu)$ then set $D_{\nu,k} = D$. If $H_{n(\nu)+k+1}^{\beta(\nu)}(\alpha \cup \{p(\nu)\}) \cap \alpha(\nu)$ is unbounded in $\alpha(\nu)$ for some $\alpha < \alpha(\nu)$ then we can choose $D_{\nu,k}$ CUB in $\alpha(\nu)$ of ordertype $< \alpha(\nu)$ so that $D_{\nu,k} \cap \alpha$ is $\Sigma^*_{n(\nu)+k+1}$-definable over $H_{n(\nu)+k}^{\beta(\nu)}(\alpha \cup \{p(\nu)\})$ uniformly for $\alpha \in \mathrm{Lim}\, D_{\nu,k}$. Otherwise define $D_{\nu,k} = \{\alpha_0, \alpha_1, \ldots\}$ where $\alpha_0 = 0$, $\alpha_{n+1} = \bigcup(\alpha(\nu) \cap H_{n(\nu)+k+1}^{\beta(\nu)}(\alpha_n \cup \{p(\nu)\}))$.

**Fine Scale Principle.** *There exist $\langle f(\nu, k, \alpha) \mid \nu \in T,\ k \in \omega, \alpha$ an infinite cardinal $< \alpha(\nu)\rangle$, $\langle C_\nu \mid \nu \in T\rangle$, $\langle D_{\nu,k} \mid \nu \in T,\ \alpha(\nu)$ an uncountable limit cardinal, $k \in \omega\rangle$ such that*

(a) *$\nu \in T \Longrightarrow \nu$ is a limit ordinal and not a cardinal, $T \cap \alpha^+$ is* CUB *in $\alpha^+$. $\alpha(\nu)$ denotes the cardinality of $\nu$.*

(b) *$f(\nu, k, \alpha) \in T \cap \alpha^+$; $C_\nu$ is* CUB *in $\nu$, ordertype $C_\nu \leq \alpha(\nu)$; $D_{\nu,k}$ is* CUB *in $\alpha(\nu)$. For $\alpha$ an uncountable limit cardinal and any $f : \alpha \longrightarrow \alpha$ such that $f(\alpha_0) < \alpha_0^+$ for every $\alpha_0 < \alpha$, there is $\nu \in T \cap \alpha^+$ such that $f(\alpha_0^+) < f(\nu, 0, \alpha_0^+)$ for sufficiently large $\alpha_0 < \alpha$.*

(c) *For any $\nu \in T$ there is $\alpha_0(\nu) < \alpha(\nu)$ such that for $\alpha_0(\nu) \leq \alpha < \alpha(\nu)$, $\alpha$ an infinite cardinal and $\bar{\nu} \in$* Lim *$C_\nu$:*

(c1) $f(\bar{\nu}, 0, \alpha) = \bigcup\{f(\bar{\bar{\nu}}, 0, \alpha) \mid \bar{\bar{\nu}} \in C_\nu \cap \bar{\nu}\}$

(c2) $\{f(\bar{\bar{\nu}}, 0, \alpha) \mid \bar{\bar{\nu}}\nu \in C_\nu \cap \bar{\nu}\} \in \widetilde{J}_\beta$ *where $\beta = T$-successor to $f(\bar{\nu}, 0, \alpha)$.*

(d) *For any* $\nu \in T$, $k \geq 0$ *there is* $\alpha_0(\nu, k) < \alpha(\nu)$ *such that for* $\alpha_0(\nu, k) \leq \alpha_0 < \alpha(\nu)$, $\alpha_0$ *an infinite cardinal and* $\alpha \in \mathrm{Lim}\, D_{\nu,k}$*:*

(d1) $f(f(\nu, k, \alpha), 1, \alpha_0) = \bigcup\{f(f(\nu, k, \bar{\alpha}), 1, \alpha_0) \mid \bar{\alpha} \in D_{\nu,k} \cap \alpha\}$

(d2) $\left\{f\left(f\left(\nu, k, \bar{\alpha}\right), 1, \alpha_0\right) \mid \bar{\alpha} \in D_{\nu,k} \cap \alpha\right\} \in \widetilde{J}_\beta$ *where* $\beta =$ $T$*-successor to* $f(f(\nu, k, \alpha), 1, \alpha_0)$.

*Proof.* Clauses (a), (b) are clear.

(c) Choose $\alpha_0(\nu)$ larger than ordertype $(C_\nu)$ if the latter is less than $\alpha(\nu)$. In this case the properties follow from the $\Sigma^*_{n(\nu)}$-elementarity of $H(\bar{\nu}, 0, \alpha)$ in $\widetilde{J}_{\beta(\bar{\nu})}$ and the $\Pi^*_{n(\nu)}$-definability of $C_{\bar{\nu}} = C_\nu \cap \bar{\nu}$. In case ordertype $(C_\nu) = \alpha(\nu)$ then note that $f(\bar{\nu}, 0, \alpha) = f(\bar{\nu}_\alpha, 0, \alpha)$ where $\bar{\nu}_\alpha \leq \bar{\nu}$ are in $C_\nu$ and $\bar{\nu}_\alpha = \alpha^{\text{th}}$ element of $C_\nu$. So the argument also works in this case.

(d) If ordertype $(D_{\nu,k}) < \alpha(\nu)$ then choose $\alpha_0(\nu, k)$ larger than this ordertype. Note that $D_{\nu,k} \cap \alpha$ is $\Sigma^*_{n(\nu)+k+1}$ or $\Pi^*_{n(\nu)+k+1}$-definable over $H(\nu, k, \alpha)$ when $\alpha < \alpha(\nu)$; so the result follows from the $\Sigma^*_{n(\nu)+k}$-elementarity of $H(\nu, k, \bar{\alpha})$ in $H(\nu, k, \alpha)$ and the fact that $n(f(\nu, k, \alpha)) = n(\nu) + k$. Also in case ordertype $(D_{\nu,k}) = \alpha(\nu)$ then note that $f(f(\nu, k, \alpha), 1, \alpha_0)$ is constant for $\alpha \geq \alpha_0^{\text{th}}$ element of $\mathrm{Lim}\, D_{\nu,k}$. □

The key clause in the Fine Scale Principle is (d). It says that $f(\nu, k+1, -)$ can be uniformly approximated by functions which differ from $f(\nu, k, -)$ only on a proper initial segment of $\alpha(\nu)$, in such a way that at limit stages $\alpha$, the $\alpha^{\text{th}}$ approximation can easily recover the $\alpha$-sequence of smaller approximations. This is a powerful tool for proving a statement for each $f(\nu, k, -)$, by induction on $(\nu, k)$. In the case of Jensen coding, extendibility of conditions can be proved in this way.

We conclude with a discussion of Gap 2 morasses. Again we begin with a description of the intended object. Let $S^0 = \{\alpha > \omega \mid \alpha$ is a limit ordinal, $\alpha$ is cardinal-correct$\}$, $S^1 = \{\nu \mid \nu$ is a limit ordinal and for some $\alpha(\nu) \in S^0$, $\widetilde{J}_\nu \vDash \alpha(\nu)$ is the largest cardinal and $\alpha(\nu)$ is regular$\}$, $S^2 = \{\mu \mid \mu$ is a limit ordinal $\mu$ is not a cardinal and for some $\nu(\mu) \in S^1$, $\widetilde{J}_\mu \vDash \nu(\mu)$ is the largest cardinal$\}$. Thus if $\mu \in S^2$ then $\widetilde{J}_\mu \vDash \alpha(\nu(\mu))$ is regular, $\nu(\mu) = \alpha(\nu(\mu)^+$ is the largest cardinal. We write $\nu_0 <_0 \nu_1$ if $\nu_0 < \nu_1$ and for some $\alpha$, $\nu_0$ and $\nu_1$ both belong to $S_\alpha = \{\nu \mid \alpha(\nu) = \alpha\}$; also we write $\mu_0 <_0 \mu_1$ if $\mu_0 < \mu_1$ and for some $\nu \in S^1$, $\mu_0$ and $\mu_1$ both belong to $S_\nu = \{\mu \mid \nu(\mu) = \nu\}$. For $\alpha \in S^0$, $\nu(\alpha)$ denotes max $S_\alpha$ (when $\alpha$ is not regular) and for $\nu \in S^1$, $\mu(\nu)$ denotes $\nu \cup \max S_\nu$ (when $\nu$ is not regular).

Now we give the main definition. For $\nu \in S^1$, $\nu$ not regular let $\beta(\nu) \geq \nu$ be the least limit ordinal such that for some least $n(\nu)$, $\rho^{\beta(\nu)}_{n(\nu)} \leq \alpha(\nu)$ and let $q(\nu) = p^{\beta(\nu)}_{n(\nu)} - \alpha(\nu)$. (Thus $q(\nu)$ is least so that $H^{\beta(\nu)}_{n(\nu)}(\alpha(\nu) \cup \{q(\nu)\}) = \widetilde{J}_{\beta(\nu)}$.) The previous, as well the definition of $\bar{\nu} <_1 \nu$ are as in the gap 1 case: $\bar{\nu} <_1 \nu$ iff there exists $\tilde{\pi}_{\bar{\nu}\nu} = \pi :$ $\widetilde{J}_{\beta(\bar{\nu})} \longrightarrow \widetilde{J}_{\beta(\nu)}$ which is $\Sigma^*_{n(\nu)}$-elementary, $n(\bar{\nu}) = n(\nu)$, $\pi \restriction \alpha(\bar{\nu}) =$ identity, $\pi(\alpha(\bar{\nu})) = \alpha(\nu)$, $\pi(q(\bar{\nu})) = q(\nu)$ and the $Q$-condition is satisfied: whenever $\varphi(x)$ is $\Sigma^*_{n(\nu)}$ in parameters $\bar{p} \in \widetilde{J}_{\beta(\bar{\nu})}$ then $\{\bar{\nu}' < \bar{\nu} \mid \widetilde{J}_{\beta(\bar{\nu})} \vDash \varphi(\bar{\nu}', \bar{p})\}$ is bounded in $\bar{\nu}$ iff

$\{\nu' < \nu \mid \widetilde{J}_{\beta(\nu)} \models \varphi(\nu', \pi(\bar{p}))\}$ is bounded in $\nu$. We write $\pi_{\bar{\nu}\nu}$ for $\pi \restriction \mu(\bar{\nu}) \cup \bar{\nu}$. Now in addition, for $\mu \in S^2$, define $\beta(\mu), \mu(\mu), q(\mu)$ in the same way, with $\alpha(\nu)$ replaced by $\nu(\mu)$. Also define $\bar{\mu} <_1 \mu$ in the same way, with $\alpha(\bar{\nu}), \alpha(\nu)$ replaced by $\nu(\bar{\mu}), \nu(\mu)$. We write $\pi_{\bar{\mu}\mu}$ for $\pi \restriction \bar{\mu}$.

Note that we defined $\pi_{\bar{\nu}\nu}$ for $\bar{\nu} <_1 \nu$ in $S^1$ to be $\pi \restriction \mu(\bar{\nu})$ and not simply $\pi \restriction \bar{\nu}$. This means that $\pi_{\bar{\nu}\nu}$ moves ordinals $\bar{\mu} \in S^2$, $\bar{\mu} < \mu(\bar{\nu})$ and raises interesting questions about how the relation $<_1$ on such ordinals is affected by applying $\pi_{\bar{\nu}\nu}$. Thus our gap 2 morass properties pertain not only to the "gap 1" relationships $\bar{\nu} <_1 \nu$, $\bar{\mu} <_1 \mu$ but also to the way in which they interact.

**Gap 2 Morass.** *There exist $\langle S_\alpha \mid \alpha \in S^0\rangle$, $\langle S_\nu \mid \nu \in S^1\rangle$, a binary relation $<_1$ on $(S^1 \times S^1) \cup (S^2 \times S^2)$, and sequences $\langle \pi_{\bar{\nu}\nu} \mid \bar{\nu} <_1 \nu$ in $S^1\rangle$, $\langle \pi_{\bar{\mu}\mu} \mid \bar{\mu} <_1 \mu$ in $S^2\rangle$ such that $S^1 = \bigcup\{S_\alpha \mid \alpha \in S^0\}$, $S^2 = \bigcup\{S_\nu \mid \nu \in S^1\}$ and:*

(a) *$S^0 \cap \kappa$ is* CUB *in $\kappa$ for each uncountable cardinal $\kappa$.*

(b) *For $\alpha \in S^0$, $S_\alpha$ is a closed subset of $(\alpha, \alpha^+]$ and for $\nu \in S^1$, $S_\nu$ is a closed subset of $(\nu, \nu^+)$. And:*

(b1) *$\alpha$ regular $\Longrightarrow S_\alpha = S^0 \cap (\alpha, \alpha^+]$*

(b2) *$\alpha$ a singular cardinal $\Longrightarrow S_\alpha$ is a proper initial segment of $S^0 \cap (\alpha, \alpha^+)$.*

(b3) *$\alpha < \alpha'$ in $S^0$, $\alpha$ not a cardinal $\Longrightarrow \bigcup S_\alpha < \alpha'$*

(b4) *$\nu < \nu'$ in $S^1$, $\nu$ not a cardinal $\Longrightarrow \bigcup S_\nu < \nu'$.*

**Notation.** For $\nu \in S^1$, $\alpha(\nu)$ denotes the $\alpha$ such that $\nu \in S_\alpha$ and for $\mu \in S^2$, $\nu(\mu)$ denotes the $\nu$ such that $\mu \in S_\nu$. We write $\nu <_0 \nu'$ if $\nu < \nu'$ and $\alpha(\nu) = \alpha(\nu')$, $\mu <_0 \mu'$ if $\mu < \mu'$ and $\nu(\mu) = \nu(\mu')$. If $\alpha \in S^0$ then $\nu(\alpha) = \max S_\alpha$ and if $\nu \in S^1$ is not a cardinal then $\mu(\nu) = \nu \cup \max S_\nu$. If $\nu$ is not $<_0$-maximal then $\nu^+$ denotes its $<_0$-successor (similarly for $\mu \in S^2$).

(c) *$<_1$ is a tree and if $\bar{\nu} <_1 \nu$ in $S^1$ then $\alpha(\bar{\nu}) < \alpha(\nu)$ and $\nu$ is not a cardinal. If $\bar{\mu} <_1 \mu$ in $S^2$ then $\nu(\bar{\mu}) < \nu(\mu)$. If $\bar{\nu} <_1 \nu$ then $\bar{\mu}$ is $<_0$-minimal, successor, limit iff $\mu$ is $<_0$-minimal, successor, limit where either $\bar{\mu} < \mu(\bar{\nu})$ and $\mu = \pi_{\bar{\nu}\nu}(\bar{\mu})$, or $(\bar{\mu}, \mu) = (\mu(\bar{\nu}), \mu(\nu))$. If $\bar{\mu} <_1 \mu$ then $\bar{\mu}$ is $<_0$-minimal, successor, limit iff $\mu$ is $<_0$-minimal, successor, limit.*

(d) *If $\bar{\nu} <_1 \nu$ then $\pi = \pi_{\bar{\nu}\nu} : \mu(\bar{\nu}) \longrightarrow \mu(\nu)$ is order-preserving, $\pi^{-1}[S_{\alpha(\nu)}] = S_{\alpha(\bar{\nu})} \cap \bar{\nu}$, $\pi^{-1}[S_\nu] = S_{\bar{\nu}}$, and $\pi(\bar{\mu}^+) = \pi(\bar{\mu})^+$ whenever $\pi(\bar{\mu})^+ < \mu(\nu)$. If $\bar{\nu}_0 <_0 \bar{\nu}$, $\nu_0 = \pi(\bar{\nu}_0)$ then $\bar{\nu}_0 <_1 \nu_0$ and $\pi_{\bar{\nu}_o\nu_0} = \pi \restriction \mu(\bar{\nu}_0)$. If $\bar{\nu}$ is a $<_0$-limit, $\lambda = \bigcup \mathrm{Range}(\pi \restriction \bar{\nu})$ then $\bar{\nu} <_1 \lambda$, and if $\mu_\lambda = \bigcup \mathrm{Range}\, \pi_{\bar{\nu}\lambda}$, $\mu = \bigcup \mathrm{Range}\, \pi_{\bar{\nu}\nu}$ then $\mu_\lambda <_1 \mu$ and $\pi_{\bar{\nu}\nu} = \pi_{\mu_\lambda\mu} \circ \pi_{\bar{\nu}\lambda}$. If $\bigcup \mathrm{Range}(\pi \restriction \bar{\nu}) = \nu$, $\alpha \in S^0$, and for each $\bar{\nu}_0 <_0 \bar{\nu}$, $\alpha = \alpha(\nu_0')$ for some $\nu_0' <_1 \pi_{\bar{\nu}\nu}(\bar{\nu}_0)$, then $\alpha = \alpha(\nu')$ for some $\nu' <_1 \nu$. Similarly for $\pi_{\bar{\mu}\mu}$ when $\bar{\mu} <_1 \mu$, with $\mu(\bar{\nu}), \mu(\nu), \alpha(\bar{\nu}), \alpha(\nu)$ replaced by $\bar{\mu}, \mu, \nu(\bar{\mu}), \nu(\mu)$ and $\pi_{\mu_\lambda\mu}$ replaced by the identity.*

(e) $\bar{\bar{\nu}} <_1 \bar{\nu} <_1 \nu \Longrightarrow \pi_{\bar{\bar{\nu}}\nu} = \pi_{\bar{\nu}\nu} \circ \pi_{\bar{\bar{\nu}}\bar{\nu}}$. *For* $\nu \in S^1$, $\{\alpha(\bar{\nu}) \mid \bar{\nu} <_1 \nu\}$ *is closed in* $\alpha(\nu)$ *and unbounded unless* $\nu$ *is* $<_0$*-maximal. If* $\{\alpha(\bar{\nu}) \mid \bar{\nu} <_1 \nu\}$ *is unbounded in* $\alpha(\nu)$ *then* $\mu(\nu) = \bigcup\{\text{Range}(\pi_{\bar{\nu}\nu}) \mid \bar{\nu} <_1 \nu\}$. *Similarly for* $\mu \in S^2$, *with* $\alpha(\nu), \mu(\nu)$ *replaced by* $\nu(\mu), \mu$.

(f) *Suppose* $\bar{\nu} <_1 \nu$. *Then* $\bar{\nu} < \mu(\bar{\nu})$ *iff* $\nu < \mu(\nu)$. *If the latter holds, then* $\mu(\bar{\nu})$ *is* $<_1$*-minimal, successor, limit iff* $\mu(\nu)$ *is* $<_1$*-minimal, successor, limit; for* $\bar{\mu}_0, \bar{\mu}_1 < \mu(\bar{\nu})$, $\bar{\mu}_0 <_1 \bar{\mu}_1$ *iff* $\pi_{\bar{\nu}\nu}(\bar{\mu}_0) <_1 \pi_{\bar{\nu}\nu}(\bar{\mu}_1)$*; and for* $\bar{\nu}_0 \in S^1$, $\bar{\nu}_0 = \nu(\bar{\mu}_0)$ *for some* $\bar{\mu}_0 <_1 \mu(\bar{\nu})$ *iff* $\pi_{\bar{\nu}\nu}(\bar{\nu}_0) = \nu(\mu_0)$ *for some* $\mu_0 <_1 \mu(\nu)$.

(g) *Suppose* $\bar{\nu} <_1 \nu, \bar{\nu} < \mu(\bar{\nu})$ *and* $\bar{\mu} = \mu(\bar{\nu})$ *is a* $<_1$*-successor. Let* $\bar{\mu}_0 <_1^* \bar{\mu}$ *denote that* $\bar{\mu}_0$ *is the* $<_1$*-predecessor to* $\bar{\mu}$. *Then if* $\mu_0 <_1^* \mu$ *we have* $\pi_{\bar{\nu}\nu}(\nu(\bar{\mu}_0)) = \nu(\mu_0)$. *And* $\pi_{\bar{\mu}_0\bar{\mu}}$ *is cofinal iff* $\pi_{\mu_0\mu}$ *is cofinal. If* $\pi_{\bar{\mu}_0\bar{\mu}}$ *is not cofinal,* $\bar{\lambda} = \bigcup\text{Range}(\pi_{\bar{\mu}_0\bar{\mu}})$ *then* $\pi_{\bar{\nu}\nu}(\bar{\mu}_0) = \mu_0$, $\pi_{\bar{\nu}\nu}(\bar{\lambda}) = \bigcup\text{Range}(\pi_{\mu_0\mu})$.

(h) *Suppose* $\bar{\nu} <_1 \nu, \bar{\nu} < \mu(\bar{\nu})$ *and* $\mu(\bar{\nu})$ *is not a* $<_1$*-limit. Then* $\pi_{\bar{\nu}\nu}$ *is cofinal iff* $\pi_{\bar{\nu}\nu} \restriction \bar{\nu}$ *is cofinal.*

(i) *Suppose* $\bar{\nu} <_1 \nu$ *and* $\bar{\nu} < \mu(\bar{\nu}) = \bar{\mu}$. *If* $\bar{\mu}_0 <_1 \bar{\mu}_1 \leq \bar{\mu}$, $\mu_0 <_1 \mu_1 = \pi_{\bar{\nu}\nu}(\bar{\mu}_1)$ *(or* $\mu_0 <_1 \mu(\nu)$ *if* $\bar{\mu}_1 = \bar{\mu}$*) and* $\nu(\mu_0) = \pi_{\bar{\nu}\nu}(\nu(\bar{\mu}_0))$ *then* $\pi_{\mu_0\mu_1^0}\pi_{\bar{\nu}\nu} \restriction \bar{\mu}_0 = \pi_{\bar{\nu}\nu^0}\pi_{\bar{\mu}_0\bar{\mu}_1}$.

**Remarks.** (1) Jensen points out that we cannot have perfect tree preservation, which would say: $\bar{\mu}_0 <_1 \bar{\mu}_1 \Longleftrightarrow \pi_{\bar{\nu}\nu}(\bar{\mu}_0) <_1 \pi_{\bar{\nu}\nu}(\bar{\mu}_1)$ for $\bar{\mu}_1 \leq \mu(\bar{\nu})$. (We take $\pi_{\bar{\nu}\nu}(\mu(\bar{\nu}))$ to be $\mu(\nu)$.) Thus in (g) we wrote only $\pi_{\bar{\nu}\nu}(\nu(\bar{\mu}_0)) = \nu(\mu_0)$ rather than $\pi_{\bar{\nu}\nu}(\bar{\mu}_0) = \mu_0$. So $\pi_{\bar{\nu}\nu}$ may send $\bar{\mu} <_1^* \mu(\bar{\nu})$ to $\mu$ where $\nu(\mu) = \nu(\mu_0)$, $\mu_0 <_1^* \mu(\nu)$, $\mu \neq \mu_0$. If $\pi_{\bar{\mu}\mu(\bar{\nu})}$ is not cofinal, however, this will not happen and we get $\mu = \mu_0$.

(2) Though $\pi_{\bar{\nu}\nu}$ may fail to preserve the relation $\bar{\mu}_0 <_1 \bar{\mu}_1$ when $\bar{\mu}_1 = \mu(\bar{\nu})$ it does preserve the following relation $\dashv$: $\bar{\mu}_0 \dashv \bar{\mu}_1$ iff there are $\bar{\bar{\mu}}_0 <_1^* \bar{\mu}_0, \bar{\bar{\mu}}_1 <_1^* \bar{\mu}_1$, $\bar{\bar{\mu}}_0 <_0 \bar{\bar{\mu}}_1$ and $\bar{\mu}_0 <_1 \pi_{\bar{\bar{\mu}}_1\bar{\mu}_1}(\bar{\bar{\mu}}_0)$. Moreover in case $\pi_{\bar{\bar{\mu}}_1\bar{\mu}_1}$ is cofinal (the troublesome case for $<_1$-preservation) then $\bar{\mu}_1$ is the direct limit of $\{\bar{\mu}_0 \mid \bar{\mu}_0 \dashv \bar{\mu}_1\}$ via natural maps (if $\sigma \dashv \tau$ then $f_{\sigma\tau}$ is $\pi_{\sigma\gamma}$ where $\bar{\sigma} <_1^* \sigma, \bar{\tau} <_1^* \tau, \gamma = \pi_{\bar{\tau}\tau}(\bar{\sigma})$).

(3) Of course one could formulate "Gap 2 Morass with $\square$" but as we know of no applications of this principle, we have elected not to do so here for the sake of simplicity.

*Proof.* (a)–(e) This is just as in the gap 1 case, with one exception: we must show that if $\bar{\nu}$ is $<_0$-limit, $\bar{\nu} < \mu(\bar{\nu})$, $\bar{\nu} <_1 \nu$, $\pi_{\bar{\nu}\nu} \restriction \bar{\nu}$ not cofinal, $\lambda = \bigcup\text{Range}(\pi_{\bar{\nu}\nu} \restriction \bar{\nu})$, $\mu_\lambda = \bigcup(\text{Range}(\pi_{\bar{\nu}\nu}))$, $\mu = \bigcup(\text{Range}(\pi_{\bar{\nu}\nu}))$ then $\mu_\lambda <_1 \mu$ and $\pi_{\bar{\nu}\nu} = \pi_{\mu_\lambda\mu} \circ \pi_{\bar{\nu}\lambda}$. Consider $H_\lambda = H_{n(\nu)}^{\beta(\nu),\sigma}(\lambda \cup \{q(\nu)\})$, $H = H_{n(\nu)}^{\beta(\nu),\sigma}(\nu \cup \{q(\nu)\})$ where $\sigma = \bigcup(\rho_{n(\nu)-1}^{\beta(\nu)} \cap H_{n(\nu)}^{\beta(\nu)}(\alpha(\bar{\nu}) \cup \{q(\nu)\}))$. Then $H_\lambda$ is $\Sigma_{n(\nu)}^*$-elementary in $H$ and after transitive collapse yields $\pi_{\mu_\lambda\mu}$. And $\pi_{\bar{\nu}\nu} = \pi_{\mu_\lambda\mu} \circ \pi_{\bar{\nu}\lambda}$ follows from the fact that $\pi_{\bar{\nu}\nu}, \pi_{\bar{\nu}\lambda}$ are obtained respectively by collapsing the inclusion of $H_{n(\nu)}^{\beta(\nu)}(\alpha(\bar{\nu}) \cup \{q(\nu)\})$ in $H$, $H_\lambda$.

(f) The fact that $\bar{\mu} > \bar{\nu}$ iff $\mu > \nu$ is clear from $\Sigma_{n(\nu)}^*$-elementarity of $\tilde{\pi}_{\bar{\nu}\nu}$. For the rest, first suppose that for some $\mu_0 <_1 \mu$, $\pi_{\mu_0\mu}$ is cofinal. If $\mu$ is a $<_1$-successor then we can take $\mu_0 <_1^* \mu$ and then we have $\nu(\mu_0) \in q(\nu)$ and therefore $\nu(\mu_0) = \pi(\bar{\nu}_0)$

for some $\bar{\nu}_0$, where $\pi = \tilde{\pi}_{\bar{\nu}\nu}$. Then $H^{\beta(\bar{\nu})}_{n(\nu)}(\bar{\nu}_0 \cup \{q(\bar{\nu})\}) = H$ is cofinal in $\bar{\mu}$ and $H \cap \bar{\nu} = \bar{\nu}_0$. So we get $\bar{\mu}_0 <^*_1 \bar{\mu}$, $\nu(\bar{\mu}_0) = \bar{\nu}_0$. If $\mu$ is a $<_1$-limit then $n(\mu) < n(\nu)$ and $\pi$ is therefore $\Sigma^*_{n(\mu)+1}$-elementary. Thus $\mathrm{Range}(\pi) \cap \{\nu(\mu_0) \mid \mu_0 <_1 \mu\}$ is unbounded in $\mathrm{Range}(\pi) \cap \nu$ so $\bar{\mu}$ is a $<_1$-limit as all maps $\pi_{\mu_0\mu}$, $\mu_0 <_1 \mu$ sufficiently large, are cofinal.

Second suppose that there is no cofinal $\pi_{\mu_0\mu}$, $\mu_0 <_1 \mu$. If $n(\mu) < n(\nu)$ then for $\mu_0 <_1 \mu$ we must have $\Sigma^*_{n(\mu)+1}$-elementarity for $\tilde{\pi}_{\mu_0\mu}$ (see the proof of (g) from the gap 1 case). Thus if $\mu$ is a $<_1$-limit then $n(\nu) \geq n(\mu)+2$ and we get that $\bar{\mu}$ is a $<_1$-limit as $\{\nu(\mu_0) \mid \mu_0 <_1 \mu\}$ is $\Pi^*_{n(\mu)+1}$. If $\mu$ is not a $<_1$-limit then $\max\{\nu(\mu_0) \mid \mu_0 <_1 \mu\}$ belongs to $\mathrm{Range}(\pi)$ as it is either in $q(\nu)$ or is 0. Thus $\bar{\mu}$ is not a $<_1$-limit and if $\mu_0 <^*_1 \mu$ then $\nu(\mu_0) = \nu(\bar{\mu}_0)$ where $\bar{\mu}_0 <^*_1 \bar{\mu}$. If $\mu$ is $<_1$-minimal then so is $\bar{\mu}$. Finally, if $n(\mu) = n(\nu)$ then $\mu$ is not a $<_1$-limit; if $\nu_0 = \bigcup\{\nu' < \nu \mid \nu' = \nu \cap H^{\beta(\nu)}_{n(\nu)}(\nu' \cup \{q(\mu)\})\}$ then $\nu_0 \in \mathrm{Range}(\pi)$ but now since $\pi$ is $Q$-elementary we get $H^{\beta(\nu)}_{n(\nu)}(\nu_0 \cup \{q(\mu)\}) \cap \mu$ bounded below $\mu^* = \bigcup(\mathrm{Range}(\pi) \cap \mu)$. It follows that $\mu_0 <_1 \mu$ iff $\mu_0 <_1 \mu^*$, so $\bar{\mu}$ is not a $<_1$-limit and if $\mu_0 <^*_1 \mu$ then $\nu(\mu_0) = \pi(\nu(\bar{\mu}_0))$ where $\bar{\mu}_0 <^*_1 \bar{\mu}$. If $\mu$ is $<_1$-minimal then so is $\bar{\mu}$.

Note that in the above argument we also verified the final statement of (f). The remaining claim in (f) is clear by $\Sigma_1$-elementarity.

(g) The argument in the proof of (f) showed that $\pi_{\bar{\nu}\nu}(\nu(\bar{\mu}_0)) = \nu(\mu_0)$ and $\pi_{\bar{\mu}_0\mu}$ cofinal iff $\pi_{\mu_0\mu}$ cofinal. Finally if $\lambda = \bigcup \mathrm{Range}(\pi_{\mu_0\mu}) < \mu$ then we get $\lambda \in \mathrm{Range}(\pi_{\bar{\nu}\nu})$ by the argument in (f), and hence $\mu_0 \in \mathrm{Range}(\pi_{\bar{\nu}\nu})$. Then we must have $\pi^{-1}(\mu_0) = \bar{\mu}_0$.

(h) If $\mu$ is not a $<_1$-limit then either $n(\mu) = n(\nu)$ and the result follows easily or $H^{\beta(\nu)}_{n(\mu)}(\nu_0 \cup \{q(\mu)\}) \cap \mu$ is bounded in $\mu$ for each $\nu_0 < \nu$, which means that $X \cap \mu$ bounded in $\mu$ iff $X \cap \nu$ bounded in $\nu$ for any $X$ which is $\Sigma^*_{n(\nu)}$-elementary in $\tilde{J}_{\beta(\nu)}$.

(i) This is clear if $\bar{\mu}_1 < \bar{\mu}$. Otherwise it follows immediately when $\pi_{\bar{\nu}\nu}(\bar{\mu}_0) = \mu_0$ and otherwise by the fact that $\pi_{\bar{\mu}_0\mu}$ is given by $H^{\beta(\bar{\mu})}_{n(\bar{\mu})}(\nu(\bar{\mu}_0) \cup \{q(\bar{\mu})\})$, $\pi_{\mu_0\mu}$ is given by $H^{\beta(\mu)}_{n(\mu)}(\nu(\mu_0) \cup \{q(\mu)\})$ and $\pi_{\bar{\nu}\nu}(\nu(\bar{\mu}_0)) = \nu(\mu_0)$. □

*Chapter 2*

# Forcing

In this chapter we introduce the method of forcing, which provides a way to construct extensions of Gödel's model $L$. Cohen [66] invented this method to demonstrate the unprovability of the continuum hypothesis (CH) in ZFC and of the axiom of choice (AC) in ZF; as AC, CH hold in $L$ we obtain in this way two striking examples of undecidable propositions.

Cohen's method was extended by Solovay [70] to provide a very general and powerful technique for enlarging any transitive ZFC model $M$, given the choice of a partial ordering $P \in M$. In the first section of this chapter we present this technique, called "set forcing" due to the fact that $P$ is required to be a set in $M$. Then we turn to "class forcing," the first example of which appears in Easton [70], where $P$ is now allowed to be a definable (or more generally, "amenable") class of $M$. An important point is that unlike with set forcing, ZFC need not be preserved by class forcing and in Section 2.2 we isolate the necessary "tameness" requirement on $P$ needed to guarantee ZFC preservation. Section 2.3 discusses four basic examples.

## 2.1 Set Forcing

Let $M$ be a transitive model of ZF, either a set or a class. The case that interests us most is when $M$ is $L$, but the forcing method does not require such a restriction. Let $P \in M$ be a partial ordering; our plan is to do the following:

(1) We define what it means for $G \subseteq P$ to be $P$-generic over $M$.

(2) We describe, for each $G \subseteq P$, a transitive $M[G] \supseteq M \cup \{G\}$.

(3) We prove that if $G$ is $P$-generic over $M$ then $M[G]$ is a model of ZF and, assuming AC holds in $M$, that AC holds in $M[G]$.

(1)–(3) summarizes the forcing method. The one thing missing is a demonstration that $G$'s as above actually exist. One solution is to make the assumption that $M$ is countable; however we mentioned that the case of greatest interest for us is when $M = L$. Chapter 3 is devoted to the question of finding $P$-generic $G$ in this case, when $P$ is a set or class of $L$.

## $P$-Generic Sets

We assume that $P = (P, \leq)$ has a greatest element, which we call $1^P$. We think of $p \leq q$ as meaning "$p$ is at least as strong as $q$."

**Definition.** $p, q$ are *compatible* if for some $r$, $r \leq p$ and $r \leq q$. $D \subseteq P$ is *dense* if $\forall p \exists q (q \leq p$ and $q \in D)$. $G \subseteq P$ is *$P$-generic over $M$* if:

(1) $p, q \in G \Longrightarrow p, q$ are compatible.

(2) $p \geq q \in G \Longrightarrow p \in G$.

(3) $D \subseteq P$, $D$ dense, $D \in M \Longrightarrow G \cap D \neq \emptyset$.

**Assumption.** We assume that for each $p \in P$ there exists $G \subseteq P$, $p \in G$, $G$ $P$-generic over $M$.

Our Assumption is vacuous if $M$ is countable as we can list the dense $D \in M$ as $D_0, D_1, \ldots$, define $p_0 = p$, $p_n \geq p_{n+1} \in D_n$ and take $G = \{p \mid p_n \leq p$ for some $n\}$. We will discuss this Assumption in more detail in Chapter 3.

## The Extension $M[G]$

We define $M[G]$ to consist of sets which have "names" in $M$, interpreted using $G$. Our approach owes much to Kunen [80] and to Shoenfield [71].

A *name* is a set $\sigma \in M$ consisting of pairs $\langle \tau, p \rangle$ where $\tau$ is a name and $p \in P$. Equivalently, a name is an element of $\cup\{\text{Name}_\alpha \mid \alpha \in \text{ORD}(M)\}$ where $\text{Name}_0 = \emptyset$, $\text{Name}_{\alpha+1}$ = All subsets of $\text{Name}_\alpha \times P$ in $M$, $\text{Name}_\lambda = \cup\{\text{Name}_\alpha \mid \alpha < \lambda\}$ for limit $\lambda$.

The interpretation of the name $\sigma$ is $\sigma^G = \{\tau^G \mid \langle \tau, p \rangle \in \sigma, p \in G\}$. Then $M[G] = \{\sigma^G \mid \sigma \text{ a name}\}$.

**Lemma 2.1.** *Suppose* $1^P \in G \subseteq P$.

(a) $M \subseteq M[G]$, $G \in M[G]$,

(b) $M[G]$ *is transitive,* $\text{ORD}(M[G]) = \text{ORD}(M)$.

(c) *If* $M \cup \{G\} \subseteq N$, $N$ *a model of* ZF *then* $M[G] \subseteq N$.

*Proof.* (a) For $a \in M$ define $\hat{a} = \{\langle \hat{b}, 1^P \rangle \mid b \in a\}$ and then $\hat{a}^G = a$. Also $G = \gamma^G$ where $\gamma = \{\langle \hat{p}, p \rangle \mid p \in P\}$.

(b) If $a \in \sigma^G \in M[G]$ then by definition $a = \tau^G \in M[G]$ for some $\tau$; so $M[G]$ is transitive. By induction on rank $\sigma$ = least $\alpha$ such that $\sigma \in \text{Name}_{\alpha+1}$, it follows that the von Neumann rank of $\sigma^G$ is at most rank $\sigma \in \text{ORD}(M)$. So $\text{ORD}(M[G]) \subseteq \text{ORD}(M)$.

(c) For each $\alpha \in \text{ORD}(M)$, the inductive definition of $\sigma^G$ for rank $\sigma < \alpha$ can be carried out in $N$. □

**Definition.** Suppose $p$ belongs to $P$, $\varphi(v_1, \ldots, v_n)$ is a formula and $\sigma_1, \ldots, \sigma_n$ are names. We write $p \Vdash \varphi(\sigma_1, \ldots, \sigma_n)$, *$p$ forces* $\varphi(\sigma_1, \ldots, \sigma_n)$, iff whenever $G \subseteq P$ is $P$-generic over $M$ and $p \in P$, we have $M[G] \models \varphi(\sigma_1^G, \ldots, \sigma_n^G)$. And we write $P \Vdash \varphi(\sigma_1, \ldots, \sigma_n)$ for $1^P \Vdash \varphi(\sigma_1, \ldots, \sigma_n)$.

The key to forcing is to establish the Definability and Truth lemmas. The Definability lemma, much like Gödel's Completeness Theorem equating nonconstructive semantical validity with semiconstructive syntactical provability, says that the forcing relation is $M$-definable for each $\varphi$ (as a property of $p, \sigma_1, \ldots, \sigma_n$). The Truth lemma says that $P$-generic $G$ are in fact "generic" in the intuitive sense: If $\varphi(\sigma_1^G, \ldots, \sigma_n^G)$ is true in $M[G]$ then for some $p \in G$, it is true in every $M[H]$, $H$ $P$-generic and containing $p$.

**Lemma 2.2** (Definability of $\Vdash$). *For any $\varphi$, the relation "$p \Vdash \varphi(\sigma_1, \ldots, \sigma_n)$" is definable in $M$.*

**Lemma 2.3** (Truth Lemma). *If $G$ is $P$-generic over $M$ then $M[G] \models \varphi(\sigma_1^G, \ldots, \sigma_n^G)$ $\Longleftrightarrow \exists p \in G\ (p \Vdash \varphi(\sigma_1, \ldots, \sigma_n))$.*

Our proof strategy for these lemmas is indirect: We define a relation $\Vdash^*$ for which Lemma 2.2 is clear, prove Lemma 2.3 for $\Vdash^*$ and finally show $\Vdash = \Vdash^*$.

**Definition** ($\Vdash^*$). We say that $D \subseteq P$ is *dense* $\leq p$ if $\forall q \leq p \exists r (r \leq q, r \in D)$.

(1) $p \Vdash^* \sigma \in \tau$ iff $\{q \mid \exists \langle \pi, r \rangle \in \tau$ such that $q \leq r, q \Vdash^* \sigma = \pi\}$ is dense $\leq p$.

(2) $p \Vdash^* \sigma = \tau$ iff for all $\langle \pi, r \rangle \in \sigma \cup \tau$, $p \Vdash^* (\pi \in \sigma \Longleftrightarrow \pi \in \tau)$.

(3) $p \Vdash^* \varphi \wedge \psi$ iff $p \Vdash^* \varphi$ and $p \Vdash^* \psi$.

(4) $p \Vdash^* \sim \varphi$ iff $\forall q \leq p (\sim q \Vdash^* \varphi)$.

(5) $p \Vdash^* \forall x \varphi$ iff for all names $\sigma$, $p \Vdash^* \varphi(\sigma)$.

Note that circularity is avoided in (1), (2) as $\max(\operatorname{rank} \sigma, \operatorname{rank} \tau)$ goes down (in at most three steps) when these definitions are applied. Also all quantifiers in (1), (2) are bounded, as $P$ is a set, so the above definition can be carried out in $M$ and Lemma 2.2 does hold for $\Vdash^*$.

**Lemma 2.4.** (a) $p \Vdash^* \varphi, q \leq p \Longrightarrow q \Vdash^* \varphi$.

(b) *If $\{q \mid q \Vdash^* \varphi\}$ is dense $\leq p$ then $p \Vdash^* \varphi$.*

(c) *If $\sim p \Vdash^* \varphi$ then $\exists q \leq p\ (q \Vdash^* \sim \varphi)$.*

*Proof.* (a) Clear, by induction on $\varphi$, as dense $\leq p \Longrightarrow$ dense $\leq q$.

(b) Again by induction on $\varphi$. The proof uses the following facts: If $\{q \mid D$ is dense $\leq q\}$ is dense $\leq p$ then $D$ is dense $\leq p$; if $\{q \mid q \Vdash^* \sim \varphi\}$ is dense $\leq p$ then $\forall q \leq p (\sim q \Vdash^* \varphi)$, using (a).

(c) is immediate by (b). $\square$

We are ready to prove Lemma 2.3 for $\Vdash^*$.

**Lemma 2.5.** *For $G$ $P$-generic over $M$:*

$$M[G] \models \varphi(\sigma_1^G, \ldots, \sigma_n^G) \iff \exists p \in G(p \Vdash^* \varphi(\sigma_1, \ldots, \sigma_n)).$$

*Proof.* By induction on $\varphi$.

$\sigma \in \tau$. ($\Longrightarrow$) If $\sigma^G \in \tau^G$ then choose $\langle \pi, r\rangle \in \tau$ such that $\sigma^G = \pi^G$ and $r \in G$. By induction we can choose $p \in G$, $p \leq r$, $p \Vdash^* \sigma = \pi$. Then $p \Vdash^* \sigma \in \tau$.

($\Longleftarrow$) If $p \in G$, $\{q \mid \exists \langle \pi, r\rangle \in \tau$ such that $q \leq r, q \Vdash^* \sigma = \pi\} = D$ is dense $\leq p$ then by genericity we can choose $q \in G$, $\langle \pi, r\rangle \in \tau$ such that $q \leq r, q \Vdash^* \sigma = \pi$; then by induction $\sigma^G = \pi^G$ and as $r \geq q \in G$ we get $r \in G$ and hence by definition of $\tau^G, \pi^G \in \tau^G$. So $\sigma^G \in \tau^G$.

$\sigma = \tau$. ($\Longrightarrow$) Suppose $\sigma^G = \tau^G$. Consider $D = \{p \mid$ Either $p \Vdash^* \sigma = \tau$ or for some $\langle \pi, r\rangle \in \sigma \cup \tau$, $p \Vdash^* \sim (\pi \in \sigma \iff \pi \in \tau)\}$. Then $D$ is dense, using the definition of $p \Vdash^* \sigma = \tau$ and Lemma 2.4(c). By genericity there is $p \in G \cap D$ and by induction it must be that $p \Vdash^* \sigma = \tau$.

($\Longleftarrow$) Suppose $p \in G$, $p \Vdash^* \sigma = \tau$. Then by induction, $\pi^G \in \sigma^G \iff \pi^G \in \tau^G$ for all $\langle \pi, r\rangle \in \sigma \cup \tau$. So $\sigma^G = \tau^G$.

$\varphi \wedge \psi$. Clear by induction, using the fact that $p, q \in G \Longrightarrow \exists r \in G(r \leq p$ and $r \leq q)$.

$\sim \varphi$. Clear by induction, using the density of $\{p \mid p \Vdash^* \varphi$ or $p \Vdash^* \sim \varphi\}$.

$\forall x \varphi$. ($\Longrightarrow$) Suppose $M[G] \models \forall x \varphi$. As in the proof of ($\Longrightarrow$) for $\sigma = \tau$, there is $p \in G$ such that either $p \Vdash^* \forall x \varphi$ or for some $\sigma$, $p \Vdash^* \sim \varphi(\sigma)$. By induction the latter is impossible so $p \Vdash^* \forall x \varphi$.

($\Longleftarrow$) Clear by induction. □

**Lemma 2.6.** $\Vdash^* = \Vdash$.

*Proof.* By Lemma 2.5, $p \Vdash^* \varphi(\sigma_1, \ldots, \sigma_n) \Longrightarrow p \Vdash \varphi(\sigma_1, \ldots, \sigma_n)$. And $\sim p \Vdash^* \varphi(\sigma_1, \ldots, \sigma_n) \Longrightarrow q \Vdash^* \sim \varphi(\sigma_1, \ldots, \sigma_n)$ for some $q \leq p$ (by Lemma 2.4(c)) $\Longrightarrow$ $\sim p \Vdash \varphi(\sigma_1, \ldots, \sigma_n)$ using our Assumption about the existence of generics. □

## ZFC and Cofinalities in $M[G]$

**Theorem 2.7.** *If $G$ is $P$-generic over $M$ then $M[G]$ is a model of* ZF. *If $M$ satisfies* AC *then so does $M[G]$.*

*Proof.* As $M[G]$ is transitive and contains $\omega$, it is a model of all ZF axioms with the possible exception of pairing, union, power and replacement.

For pairing, given $\sigma_1^G, \sigma_2^G$ consider $\sigma = \{\langle \sigma_1, 1^P\rangle, \langle \sigma_2, 1^P\rangle\}$. Then $\sigma^G = \{\sigma_1^G, \sigma_2^G\}$.

For union, given $\sigma^G$ consider $\pi = \{\langle\tau, p\rangle \mid p \Vdash \tau \in \bigcup\sigma, \operatorname{rank}\tau < \operatorname{rank}\sigma\}$. By the Truth Lemma, $\pi^G = (\bigcup\sigma^G) \cap \{\tau^G \mid \operatorname{rank}\tau < \operatorname{rank}\sigma\}$. As any element of $\bigcup\sigma^G$ is of the form $\tau^G$, $\operatorname{rank}\tau < \operatorname{rank}\sigma$ we get $\pi^G = \bigcup\sigma^G$.

For power, given $\sigma^G$ consider $\pi = \{\langle\tau, p\rangle \mid p \Vdash \tau \subseteq \sigma, \operatorname{rank}\tau \leq \operatorname{rank}\sigma\}$. Then $\pi^G = \mathcal{P}(\sigma^G) \cap \{\tau^G \mid \operatorname{rank}\tau \leq \operatorname{rank}\sigma\}$. Now suppose that $\tau^G \subseteq \sigma^G$, with no restriction on $\operatorname{rank}\tau$. Form the name $\tau^*$ by replacing each $\langle\tau_0, p\rangle \in \tau$ by all of the $\langle\tau_0^*, q\rangle$ such that $\operatorname{rank}\tau_0^* < \operatorname{rank}\sigma, q \leq p, q \Vdash \tau_0^* = \tau_0$. Then $\operatorname{rank}\tau^* \leq \operatorname{rank}\sigma$ and $\tau^{*G} = \tau^G$ since if $\langle\tau_0, p\rangle \in \tau, p \in G$ then $\tau_0^G \in \sigma^G$ and hence there is $q \leq p, q \in G$, $q \Vdash \tau_0 = \tau_0^*$ where $\operatorname{rank}\tau_0^* < \operatorname{rank}\sigma$; conversely, if $q \leq p, q \Vdash \tau_0^* = \tau_0$ and $q \in G$ then $p \in G$ and $\tau_0^{*G} = \tau_0^G$. So we conclude that $\pi^G = \mathcal{P}(\sigma^G) \cap M[G]$.

For replacement, given $f : \sigma^G \longrightarrow M[G]$, $f$ definable (with parameters) in $M[G]$ consider $\pi_\alpha = \{\langle\tau, p\rangle \mid \operatorname{rank}\tau < \alpha$ and for some $\sigma_0$, $\operatorname{rank}\sigma_0 < \operatorname{rank}\sigma, p \Vdash \sigma_0 \in \sigma \wedge f(\sigma_0) = \tau\}$. Then $\pi_\alpha^G = \operatorname{Range}(f) \cap \{\tau^G \mid \operatorname{rank}\tau < \alpha\}$. Now choose $\alpha \in \operatorname{ORD}(M)$ so large that if $p \in P$, $\operatorname{rank}\sigma_0 < \operatorname{rank}\sigma$ and $p \Vdash f(\sigma_0) = \tau$ for some $\tau$, then there is such a $\tau$ of rank $< \alpha$. This is possible by replacement in $M$. Then $\pi_\alpha^G = \operatorname{Range}(f)$.

Finally if $M$ satisfies AC, we can well-order $\sigma^G$ in $M[G]$ by first choosing a well-ordering of names of rank $< \operatorname{rank}\sigma$ in $M$, and then comparing $x, y \in \sigma^G$ by comparing the least names $\sigma_x, \sigma_y$ such that $\sigma_x^G = x, \sigma_y^G = y$. □

It does not follow from Theorem 2.7 that $M$, $M[G]$ have the same cardinals. We now turn to conditions on $P$ which guarantee that cardinals (indeed, cofinalities) are preserved. Assume that AC holds in $M$ and hence also in $M[G]$.

**Definition.** An *antichain* is a set $A \subseteq P$ such that $p \neq q$ in $A \Longrightarrow p, q$ are incompatible. For regular, uncountable $\kappa$, $P$ is $\kappa$-cc (*$\kappa$-chain condition*) if every antichain has cardinality $< \kappa$.

**Lemma 2.8.** *If $P$ is $\kappa$-cc in $M$ and* $\operatorname{cof}(\alpha) \geq \kappa$ *in $M$ then* $\operatorname{cof}(\alpha) \geq \kappa$ *in $M[G]$.*

*Proof.* It suffices to show that if $f : \beta \longrightarrow \gamma$ belongs to $M[G]$ then there is $g : \beta \longrightarrow P(\gamma)$ in $M$ such that for each $\beta_0 < \beta$, $f(\beta_0) \in g(\beta_0)$, $\operatorname{card}(g(\beta_0)) < \kappa$ in $M$. Let $\sigma^G = f$ and define $g$ by $g(\beta_0) = \{\gamma_0 < \gamma \mid p \Vdash \sigma$ is a function and $\sigma(\hat{\beta}_0) = \hat{\gamma}_0$, for some $p\}$. □

**Definition.** If $D \subseteq P$ and $p \in P$ then we say that $p$ *meets* $D$ if $p \leq q \in D$ for some $q$. For regular, uncountable $\kappa$, $P$ is *$\kappa$-distributive* if whenever $p \in P$ and $\langle D_i \mid i < \beta\rangle$ are dense subsets of $P$, $\beta < \kappa$ then $\exists q \leq p$ ($q$ meets each $D_i$).

**Lemma 2.9.** *If $P$ is $\kappa$-distributive in $M$ and* $\operatorname{cof}(\alpha) \geq \kappa$ *in $M$ then* $\operatorname{cof}(\alpha) \geq \kappa$ *in $M[G]$.*

*Proof.* It suffices to show that if $f : \beta \longrightarrow \gamma$, $\beta < \kappa$ belongs to $M[G]$ then it belongs to $M$. Let $\sigma^G = f$ and note that for each $\beta_0 < \beta$, $D_{\beta_0} = \{q \mid$ For some $\gamma_0 < \gamma, q \Vdash \sigma$ a total function $\Longrightarrow \sigma(\hat{\beta}_0) = \hat{\gamma}_0\}$ is dense. If $p \in G$, $p \Vdash \sigma$ total and $p$ meets each $D_{\beta_0}$ then $f(\beta_0) =$ unique $\gamma_0$, $p \Vdash \sigma(\hat{\beta}_0) = \hat{\gamma}_0$; so $f \in M$. □

There is one more condition for cofinality preservation to consider, which is best motivated by an example. Suppose that $\kappa$ is regular and that the ground model $M$ is $L$. Let $P$ consist of all functions $p$ on $I = \{0\} \cup$ All infinite cardinals $< \kappa$ such that for all $\alpha \in I$, $p(\alpha)$ is a bounded subset of $\alpha^+$ (we take $0^+ = \omega$). Order $P$ by $p \leq q \iff$ For each $\alpha \in I$, $q(\alpha)$ is an initial segment of $p(\alpha)$. For inaccessible $\kappa$, $P$ is neither $\kappa^+$-cc nor $\kappa^+$-distributive, yet "cofinality $> \kappa$" is preserved when forcing with $P$. This is because $P$ is $\Delta$-distributive at $\kappa$, a concept that we now define.

**Definition.** Let $\kappa$ be regular. We say that $d \subseteq P$ is *predense* $\leq p$ if $q \leq p \Longrightarrow q$ is compatible with an element of $d$. If $D \subseteq P$ is dense then $p$ *$\alpha^+$-reduces* $D$ if there exists $d \subseteq D$, $\operatorname{card}(d) \leq \alpha^+$, $d$ predense $\leq p$. $P$ is *$\Delta$-distributive at* $\kappa$ if whenever $\langle D_i \mid i < \kappa \rangle$ are dense subsets of $P$ and $p \in P$, there is $q \leq p$, $q$ $i^+$-reduces $D_i$ for each $i$. (We take $i^+ = \omega$ for finite $i$.)

**Lemma 2.10.** *If $P$ is $\Delta$-distributive at $\kappa$ in $M$ and* $\operatorname{cof}(\alpha) \geq \kappa^+$ *in $M$ then* $\operatorname{cof}(\alpha) \geq \kappa^+$ *in $M[G]$.*

*Proof.* It suffices to show that if $f : \kappa \longrightarrow \gamma$ belongs to $M[G]$ then there is $g : \kappa \longrightarrow \mathcal{P}(\gamma)$ in $M$ such that $\operatorname{card}(g(i)) \leq \kappa$, $f(i) \in g(i)$ for each $i < \kappa$. Let $\sigma^G = f$ and note that $D_i = \{p \mid \text{For some } \bar{\gamma} < \gamma,\ p \Vdash \sigma \text{ total} \Longrightarrow \sigma(\hat{i}) = \hat{\bar{\gamma}}\}$ is dense for each $i$. Let $p \in G$, $p \Vdash \sigma$ total, $p$ $i^+$-reduces $D_i$ for each $i$. Then the desired $g$ is $g(i) = \{\bar{\gamma} < \gamma \mid q \Vdash \sigma(\hat{i}) = \hat{\bar{\gamma}} \text{ for some } q \leq p\}$. □

**Corollary 2.11.** *If for some $\kappa$, $P$ is either both $\kappa$-distributive and $\kappa^+$-cc, or both $\Delta$-distributive at $\kappa$ and $\kappa^{++}$-cc then $P$ preserves cofinalities.*

Lemmas 2.8, 2.9, 2.10 are the basic tools for proving cofinality preservation in the basic examples of class forcing introduced later in this chapter. We end this discussion of cofinality-preservation for set forcing with a characterization of this property, which serves to motivate our discussion of tameness for class forcing in the next section.

**Proposition 2.12.** *The following are equivalent:*

(a) *For every $P$-generic $G$ and $\alpha \in \mathrm{ORD}(M)$,* $\operatorname{cof}(\alpha)$ *in* $M[G] = \operatorname{cof}(\alpha)$ *in $M$.*

(b) *In $M$: For each $p \in P$ and $\kappa < \lambda$, $\lambda$ regular, if $\langle D_i \mid i < \kappa \rangle$ are dense and $f_i : D_i \longrightarrow \lambda$ for each $i < \kappa$ then there is $q \leq p$ and $\langle d_i \mid i < \kappa \rangle$ such that for each $i$, $d_i \subseteq D_i$, $d_i$ is predense $\leq q$ and* $\operatorname{Range}(f_i \restriction d_i)$ *is bounded in $\lambda$.*

*Proof.* (a) $\Longrightarrow$ (b). Given the hypothesis of (b) choose a $P$-generic $G$, $p \in G$ and define $f_G : \kappa \longrightarrow \lambda$ by $f_G(i) = \min\{f_i(q) \mid q \in G \cap D_i\}$. Then there is $q \leq p$, $\lambda_0 < \lambda$, $q \Vdash \operatorname{Range}(f) \subseteq \hat{\lambda}_0$ (since $G$ preserves cofinalities). Then let $d_i = \{r \in D_i \mid f_i(r) < \lambda_0\}$.

(b) $\Longrightarrow$ (a). Given a $P$-generic $G$ and $f \in M[G]$, $f : \kappa \longrightarrow \lambda$, $\kappa < \lambda$, $\lambda$ regular in $M$, let $\sigma^G = f$ and define $D_i = \{q \mid \text{If } q \Vdash \sigma \text{ a total function then for some } \lambda^i_q < \lambda, q \Vdash \sigma(\hat{i}) = \hat{\lambda}^i_q\}$, $f_i(q) = \lambda^i_q$ (or 0 if $\sim q \Vdash \sigma$ total). By (b), $\exists q \in G$ ($q \Vdash \operatorname{Range}(f)$ bounded in $\hat{\lambda}$). □

## GCH Preservation

Given that cofinalities are preserved, we can ask what further conditions we need on $P$ to guarantee that GCH, if true in $M$, will remain true in $M[G]$ for $P$-generic $G$. The basic fact is the following.

**Lemma 2.13.** *If $M \vDash 2^\kappa = \kappa^+$, $P \in M$ and either $P$ is $\kappa^+$-distributive or $P$ is $\kappa^+$-preserving,* $\mathrm{card}(P) \leq \kappa^+$ *then $G$ $P$-generic over $M \Longrightarrow M[G] \vDash 2^\kappa = \kappa^+$.*

*Proof.* This is clear if $P$ is $\kappa^+$-distributive as then $\mathcal{P}(\kappa)$ in $M[G] = \mathcal{P}(\kappa)$ in $M$. Now if $P$ is a $\kappa^+$-preserving forcing of cardinality $\leq \kappa^+$ choose $f : P \overset{1-1}{\longrightarrow} \kappa^+$ and let $P_\alpha = f^{-1}[\alpha]$ for $\alpha < \kappa^+$. If $\sigma^G \subseteq \kappa$ then there is $\alpha < \kappa^+$ such that for all $i < \kappa, i \in \sigma^G \Longleftrightarrow \exists p \in P_\alpha \cap G(p \Vdash \hat{i} \in \sigma)$. Thus $\sigma^G$ is uniquely determined by $\alpha$, $\langle S_i \mid i < \kappa \rangle$ where $\alpha < \kappa^+$, $S_i = \{p \in P_\alpha \mid p \Vdash \hat{i} \in \sigma\}$ and hence in $M[G]$ there are at most $\kappa^+$-many such $\sigma^G$. □

## Cohen's Results

**Theorem 2.14.** *If* ZF *is consistent then so is* ZFC + $\sim$ CH.

*Proof.* First suppose that ZF has a countable transitive model $N$; then so does ZFC for we can take $M = (L)^N$. Now take $P \in M$ to consist of all $p : F_p \longrightarrow 2$, $F_p$ a finite subset of $\omega \times \aleph_2^M$, ordered by $p \leq q \Longleftrightarrow p$ extends $q$ as a function. If $G$ is $P$-generic over $M$ (such $G$ exist by the assumption that $M$ is countable) then $\cup G : \omega \times \aleph_2^M \longrightarrow 2$, since for each $(n, \alpha) \in \omega \times \aleph_2^M$ the set $D = \{p \mid (n, \alpha) \in F_p\}$ is dense. Also $\alpha < \beta < \aleph_2^M \Longrightarrow G_\alpha \neq G_\beta$ where $G_\alpha(n) = (\cup G)(n, \alpha)$. So $M[G] \vDash \mathrm{ZFC} + 2^{\aleph_0} \geq \aleph_2^M$. Thus to get $\sim$ CH in $M[G]$ we only need $\aleph_2^M = \aleph_2^{M[G]}$, which will follow by Lemma 2.8 if we can show that $P$ is $\aleph_1$-cc in $M$.

**Claim.** $P$ is $\aleph_1$-cc in $M$.

*Proof of Claim.* Suppose $A$ were an uncountable antichain and choose $F$ maximal so that $F \subseteq F_p$ for uncountably many $p \in A$. We may assume that $p \restriction F$ is constant for $p \in A$. But then for any $p \in A$ choose $p \neq q \in A$ such that $F_q \cap F_p = F$ and we see that $p, q$ are compatible, contradiction. □ (Claim)

Now to prove the theorem notice the following: The above shows that if $\mathrm{ZF}_{n+17}$(= ZF with only $\Sigma_{n+17}$ Replacement) has a countable transitive model then so does $\mathrm{ZF}_n$ + AC + $\sim$ CH. But in ZF we can prove that $\mathrm{ZF}_{n+17}$ has a countable transitive model, so if ZF + AC + $\sim$ CH were inconsistent we would get an inconsistency in ZF. □

**Theorem 2.15.** *If* ZF *is consistent then so is* ZF + $\sim$ AC.

*Proof.* As in the previous theorem, it will suffice to show that if ZF $+V = L$ has a countable transitive model $M$ then so does ZF$+ \sim$ AC. Let $P \in M$ be the partial ordering of all $p : F_p \longrightarrow 2$ where $F_p$ is a finite subset of $\omega \times \omega$, ordered by $p \leq q \Longleftrightarrow p$ extends $q$. If $G$ is $P$-generic over $M$ then $\bigcup G : \omega \times \omega \longrightarrow 2$ and $n \neq m \Longrightarrow G_n \neq G_m$ where $G_n(i) = (\bigcup G)(i, n)$.

For any $m, n \in \omega$ define $\pi_{mn} : P \longrightarrow P$ as follows: if $p \in P$ then $\pi_{mn}(p)$ agrees with $p$ except it sends $(i, m)$ to $p(i, n)$ and $(i, n)$ to $p(i, m)$. Then $G_{mn} = \{\pi_{mn}(p) \mid p \in G\}$ is $P$-generic over $M$ and $M[G] = M[G_{mn}]$. It follows that if $f : \omega \longrightarrow S = \{G_n \mid n \in \omega\}$ is definable in $M[G]$ with parameters from $M \cup \{S, G_0, G_1, \dots\}$ then Range$(f)$ is finite: If the formula $\varphi$ defining $f$ does not have $f(k) = G_m$ as a parameter, choose $p \in G$, $p \Vdash f$ is a function, $f(k) = G_m$; then for large enough $n \geq m$, $p$ and $\pi_{mn}(p)$ are compatible and together force $f(k)$ to equal both $G_m$ and $G_n$, contradiction.

Let $N = \bigcup\{t \in M[G] \mid t$ transitive and $x \in t \Longrightarrow x$ is definable in $M[G]$ with parameters from $M \cup \{S, G_0, G_1, \dots\}\}$. We have shown that $f : \omega \longrightarrow S$, $f \in N \Longrightarrow$ Range$(f)$ finite and clearly $S \in N$. So we need only show that $N$ is a model of ZF. Note that $N$ is a transitive, definable (with parameter $G$) subclass of $M[G]$, since by the Reflection Principle, $N = \bigcup\{t \in M[G] \mid t$ transitive and $x \in t \Longrightarrow$ for some $\alpha \in \mathrm{ORD}(M)$, $x$ is definable in $V_\alpha^{M[G]}$ with parameters from $M \cup \{S, G_0, G_1, \dots\}\}$. The axioms of extensionality, foundation, empty and infinity obviously hold in $N$. Pairing and union hold as these are definable operations and the transitive closure ($TC$) of $\{x, y\}$ is $TC\{x\} \cup TC\{y\}$, $TC(\bigcup x) \subseteq TC(x)$. For power, use the definability of $N$ to get $x \in N \Longrightarrow P(x) \cap N \in N$. Finally, for replacement use replacement in $M[G]$ and the definability of $N$. □

## 2.2 Class Forcing

We now relax the requirement that the forcing partial ordering $P$ be a set in $M$ and investigate to what extent the forcing method is still applicable. Again $M$ is a transitive set or class modelling ZF, but now we also have $A \subseteq M$. We say that $\langle M, A\rangle$ is a *model of* ZF if $M$ is a model of ZF and the scheme of replacement holds in $M$ for formulas which mention $A$ as a predicate. In addition we require $\langle M, A\rangle$ to be a it ground model, which means that $\langle M, A\rangle$ satisfies: $V = L(A) = \bigcup\{L(A \cap V_\alpha) \mid \alpha \in \mathrm{ORD}\}$. Any ZF model $\langle M, A\rangle$ is easily modified to a ground model $\langle M, A^*\rangle$ (with the same definable predicates) by taking $A^*$ to be $\{\langle 0, x\rangle \mid x \in A\} \cup \{\langle 1, V_\alpha^M\rangle \mid \alpha \in \mathrm{ORD}(M)\}$. This "minimality" property of $M$ relative to $A$ is needed to guarantee that $M$ is definable as a predicate in all of its extensions $\langle M[G], A, G\rangle$.

A partial ordering $P$ is a *class forcing* for $M$ (or an $M$*-forcing*) if for some ground model $\langle M, A\rangle$, $P$ (with its ordering) is definable with parameters over $\langle M, A\rangle$. Assume that this is the case and that $P$ has a greatest element $1^P$. Compatibility of $p, q \in P$ and density of $D \subseteq P$ are defined as in Section 2.1.

**Definition.** $G \subseteq P$ is $P$-generic over $\langle M, A\rangle$ iff:

(1) $p, q \in G \Longrightarrow p, q$ are compatible.

(2) $p \geq q \in G \Longrightarrow p \in G$.

(3) If $D \subseteq P$ is dense and $\langle M, A\rangle$-definable (with parameters) then $G \cap D \neq \emptyset$.

We make the same Assumption as in Section 2.1, that for each $p \in P$ there exists $G$ such that $p \in G$ and $G$ is $P$-generic over $\langle M, A\rangle$. Again this is provable when $M$ is countable. We will discuss (and dispense with) this Assumption in Chapter 3.

Define names and $M[G]$ as in Section 2.1. We have the following:

**Lemma 2.16.** (a) *$M \subseteq M[G]$ and $M[G]$ is transitive,* $\mathrm{ORD}(M[G]) = \mathrm{ORD}(M)$.

(b) *$G \cap V_\alpha \in M[G]$ for each $\alpha \in \mathrm{ORD}(M)$ and if $M \subseteq N$, $\langle N, G\rangle$ is amenable and $N$ is a model of* ZF *then $M[G] \subseteq N$ and $M$ is definable over $\langle N, A\rangle$.*

*Proof.* (a) Exactly as in Section 2.1.

(b) For each $\alpha \in \mathrm{ORD}(M)$, $G \cap V_\alpha = \gamma_\alpha^G$ where $\gamma_\alpha = \{\langle \hat{p}, p\rangle \mid p \in P \cap V_\alpha\}$, so $G \cap V_\alpha \in M[G]$. Under the assumptions on $N$ we can define $\sigma^G$ as an element of $N$, for each name $\sigma$; $M$ is definable over $\langle N, A\rangle$ as it equals $L(A)^N$. □

Define $\Vdash$ and $\Vdash^*$ as in Section 2.1. We would like to carry out the argument of Section 2.1 to show that the Truth and Definability lemmas hold for $\Vdash$. But we immediately run into trouble: We do not know that the Definability lemma holds for $\Vdash^*$. The problem is in (1), (2) of the definition of $\Vdash^*$:

(1) $p \Vdash^* \sigma \in \tau$ iff $\{q \mid \exists \langle \pi, r\rangle \in \tau$ such that $q \leq r, q \Vdash^* \sigma = \pi\}$ is dense $\leq p$.

(2) $p \Vdash^* \sigma = \tau$ iff for all $\langle \pi, r\rangle \in \sigma \cup \tau$, $p \Vdash^* (\pi \in \sigma \Longleftrightarrow \pi \in \tau)$ iff for all $\langle \pi, r\rangle \in \sigma \cup \tau$, $\{q \mid q \Vdash^* (\pi \in \sigma \wedge \pi \in \tau)$ or $q \Vdash^* (\pi \notin \sigma \wedge \pi \notin \tau)\}$ is dense $\leq p$.

As $P$ may now be a proper class these clauses now involve unbounded quantifiers, and therefore lead to definitions of $p \Vdash^* \sigma \in \tau$, $p \Vdash^* \sigma = \tau$ whose quantifier complexity may increase with the ranks of $\sigma, \tau$.

By introducing a further condition on $P$ we can control the quantifier complexity of the relations $p \Vdash^* \sigma \in \tau$, $p \Vdash^* \sigma = \tau$ and therefore obtain the Definability lemma for $\Vdash^*$. The following is reminiscent of Proposition 2.12. By "definable" we always mean "definable with parameters" unless we say otherwise.

**Tameness Condition, Part One.** $P$ is *pretame* iff whenever $\langle D_i \mid i \in a\rangle$ is an $\langle M, A\rangle$-definable sequence of dense classes, $a \in M$ and $p \in P$ then there is $q \leq p$ and $\langle d_i \mid i \in a\rangle \in M$ such that $d_i \subseteq D_i$ and $d_i$ is predense $\leq q$ for each $i$.

**Proposition 2.17.** *Suppose that for each $p \in P$ there is $G \subseteq P$ such that $p \in G$, $G$ is $P$-generic over $\langle M, A\rangle$ and $\langle M[G], A, G\rangle$ is a model of* ZF − Power. *Then $P$ is pretame.*

*Proof.* Given $\langle D_i \mid i \in a\rangle$ and $p$ as in the statement of pretameness choose $G$ such that $p \in G$, $G$ $P$-generic over $\langle M, A\rangle$ and consider $f(i) =$ least rank of an element of $G \cap D_i$. If pretameness failed for $p$, $\langle D_i \mid i \in a\rangle$ then for every $q \leq p$ and $\alpha \in \mathrm{ORD}(M)$ there would be $r \leq q$ and $i \in a$ with $r$ incompatible with each element of $D_i \cap V_\alpha$. But then by genericity, no ordinal of $M$ can bound the range of $f$, so replacement fails in $\langle M[G], A, G, M\rangle$. As $\langle M, A\rangle$ is a ground model, replacement fails in $\langle M[G], A, G\rangle$. □

Thus pretameness is necessary for a reasonable notion of class forcing. We now prove the Definability lemma for $\Vdash^*$ assuming pretameness. By *formula* we now mean a formula in the language of set theory with the addition of the unary predicate symbols $\underline{A}, \underline{G}$. Of course $\langle M[G], A, G\rangle \vDash \underline{A}(\sigma^G)$ iff $\sigma^G \in A$, $\langle M[G], A, G\rangle \vDash \underline{G}(\sigma^G)$ iff $\sigma^G \in G$. And extend the definition of $\Vdash^*$ by adding:

(6) $p \Vdash^* \underline{A}(\sigma)$ iff $p \Vdash^* \sigma \in \hat{a}_\alpha$, where $a_\alpha = A \cap V_\alpha$, $\alpha = \operatorname{rank}\sigma + 1$

(7) $p \Vdash^* \underline{G}(\sigma)$ iff $p \Vdash^* \sigma \in \gamma_\alpha$, where $\gamma_\alpha = \{\langle \hat{p}, p\rangle \mid p \in P \cap V_\alpha\}$, $\alpha = \operatorname{rank}\sigma + 1$.

**Theorem 2.18.** *If $P$ is pretame then for any formula $\varphi$, the relation "$p \Vdash^* \varphi(\sigma_1, \ldots, \sigma_n)$" of $p, \sigma_1, \ldots, \sigma_n$ is $\langle M, A\rangle$-definable.*

*Proof.* It suffices to show that the relations $p \Vdash^* \sigma \in \tau$ and $p \Vdash^* \sigma = \tau$ are $\langle M, A\rangle$-definable, for then we may induct on teh structure of $\varphi$. Note that by modifying $A$ if necessary, we may assume that the relations "$x = V_\alpha^M$," "$p, q$ are compatible," "$d$ is predense below $p$," as well as $(P, \leq)$, are $\Delta_1$-definable over $\langle M, A\rangle$. Also note that Lemma 2.4 holds in this context as its proof made no use of Definability for $\Vdash^*$.

Using pretameness we shall define a function $F$ from pairs $(p, \sigma \in \tau)$, $(p, \sigma = \tau)$ into $M$ such that:

(1) $F(p, \sigma \in \tau) = (i, d)$ where $d \in M$ is a nonempty subset of $P(\leq p) = \{q \in P \mid q \leq p\}$ and either ($i = 1$ and $q \Vdash^* \sigma \in \tau$ for $q \in d$) or ($i = 0$ and $q \Vdash^* \sigma \notin \tau$ for $q \in d$).

(2) The same holds for $\sigma = \tau, \sigma \neq \tau$ instead of $\sigma \in \tau, \sigma \notin \tau$.

(3) $F$ is $\Sigma_1$-definable over $\langle M, A\rangle$.

Given this we can define $p \Vdash^* \sigma \in \tau$ by: $p \Vdash^* \sigma \in \tau$ iff for all $q \leq p$, $F(q, \sigma \in \tau) = (1, d)$ for some $d$. This holds because by Lemma 2.4, $p \Vdash^* \sigma \in \tau$ iff $\{q \mid q \Vdash^* \sigma \in \tau\}$ is dense $\leq p$. Similarly we can define $p \Vdash^* \sigma = \tau$.

Now define $F$ by induction on $\sigma \in \tau, \sigma = \tau$. We consider the two cases separately.

$\sigma \in \tau$. Given $p$, search for $\langle \pi, r\rangle \in \tau$ and $q \leq p$, $q \leq r$ such that $F(q, \sigma = \pi) = (1, d)$ for some $d$. If such exist, let $F(p, \sigma \in \tau) = (1, e)$ where $e$ is the union of all such $d$ which appear by the least possible stage $\alpha$ (i.e., this $\Sigma_1$ property is true in $\langle V_\alpha^M, A \cap V_\alpha^M\rangle$, $\alpha$ least). If not then for each $\langle \pi, r\rangle \in \tau$, $D(\pi, r) = \cup\{d \mid$ For some $q \le r$, $F(q, \sigma = \pi) = (0, d)\} \cup \{q \mid q$ incompatible with $r\}$ is dense below $p$. So also search for $\langle d(\pi, r) \mid \langle \pi, r\rangle \in \tau\rangle \in M$ and $q \le p$ such that $d(\pi, r) \subseteq D(\pi, r)$ for each $\langle \pi, r\rangle$ and each $d(\pi, r)$ is predense $\le q$; if this latter search terminates then set $F(p, \sigma \in \tau) = (0, e)$, where $e$ consists of all such $q$ witnessed by the least possible stage $\alpha$. One of these searches must terminate (by pretameness) and hence $F(p, \sigma \in \tau)$ is defined and either of the form $(1, e)$ where $q \in e \Longrightarrow q \le p, q \Vdash^* \sigma \in \tau$, or of the form $(0, e)$ where $q \in e \Longrightarrow q \le p, q \Vdash^* \sim (\sigma \in \tau)$.

$\sigma = \tau$. Given $p$, search for $\langle \pi, r\rangle \in \sigma \cup \tau$ and $q \le p, q \le r$ such that for some $i, d, q'$ and $e$, $F(q, \pi \in \sigma) = (i, d), q' \in d$, $F(q', \pi \in \tau) = (1 - i, e)$. If this search terminates then set $F(p, \sigma = \tau) = (0, f)$ where $f$ is the union of all such $e$ which appear by the least possible stage $\alpha$. If this search fails then for each $\langle \pi, r\rangle \in \sigma \cup \tau$, $D(\pi, r) = \cup \{e \mid$ For some $q \le r$, some $i, d, q'$, $F(q, \pi \in \sigma) = (i, d), q' \in d$, $F(q', \pi \in \tau) = (i, e)\} \cup \{q \mid q$ is incompatible with $r\}$ is dense $\le p$. So also search for $\langle d(\pi, r) \mid \langle \pi, r\rangle \in \sigma \cup \tau\rangle \in M$ and $q \le p$ such that for each $\langle \pi, r\rangle$, $d(\pi, r) \subseteq D(\pi, r)$ and $d(\pi, r)$ is predense $\le q$. If this latter search terminates then $q \Vdash^* \sigma = \tau$ for all such $q$ and let $F(p, \sigma = \tau) = (1, f)$, where $f$ consists of all such $q \le p$ witnessed to obey the above by the least stage $\alpha$. □

Theorem 2.18 was proved independently by M. Stanley (see Stanley [97]). The author does not know if the assumption of pretameness is necessary for Theorem 2.18.

Now that we have the Definability lemma for $\Vdash^*$ we can prove the Truth lemma for $\Vdash^*$ as we did in Lemma 2.5; the two new clauses (f), (g) cause no difficulty. Then we infer that $\Vdash = \Vdash^*$ as in Lemma 2.6.

Pretameness is sufficient to verify that ZF – Power is preserved.

**Lemma 2.19.** *If $P$ is pretame and $G$ is $P$-generic over $\langle M, A\rangle$ then $\langle M[G], A, G\rangle$ is a model of* ZF – Power. *If $M$ is a model of* AC *then so is $M[G]$.*

*Proof.* This is exactly like Theorem 2.7, except for the verifications of replacement, union. For replacement, suppose $f : \sigma^G \longrightarrow M[G]$, $f$ definable (with parameters) in $\langle M[G], A, G\rangle$ and choose $p \in G$, $p \Vdash f$ is a total function on $\sigma$. Then for each $\sigma_0$ of rank $<$ rank $\sigma$, $D(\sigma_0) = \{q \mid$ For some $\tau$, $q \Vdash \sigma_0 \in \sigma \Longrightarrow f(\sigma_0) = \tau\}$ is dense $\le p$. Thus by pretameness we get that for each $q \le p$ there is $r \le q$ and $\alpha \in \mathrm{ORD}(M)$ such that $D_\alpha(\sigma_0) = \{s \mid s \in V_\alpha$ and for some $\tau$ of rank $< \alpha$, $s \Vdash \sigma_0 \in \sigma \Longrightarrow f(\sigma_0) = \tau\}$ is predense $\le r$ for each $\sigma_0$ of rank $<$ rank $\sigma$. By genericity there is $q \in G$ and $\alpha \in \mathrm{ORD}(M)$ such that $q \le p$, $D_\alpha(\sigma_0)$ is predense $\le q$ for each $\sigma_0$ of rank $<$ rank $\sigma$. Thus $\mathrm{Range}(f) = \pi^G$ where $\pi = \{\langle \tau, r\rangle \mid \mathrm{rank}\, \tau < \alpha, r \in V_\alpha, r \Vdash \tau \in \mathrm{Range}(f)\}$. So $\mathrm{Range}(f) \in M[G]$.

For union, given $\sigma^G$ consider $\pi = \{\langle \tau, p\rangle \mid p \Vdash \tau \in \cup\sigma\}$. This is not a set, but for each $\alpha$ we may consider $\pi_\alpha = \pi \cap V_\alpha^M$. By replacement in $\langle M[G], A, G\rangle$, $\pi_\alpha^G$ is constant for sufficiently large $\alpha \in \mathrm{ORD}(M)$. For such $\alpha$ we have $\pi_\alpha^G = \cup\sigma^G$. □

Thus pretameness is equivalent to ZF – Power preservation.

To handle the power set axiom we need to introduce tameness.

**Definition.** A *predense* $\leq p$ *partition* is a pair $(D_0, D_1)$ such that $D_0 \cup D_1$ is predense $\leq p$ and $p_0 \in D_0$, $p_1 \in D_1 \Longrightarrow p_0, p_1$ are incompatible. Suppose $\langle (D_0^i, D_1^i) \mid i \in a\rangle$, $\langle (E_0^i, E_1^i) \mid i \in a\rangle$ are sequences of predense $\leq p$ partitions. We say that they are *equivalent* $\leq p$ if for each $i \in a$, $\{q \mid q \text{ meets } D_0^i \Longleftrightarrow q \text{ meets } E_0^i\}$ is dense $\leq p$. When $p = 1^P$ we omit "$\leq p$".

To each sequence of predense $\leq p$ partitions $\vec{D} = \langle (D_0^i, D_1^i) \mid i \in a\rangle \in M$ and $G$ $P$-generic over $\langle M, A\rangle$, $p \in G$ we can associate the function $f_{\vec{D}}^G : a \longrightarrow 2$ defined by $f(i) = 0 \Longleftrightarrow G \cap D_0^i \neq \emptyset$. Then two such sequences are equivalent $\leq p$ exactly if their associated functions are equal, for each choice of $G$.

**Tameness Condition, Part Two.** $P$ is *tame* iff $P$ is pretame and for each $a \in M$ and $p \in P$ there is $q \leq p$ and $\alpha \in \mathrm{ORD}(M)$ such that whenever $\vec{D} = \langle (D_0^i, D_1^i) \mid i \in a\rangle \in M$ is a sequence of predense $\leq q$ partitions, $\{r \mid \vec{D}$ is equivalent $\leq r$ to some $\vec{E} = \langle (E_0^i, E_1^i) \mid i \in a\rangle$ in $V_\alpha^M\}$ is dense below $q$.

**Proposition 2.20.** *Suppose that for each $p \in P$ there is $G \subseteq P$ such that $p \in G$, $G$ is $P$-generic over $\langle M, A\rangle$ and $M[G] \vDash ZF$. Then $P$ is tame.*

*Proof.* If not then choose $a \in M$ and $p \in P$ such that for every $\alpha \in \mathrm{ORD}(M)$ and $q \leq p$ there are $r \leq q$ and $\vec{D}$ as in the statement of tameness such that $s \leq r \Longrightarrow \vec{D}$ is not equivalent $\leq s$ to any $\vec{E}$ in $V_\alpha^M$. Choose $G$ to be $P$-generic over $\langle M, A\rangle$, $p \in G$ and as the condition on $r$ is dense below $p$, choose $r_\alpha \in G$ and $\vec{D}_\alpha$ such that $s \leq r_\alpha \Longrightarrow \vec{D}_\alpha$ is not equivalent $\leq s$ to any $\vec{E}$ in $V_\alpha^M$. Then $r_\alpha \Vdash f_{\vec{D}_\alpha}^G \neq f_{\vec{E}}^G$ for each $\vec{E}$ in $V_\alpha^M$. It follows that $\{f \in M[G] \mid f : a \longrightarrow 2\} = X$ is a proper class in $M[G]$, as each $f_{\vec{D}_\alpha}^G$ belongs to $X$. So $M[G]$ does not satisfy the power set axiom. □

**Theorem 2.21.** *If $P$ is tame and $G \subseteq P$ is $P$-generic over $\langle M, A\rangle$ then $\langle M[G], A, G\rangle$ is a model of* ZF. *If $M \vDash$* AC *then $M[G] \vDash$* AC.

*Proof.* We only need to verify the power set axiom in $M[G]$. It suffices to show that $\mathcal{P}(a)$ exists in $M[G]$ for each $a \in M$, because if $a \in M[G]$ then in $M[G]$ there is a function from $\bar{a} = \{\sigma \mid \sigma \text{ a name, rank } \sigma < \text{(von Neumann) rank } a\}$ onto $a$, and hence from $\mathcal{P}(\bar{a})$ onto $\mathcal{P}(a)$.

So suppose $a \in M$ and by tameness and genericity choose $p \in G$ and $\alpha \in \mathrm{ORD}(M)$ such that whenever $\vec{D} = \langle (D_0^i, D_1^i) \mid i \in a\rangle \in M$ is a sequence of predense $\leq p$ partitions, there is $r \in G$, $r \leq p$ and $\vec{E} \in V_\alpha^M$ such that $\vec{D}, \vec{E}$ are equivalent $\leq r$. Now for any $\sigma^G \subseteq a$ consider $\vec{D}(\sigma)$ defined by $D(\sigma)_0^i = \{q \mid q \Vdash \hat{i} \notin \sigma\}$, $D(\sigma)_1^i = \{q \mid q \Vdash \hat{i} \in \sigma\}$. By pretameness we can assume that there is a predense $\leq p$ partition $\vec{d}^i = \langle (d_0^i, d_1^i) \mid i \in a\rangle \in M$ such that $d_0^i \subseteq D_0^i$, $d_1^i \subseteq D_1^i$ for each $i \in a$. Then we see that for some $\vec{E} \in V_\alpha^M$, $i \in \sigma^G \iff G \cap E_1^i \neq \emptyset$ and therefore $\sigma^G = \sigma_0^G$ where $\sigma_0 = \{\langle \hat{i}, p\rangle \mid p \in E_1^i\}$. Thus $\mathcal{P}(a) \cap M[G] = \pi^G$ where $\pi = \{\langle \pi_0, 1^P\rangle \mid \pi_0$ of the form $\{\langle \hat{i}, p\rangle \mid p \in E, i \in a\}$ for some $E \in V_\alpha^M\}$. So $\mathcal{P}(a) \cap M[G] \in M[G]$. □

Thus tameness is equivalent to ZF-preservation.

Fortunately there are some simple and useful sufficient conditions for tameness, which in addition can be used to demonstrate cofinality and GCH preservation. Assume now that $M$ satisfies AC.

**Definition.** For regular $\kappa > \omega$, $P$ is *$\kappa$-distributive* if whenever $p \in P$ and $\langle D_i \mid i < \beta\rangle$ is an $\langle M, A\rangle$-definable sequence of dense classes, $\beta < \kappa$ then there is $q \leq p$ meeting each $D_i$. $P$ is *tame below $\kappa$* if the tameness conditions hold for $P$ provided we impose the added restriction that $\mathrm{Card}(a) < \kappa$.

**Lemma 2.22.** *If $P$ is $\kappa$-distributive then $P$ is tame below $\kappa$ and if $P$ is $\kappa$-distributive and pretame then whenever $G \subseteq P$ is $P$-generic over $\langle M, A\rangle$ and* $\mathrm{cof}(\alpha) \geq \kappa$ *in $M$ we have* $\mathrm{cof}(\alpha) \geq \kappa$ *in $M[G]$.*

*Proof.* Pretameness below $\kappa$ is clear as we may take $q \leq p$ to meet each $D_i$ and let $d_i = \{r_i\}$, where $q \leq r_i \in D_i$. For tameness below $\kappa$, first note that for any $b \subseteq a \in M$ we can form a sequence of predense partitions $\vec{E}(a, b) = \langle (E_0^i, E_1^i) \mid i \in a\rangle$ by taking $E_0^i = \{1^P\}$ if $i \in b$, $E_0^i = \emptyset$ if $i \notin b$ (and dually for $E_1^i$). Then $\vec{E}(a, b) \in V_\kappa^M$ if $a \in V_\kappa^M$. Now for any $\langle M, A\rangle$-definable sequence $\vec{D} = \langle (D_0^i, D_1^i) \mid i \in a\rangle$ of predense partitions, $a \in V_\kappa^M$, $\mathrm{Card}(a) < \kappa$, we see that $\{r \mid \vec{D}$ is equivalent $\leq r$ to some $\vec{E} \in V_\kappa^M\}$ is dense, because by $\kappa$-distributivity we can extend any $q$ to an $r$ such that $r$ meets $D_0^i \cup D_1^i$ for each $i \in a$; then $\vec{D}$ is equivalent $\leq r$ to $\vec{E}(a, b)$ where $b = \{i \in a \mid r$ meets $D_0^i\}$. So tameness holds for each $a \in V_\kappa^M$, $\mathrm{Card}(a) < \kappa$ and therefore for each $a$ of cardinality $< \kappa$, since we can assume that $a$ is a cardinal. The second statement of the lemma is clear, as pretameness gives us the definability of $\Vdash$ and so we can use the proof of Lemma 2.9. □

We must be careful when defining $\Delta$-distributivity for class forcing.

**Definition.** Let $\kappa \geq \omega$ be regular and $I = \{0\} \cup$ All infinite cardinals $< \kappa$. If $D \subseteq P$ is dense and $p, q \in P$ are compatible then we say that *$p \wedge q$ meets $D$* if every $r$ below both $p$ and $q$ meets $D$. Now fix $\langle P_\alpha \mid \alpha \in I\rangle \in M$, where each $P_\alpha$ is a subset of $P$. If $D \subseteq P$ is dense then *$p$ $\alpha^+$-reduces $D$* if there exists $\bar{d} \subseteq P_\alpha$ of cardinality $\leq \alpha^+$ such that $\bar{d}$ is predense $\leq p$ and $q \in \bar{d} \implies q, p$ are compatible, $p \wedge q$ meets $D$. $P$ is *$\Delta$-distributive at $\kappa$* if there is $\langle P_\alpha \mid \alpha \in I\rangle \in M$ as above such that whenever $\langle D_i \mid i < \kappa\rangle$ is an $\langle M, A\rangle$-definable sequence of dense classes and $p \in P$ there exists $q \leq p$ such that $q$ $i^+$-reduces each $D_i$. (We take $i^+$ to be $\omega$ for finite $i$.)

Note that if $P$ is a set forcing then the above definition is equivalent to our earlier one.

**Lemma 2.23.** *If $P$ is $\Delta$-distributive at $\kappa$ then $P$ is tame below $\kappa^+$ and if $P$ is $\Delta$-distributive at $\kappa$ and pretame then whenever $G \subseteq P$ is $P$-generic over $\langle M, A\rangle$ and* $\mathrm{cof}(\alpha) \geq \kappa^+$ *in $M$, we have* $\mathrm{cof}(\alpha) \geq \kappa^+$ *in $M[G]$.*

*Proof.* For pretameness below $\kappa^+$, given $p \in P$ and $\langle D_i \mid i < \kappa\rangle$ an $\langle M, A\rangle$-definable sequence of dense classes, choose $q \leq p$ to $i^+$-reduce $D_i$ for each $i$. Thus for each $i$ there is $\bar{d}_i \subseteq P_i$ such that $\bar{d}_i$ is a set, $\bar{d}_i$ is predense $\leq q$ and $r \in \bar{d}_i \implies r, q$ are compatible, $q \wedge r$ meets $D_i$. We can choose a set $d_i \subseteq D_i$ such that $r \in \bar{d}_i \implies q \wedge r$ meets $d_i$; thus $d_i$ is predense $\leq q$ and pretameness below $\kappa^+$ is proved.

Our careful definition of $\Delta$-distributivity was designed to facilitate a proof of tameness below $\kappa^+$. Note that if $\vec{D} = \langle (D_0^i, D_1^i) \mid i < \kappa\rangle$ is a sequence of predense partitions and $p$ $i^+$-reduces $\{q \mid q \text{ meets } D_0^i \cup D_1^i\}$ for each $i$, then $\vec{D}$ is equivalent $\leq p$ to $\vec{E} = \langle (E_0^i, E_1^i) \mid i < \kappa\rangle$, where $E_0^i = \{q \in P_{i^+} \mid p, q$ are compatible, $p \wedge q$ meets $D_0^i\}$, $E_1^i$ defined similarly. Moreover $\vec{E} \in V_\alpha^M$ where $\alpha$ is chosen so that $\langle P_\beta \mid \beta \in I\rangle \in V_\alpha^M$. Thus tameness below $\kappa^+$ is satisfied using this fixed $\alpha$. The second statement of the lemma is clear, as pretameness gives us the definability of $\Vdash$ and therefore we can use the proof of Lemma 2.10. □

In the next section we will use the second statements of Lemmas 2.22, 2.23 to demonstrate cofinality preservation for the basic examples of class forcing. For each class forcing example $P$ and a regular $\kappa \geq \omega$, we will show that $P$ can be "factored" as $P_0 * P_1$ where each $P_i$ is either a $\kappa^+$-cc set forcing, is $\kappa^+$-distributive or is $\Delta$-distributive at $\kappa$. Then cofinality preservation will follow by Lemma 2.8 and the second statements of Lemmas 2.22, 2.23.

**GCH Preservation.** $\Delta$-distributivity must be strengthened if we wish to preserve GCH.

**Definition.** $P$ is *strongly $\Delta$-distributive at $\kappa$* if it obeys the definition of $\Delta$-distributive at $\kappa$, with the added condition that $\mathrm{Card}(P_\alpha) \leq \alpha^+$.

**Lemma 2.24.** *If $M \models 2^\kappa = \kappa^+$, $P$ is a pretame class forcing and either $P$ is $\kappa^+$-distributive or strongly $\Delta$-distributive at $\kappa$ then $M[G] \models 2^\kappa = \kappa^+$ for $G$ $P$-generic over $\langle M, A\rangle$.*

*Proof.* If $P$ is $\kappa^+$-distributive then $\mathcal{P}(\kappa)$ in $M[G] = \mathcal{P}(\kappa)$ in $M$ so the result is clear. Now suppose that $P$ is strongly $\Delta$-distributive at $\kappa$ and $\sigma^G \subseteq \kappa$, $G$ $P$-generic over $\langle M, A\rangle$. Consider the $\kappa$-sequence of predense partitions $\vec{D} = \langle (D_0^i, D_1^i) \mid i < \kappa\rangle$ defined by $D_0^i = \{p \mid p \Vdash \hat{i} \in \sigma\}$, $D_1^i = \{p \mid p \Vdash \hat{i} \notin \sigma\}$ and choose $p \in G$ so that $p$ $i^+$-reduces $D_0^i \cup D_1^i$ for each $i < \kappa$, relative to $\langle P_\alpha \mid \alpha \in I\rangle$ where $\mathrm{card}(P_\alpha) \leq \alpha^+$. Then $i \in \sigma^G$ iff $G \cap E_0^i \neq \emptyset$ where $E_0^i = \{q \in P_i \mid \forall r(r \leq p, q \implies r \Vdash \hat{i} \in \sigma)\}$; thus $\sigma^G$ is uniquely determined by $\langle E_0^i \mid i < \kappa\rangle$ where $E_0^i \subseteq \cup\{P_\alpha \mid \alpha < \kappa\}$ and there are at most $(2^\kappa)^\kappa = \kappa^+$ such sequences in $M$. So $M[G] \models 2^\kappa = \kappa^+$. □

In the next section we will get GCH preservation via Lemmas 2.13, 2.24 for the basic examples of class forcing.

We close this section by describing two examples of untame forcings.

**Proposition 2.25.** *There exist $L$-definable forcings $P$ such that*

(a) *The relation "$p \Vdash \varphi(\sigma_1, \dots, \sigma_n)$" for $P$ is $L$-definable for each $\varphi$, but $P$ is not pretame.*

(b) *$P$ is pretame but not tame.*

*Proof.* (a) A condition is an increasing $p : n \longrightarrow \mathrm{ORD}, n \in \omega$; ordered by $p \leq q \iff p$ extends $q$. Clearly this forcing is not pretame. To prove the definability of "$p \Vdash \varphi(\sigma_1, \dots, \sigma_n)$" for each $\varphi$ it suffices, as in the proof of Theorem 2.18 to define "$p \Vdash \sigma \in \tau$" and "$p \Vdash \sigma = \tau$." But if $\operatorname{rank} \sigma$, $\operatorname{rank} \tau < \alpha$, $p : n \longrightarrow \alpha$ then $p \Vdash \sigma \in \tau$ iff $\sigma^g \in \tau^g$ for every $g$ of the form $\{q \restriction m \mid 0 \leq m \leq \mathrm{Dom}(q)\}$ where $q \leq p, q : \mathrm{Dom}(q) \longrightarrow \alpha$ (as $G \cap V_\alpha^M$ is of this form for generic $G$). Similarly for $p \Vdash \sigma = \tau$.

(b) A condition is of the form $p : F \longrightarrow 2$, $F$ a finite set of ordinals, ordered by $p \leq q$ iff $p$ extends $q$. A generic adds a proper class of reals, but this forcing is $\aleph_1$-cc (as in the Claim of the proof of Theorem 2.14) and hence is pretame. □

## 2.3 Examples

In this section we describe the four basic examples of tame class forcing: Easton, Long Easton, Reverse Easton and Amenable forcing. We introduce the theory of product forcing to facilitate an analysis of Easton and Long Easton forcing. For our discussion of Reverse Easton forcing we will need the theory of iterated set forcing. (Iterated *class* forcing is discussed in Chapter 8.)

We now fix our ground model $\langle M, A \rangle$ to just be $\langle L, \emptyset \rangle$, and maintain the Assumption that for each forcing $P$ considered, $P$-generic classes exist containing any given condition in $P$ (where $P$-generic means $P$-generic over $\langle L, \emptyset \rangle$). The next chapter is devoted to the question of generic class existence and will show how to eliminate this Assumption, when establishing first-order properties of $P$-generic class

### Easton Forcing

Building on work of Solovay [65], Easton [70] extended Cohen's independence proof for CH to all regular cardinals, showing that the function $f(\kappa) = 2^\kappa$ can exhibit any reasonable behavior for regular $\kappa$. To do so he developed a class forcing for adding generic subsets to all regular $\kappa$ simultaneously. We describe here a version of his technique, where we explicitly add only one generic subset to each regular $\kappa$, thereby preserving GCH.

A condition in $P$ is a function $p : \alpha(p) \longrightarrow L$ where $\alpha(p) \in$ ORD and for $\alpha < \alpha(p)$, $p(\alpha) = \emptyset$ unless $\alpha$ is infinite and regular, in which case $p(\alpha) \in 2^{<\alpha} = \{f : \beta \longrightarrow 2 \mid \beta < \alpha\}$. In addition we require that $p$ has *Easton support* which means that for inaccessible $\kappa$, $\{\alpha < \kappa \mid p(\alpha) \neq \emptyset\}$ is bounded in $\kappa$. Extension is defined by: $p \leq q$ iff $\alpha(p) \geq \alpha(q)$, $\alpha < \alpha(q) \Longrightarrow p(\alpha)$ extends $q(\alpha)$. The key to analyzing $P$ is to observe that for each infinite regular $\kappa$, $P$ is isomorphic to $P(\leq \kappa) \times P(> \kappa)$, where $P(\leq \kappa) = \{p \upharpoonright [0, \kappa] \mid p \in P\}$, $P(> \kappa) = \{p \upharpoonright (\kappa, \infty) \mid p \in P\}$, ordered in the natural way. Note that $P(\leq \kappa)$ is $\kappa^+$-cc (indeed has cardinality $\kappa$) and $P(> \kappa)$ is $\kappa^+$-distributive (indeed is $\kappa^+$*-closed*: decreasing $\kappa$-sequences of conditions have lower bounds).

**Lemma 2.26.** *$P$ is pretame.*

*Proof.* Suppose $p \in P$, $\langle D_i \mid i < \kappa\rangle$ is a definable sequence of classes which are dense $\leq p$, $\kappa$ regular. Let $\langle q_i \mid i < \kappa\rangle$ list all elements of $P(\leq \kappa)$. View each $i < \kappa$ as a pair $\langle i_0, i_1\rangle$ of ordinals $< \kappa$ and define: $p_0 = p$, $p_{i+1}$ = least $r \leq p_i$ such that $r \upharpoonright [0, \kappa] = p_i \upharpoonright [0, \kappa]$ and $q_{i_0} \cup r$ is a condition below some $r_i \in D_{i_1}$, if possible; $p_{i+1} = p_i$ otherwise. For limit $\lambda \leq \kappa$, $p_\lambda$ = greatest lower bound $\langle p_i \mid i < \lambda\rangle$. Then $p^* = p_\kappa \leq p$ has the property: $r \leq p^*$, $r$ meets $D_i \Longrightarrow r$ extends $r_j \in D_i$ for some $j < \kappa$. Thus $d_i = \{r_j \mid r_j \in D_i\}$ is predense $\leq p^*$ for each $i$, proving pretameness. □

The following general lemma reduces the analysis of $P$ to an analysis of its factors, $P(\leq \kappa)$ and $P(> \kappa)$. For a partial order $P$ defined over a ground model $\langle M, A\rangle$, we say that *$P$-forcing is definable* if the relation "$p \Vdash \varphi(\sigma_1, \ldots, \sigma_n)$" is definable over $\langle M, A\rangle$ for each $\varphi$.

**Lemma 2.27** (Product Lemma). *Suppose that $P = P_0 \times P_1$ where $P_0$ and $P_1$ are definable over the ground model $\langle M, A\rangle$.*

(a) *If $G_0$ is $P_0$-generic over $\langle M, A\rangle$ and $G_1$ is $P_1$-generic over $\langle M[G_0], A, G_0\rangle$ then $G_0 \times G_1$ is $P$-generic over $\langle M, A\rangle$.*

(b) *If $G$ is $P$-generic over $\langle M, A\rangle$ then $G = G_0 \times G_1$ where $G_0$ is $P_0$-generic over $\langle M, A\rangle$. If in addition $P_0$-forcing is definable then $G_1$ is $P_1$-generic over $\langle M[G_0], A, G_0\rangle$.*

*Proof.* (a) Suppose that $D \subseteq P$ is dense and $\langle M, A\rangle$-definable. Then $D_1 = \{p_1 \mid \exists p_0 \in G_0\ (p_0, p_1)$ meets $D\}$ is $\langle M[G_0], A, G_0\rangle$-definable; we claim that it is dense on $P_1$: Given $p_1 \in P_1$ form $D_0(p_1) = \{p_0 \mid (p_0, p_1')$ meets $D$ for some $p_1' \leq p_1\}$. Then $D_0(p_1)$ is dense since $D$ is, so $G_0 \cap D_0(p_1) \neq \emptyset$. Thus $(p_0, p_1')$ meets $D$ for some $p_0 \in G_0$ and some $p_1' \leq p_1$ and therefore $p_1'$ is an extension of $p_1$ in $D_1$.

As $D_1$ is dense we can choose $p_1 \in G_1 \cap D_1$ and so we get $(p_0, p_1) \in G_0 \times G_1$ where $(p_0, p_1)$ meets $D$. As $G_0 \times G_1$ is compatible and closed upwards (since $G_0, G_1$ are) we have shown that $G_0 \times G_1$ is $P$-generic over $\langle M, A\rangle$.

(b) Let $G_0 = \{p_0 \in P_0 \mid (p_0, p_1) \in G$ for some $p_1\}$, $G_1 = \{p_1 \mid (p_0, p_1) \in G$ for some $p_0\}$. Clearly $G \subseteq G_0 \times G_1$ and conversely if $(p_0, p_1) \in G_0 \times G_1$ then $(p_0, p_1)$ is compatible with every element of $G$ and hence by genericity of $G$, $(p_0, p_1) \in G$. If $D_0 \subseteq P_0$ is dense and $\langle M, A\rangle$-definable then $D = \{(p_0, p_1) \mid p_0 \in D_0\} \subseteq P$ is dense and $\langle M, A\rangle$-definable and since $G$ meets $D$, we get that $G_0$ meets $D_0$. So $G_0$ is $P_0$-generic over $\langle M, A\rangle$, as compatibility and upward closure for $G_0$ follow from these properties for $G$.

Suppose that $D_1 \subseteq P_1$ is $\langle M[G_0], A, G_0\rangle$-definable and dense. Then $D = \{(p_0, p_1) \mid p_0 \Vdash \hat{p}_1 \in D_1\}$ is $\langle M, A\rangle$-definable by the definability of $P_0$-forcing (where "$\hat{p}_1 \in D_1$" is expressed using a defining formula for $D_1$). Also $D$ is dense $\leq (p_0, p_1)$ whenever $p_0 \Vdash D_1$ is dense. As $G_0$ is $P_0$-generic over $\langle M, A\rangle$ we can choose $p_0 \in G_0$, $p_0 \Vdash D_1$ is dense and then the genericity of $G$ over $\langle M, A\rangle$ produces $(p_0', p_1) \in G$, $p_0' \Vdash \hat{p}_1 \in D_1$; then $p_1 \in G_1 \cap D_1$ and as compatibility, upward closure for $G_1$ follow from these properties for $G$, we have shown that $G_1$ is $P_1$-generic over $\langle M[G_0], A, G_0\rangle$. □

In the case of Easton forcing, if $G \subseteq P$ is $P$-generic then $L[G] = L[G(>\kappa)][G(\leq \kappa)]$ where $G(>\kappa)$ is $P(>\kappa)$- generic, $G(\leq \kappa)$ is $P(\leq \kappa)$-generic over $\langle L[G(>\kappa)], G(>\kappa)\rangle$; (b) of the Product Lemma applies as the proof of Lemma 2.26 shows that $P(>\kappa)$ is pretame and hence $P(>\kappa)$-forcing is definable. By Lemma 2.22, $\mathrm{cof}(\alpha) > \kappa$ in $L \Longrightarrow \mathrm{cof}(\alpha) > \kappa$ in $L[G(>\kappa)]$. If $\kappa$ is regular then Lemma 2.8 applies, showing that $\mathrm{cof}(\alpha) > \kappa$ in $L[G(>\kappa)] \Longrightarrow \mathrm{cof}(\alpha) > \kappa$ in $L[G]$. So $P$ preserves all cofinalities. Lemma 2.24 gives us that $2^\kappa = \kappa^+$ in $L[G(>\kappa)]$. Lemma 2.13 gives us $2^\kappa = \kappa^+$ in $L[G(>\kappa)][G(\leq\kappa)]$ as $G(\leq\kappa)$ is $\kappa^+$-preserving and has cardinality $\leq \kappa^+$.

## Long Easton Forcing

This is like Easton forcing, except we drop the Easton support requirement. There are two types of Long Easton forcing, depending upon whether or not the forcing is trivial at inaccessibles. We begin with the simpler case, called Long Easton forcing at Successors. We treat $\omega$ as a successor cardinal in this discussion: $0^+ = \omega$.

A condition in $P$ is a function $p : \alpha(p) \longrightarrow L$, $\alpha(p) \in \mathrm{ORD}$ where $p(\alpha) = \emptyset$ unless $\alpha$ is a successor cardinal, in which case $p(\alpha) \in 2^{<\alpha}$. Extension is defined by $p \leq q$ iff $\alpha(p) \geq \alpha(q)$ and for each $\alpha < \alpha(q)$, $p(\alpha)$ extends $q(\alpha)$. For any infinite regular $\kappa$ we can factor $P$ as $P(\leq\kappa) \times P(>\kappa)$ and $P(>\kappa)$ is $\kappa^+$-distributive. However if $\kappa$ is inaccessible, $P(\leq\kappa) = P(<\kappa)$ is not $\kappa^+$-cc so the analysis we used for Easton forcing does not apply here.

**Lemma 2.28.** *For every infinite regular $\kappa$, $P$ is strongly $\Delta$-distributive at $\kappa$.*

*Proof.* For each $\alpha \in \{0\} \cup$ Infinite Cardinals define $P_\alpha = \{p \in P \mid \alpha(p) \leq \alpha^+ + 1\}$; then $\mathrm{card}(P_\alpha) = \alpha^+$ (where $0^+$ is $\omega$). Now if $\kappa$ is infinite and regular and $\langle D_i \mid i < \kappa\rangle$ is an $L$-definable sequence of dense classes, $p \in P$ we define a sequence $p = p_0 \geq p_1 \geq \ldots$ of length $\kappa + 1$ such that $p_\kappa$ $i^+$-reduces each $D_i$, relative to the

$P_\alpha$'s defined above. First select a list $\langle (i_\beta, q_\beta) \mid \beta < \kappa\rangle$ of all pairs $(i, q)$ with $i < \kappa$, $q \in P_{<\kappa} = \bigcup\{P_\alpha \mid \alpha < \kappa\}$ such that for successor cardinals $\alpha^+ \leq \kappa$, all pairs $(i, q)$ with $i < \alpha^+, q \in P_{\alpha^+}$ are listed at $(i_\beta, q_\beta)$ for some $\beta < \alpha^+$.

Now define $p_{\beta+1}$ to extend $p_\beta$ and to have the property that $p_{\beta+1}(\alpha) = p_\beta(\alpha)$ for $\alpha \leq \beta^+$, and if possible, such that $p_{\beta+1} \wedge q_\beta$ meets $D_{i_\beta}$ (where $\wedge$ denotes greatest lower bound). At limit stages $\lambda \leq \kappa$, let $p_\lambda$ be the greatest lower bound to $\langle p_i \mid i < \lambda\rangle$, which exists since we have properly extended $p_i(\alpha^+)$ only for $i < \alpha$. Then if $p^* = p_\kappa$ we have: If $r \leq p^*$ meets $D_i$ then so does $p^* \wedge q$ where $q = r \restriction (i^+ + 1)$. So $\{q \in P_{\mathrm{card}(i)} \mid p^*, q \text{ are compatible}, p^* \wedge q \text{ meets } D_i\}$ is predense $\leq p^*$ for each $i < \kappa$ and therefore $p^*$ $i^+$-reduces $D_i$ for all $i$. □

Then Lemma 2.23 tells us that $P$ is tame below $\kappa^+$ for each $\kappa$ and therefore tame, and that $P$ preserves cofinalities. GCH preservation follows by Lemma 2.24.

Now we consider (unrestricted) Long Easton forcing, where we redefine $P$ so as to allow $p(\alpha) \in 2^{<\alpha}$ for any infinite regular $\alpha$, not just successor cardinals. Then for any infinite *successor* cardinal $\kappa$ we can factor $P$ as $P(\leq \kappa) \times P(> \kappa)$ and the analysis of Easton forcing shows that $P$ is tame and preserves "cofinality $> \kappa$" for *successor* cardinals $\kappa$. However $P$ is *not* cofinality-preserving in general. A cardinal $\kappa$ is *Mahlo* if $\kappa$ is inaccessible and $\{\alpha < \kappa \mid \alpha \text{ inaccessible}\}$ is stationary in $\kappa$.

**Theorem 2.29.** *Suppose $G$ is $P$-generic over $L$ and $\kappa$ is $L$-regular. Then $(\kappa^+)^{L[G]} = (\kappa^+)^L$ iff $\kappa$ is not Mahlo in $L$.*

*Proof.* Let $G = \langle G_\alpha \mid \alpha \text{ infinite, regular}\rangle$ be $P$-generic. For each $\alpha < \kappa$ consider $A_\alpha \subseteq \kappa$ defined by: $\beta \in A_\alpha \iff \alpha \in G_\beta$.

**Claim.** Suppose $\kappa$ is Mahlo. Then $\{A_\alpha \mid \alpha < \kappa\} \subseteq L$ but for no $\gamma < (\kappa^+)^L$ do we have $\{A_\alpha \mid \alpha < \kappa\} \subseteq L_\gamma$.

*Proof of Claim.* For any $\alpha < \kappa$ and condition $p$, we can extend $p$ to $q$ so that $\alpha < \bar{\kappa} < \kappa, \bar{\kappa} \text{ regular} \implies p(\bar{\kappa})$ has length greater than $\alpha$. Thus $A_\alpha$ is forced to belong to $L$.

Given $\gamma < (\kappa^+)^L$ and a condition $p$, define $f(\bar{\kappa}) = \mathrm{length}(p(\bar{\kappa}))$ for regular $\bar{\kappa} < \kappa$. As $\kappa$ is Mahlo, $f$ has stationary domain and hence by Fodor's Theorem we may choose $\alpha < \kappa$ such that $\mathrm{length}(p(\bar{\kappa}))$ is less than $\alpha$ for stationary many regular $\bar{\kappa} < \kappa$. Then $p$ can be extended so that $A_\alpha$ is guaranteed to be distinct from the $\kappa$-many subsets of $\kappa$ in $L_\gamma$. □ (Claim)

Thus $\kappa^+$ is collapsed if $\kappa$ is Mahlo. Conversely, if $\kappa$ is not Mahlo, then choose a CUB $C \subseteq \kappa$ consisting of cardinals which are not inaccessible (we may assume that $\kappa$ is a limit cardinal). Suppose that $\langle D_\alpha \mid \alpha \in C\rangle$ is a definable sequence of dense classes. Given $p$ we can successively extend $p(\geq \alpha^+)$, $\alpha \in C$ so that $\{q \leq p \mid q, p \text{ agree} \geq \alpha^+, q \in D_\alpha\}$ is predense $\leq p$. There is no difficulty in obtaining a condition at a limit stage less than $\kappa$ precisely because conditions are trivial at limit points of $C$. Thus we have shown that $P(< \kappa) \times P(> \kappa)$ preserves $\kappa^+$ as $\kappa$-many dense classes can be simultaneously reduced to predense subsets of size $< \kappa$. Finally $P \simeq P(<\kappa) \times P(>\kappa) \times P(\kappa)$ and $P(\kappa)$ preserves $\kappa^+$ as it has size $\kappa$. □

**Remark.** The previous proof shows that Long Easton forcing preserves the cofinality of any ordinal $\alpha$ unless $\alpha$ has $L$-cofinality $\kappa^+$, where $\kappa$ is Mahlo in $L$; in that case the cofinality of $\alpha$ becomes $\kappa$ after forcing with $P$. Full cofinality-preservation does hold for *Thin* Easton forcing, defined like Long Easton forcing but with the requirement that for inaccessible $\kappa$, $\{\alpha < \kappa \mid p(\alpha) \neq \emptyset\}$ is *nonstationary* in $\kappa$. The argument of the proof of Lemma 2.28 applies in this case.

## Reverse Easton Forcing

Our third class forcing example is a type of iteration of set forcings first considered by Silver. We begin by discussing iterated forcing in general and then specialize to the particular kind of iteration that we wish to consider. First we consider two-step iterations.

**Definition.** Suppose that $P$ is a tame class forcing, defined over the ground model $\langle M, A\rangle$. A *named formula* is a formula $\varphi(x_1, \dots, x_n, \sigma_1, \dots, \sigma_m)$ where $\varphi$ is in the language of set theory augmented by the unary predicate symbols $\underline{A}$, $\underline{G}$ and where $\sigma_1, \dots, \sigma_m$ are names (built using $P$).

We would like to define the two-step iteration $P * Q$ to consist of pairs $(p, q)$ where $p \in P$, $p \Vdash q \in Q$, where $q$ is a $P$-name and $Q$ is a named formula, such that $P \Vdash Q$ is a partial ordering. However, we note that the resulting natural ordering on $P * Q$ is reflexive and transitive but not necessarily antisymmetric, as $p \Vdash q_0 = q_1$ does not imply that $q_0, q_1$ are the same name. Thus we drop the antisymmetry condition on $P$, $Q$ and only require that $P$ is a *preorder* (a reflexive, transitive relation) and that $P \Vdash Q$ is a preorder. Our earlier work in this chapter made no use of the antisymmetry condition, so we may safely work with preorders throughout.

**Definition.** Suppose $P$ is a tame forcing defined over the ground model $\langle M, A\rangle$, $Q$ is a named formula and $P \Vdash Q$ is a preorder with greatest element $1^Q$. Then $P * Q$ is the $\langle M, A\rangle$-definable preorder consisting of pairs $(p, q)$ where $p \in P$, $q$ is a $P$-name and $p \Vdash q \in Q$. (The latter means that, viewing $Q$ as a formula defining the ordering $q_0 \leq q_1$ of $Q$, we have $p \Vdash Q\ (q, q)$.) The ordering on $P * Q$ is given by: $(p_0, q_0) \leq (p_1, q_1)$ iff $p_0 \leq p_1$, $p_0 \Vdash q_0 \leq q_1$ in $Q$.

We need a version of the Product Lemma 2.27 for $P * Q$. The proof is almost the same. Note that we have assumed that $P$ is tame.

**Lemma 2.30.** *Suppose that $P * Q$ is a two-step iteration defined over $\langle M, A\rangle$ as above.*

(a) *If $G$ is $P$-generic over $\langle M, A\rangle$ and $H^G$ is $Q^G$-generic over $\langle M[G], A, G\rangle$ where $Q^G = \{q^G \mid \exists p \in G\ p \Vdash q \in Q\}$ (ordered by $q_0^G \leq q_1^G$ iff $\exists p \in G\ p \Vdash$*

*$Q(q_0, q_1)$) then $(G \times H) \cap (P * Q)$ is $P * Q$-generic over $\langle M, A\rangle$, where $H = \{q \mid q^G \in H^G\}$.*

(b) *If $K \subseteq P * Q$ is $P * Q$-generic over $\langle M, A\rangle$ then $K = (G \times H) \cap (P * Q)$ where $G$ is $P$-generic over $\langle M, A\rangle$, $H^G$ is $Q^G$-generic over $\langle M[G], A, G\rangle$ and $Q^G = \{q^G \mid \exists p \in G\ p \Vdash q \in Q\}$, $H^G = \{q^G \mid q \in H\}$.*

*Proof.* (a) First note that $K = (G \times H) \cap (P * Q)$ is compatible, for if $(p_0, q_0)$, $(p_1, q_1)$ both belong to $K$ then by the compatibility of $G$ we may assume $p_0 = p_1$ and then by the compatibility of $H^G$ in $Q^G$ we can choose $p \leq p_0$ in $G$ and $q$ such that $p \Vdash q \leq q_0, q_1$ (using the truth lemma for $P$). Second note that $K$ is upward closed as if $(p, q) \leq (p_0, q_0)$, $(p, q) \in K$ then $p_0 \in G$ by the upward closure of $G$ and $q_0 \in H$ as $q^G \leq q_0^G$ and $H^G$ is upward closed.

Now suppose that $D \subseteq P * Q$ is dense and $\langle M, A\rangle$-definable. Then $D_1 = \{q^G \mid \exists p \in G\ (p, q)$ meets $D\}$ is $\langle M[G], A, G\rangle$-definable; we claim that it is dense on $Q^G = \{q^G \mid \exists p \in G\ p \Vdash q \in Q\}$: Given $q^G \in Q^G$ consider $D_0(q) = \{p \in P \mid p \Vdash q \notin Q$ or $(p, q_0)$ meets $D$ for some $(p, q_0) \leq (p, q) \in P * Q\}$. Then $D_0(q) \subseteq P$ is dense so $G \cap D_0(q) \neq \emptyset$. Thus $(p, q_0)$ meets $D$ for some $p \in G$, $(p, q_0) \leq (p, q) \in P * Q$ so $q_0^G \leq q^G$ shows that $q^G$ can be extended into $D_1$. Lastly, the density of $D_1$ implies that $(p, q)$ meets $D$ for some $p \in G$, $q^G \in H^G$ so $K$ meets $D$.

(b) Let $G = \{p \in P \mid (p, q) \in K$ for some $q\}$, $H = \{q \mid (p, q) \in K$ for some $p\}$. Then clearly $K \subseteq G \times H$ and conversely if $(p, q) \in (G \times H) \cap (P * Q)$ then we can choose $(p^*, q^*) \leq (p, q_0)$, $(p_0, q)$ where all three of these conditions belong to $K$; then $(p^*, q^*) \leq (p, q)$ so $(p, q) \in K$. If $D_0 \subseteq P$ is dense and $\langle M, A\rangle$-definable then $D = \{(p, q) \in P * Q \mid p$ meets $D_0\}$ is dense so as $K$ meets $D$, we get that $G$ meets $D_0$. The genericity of $K$ implies that $G$ is upward closed (as $(p, 1^Q) \in P * Q$ for all $p \in P$) and compatible, so $G$ is $P$-generic over $\langle M, A\rangle$.

Suppose that $D_1 \subseteq Q^G$ is dense and $\langle M[G], A, G\rangle$-definable. Then by the definability of $P$-forcing, $D = \{(p, q) \in P * Q \mid p \Vdash q \in D_1\}$ is $\langle M, A\rangle$-definable. Also $D$ is dense $\leq (p, q)$ provided $p \Vdash D_1$ is dense on $Q$. By the truth lemma for $P$ there is such a $(p, q) \in K$ so by the genericity of $K$ there is $(p, q) \in K$, $p \Vdash q \in D_1$. So $H^G$ meets $D_1$. Upward closure for $H^G$ follows from upward closure for $K$: If $q^G \leq q_0^G$ then choose $p \in G$, $p \Vdash q \leq q_0$ in $Q$ and as $(p, q) \leq (p, q_0)$, $(p, q) \in K$ we get $(p, q_0) \in K$ so $q_0^G \in H^G$. Compatibility for $H^G$ follows from compatibility for $K$: if $q_0^G, q_1^G \in H^G$ then for some $(p, q) \in K$, $p \Vdash q \leq q_0, q_1$ so $q^G \leq q_0^G, q_1^G$. So $H^G$ is $Q^G$-generic over $\langle M[G], A, G\rangle$. □

We also have a Product Lemma for tameness:

**Lemma 2.31.** *Suppose that $P * Q$ is a two-step iteration defined over $\langle M, A\rangle$ and $\kappa > \omega$ is regular. If $Q^G$ is pretame below $\kappa$ in $\langle M[G], A, G\rangle$ for each $G$ $P$-generic over $\langle M, A\rangle$ then $P * Q$ is pretame below $\kappa$. If $Q^G$ is both pretame and tame below $\kappa$ in $\langle M[G], A, G\rangle$ for each $G$ $P$-generic over $\langle M, A\rangle$ then $P * Q$ is both pretame and tame below $\kappa$.*

*Proof.* Let $(p, q) \in P * Q$ and let $\langle D^i \mid i < \gamma \rangle$ be an $\langle M, A \rangle$-definable sequence of dense subclasses of $P * Q$, $\gamma < \kappa$. Let $G \subseteq P$ be $P$-generic over $\langle M, A \rangle$, $p \in G$ and let $D_1^i = \{q^G \mid (p^*, q)$ meets $D^i$ for some $p^* \in G\}$. Then $D_1^i$ is dense on $Q^G$ for each $i$, using the density of $D^i$ and the genericity of $G$. By the pretameness below $\kappa$ of $Q^G$ there are $q^{*G} \leq q^G$ and $\langle d_1^i \mid i < \gamma \rangle \in M[G]$ such that $d_1^i \subseteq D_1^i$ for each $i$ and $d_1^i$ is predense below $q^{*G}$ for each $i$. Using the truth lemma for $P$-forcing (which is guaranteed by our assumption that $P$ is tame) we may choose $p^* \leq p$ in $G$ such that $p^*$ forces the above property of $q^*$ and (a $P$-name for) the $d_1^i$'s, as well as "$d_1^i \subseteq V_{\hat{\alpha}}^{M[G]}$" for some fixed $\alpha \in \mathrm{ORD}(M)$ independent of $i < \gamma$.

For each pair $\langle i, q^{**} \rangle$ with $i < \gamma$, $q^{**}$ a $P$-name of rank $< \alpha$ the class $D_0(i, q^{**}) = \{p^{**} \mid p^{**} \Vdash q^{**} \in d_1^i \Longrightarrow (p^{**}, q^{**})$ meets $D^i\}$ is dense $\leq p^*$ so by pretameness of $P$ we can choose $p^{**} \leq p^*$ and $\langle d_0(i, q^{**}) \mid i < \gamma, q^{**}$ of rank $< \alpha \rangle \in M$ such that $d_0(i, q^{**}) \subseteq D_0(i, q^{**})$ and $d_0(i, q^{**})$ is predense $\leq p^{**}$ for each $(i, q^{**})$. Then if we let $d^i = D^i \cap V_\beta^M$ where $\alpha \leq \beta$, $\langle d_0(i, q^{**}) \mid i < \gamma, q^{**}$ of rank $< \alpha \rangle \in V_\beta^M$ we get that $d^i \subseteq D^i$ is predense $\leq (p^{**}, q^*)$ for each $i < \gamma$ and $(p^{**}, q^*) \leq (p, q)$. So we have established the pretameness below $\kappa$ of $P * Q$. The last statement of the lemma now follows, as for a pretame forcing, tameness below $\kappa$ is equivalent to preservation of the Power Set axiom below $\kappa$. □

Now we consider transfinite iterations.

**Definition.** $\langle P(< i) \mid i \leq \alpha \rangle, \alpha \leq \infty = \mathrm{ORD}$ is a *tame iteration of length* $\alpha$ over $\langle M, A \rangle$ if $\langle P(< i) \mid i \leq \alpha \rangle$ is an $\langle M, A \rangle$-definable sequence of preorders (with greatest elements $1^{P(<i)}$) such that

(1) $P(< 0)$ is the trivial forcing $\{\emptyset\}$ and each $P(< i)$ consists of functions $p : i \longrightarrow M$.

(2) For $i + 1 \leq \alpha$, $P(\leq i) \simeq P(< i) * P(i)$ via the isomorphism $p \longmapsto (p(< i), p(i))$, where $p(< i) = p \restriction i$, $P(< i)$ is tame, $P(< i) \Vdash P(i)$ is tame.

(3) For limit $\lambda \leq \alpha$, Direct Limit $\langle P(< i) \mid i < \lambda \rangle \subseteq P(< \lambda) \subseteq$ Inverse Limit $\langle P(< i) \mid i < \lambda \rangle$ and $P(< \lambda)$ is tame.

Clause (3) requires some further explanation. Inverse Limit $\langle P(< i) \mid i < \lambda \rangle$ consists of all $p : \lambda \longrightarrow M$ such that $p(< i) \in P(< i)$ for all $i < \lambda$. For the Direct Limit $\langle P(< i) \mid i < \lambda \rangle$ we add the requirement that $p(i) = 1^{P(i)}$ for sufficiently large $i < \lambda$. (Thus we assume that we have canonical names $1^{P(i)}$ for the greatest element of $P(i)$, uniformly for all $i$.) Both types of Limit are ordered by: $p \leq q$ iff $p(< i) \leq q(< i)$ for each $i < \lambda$.

It is not difficult to produce examples of tame iterations (of arbitrary length) by choosing the $P(i)$'s to be *set* forcings; as long as this is done $\langle M, A \rangle$-definably we are guaranteed to have tameness at each stage. Iterated *class* forcing is more problematic and will be discussed in Chapter 8.

However, even with iterated set forcing there may be difficulty in obtaining tameness for $P(< \infty) =$ Direct Limit $\langle P(< i) \mid i < \infty \rangle$. Reverse Easton forcing is one type of iteration for which this difficulty can be overcome, and which in addition has the following useful Factoring Property: For any $\alpha < \beta \leq \infty$, $P(< \beta)$ is isomorphic to a dense subordering of $P(< \alpha) * P[\alpha, \beta)$, where $P(< \alpha) \Vdash P[\alpha, \beta)$ is a tame iteration of length $\beta - \alpha$. We turn now to this example.

Define the iteration $\langle P(< i) \mid i \leq \infty \rangle$ in $L$ by: $P(< 0) = \{\emptyset\}$, the trivial forcing; $P(\leq i) \simeq P(< i) * P(i)$ where $P(i)$ is (a formula for) the trivial forcing unless $i \geq \omega$ is regular, in which case $P(i)$ is (a formula for) the forcing $2^{<i} = \{p : \alpha \longrightarrow 2 \mid \alpha < i\}$, ordered by extension; for $i$ limit we take $P(< i) =$ Inverse Limit $\langle P(< j) \mid j < i \rangle$ if $i$ is singular and Direct Limit $\langle P(< j) \mid j < i \rangle$ if $i$ is regular (or if $i = \infty$).

**Lemma 2.32.** *For each $i < \infty$, $P(\leq i)$ has a dense subordering which is a set of cardinality $\leq i^+$ (by convention, $0^+ = \omega$).*

*Proof.* Define $P^*(\leq i)$ to consist of all $p \in P(\leq i)$ in $L_{i^+}$. By induction on $i$ we show that $P^*(\leq i)$ is dense in $P(\leq i)$. This is clear for $i = 0$ and clear by induction for $i$ unless $i$ is a limit. Given that $i$ is regular and $P^*(< i) = \cup\{P^*(\leq i_0) \mid i_0 < i\}$ is dense in $P(< i)$ then any condition $(p, q_0)$ in $P(< i) * P(i)$ can be extended to one of the form $(p, q)$ where $q$ is a name contained in $\{\check{j} \mid j < i\} \times P^*(< i)$. Thus $(P(< i) * P(i)) \cap L_{i^+}$ is dense in $P(< i) * P(i)$. For $i$ singular, note that any $p \in$ Inverse Limit $\langle P(< j) \mid j < i \rangle$ can be successively extended to $p_0 \geq p_1 \geq p_2 \geq \cdots$ so that $p_j(< j) \in P_j^*$ and in addition, $j_0 < j_1 \Longrightarrow p_{j_1}(< j_0) = p_{j_0}(< j_0)$. So $q =$ (greatest lower bound to the $p_j$'s) exists and $q \in P^*(< i)$. □

**Lemma 2.33.** *For $\kappa$ regular and infinite, $P(\leq \kappa)$ is $\kappa^+$-cc.*

*Proof.* Note that $P(\leq \kappa) \simeq P(< \kappa) * P(\kappa)$ where $P(< \kappa) =$ Direct Limit $\langle P(< i) \mid i < \kappa \rangle$. By Lemma 2.32, $P(< \kappa)$ has a dense subordering of cardinality $\leq \kappa$ and hence $P(< \kappa)$ is $\kappa^+$-cc; also $P(< \kappa) \Vdash P(\kappa)$ is $\kappa^+$-cc. It follows that $P(< \kappa) * P(\kappa)$ is $\kappa^+$-cc: If $D \subseteq P(< \kappa) * P(\kappa)$ is dense, $D \in L$ and $G(< \kappa) \subseteq P(< \kappa)$ is $P(< \kappa)$-generic over $L$ then $D^{G(<\kappa)} = \{q \in P(\kappa)^{G(<\kappa)} \mid (p, q)$ meets $D$ for some $p \in G(< \kappa)\}$ is dense on $P(\kappa)^{G(<\kappa)}$, so as $P(< \kappa) \Vdash P(\kappa)$ is $\kappa^+$-cc there is $d \subseteq \hat{D}$ of cardinality $\leq \kappa$ such that $d^{G(<\kappa)}$ (defined as above) is predense on $P(\kappa)^{G(<\kappa)}$. As $P(< \kappa)$ is $\kappa^+$-cc we may choose $d \in L$ of cardinality $\leq \kappa$. Thus $\{p \in P(< \kappa) \mid$ for some $d \subseteq D$ of cardinality $\leq \kappa$, $d$ is predense below $(p, 1^{P(\kappa)})\}$ is dense on $P(< \kappa)$, so again since $P(< \kappa)$ is $\kappa^+$-cc, $D$ has a predense subset of cardinality $\leq \kappa \cdot \kappa = \kappa$. □

We can now prove the Factoring Property. For $\alpha \leq \beta \leq \infty$ we let $P[\alpha, \beta)$ be (a formula for) the iteration of length $\beta - \alpha$ stages defined just like $P$, except beginning at index $\alpha$ and ending after $\beta - \alpha$ stages. Then $P(< \alpha) * P[\alpha, \beta)$ consists of pairs $(p, q)$ where $p \in P(< \alpha)$ and $q$ is a $P(< \alpha)$-name for a condition in the iteration $P[\alpha, \beta)$.

**Lemma 2.34** (Factoring Property). *$P(< \beta)$ is isomorphic to $P(< \alpha) * P[\alpha, \beta)$.*

*Proof.* By induction on $\beta \geq \alpha$ we describe the isomorphism. If $\beta = \alpha$ then we can use the identity. Suppose that $\beta = \gamma + 1 \geq \alpha$ and we have defined $\pi : P(< \gamma) \simeq P(< \alpha) * P[\alpha, \gamma)$. Now $P(\leq \gamma) \simeq P(< \gamma) * P(\gamma) \simeq (P(< \alpha) * P[\alpha, \gamma)) * P(\gamma)$. We claim that in general, $(P * Q) * R$ is isomorphic to $P * (Q * R)$, where $Q * R$ is the formula defining all pairs $(q, r)$ such that $q \in Q^{\underline{G}}$ and $r$ is a $Q^{\underline{G}}$-name for an element of $R^{\underline{G}*\underline{H}}$ ($\underline{G}, \underline{H}$ denoting the $P$-generic, $Q^{\underline{G}}$-generic, respectively). To see this, first note that if $r$ is a $P * Q$-name and $G \subseteq P$ is $P$-generic then we can form a $Q^G$-name $r^G$ by taking $r^G = \{\langle s^G, q^G\rangle \mid q^G \in Q^G$ and for some $p \in G$, $\langle s, \langle p, q\rangle\rangle \in r\}$. Thus we get a $P$-name $\tilde{r}$ for a $Q^{\underline{G}}$-name by taking $\{\langle \pi, p\rangle \mid \operatorname{rank} \pi < \operatorname{rank} r$ and $p \Vdash \pi \in r^{\underline{G}}\}$. Conversely, given a $P$-name $\tilde{r}$ for a $Q^{\underline{G}}$-name we get a $P * Q$-name $r$ by taking $\{\langle \tilde{\pi}, \langle p, q\rangle\rangle \mid \operatorname{rank} \tilde{\pi} < \operatorname{rank} \tilde{r}$ and $\langle p, q\rangle \Vdash \tilde{\pi} \in \left((\tilde{r})^{\underline{G}}\right)^{\underline{H}}\}$. The isomorphism from $(P * Q) * R$ to $P * (Q * R)$ is given by: $\langle\langle p, q\rangle, r\rangle \longmapsto \langle p, [q, \tilde{r}]\rangle$, $\tilde{r}$ defined as above, where $[\sigma, \tau]$ is a canonical $P$-name for the ordered pair $\langle \sigma, \tau\rangle$.

Thus we get $(P(< \alpha) * P[\alpha, \gamma)) * P(\gamma) \simeq P(< \alpha) * (P[\alpha, \gamma) * P(\gamma)) = P(< \alpha) * P[\alpha, \gamma + 1)$. This completes the successor step.

For limit $\beta > \alpha$ we easily get $P(< \beta) \simeq P(< \alpha) * P[\alpha, \beta)$ by induction, provided $\beta$ regular $\Longrightarrow P(< \alpha) \Vdash \beta$ regular (for then we are taking the same type of Limit in forming $P(< \beta)$, $P[\alpha, \beta)$; also note that a $P(< \alpha)$-name for a $(\beta - \alpha)$-sequence can be equivalently viewed as a $(\beta - \alpha)$-sequence of $P(< \alpha)$-names). By Lemma 2.33 this condition is met. □

**Lemma 2.35.** *For $\kappa$ regular and infinite, $P(\leq \kappa) \Vdash P[\kappa + 1, \infty)$ is $\kappa^+$-closed (descending sequences of length $\leq \kappa$ have lower bounds).*

*Proof.* If $p_0 \geq p_1 \geq \cdots$ is a sequence of length $\lambda \leq \kappa$ in $P[\kappa + 1, \infty)^G$ where $G$ is $P(\leq \kappa)$-generic then define $p(j) = (P[\kappa + 1, j)$-name for) the greatest lower bound to $\langle p_i(j) \mid i < \lambda\rangle$. Note that for regular $\delta > \kappa$, $p(j) = 1^{P(j)}$ for sufficiently large $j < \delta$ so $p$ is indeed a condition. □

Now we have all the necessary facts to analyze $P =$ Direct Limit $\langle P(< i) \mid i < \infty\rangle$. $P$ is tame: By Lemma 2.34, for regular $\kappa$ we can write $P \simeq P(\leq \kappa) * P[\kappa + 1, \infty)$ and by Lemmas 2.32, 2.35 $P(\leq \kappa)$ is (equivalent to) a set forcing, $P(\leq \kappa) \Vdash P[\kappa + 1, \infty)$ is $\kappa^+$-distributive. By Lemma 2.22, $P(\leq \kappa) \Vdash P[\kappa + 1, \infty)$ is tame below $\kappa^+$. By Lemma 2.31, $P$ is tame below $\kappa^+$ for each $\kappa$ and hence tame.

$P$ preserves cofinalities: By Lemmas 2.8, 2.33, $P(\leq \kappa)$ preserves cofinality $\geq \kappa^+$ and by Lemmas 2.22, 2.35 $P(\leq \kappa) \Vdash P[\kappa + 1, \infty)$ preserves cofinality $\geq \kappa^+$. By the Product Lemma 2.30 we get that $P$ preserves cofinality $\geq \kappa^+$, and since $\kappa$ is arbitrary $P$ preserves cofinalities. GCH-preservation is similar, using Lemma 2.13.

## Amenable Class Forcing

Our fourth and final basic example of class forcing is where one has $\kappa$-distributivity for every $\kappa$. Tameness and preservation of cofinalities follow easily. Note that in this case one adds a generic class but no new sets, so GCH preservation is trivial.

A simple example is $P =$ all functions $p : \alpha \longrightarrow 2$, $\alpha \in$ ORD, ordered by extension. Another is $P =$ all closed sets of ordinals, ordered by end extension.

This completes our discussion of the basics of class forcing. In the next chapter we pose the question: For which class forcings $P$ defined in $L$ can we construct $P$-generic classes? We will make sense of this question using Silver's theory of indiscernibles for $L$, which will lead us to some unexpected answers.

*Chapter 3*

# Construction of Generic Classes

Recall that in Chapter 2 we imposed the Assumption that $P$-generic classes exist for any class forcing defined over a ground model $\langle M, A\rangle$. This is true when $M$ is countable, but not in general. In this chapter we drop this Assumption and study in detail, for the case of forcings defined over $L$, the problem of generic class existence. We will see that there is a natural condition, *$L$-rigidity*, shared by all tame class-generic extensions of $L$ and if this property fails in $V$ then there is a least inner model in which it fails, $L[0^\#]$. We then use $L[0^\#]$ to provide a criterion for deciding which class forcings $P$ defined over $L$ have generic classes, by defining such a $P$ to be *relevant* if it has a generic definable in $L[0^\#]$. Finally, we determine which of the basic class-forcing examples from Chapter 2 are relevant, using properties of the $L$-indiscernibles provided by $0^\#$.

First we must verify that if a $P$-generic $G$ exists then the model $\langle M[G], A, G\rangle$ does behave as described in Chapter 2, under the various hypotheses on $P$ discussed there. This is not immediate as we in fact do need the Assumption to prove some of the basic facts about $\Vdash$ such as the fact that $\Vdash$ and $\Vdash^*$ coincide, as well as the Definability and Truth lemmas for $\Vdash$.

However note the following:

**Proposition 3.1.** *Suppose $\varphi$ is a first-order property true in $\langle M[G], A, G\rangle$ whenever $M$ is countable and $G$ is $P$-generic over $\langle M, A\rangle$ for a forcing $P$ definable over $\langle M, A\rangle$. Then $\varphi$ is true for all such $\langle M[G], A, G\rangle$, without the assumption that $M$ is countable.*

*Proof.* Given an arbitrary $\langle M[G], A, G\rangle$ let $\langle \bar{M}[\bar{G}], \bar{A}, \bar{G}\rangle$ be the transitive collapse of a sufficiently elementary countable submodel and apply the hypothesis about $\varphi$ and elementarity to conclude that $\varphi$ holds in $\langle M[G], A, G\rangle$. □

Thus when establishing first-order properties of $\langle M[G], A, G\rangle$ for $P$-generic $G$, we may in fact use our earlier Assumption. As a consequence we get the following results, summarizing the work of Chapter 2.

**Theorem 3.2.** *Suppose $P$ is a forcing defined over the ground model $\langle M, A\rangle$ and $G$ is $P$-generic over $\langle M, A\rangle$.*

(a) *If $P$ is tame then $\langle M[G], A, G\rangle$ is a model of $ZF$. If $M \vDash AC$ then $M[G] \vDash AC$.*

*Now assume that $P$ is tame.*

(b) *If $\kappa$ is regular and uncountable in $M$ and $P$ is $\kappa$-cc, $\kappa$-distributive or $\Delta$-distributive at $\kappa$ in $\langle M, A\rangle$ then* $\mathrm{cof}(\alpha) \geq \kappa$ *in* $M \Longrightarrow \mathrm{cof}(\alpha) \geq \kappa$ *in* $M[G]$.

(c) *If $\kappa$ is an infinite cardinal of $M$, $M \vDash 2^\kappa = \kappa^+$ and $P$ is $\kappa^+$-preserving of cardinality $\leq \kappa^+$, $\kappa^+$-distributive or strongly $\Delta$-distributive at $\kappa$ then $M[G] \vDash 2^\kappa = \kappa^+$.*

**Theorem 3.3.** *Suppose $P = P_0 * P_1$ is defined over $\langle M, A\rangle$ with $P_0$ tame.*

(a) *If $G_0$ is $P_0$-generic over $\langle M, A\rangle$ and $G_1^{G_0}$ is $P_1^{G_0}$-generic over $\langle M[G_0], A, G_0\rangle$ where $P_1^{G_0} = \left\{p_1^{G_0} \mid \exists\, p_0 \in G_0\; p_0 \Vdash p_1 \in P_1\right\}$ then $(G_0 \times G_1) \cap P$ is $P$-generic over $\langle M, A\rangle$, where $G_1 = \left\{p_1 \mid p_1^{G_0} \in G_1^{G_0}\right\}$. Conversely, if $G$ is $P$-generic over $\langle M, A\rangle$ then $G = (G_0 \times G_1) \cap P$ where $G_0$ is $P_0$-generic over $\langle M, A\rangle$ and $G_1^{G_0}$ is $P_1^{G_0}$-generic over $\langle M[G_0], A, G_0\rangle$, where $P_1^{G_0} = \left\{p_1^{G_0} \mid \exists p_0 \in G_0,\, p_0 \Vdash p_1 \in P_1\right\}$ and $G_1^{G_0} = \left\{p_1^{G_0} \mid p_1 \in G_1\right\}$.*

(b) *If $P_0 \Vdash P_1$ is pretame below $\kappa$ then $P$ is pretame below $\kappa$. If $P_0 \Vdash P_1$ is tame below $\kappa$ and pretame then $P$ is tame below $\kappa$.*

**Theorem 3.4.** *If $P$ is one of the basic examples of class forcing over $L$ (Easton, Long Easton at Successors, Reverse Easton, Amenable) then $P$ is tame and preserves both cofinalities and the* GCH.

## 3.1 Rigidity

Which forcings $P$ defined in $L$ have generic classes? Of course if $V = L$ then for no nontrivial $P$ does there exist a $P$-generic class, however we declare this hypothesis to be too restrictive. A necessary condition for every $p \in P$ to belong to a $P$-generic, as we have seen in Chapter 2, is that $P$ be tame, and for any such $P$ it is consistent that a $P$-generic class exists. However, the possibility that a $P$-generic class exist for every tame $P$ which is $L$-definable without parameters is ruled out by the following result.

**Proposition 3.5.** *There exist tame forcings $P_0, P_1$ which are $L$-definable without parameters such that if $G_0, G_1$ are $P_0, P_1$-generic over $L$, respectively, then $\langle L[G_0, G_1], G_0, G_1\rangle$ is not a model of $ZF$.*

*Proof.* For any ordinal $\alpha$, let $n(\alpha)$ be the least $n$ such that $L_\alpha$ is not a model of $\Sigma_n$-replacement, if such an $n$ exists. Let $S_0 = \{\alpha \mid n(\alpha) \text{ exists and is even}\}$. $P_0$ consists of all closed $p$ such that $p \subseteq S_0$, ordered by $p \leq q$ iff $q$ is an initial segment of $p$.

Note that $S_0$ is unbounded in ORD: Given $\alpha$, let $\beta$ be least such that $\beta > \alpha$ and $L_\beta \vDash \Sigma_1$-Replacement. Then $n(\beta) = 2$ so $\beta \in S_0$. If $G_0 \subseteq P_0$ is $P_0$-generic over $L$ then $\cup G_0$ is therefore a closed unbounded subclass of ORD contained in $S_0$. To show that $P_0$ is tame, it suffices to show that it is $\kappa$-distributive for every $L$-regular $\kappa$ : If $\langle D_i \mid i < \kappa\rangle$ is an $L$-definable sequence of classes dense on $P_0$ and $p \in P_0$ then choose $n$ odd so that $\langle D_i \mid i < \kappa\rangle$ is $\Sigma_n$ definable and choose $\langle \alpha_i \mid i < \kappa\rangle$ to be first $\kappa$-many $\alpha$

such that $L_\alpha$ is $\Sigma_n$-elementary in $L$ and $\kappa, p, x \in L_\alpha$ where $x$ is the defining parameter for $\langle D_i \mid i < \kappa\rangle$. We can define $p \geq p_0 \geq p_1 \geq \dots$ so that $p_{i+1}$ meets $D_i$ and $\max(p_i) = \alpha_i$, using the $\Sigma_n$-elementarity of $L_{\alpha_i}$ in $L$. As $n(\alpha_i) = n+1$ and $n+1$ is even, we may define $p_\lambda$ to be $\bigcup\{p_i \mid i < \lambda\} \cup \{\alpha_\lambda\}$ for limit $\lambda \leq \kappa$ and we see that $q = p_\kappa \leq p$ meets each $D_i$.

Now define $P_1$ in the same way, but using $S_1 = \{\alpha \mid n(\alpha) \text{ is defined and odd}\}$. Then $P_1$ is also tame yet if $G_0, G_1$ are $P_0, P_1$-generic over $L$ (respectively) then $\bigcup G_0, \bigcup G_1$ are disjoint CUB subclasses of ORD. □

So we need a criterion for choosing $L$-definable forcings for which we can have a generic. Our approach is to isolate a "property of transcendence" (#) such that:

(1) In tame class-generic extensions of $L$, (#) fails.

(2) If (#) is true in $V$ then there is a least inner model $L(\#)$ satisfying (#).

Then our criterion for generic class existence is: $P$ has a generic iff it has one definable over $L(\#)$.

**Definition.** An amenable $\langle L, A\rangle$ is *rigid* if there is no nontrivial elementary embedding $\langle L, A\rangle \longrightarrow \langle L, A\rangle$. $L$ is *rigid* if $\langle L, \emptyset\rangle$ is rigid.

We shall take (#) to be: $L$ is not rigid. First we demonstrate property (b) above, i.e., that there is a least model in which $L$ is not rigid (if there is one at all).

**Theorem 3.6** (Kunen). *If $L$ is not rigid then there exists a* CUB *class $C$ of ordinals which are $L$-indiscernible: If $\varphi$ is an $n$-ary formula, $\alpha_1, \dots, \alpha_n$ and $\beta_1, \dots, \beta_n$ are increasing $n$-tuples from $C$ then $L \models \varphi(\alpha_1, \dots, \alpha_n) \iff \varphi(\beta_1, \dots, \beta_n)$.*

*Proof.* We need a lemma.

**Lemma 3.7.** *Suppose there exists $j : L \longrightarrow L$. Then there exists such a $j$ which is definable (with parameters) and such that every cardinal $\lambda$ of $L$-cofinality greater than $\kappa$ satisfying $\bar{\lambda} < \lambda \Longrightarrow \mathrm{Card}(\bar{\lambda}^\kappa)^L < \lambda$ is a fixed point of $j$, where $\kappa = \mathrm{crit}(j) =$ least $\alpha$ such that $j(\alpha) \neq \alpha$. ("crit" stands for "critical point".)*

*Proof of Lemma* 3.7. We use the ultrapower construction. Define an ultrafilter $U$ on $\mathcal{P}(\kappa) \cap L$ by: $X \in U$ iff $\kappa \in j(X)$. Then there is an elementary embedding $k : L \longrightarrow \mathrm{Ult}(L, U)$ where $\mathrm{Ult}(L, U)$ is the ultrapower $L^\kappa / U$ defined using functions $f : \kappa \longrightarrow L$ which belong to $L$. Thus an element of $\mathrm{Ult}(L, U)$ is $[f] = \{g : \kappa \longrightarrow L \mid g \in L$ and for some $X \in U, \alpha \in X \Longrightarrow g(\alpha) = f(\alpha)\}$, with $E = \in$-relation of $\mathrm{Ult}(L, U)$ defined in the natural way: $[f]E[g]$ iff $\{\alpha \mid f(\alpha) \in g(\alpha)\} \in U$.

The map $[f] \longmapsto j(f)(\kappa)$ gives an elementary embedding from $\mathrm{Ult}(L, U)$ into $L$ and hence $\mathrm{Ult}(L, U)$ is well-founded and isomorphic to $L$. If $h : \mathrm{Ult}(L, U) \simeq L$ then $j^* = h \circ k : L \longrightarrow L$ is definable with parameters $\kappa, U$. If $\lambda$ has $L$-cofinality greater than $\kappa$ then $k$ (and hence $j^*$) is continuous at $\lambda$ since any constructible $f : \kappa \longrightarrow \lambda$ is bounded by the constant function $c_{\bar{\lambda}}$ with value $\bar{\lambda}$ for some $\bar{\lambda} < \lambda$ (hence

$[f]E[c_\lambda] \longrightarrow [f]E[c_{\bar{\lambda}}]$ for some $\bar{\lambda} < \lambda$). But if $[f]E[c_{\bar{\lambda}}]$ then $\{[g] \mid [g]E[f]\}$ has size at most $\mathrm{Card}(\bar{\lambda}^\kappa)^L$, and if this is smaller than $\lambda$ then $j^*[\lambda] \subseteq \lambda$ and hence by continuity $j^*(\lambda) = \lambda$. □ (Lemma)

Lemma 3.7 shows that if $L$ is not rigid then there is $j : L \longrightarrow L$ with critical point $\kappa$ such that every limit cardinal of cofinality $> \kappa$ is a fixed point of $j$. It follows that if $F = \{\alpha \mid \alpha$ a limit cardinal of cofinality $> \kappa\}$ then $\kappa \notin \mathrm{Hull}(\kappa \cup F)$ where Hull denotes the Skolem hull in $L$.

For any class of ordinals $G$ let $G^*$ denote $\{\alpha \in G \mid \alpha = \mathrm{ordertype}(\alpha \cap G)\}$. Then define inductively: $F_0 = F$, $F_{\alpha+1} = (F_\alpha)^*$, $F_\lambda = (\bigcap\{F_\alpha \mid \alpha < \lambda\})^*$ for limit $\lambda$. For any $\alpha$, $H_\alpha$ denotes $\mathrm{Hull}(\kappa \cup F_\alpha)$. And $\langle \kappa_\alpha \mid \alpha \in \mathrm{ORD}\rangle$ is defined by: $\kappa_0 = \kappa$, $\kappa_{\alpha+1} = \min(H_\alpha - \kappa)$, $\kappa_\lambda = \cup\{\kappa_\alpha \mid \alpha < \lambda\}$ for limit $\lambda$.

**Claim.** For every $\alpha$, $\kappa_\alpha < \kappa_{\alpha+1}$.

*Proof.* We may assume that $\alpha$ is not 0. As $\kappa_{\alpha+1}$ belongs to $\mathrm{Hull}(\kappa \cup F_\alpha)$ it is a fixed point of the isomorphism $L \simeq H_{<\alpha} = \mathrm{Hull}(\kappa \cup \bigcap\{F_\beta \mid \beta < \alpha\})$. But $H_{<\alpha} \cap [\kappa, \kappa_\alpha) = \emptyset$, so $\kappa_\alpha$ is *not* a fixed point of this isomorphism, using $\kappa < \kappa_\alpha$. □

**Claim.** Let $\pi_{\alpha\beta} : L \simeq \mathrm{Hull}(\kappa_\alpha \cup F_\beta)$. Then $\pi_{\alpha\beta}$ fixes $\kappa_\gamma$ when $\gamma < \alpha$ or when $\gamma$ is a successor ordinal $> \beta + 1$. Also $\pi_{\alpha\beta}(\kappa_\alpha) = \kappa_{\beta+1}$.

*Proof.* $\gamma < \alpha \Longrightarrow \kappa_\gamma < \kappa_\alpha$, so clearly $\pi_{\alpha\beta}$ fixes $\kappa_\gamma$. If $\beta + 1 < \gamma$, $\gamma$ successor then $\kappa_\gamma \in \mathrm{Hull}(\kappa \cup F_{\gamma-1})$, so $\kappa_\gamma$ is a fixed point of $\pi_{\alpha\beta}$.

As $\kappa_{\beta+1} \in \mathrm{Hull}(\kappa_\alpha \cup F_\beta) = H$, we have $\kappa_\alpha \leq \pi_{\alpha\beta}(\kappa_\alpha) \leq \kappa_{\beta+1}$. Conversely, suppose that $\kappa_\alpha \leq \delta < \kappa_{\beta+1}$, $\delta \in H$; we derive a contradiction. Write $\delta = t(\vec{\xi}, \vec{\eta})$ where the components of $\vec{\xi}$ are less than $\kappa_\alpha$ and the components of $\vec{\eta}$ belong to $F_\beta$. Choose $\bar{\alpha} + 1 \leq \alpha$ least so that the components of $\vec{\xi}$ are less than $\kappa_{\bar{\alpha}+1}$. Then $L \models \exists \vec{\xi}$ with components $< \kappa_{\bar{\alpha}+1} (\kappa_{\bar{\alpha}+1} \leq t(\vec{\xi}, \vec{\eta}) < \kappa_{\beta+1})$. Let $\pi : L \simeq \mathrm{Hull}(\kappa \cup F_{\bar{\alpha}})$. Then $\pi(\kappa) = \kappa_{\bar{\alpha}+1}$, $\pi(\vec{\eta}) = \eta$, $\pi(\kappa_{\beta+1}) = \kappa_{\beta+1}$. So $L \models \exists \vec{\xi}$ with components $< \kappa (\kappa \leq t(\vec{\xi}, \vec{\eta}) < \kappa_{\beta+1})$, contradicting the defintion of $\kappa_{\beta+1}$. □

Now for any two increasing $n$-tuples $\alpha_1, \ldots, \alpha_n$ and $\beta_1, \ldots, \beta_n$ with $\alpha_n < \beta_1$ we can obtain $\pi : L \longrightarrow L$ such that $\pi(\kappa_{\alpha_i}) = \kappa_{\beta_i+1}$ for all $i$, by taking $\pi_{\alpha_1\beta_1} \circ \cdots \circ \pi_{\alpha_n\beta_n}$. This implies that $C = \{\kappa_\alpha \mid \alpha \in \mathrm{ORD}\}$ is a class of $L$-indiscernibles. □

Now we introduce $0^\#$. As before, Hull denotes Skolem hull in $L$.

**Theorem 3.8** (Silver [71], Solovay [67]). *Suppose L is not rigid. Then there is a unique* CUB *class I of L-indiscernibles which* generate *L in the sense that* $L = \mathrm{Hull}(I)$. *Moreover I is unbounded in every uncountable cardinal and if* $0^\#$ *= First-Order theory of* $\langle L, \in, i_1, i_2, \ldots\rangle$ *(where the first* $\omega$ *elements* $i_1, i_2, \ldots$ *of I are introduced as constants) then we have the following:*

(a) $0^\# \in L[I]$, *I is* $\Delta_1(L[0^\#])$ *in the parameter* $0^\#$ *and I is unbounded in* $\alpha$ *whenever* $L_\alpha[0^\#] \models \Sigma_1$ *replacement.*

(b) $0^\#$, *viewed as a real, is the unique solution to a* $\Pi^1_2$ *formula (i.e., a formula of the form* $\forall x \exists y \psi$, *where* $x, y$ *vary over reals and* $\psi$ *is arithmetical).*

(c) *If* $f : I \longrightarrow I$ *is increasing,* $f \neq$ *identity then there is a unique* $j : L \longrightarrow L$ *extending* $f$ *with critical point in* $I$, *and every* $j : L \longrightarrow L$ *is of this form.*

(d) *If* $\langle L, A\rangle$ *is amenable then* $A$ *is* $\Delta_1(L[0^\#])$, $\langle L, A\rangle$ *is not rigid and a final segment of* $I$ *is a class of* $\langle L, A\rangle$*-indiscernibles.*

**Remarks.** (1) As $I$ is closed and is unbounded in every uncountable cardinal it follows that every uncountable cardinal belongs to $I$ and $0^\# =$ First-Order theory of $\langle L, \in, \aleph_1, \aleph_2, \dots\rangle$.

(2) The $\Sigma^1_2$-absoluteness of $L$ (Shoenfield [61]) implies that the unique solution to a $\Sigma^1_2$ formula is constructible; so in a sense (b) is best possible.

(3) $I$ is a class of *strong* indiscernibles: If $\vec{i}, \vec{j}$ are increasing tuples from $I$ of the same length and $x < \min(\vec{i}), \min(\vec{j})$ then for any $\varphi$, $L \models \varphi(x, \vec{i}) \Longleftrightarrow \varphi(x, \vec{j})$. In fact the proof below shows that any unbounded class $I$ of $L$-indiscernibles such that $I \cap \operatorname{Lim} I \neq \emptyset$ is necessarily a class of strong indiscernibles.

*Proof.* By Theorem 3.6 there exists a CUB class $C$ of $L$-indiscernibles. Let $\pi : \mathrm{Hull}(C) \simeq L$ and we see that $I = \pi[C]$ is a CUB class of generating $L$-indiscernibles. Note that $\alpha \in I \Longrightarrow L_\alpha \prec L$ and therefore $L = \Sigma_1$-Hull$(I)$. For any $\Sigma_1$ $\varphi(x, y_1, \dots, y_n)$ let $t_\varphi$ be the term $\mu x \varphi(x, y_1, \dots, y_n)$, intended to name the $L$-least $x$ such that $L \models \varphi(x, y_1, \dots, y_n)$, if it exists, and 0 otherwise. Then $L$ is described as the Ehrenfeucht–Mostowski model consisting of all terms $t_\varphi(j_1, \dots, j_n)$ (with $j_1, \dots, j_n \in I$ substituting for the variables $y_1, \dots, y_n$), with terms identified as dictated by Thy$\langle L, \in, i_1, i_2, \dots\rangle =$ First-Order theory of $\langle L, \in, i_1, i_2, \dots\rangle$. Thus $I$ is uniquely determined by Thy$\langle L, \in, i_1, i_2, \dots\rangle$. But if $I^*$ is another CUB class of generating $L$-indiscernibles we get $I \cap I^*$ infinite (and in fact CUB), hence Thy$\langle L, \in, i_1, i_2, \dots\rangle =$ Thy$\langle L, \in, i^*_1, i^*_2, \dots\rangle$. So $I$ is unique. Also note that $I$ is a class of *strong* $L$-indiscernibles in the sense that $x < \min(\vec{i}), \min(\vec{j})$, $\vec{i}$ and $\vec{j}$ of the same length from $I$ implies that $L \models \varphi(x, \vec{i}) \Longleftrightarrow \varphi(x, \vec{j})$ for any formula $\varphi$; if not then we get $\vec{i} < \min(\vec{j})$ with $\{x < \min(\vec{i}) \mid L \models \varphi(x, \vec{i})\} \neq \{x < \min(\vec{i}) \mid L \models \varphi(x, \vec{j})\}$ and $\min(\vec{i})$ a limit point of $I$. But then we can get $\vec{i} < \vec{j}_0 < \vec{j}_1 < \cdots$ of length ORD with $\alpha < \beta \Longrightarrow \{x < i_0 \mid L \models \varphi(x, \vec{j}_\alpha)\} \neq \{x < i_0 \mid L \models \varphi(x, \vec{j}_\beta)\}$; this is absurd because there are only set-many choices for subsets of $i_0$.

It follows from the strong indiscernibility of $I$ that $t(\vec{i}, \vec{j}) < \min(\vec{j})$ implies $t(\vec{i}, \vec{j}) < I$-successor to $\max(\vec{i})$. Hence for all $i \in I \cup \{0\}$, Hull$(i \cup \{i, j_1, j_2 \dots\}) \supseteq L_{i^*}$ where $i < i^* \leq j_1 < j_2 < \cdots$ are $\omega$-many elements of $I$, $i^* = I$-successor to $i$. So Card$(L_{i^*}) =$ Card$(i)$ and it follows that uncountable cardinals belong to Lim $I$. Moreover if $i \in \operatorname{Lim} I$ then $L_i =$ Hull$(I \cap i)$ and $L_i$ is isomorphic to the natural Ehrenfeucht–Mostowski model built from $I \cap i$, using $0^\# =$ Thy$\langle L, \in, i_1, i_2, \dots\rangle$ to determine when to identify two terms $t_{\varphi_0}(\vec{i}_0), t_{\varphi_1}(\vec{i}_1)$. We now verify (a)–(d).

(a) Clearly $0^\# \in L[I]$ as $0^\# = \text{Thy}\langle L_{i_\omega}, \in, i_1, i_2, \dots\rangle$ where $i_n = n^{\text{th}}$ indiscernible. If $\alpha$ is $0^\#$-admissible (i.e., $L_\alpha[0^\#] \models \Sigma_1$ replacement) then for any limit $\lambda < \alpha$, $L_{i_\lambda} \simeq$ Ehrenfeucht–Mostowski model $M(0^\#, \lambda)$ built from $\lambda$ indiscernibles and therefore belongs to $L_\alpha[0^\#]$, as $\Sigma_1$-replacement gives us the Mostowski collapse. So $\alpha = i_\alpha = \alpha^{\text{th}}$ indiscernible and $\lambda \longmapsto \langle L_{i_\lambda}, \{i_\beta \mid \beta < \lambda\}\rangle$ is $\Delta_1(L_\alpha[0^\#])$. Hence $I$ is $\Delta_1(L[0^\#])$ (with parameter $0^\#$).

(b) $0^\# = \text{Thy}\langle L, \in, i_1, i_2, \dots\rangle$ has the property that for every countable limit $\lambda$, $M(0^\#, \lambda)$ is well-founded and if $\pi : M(0^\#, \lambda) \simeq L_{i_\lambda}$, $\pi(\beta^{\text{th}}$ indiscernible in $M(0^\#, \lambda)) = i_\beta$ then $\{i_\beta \mid \beta < \lambda\}$ is CUB in $i_\lambda$. This is a $\Pi^1_2$ property as it says $\forall$ relation $R$ on $\omega$ ($R$ a well-ordering $\Longrightarrow M(0^\#, <_R)$ is well-founded and is a model of $\varphi$) where $\varphi$ is first-order. But if $0^*$ obeys this property then $M(0^*, \text{ORD}) \simeq L$ and $0^* = \text{Thy}\langle L, \in, i^*_1, i^*_2, \dots\rangle$ where $I^* = \{i^*_\beta \mid \beta \in \text{ORD}\}$ is a CUB class of generating $L$-indiscernibles. We have seen that $I = I^*$ and so $0^* = 0^\#$.

(c) If $f : I \longrightarrow I$ is increasing, $f \neq$ identity then define $j : L \longrightarrow L$ by $j(t_\varphi(j_1, \dots, j_n)) = t_\varphi(f(j_1), \dots, f(j_n))$. This is well-defined since $I$ is a class of $L$-indiscernibles. $j$ must be the identity on $i$ = the critical point of $f$ = the least $i$, $f(i) > i$, as $t_\varphi(j_1, \dots, j_n, k_1, \dots, k_m) = t_\varphi(j_1, \dots, j_n, f(k_1), \dots, f(k_m))$ when $t_\varphi(j_1, \dots, j_n, k_1, \dots, k_m) < k_1$. So the critical point of $j$ = the critical point of $f$ belongs to $I$. Clearly $j$ is unique, given $f$. If $j : L \longrightarrow L$ is arbitrary then $\alpha$ = the critical point of $j$ belongs to $I$, as by Lemma 3.7, $\alpha$ = critical point of $j^*$ where $j^*(i) = i$ for unboundedly many $i \in I$ and thus if $\alpha \notin I$ we get $\alpha = t_\varphi(x, \vec{i})$, $x < \alpha < \vec{i}$, $j^*(\vec{i}) = \vec{i}$ and thus $j^*(\alpha) = \alpha$, contradicting $\alpha$ = critical point of $j^*$. Now note that if $i \in I$ then $j(i)$ is the critical point of some $j^* : L \longrightarrow L$ as $i \notin \text{Hull}(i \cup (I - (i+1)))$ implies $j(i) \notin \text{Hull}(j(i) \cup J)$ where $J = j[I - (i+1)]$ so $k : L \simeq \text{Hull}(j(i) \cup J)$ has critical point $j(i)$. So $j(i) \in I$.

(d) If $\langle L, A\rangle$ is amenable then for each $i \in I$ we may write $A \cap i = t_{\varphi_i}(\vec{j}_i, i, \vec{k}_i)$ where $\vec{j}_i < i < \vec{k}_i$ are all from $I$. By Fodor's Theorem $(\varphi_i, \vec{j}_i)$ is constant on an unbounded subclass of $I$ and hence by indiscernibility we may assume that $A \cap i = t_\varphi(\vec{j}, i, \vec{k}_i)$ for all $i \in I$, $i > \max(\vec{j})$ where the choice of $\vec{k}_i \in I - (i+1)$ does not matter. Thus $I - (\max(\vec{j}) + 1)$ is a class of $\langle L, A\rangle$-indiscernibles and $A$ is $\Delta_1(L[0^\#])$ in parameters $\vec{j}, 0^\#$. We get $j : \langle L, A\rangle \longrightarrow \langle L, A\rangle$ by shifting $I$ above $\vec{j}$. □

In case the conclusion of Theorem 3.8 holds (i.e. in case $L$ is not rigid) we say that "$0^\#$ exists" and refer to $I$ as the *Silver Indiscernibles*. Note that Theorem 3.8 implies that if $L$ is not rigid then $L[0^\#]$ is the smallest inner model in which $L$ is not rigid.

The next theorem shows that $L$ is rigid in its tame class-generic extensions.

**Theorem 3.9.** *Suppose that $G$ is $P$-generic over $\langle L, A\rangle$ and $P$ is tame. Then $L[G] \models$ $0^\#$ does not exist.*

*Proof.* Suppose $p_0 \in P$, $p_0 \Vdash I$ = Silver indiscernibles is unbounded and $i < j$ in $I \Longrightarrow L_i \prec L_j$. Suppose that $p \leq p_0$, $p \Vdash \hat{\alpha} \in I$. Then $L_\alpha \prec L$ as this is true in any $P$-generic extension $\langle L[G], A, G\rangle$, $p \in G$. (By Löwenheim–Skolem we can assume that such a $G$ exists for the sake of this argument.) Thus an $L$-Satisfaction predicate is definable over $\langle L, A\rangle$ as $L \models \varphi(x)$ iff for some $p \in P$ below $p_0$, some $\alpha$ with $x \in L_\alpha$, $p \Vdash \varphi(\hat{x})$ is true in $L_\alpha$. This is a contradiction if $A = \emptyset$, for then $L$-satisfaction would be $L$-definable. But note that for any $A$ such that $\langle L, A\rangle$ is amenable we can apply the same argument, using the fact that by Theorem 3.8(d), $\langle L_\alpha, A \cap L_\alpha\rangle \prec \langle L, A\rangle$ for $\alpha$ in a final segment of $I$. □

The previous result was proved independently by A. Beller in Beller–Jensen–Welch [85]. In Chapter 5 we will produce a real $R$, *not* constructing $0^\#$, which fails to belong to any tame class-generic extension of $L$.

The most important sufficient condition for the existence of $0^\#$ is expressed by Jensen's Covering Theorem, to which we turn next. A set $X$ is *covered in* $L$ if there is a constructible $Y$ such that $X \subseteq Y$, $\operatorname{Card} Y = \operatorname{Card} X$.

**Theorem 3.10** (Jensen). *Suppose there exists an uncountable set of ordinals which is not covered in* $L$. *Then* $0^\#$ *exists.*

*Proof.* Suppose $X$ is an uncountable subset of $L_\kappa$ which cannot be covered in $L$, where $\kappa$ is least possible. Then any constructible $Y \supseteq X$ has $L$-cardinality $\geq \kappa$, lest we contradict the leastness of $\kappa$. We can assume that $X$ is $\Sigma_1$-elementary in $L_\kappa$ and therefore get a $\Sigma_1$-elementary $j : L_{\bar\kappa} \longrightarrow L_\kappa$ with $\operatorname{Range}(j) = X$.

If $\bar\kappa$ is an $L$-cardinal we (almost) obtain $0^\#$ as follows. We can naturally express $L$ as the union $L = \cup\{H(\alpha, P, \beta) \mid \alpha < \bar\kappa \leq \beta, P \text{ a finite subset of } L_\beta\}$ where $H(\alpha, P, \beta) =$ Skolem hull of $\alpha \cup P$ in $L_\beta$. Also write $(\alpha_0, P_0, \beta_0) \leq (\alpha_1, P_1, \beta_1)$ if $\alpha_0 \leq \alpha_1$, $P_0 \subseteq P_1, \beta_0 \leq \beta_1$ and $\beta_0 < \beta_1 \implies P_0 \cup \{\beta_0\} \subseteq P_1$; then of course $H(\alpha_0, P_0, \beta_0) \subseteq H(\alpha_1, P_1, \beta_1)$. If we transitively collapse the $H(\alpha, P, \beta)$ to $\bar H(\alpha, P, \beta)$ and define $\sigma : \bar H(\alpha_0, P_0, \beta_0) \longrightarrow \bar H(\alpha_1, P_1, \beta_1)$ to be the collapse of the inclusion map $H(\alpha_0, P_0, \beta_0) \subseteq H(\alpha_1, P_1, \beta_1)$, we obtain a direct limit system $\Pi(\bar\kappa, \infty)$ with direct limit isomorphic to $L$. The key point is that as $\bar\kappa$ is an $L$-cardinal, each element $\bar H(\alpha_0, P_0, \beta_0)$ of $\Pi(\bar\kappa, \infty)$ and each map $\sigma : \bar H(\alpha_0, P_0, \beta_0) \longrightarrow \bar H(\alpha_1, P_1, \beta_1)$ of $\Pi(\bar\kappa, \infty)$ belongs to $L_{\bar\kappa}$.

So we may define a direct limit system $j(\Pi(\bar\kappa, \infty)) = \Pi$ by applying $j$ to each element and map of $\Pi(\bar\kappa, \infty)$. If $\Pi$ has a well-founded direct limit (and this is a big "if"!) then $\operatorname{Lim} \Pi$ is isomorphic to $L$ and so we get an elementary embedding $k : L \longrightarrow L$ by virtue of the map $j : L_{\bar\kappa} \longrightarrow L_\kappa$ carrying $\Pi(\bar\kappa, \infty)$ to $\Pi$. Moreover if $x \in L_\alpha, \alpha < \bar\kappa$ then $j(x) \in j(\bar H(\alpha, \emptyset, \bar\kappa))$ represents itself in $\operatorname{Lim} \Pi$, so $k(x) = j(x)$ and we have shown that $k : L \longrightarrow L$ extends $j$. By Theorem 3.8, $0^\#$ exists.

We can refine the $\Pi(\bar\kappa, \infty)$ construction as follows: Suppose $\bar\kappa \leq \gamma$, $0 < n < \omega$ and for each $\alpha < \bar\kappa, \alpha < \beta < \gamma$, $P$ a finite subset of $L_\beta$, $H_n(\alpha, P, \beta)$ collapses to $\bar H_n(\alpha, P, \beta)$ of height less than $\bar\kappa$, where $H_n(\alpha, P, \beta) = \Sigma_n$ Skolem hull of $\alpha \cup P$ in $L_\beta$. (For any $Z \subseteq L_\beta$, $\Sigma_n$ Skolem hull of $Z$ in $L_\beta$ is the smallest $\Sigma_n$-elementary submodel of $L_\beta$ containing $Z$ as a subset; it is equal to $\{a \in L_\beta \mid a \text{ is the } <_{L_\beta}\text{-least solution to a } \Sigma_n \text{ formula with parameters from } Z\}$.) Then define $(\alpha_0, P_0, \beta_0) \leq (\alpha_1, P_1, \beta_1)$ as we did for $\Pi(\bar\kappa, \infty)$. The resulting direct limit system has direct limit $L_\gamma$ (if $\gamma$ limit;

$L_{\gamma-1}$, otherwise). This system is denoted by $\Pi(\bar{\kappa}, \gamma, n)$. If the above condition on $\gamma, n$ holds for all $n$ then we can also define $\Pi(\bar{\kappa}, \gamma, \omega)$ by letting $n$ vary: Elements of the direct limit are 4-tuples $(\alpha, P, \beta, n)$ and $(\alpha_0, P_0, \beta_0, n_0) \leq (\alpha_1, P_1, \beta_1, n_1)$ if $\alpha_0 \leq \alpha_1$, $P_0 \subseteq P_1$ and either $\beta_0 = \beta_1, n_0 \leq n_1$ or $\beta_0 < \beta_1, \beta_0 \in P_1$. And now we have $\sigma : \bar{H}_{n_0}(\alpha_0, P_0, \beta_0) \longrightarrow \bar{H}_{n_1}(\alpha_1, P_1, \beta_1)$.

Again elements and maps from $\Pi(\bar{\kappa}, \gamma, m), m \leq \omega$ belong to $L_{\bar{\kappa}}$ and so we can define $\Pi = j(\Pi(\bar{\kappa}, \gamma, m))$. We continue to suppose that $\Pi$ is well-founded and therefore has direct limit isomorphic to an initial segment of $L$. Now note that if $\gamma = \delta + 1$, $m$ finite then we get $k : L_\delta \longrightarrow L_{\delta^*}$ extending $j$, and as $\Pi(\bar{\kappa}, \gamma, m)$ maps cofinally into $\Pi$, $k$ is $\Sigma_{m+1}$- elementary. If $\gamma = \delta + 1, m = \omega$ we get $k : L_\gamma \longrightarrow L_{\gamma^*}$ $\Sigma_1$-elementary and extending $j$. And if $\gamma$ limit we also get $k : L_\gamma \longrightarrow L_{\gamma^*}$ $\Sigma_1$-elementary and extending $j$.

We are ready to consider the case where $\bar{\kappa}$ is not an $L$-cardinal. Then there is a least $\gamma \geq \bar{\kappa}$ such that for some least positive integer $n$, $H_n(\alpha, P, \gamma)$ collapses to a model of height $\geq \bar{\kappa}$ for some $\alpha < \bar{\kappa}$, $P$ a finite subset of $L_\gamma$. If $n > 1$ then let $\bar{\Pi} = \Pi(\bar{\kappa}, \gamma+1, n-1)$. If $n = 1$ then let $\bar{\Pi} = \Pi(\bar{\kappa}, \gamma, \omega)$. By the previous paragraph we get $k : L_\gamma \longrightarrow L_{\gamma^*}$ $\Sigma_n$-elementary and extending $j$. But note that $H_n(\alpha, P, \gamma)$ is isomorphic to $L_\gamma$. So in fact $H_n(\alpha, P, \gamma) = L_\gamma$ for some $\alpha < \bar{\kappa}$, $P$ a finite subset of $L_\gamma$. But then $\text{Range}(j) = X$ is exactly $H_n(\text{Range}(j \restriction \alpha), k(P), \gamma^*) \cap L_\kappa$ and therefore is included in the constructible set $H_n(j(\alpha), k(P), \gamma^*) = Y$, of $L$-cardinality $< \kappa$. This contradicts our choice of $X$.

It only remains to justify the assumption that $j(\Pi(\bar{\kappa}, \gamma, n))$ has a well-founded direct limit whenever $\Pi(\bar{\kappa}, \gamma, n)$ is defined. (This implies also that $j(\Pi(\bar{\kappa}, \infty))$ has a well-founded direct limit when $\bar{\kappa}$ is an $L$-cardinal.) If not, then choose $\gamma_0$ least such that for some least $\bar{\kappa}_0$ and least $n_0$ (for these choices of $\gamma_0, \bar{\kappa}_0$) there is a direct limit system $\bar{\Pi}$ contained cofinally in $\Pi(\bar{\kappa}_0, \gamma_0, n_0)$ with $\Pi = j(\bar{\Pi})$ having an ill-founded limit. Choose $a_0 < a_1 < \cdots$ in $\Pi$ and $\beta_i \in a_i$ such that $\sigma_{a_i a_{i+1}}(\beta_i) > \beta_{i+1}$ where $\sigma_{a_i a_{i+1}} : a_i \longrightarrow a_{i+1}$ is the canonical map given by $\Pi$. Let $X^* =$ Skolem hull of $X \cup \{\beta_0, \beta_1, \ldots\}$ in $L_\kappa$. Then for any suitable direct limit system $\Pi^* \subseteq X$ (i.e., such that $j^{-1}(\Pi^*)$ is cofinally contained in some $\Pi(\bar{\kappa}_1, \gamma_1, n_1)$), if $\Pi^*$ has an ill-founded limit then there is a witness $\{\beta_0^*, \beta_1^*, \ldots\} \subseteq X^*$ to its ill-foundedness. For by the leastness of $\bar{\kappa}_0, \gamma_0, n_0$, $\bar{\Pi}$ embeds into $\bar{\Pi}^* = j^{-1}(\Pi^*)$ and hence $\Pi$ embeds into $\Pi^*$, sending $\langle \beta_i \mid i \in \omega \rangle$ to an appropriate $\langle \beta_i^* \mid i \in \omega \rangle$ contained in $X^*$.

Now define $X_0 = X$, $X_{\alpha+1} = X_\alpha^*$, $X_\lambda = \cup\{X_\alpha \mid \alpha < \lambda\}$ for limit $\lambda$. If $\Pi \subseteq X_{\omega_1}$ is a suitable ill-founded system then some countable $\Pi_0 \subseteq \Pi \cap X_\alpha$ for some $\alpha < \omega_1$ is ill-founded and therefore $X_{\omega_1}$ contains a witness to the ill-foundedness of $\Pi$. So $X_{\omega_1}$ has the desired property that $j : L_{\bar{\kappa}} \longrightarrow L_\kappa$ with range $X_{\omega_1}$ preserves the well-foundedness of suitable direct limit systems, and $X_{\omega_1}$ has cardinality $\leq \omega_1 + \text{card}(X) = \text{card}(X)$. □

We can extract more information from the above proof. The following result is due to Carlson and is closely related to work of Shelah [82].

**Corollary 3.11** (Strong Covering). *Assume that $0^\#$ does not exist. There is a class $C$ definable over $L$ such that:*

(a) *Every uncountable set of ordinals is covered by an element of* $C$.

(b) *If* $X_0 \subseteq X_1 \subseteq \ldots$ *is a sequence from* $C$ *of length* $\lambda$ *and* $\mathrm{cof}(\lambda) > \omega$ *then* $\cup\{X_i \mid i < \lambda\}$ *belongs to* $C$.

*Proof.* Take $C$ to consist of all $X$ which are $\Sigma_1$-elementary in some $L_\alpha$ and in addition preserve well-foundedness of direct limit systems: If $\Pi$ is a direct limit system (of the type considered in the proof of Theorem 3.10) with elements and maps in $X$ and $\mathrm{Lim}\,\Pi$ is ill-founded then this is witnessed by some $\{\beta_0, \beta_1, \ldots\} \subseteq X$ (as in the proof of Theorem 3.10). We claim that $C$ is contained in $L$. For the proof of Theorem 3.10 shows that if $X$ is in $C$ then for some $n$, $P$, $\gamma$ we have $X = H_n(\bar{X}, P, \gamma) = \Sigma_n$ Skolem hull of $\bar{X} \cup P$ in $L_\gamma$, where $\bar{X}$ is a proper initial segment of $X$ (i.e., $\bar{X} = X \cap L_{\bar{\alpha}}$ for some $\bar{\alpha}$). But note that any initial segment of an element of $C$ is an element of $C$ so we can prove by induction on $\cup(X \cap \mathrm{ORD})$ that $X$ is in $L$. The proof of Theorem 3.10 also shows that any uncountable set of ordinals is a subset of an element of $C$ of the same cardinality. And property (b) is clear.

It only remains to show that $C$ is $L$-definable. But again as in the proof of Theorem 3.10, if $X \in L$ and $\Pi$ is a counterexample to $X \in C$ then $\Pi$ can be embedded into a $\Pi^*$ which is also a counterexample and belongs to $L$ (take $\Pi^*$ to be $j(\Pi(\bar{\kappa}, \gamma, n))$ for appropriate $\gamma$, $n$ where $j : L_{\bar{\kappa}} \longrightarrow L_\kappa$ has $X$ as its range). So the same $C$ results if we confine its definition to $L$. □

The next result is from Magidor [90].

**Corollary 3.12.** *Assume that* $0^\#$ *does not exist. Then any set* $X$ *which is* $\Sigma_1$*-elementary in some* $L_\alpha$ *is the countable union of constructible sets.*

*Proof.* Again let $j : L_{\bar{\kappa}} \longrightarrow L_\kappa$ be $\Sigma_1$-elementary with range $X$, $\kappa$ least so that $X \subseteq L_\kappa$. Assume that $\kappa$ has been minimized, so that $\mathrm{cof}(\kappa) > \omega$. Now redefine the canonical systems $\Pi(\bar{\kappa}, \gamma, n)$ to consist of $\sigma : \bar{H}_n(\alpha_0, P_0, \beta_0) \longrightarrow \bar{H}_n(\alpha_1, P_1, \beta_1)$ where $P_0$, $P_1$ are now *countable* sets of parameters (so we must assume that for any $\alpha < \bar{\kappa}$, $P$ countable, $\beta < \gamma$, $\bar{H}_n(\alpha, P, \beta)$ has height less than $\bar{\kappa}$). Note that even though $\sigma$ may fail to belong to $L_{\bar{\kappa}}$ we may still define a system $\Pi = j(\Pi(\bar{\kappa}, \gamma, n))$ by sending $\sigma$ to the natural map from $j(\bar{H}_n(\alpha_0, P_0, \beta_0))$ to $j(\bar{H}_n(\alpha_1, P_1, \beta_1))$ defined using $j(\alpha_0)$, $j[P_0]$, $j(\beta_0)$ and $j(\alpha_1)$, $j[P_1]$, $j(\beta_1)$. Now choose $\gamma$ least and $n$ least for this $\gamma$ such that for some $\alpha < \bar{\kappa}$, countable $P \subseteq L_\gamma$, $\Sigma_n$ Skolem hull$(\alpha \cup P)$ in $L_\gamma$ collapses to a model of height $\geq \bar{\kappa}$. Then we consider $\bar{\Pi} = \Pi(\bar{\kappa}, \gamma+1, n-1)$ if $n > 1$, $\Pi(\bar{\kappa}, \gamma, 1)$ if $\gamma$ limit and $n = 1$. (The case $\gamma$ successor, $n = 1$ cannot occur.)

$\bar{\Pi}$ is countable directed and therefore so is $\Pi = j(\bar{\Pi})$. So $\Pi$ has a well-founded limit and we can extend $j$ to $j^* : L_\gamma \longrightarrow L_{\gamma^*}$, $j^*$ $\Sigma_n$-elementary. Thus $X = L_\kappa \cap \Sigma_n$ Skolem hull $((X \cap L_{j(\alpha)}) \cup j[P])$ in $L_{\gamma^*}$ for some countable $P \subseteq L_\gamma, \alpha < \bar{\kappa}$. By induction, $X \cap L_{j(\alpha)}$ is the countable union of constructible sets, so $X$ is too. □

**Theorem 3.13.** *Each of the following is equivalent to the existence of* $0^\#$:

(a) *$L$ is not rigid.*

(b) $\langle L, A\rangle$ *is not rigid for every* $A$ *such that* $\langle L, A\rangle$ *is amenable.*

(c) *Some uncountable set of ordinals is not a subset of a constructible set of the same cardinality.*

(d) *Some singular cardinal is regular in* $L$.

(e) $\kappa^+ \neq (\kappa^+)^L$ *for some singular cardinal* $\kappa$.

(f) *Every constructible subset of* $\omega_1$ *either contains or is disjoint from a closed, unbounded subset of* $\omega_1$.

(g) $\{\alpha \mid \alpha$ *is an* $L$*-cardinal*$\}$ *is* $\Delta_1$*-definable with parameters.*

(h) *There exists* $j : L_\alpha \longrightarrow L_\beta$, $\mathrm{crit}(j) = \kappa, \kappa^+ \leq \alpha$.

(i) *There exists* $j : L_\alpha \longrightarrow L_\beta$, $\mathrm{crit}(j) = \kappa$, $(\kappa^+)^L \leq \alpha, \alpha \geq \omega_2$.

*Proof.* It is straightforward to show that these all follow from the existence of $0^\#$; using Theorem 3.8. Also (a), (b) imply the existence of $0^\#$ by Theorem 3.8. Conditions (d), (e) each easily imply (c), and we get $0^\#$ from (c) via Theorem 3.10. Condition (f) implies (a), since $L \longrightarrow L \simeq \mathrm{Ult}(L, U)$, where $U$ consists of all constructible subsets of $\omega_1$ containing a closed unbounded subset. (g) implies that $(\kappa^+)^L < \kappa^+$ for $\kappa$ a sufficiently large cardinal; by taking $\kappa$ singular we get $0^\#$ via condition (e). To see that (h) implies the existence of $0^\#$, define an ultrafilter $U$ on constructible subsets of $\kappa$ by: $X \in U$ iff $\kappa \in j(X)$. Then $\mathrm{Ult}(L, U)$ is well-founded, for if not then by Löwenheim–Skolem there would be an infinite descending chain in $\mathrm{Ult}(L_{\kappa^+}, U)$ which contradicts $\kappa^+ \leq \alpha$.

Finally we show that (i) implies the existence of $0^\#$. Define $U$ as before by: $X \in U$ iff $\kappa \in j(X)$. First suppose that $\kappa$ is at least $\omega_2$. We shall argue that $U$ is *countably complete*, i.e. that if $\langle X_n \mid n \in \omega\rangle$ belong to $U$ then $\cap\{X_n \mid n \in \omega\}$ is nonempty. (This gives $0^\#$ as it implies that $\mathrm{Ult}(L, U)$ is well-founded.) By the Covering Theorem 3.10, there is $F \in L$ of cardinality $\omega_1$ such that $X_n \in F$ for each $n$. Then as we have assumed that $\kappa \geq \omega_2$, $F$ has $L$-cardinality less than $\kappa$. We may assume that $F$ is a subset of $\mathcal{P}(\kappa) \cap L$, and hence as $\alpha$ is an $L$-cardinal, $F$ belongs to $L_\alpha$ and there is a bijection $h : F \Longleftrightarrow \gamma$ for some $\gamma < \kappa, h \in L_\alpha$. But then $F^* = \{X \in F \mid \kappa \in j(X)\}$ belongs to $L_\alpha$ as $X \in F^* \Longleftrightarrow \kappa \in j(h^{-1})(h(X))$ and $F^*$ has nonempty intersection as $j(F^*) = \mathrm{Range}(j \restriction F^*)$ and $\kappa \in \cap j(F^*)$. Thus $\{X_n \mid n \in \omega\}$ has nonempty intersection since it is a subset of $F^*$. If $\kappa$ is less than $\omega_2$ then we have $\alpha \geq \omega_2 \geq \kappa^+$ so we have a special case of (h). □

The author does not know if "$\omega_2$" can be replaced by "$\omega_1$" in (i) of the previous theorem.

## 3.2 Relevant Forcing

In Section 3.1 we showed that $L$ is rigid in its tame class generic extensions and that if $L$ is not rigid then there is a least inner model $L[0^\#]$ in which $L$ is not rigid. We now use this fact to provide a criterion for generic class existence for class forcings over $L$.

**Definition.** A forcing $P$ defined over a ground model $\langle L, A\rangle$ is *relevant* if there is a $G$ $P$-generic over $\langle L, A\rangle$ which is definable (with parameters) over $L[0^\#]$. $P$ is *totally relevant* if for each $p \in P$ the same is true for $P(\leq p) = P$ restricted to conditions extending $p$.

Assume that $0^\#$ exists. Then any $L[0^\#]$-*countable* $P \in L$ is totally relevant, as there are only countably many constructible subsets of $P$ (using the fact that $\omega_1$ is inaccessible in $L$). Note that this includes the case of any forcing $P \in L$ definable in $L$ without parameters.

The situation is far less clear for uncountable $P \in L$. The next result treats the case of $\kappa$-Cohen forcing.

**Proposition 3.14.** *Suppose $\kappa$ is $L$-regular and let $P(\kappa)$ denote $\kappa$-Cohen forcing in $L$: Conditions are constructible $p : \alpha \longrightarrow 2$, $\alpha < \kappa$ and $p \leq q$ iff $p$ extends $q$.*

(a) *If $\kappa$ has cofinality $\omega$ in $L[0^\#]$ then $P(\kappa)$ is totally relevant.*

(b) *If $\kappa$ has uncountable cofinality in $L[0^\#]$ then $P(\kappa)$ is not relevant.*

*Proof.* Let $j_n$ denote the first $n$ Silver indiscernibles $\geq \kappa$.

(a) We use the fact that $P(\kappa)$ is $\kappa$-distributive in $L$. Let $\kappa_0 < \kappa_1 < \cdots$ be an $\omega$-sequence in $L[0^\#]$ cofinal in $\kappa$. Then any $D \subseteq P(\kappa)$ in $L$ belongs to $\text{Hull}(\kappa_n \cup j_n)$ for some $n$, where Hull denotes Skolem hull in $L$. As $\text{Hull}(\kappa_n \cup j_n)$ is constructible of $L$-cardinality $< \kappa$ we can use the $\kappa$-distributivity of $P(\kappa)$ to choose $p_0 \geq p_1 \geq \cdots$ successively below any $p \in P(\kappa)$ to meet all dense $D \subseteq P(\kappa)$ in $L$.

(b) Note that in this case $\kappa \in \text{Lim}\, I$, as otherwise $\kappa = \bigcup\{\kappa_n \mid n \in \omega\}$ where $\kappa_n = \bigcup(\kappa \cap \text{Hull}(\bar{\kappa} + 1 \cup j_n)) < \kappa$, $\bar{\kappa} = \max(I \cap \kappa)$, and hence $\kappa$ has $L[0^\#]$-cofinality $\omega$. Suppose $G \subseteq P(\kappa)$ were $P(\kappa)$-generic over $L$. For any $p \in P(\kappa)$ let $\alpha(p)$ denote the domain of $p$. Define $p_0 \geq p_1 \geq \cdots$ in $G$ so that $\alpha(p_{n+1}) \in I$ and $p_{n+1}$ meets all dense $D \subseteq P(\kappa)$ in $\text{Hull}(\alpha(p_n) \cup j_n)$. Then $p = \bigcup\{p_n \mid n \in \omega\}$ meets all dense $D \subseteq P(\kappa)$ in $\text{Hull}(\alpha \cup j)$ where $\alpha = \bigcup\{\alpha(p_n) \mid n \in \omega\} \in I$, $j = \bigcup\{j_n \mid n \in \omega\}$. But then $p$ is $P(\alpha)$-generic over $L$, as every constructible dense $\bar{D} \subseteq P(\alpha)$ is of the form $D \cap P(\alpha)$ for some $D$ as above. So $p$ is not constructible, contradicting $p \in G$. □

As a consequence of Proposition 3.14(b) we see that the basic class forcing examples of Easton and Long Easton forcing are not relevant. However, we can rescue these forcings by restricting to successor cardinals, thereby not adding $\kappa$-Cohen sets for $\kappa$ of uncountable $L[0^\#]$-cofinality. *Easton forcing at Successors* is defined as follows: Conditions are constructible $p : \alpha(p) \longrightarrow L$ where for $\alpha < \alpha(p)$, $p(\alpha) = \emptyset$ unless $\alpha$ is a successor cardinal of $L$, in which case $p(\alpha) \in \alpha$-Cohen forcing; we also require that if $\alpha$ is $L$-inaccessible then $\{\beta < \alpha \mid p(\beta) \neq \emptyset\}$ is bounded in $\alpha$. Extension is defined in the natural way: $p \leq q$ iff $p(\alpha)$ extends $q(\alpha)$ for each $\alpha < \alpha(q)$.

**Theorem 3.15.** *Let $P$ be Easton forcing at Successors. Then $P$ is totally relevant.*

*Proof.* First we prove relevance. By recursion on $i \in I$ = Silver indiscernibles we define $G(< i)$ to be $P(< i)$-generic over $L$, where $P(< i)$ = Easton forcing at Successors restricted to $L_i$. For $i = \min I$ take $G(< i)$ to be any $P(< i)$-generic (note that $P(< i)$ is countable in $L[0^\#]$). If $G(< i)$ has been defined, we define $G(< i^*)$ as follows (where $i < i^*$ are adjacent in $I$) : $P(< i^*)$ factors as $P(< i) \times P(i, i^*)$ where $P(i, i^*)$ is $i^+$-closed in $L$, so it suffices to define a $P(i, i^*)$-generic $G(i, i^*)$ and then $G(< i^*) = G(< i) \times G(i, i^*)$ is $P(< i^*)$-generic. To obtain $G(i, i^*)$, successively choose $p_0 \geq p_1 \geq \cdots$ in $G(i, i^*)$ so that $p_{n+1}$ meets all dense $D \subseteq P(i, i^*)$ in $\mathrm{Hull}(i \cup j_n)$ where $j_n$ = first $n$ Siver indiscernibles $\geq i$. Then $\{p \mid p \geq p_n$ for some $n\} = G(i, i^*)$.

Finally if $i \in \mathrm{Lim}\, I$, let $G(< i) = \bigcup\{G(< j) \mid j \in I \cap i\}$. Note that if $D \subseteq P(< i)$ is dense and constructible then for some $j \in I \cap i$, $D \cap P(< j)$ is dense and constructible and hence is met by $G(< j) \subseteq G(< i)$. So $G(< i)$ is $P(< i)$-generic. Similarly, $G = \bigcup\{G(< i) \mid i \in I\}$ is $P$-generic over $L$ (and in fact meets all $L$-amenable dense $D \subseteq P$).

For total relevance, we can repeat the above construction below any condition $p \in P$, careful to guarantee that $p \restriction \alpha_i \in G(< i)$ for each $i \in I$, where $\alpha_i < i$ is chosen large enough so that $p(\alpha) = \emptyset$ for $\alpha_i \leq \alpha < i$. □

**Theorem 3.16.** *Let $P$ be Reverse Easton forcing. Then $P$ is totally relevant.*

*Proof.* First we prove relevance. Recall that $P(< \alpha)$ has a dense subset of $L$-cardinality $\leq (\alpha^+)^L$ for each $\alpha$. By induction on $i \in I$ we define $G(\leq i) = G(< i) * G(i)$ to be $P(\leq i)$-generic over $L$, where $P(\leq i) = P(< i) * P(i)$, the first $i + 1$ stages in the iteration defining $P$. We will have: $i \leq j$ in $I \Longrightarrow G(j)$ extends $G(i)$; this will enable us to get through limit stages. For $i = \min I$, take $G(\leq i)$ to be any $P(\leq i)$-generic in $L[0^\#]$. If $G(\leq i)$ has been defined and $i^*$ = $I$-successor to $i$, then write $P(< i^*)$ as $P(\leq i) * P[i+1, i^*)$ and as $P(\leq i) \Vdash P[i+1, i^*)$ is $i^+$-closed we can select $G[i+1, i^*)$ to be $P[i+1, i^*)^{G(\leq i)}$-generic over $L[G(\leq i)]$ (the collection of dense sets that must be met is the countable union of subcollections of size $i$ in $L[G(\leq i)]$, using the $\mathrm{Hull}(i \cup j_n)$'s as in the previous proof). Then $G(< i^*) = G(\leq i) * G[i + 1, i^*)$ is $P(< i^*)$-generic over $L$. We also choose $G(i^*)$ to be $P(i^*)^{G(<i^*)}$-generic over $L[G(< i^*)]$, extending the condition $G(i)$ in this forcing.

For $i \in \mathrm{Lim}\, I$ take $G(< i)$ to be $\bigcup\{G(< j) \mid j \in I \cap i\}$, as in the previous proof $G(< i)$ is $P(< i)$-generic over $L$. And we take $G(i) = \bigcup\{G(j) \mid j \in I \cap i\}$, which by our construction extends each $G(j)$, $j \in I \cap i$. Again we get genericity for $G(\leq i)$ from that of $G(\leq j)$, $j \in I \cap i$, as $G(< i)$, $G(i)$ extend $G(< j)$, $G(j)$ respectively for each $j \in I \cap i$.

Total relevance can be handled as in the previous proof. □

Before turning to *Long Easton forcing at Successors* (obtained from Easton forcing at Successors by dropping the support condition that $\{\beta < \alpha \mid p(\beta) \neq \emptyset\}$ be bounded in $\alpha$ for $L$-inaccessible $\alpha$), we establish the total relevance of *Thin Easton forcing at Successors*. The latter is obtained by weakening the support condition in Easton forcing at Successors to: $\{\beta < \alpha \mid p(\beta^+) \neq \emptyset\}$ is nonstationary in $\alpha$ for $L$-inaccessible $\alpha$.

**Theorem 3.17.** *$P$ = Thin Easton forcing at Successors is totally relevant.*

*Proof.* Factor $P$ as $P(\leq \gamma) \times P(> \gamma)$ for each $L$-cardinal $\gamma$; if $\gamma$ is a limit $L$-cardinal then $P(\leq \gamma)$ can be identified with $P(< \gamma)$. Let $i$ be any indiscernible and for any $n$ let $j_n$ be the first $n$ indiscernibles $\geq i$. We can define $p_0^i \geq p_1^i \geq \cdots$ in $P(\leq i^+)$ such that if $D \subseteq P(\leq i^+)$ is dense and belongs to $\text{Hull}(\gamma^+ \cup j_n)$ then $p_{n+1}^i$ reduces $D$ below $\gamma^+$ for any $L$-cardinal $\gamma < i$. This is possible by successively extending on $[\gamma^{++}, i^+]$ (without violating the nonstationary support requirement). Let $G_0^i = \{p \in P(\leq i^+) \mid p \geq p_n^i$ for some $n\}$.

$G_0^i$ is *not* $P(\leq i^+)$-generic over $L$ as $p \in G_0^i \Longrightarrow p(j^+) = \emptyset$ for all $j \in I \cap i$. Notice that for $i_0 < i_1 < \cdots < i_n \leq i$ in $I$, $G_0^{i_0} \cup \cdots \cup G_0^{i_n}$ is a compatible set of conditions. We take $G(\leq i^+) = \{p \in P(\leq i^+) \mid p \geq q_0 \wedge \cdots \wedge q_n$ for some $q_l \in G_0^{i_l}, i_0 < \cdots < i_n \leq i$ in $I\}$. Now we claim that $G(\leq i^+)$ is $P(\leq i^+)$- generic over $L$. Indeed if $D \subseteq P(\leq i^+)$ is dense and belongs to $\text{Hull}(\{k_0, \ldots, k_m\} \cup j_n)$ with $k_0 < \cdots < k_m < i$ in $I$ then $p_{n+1}^i$ reduces $D$ below $k_m^+$, $p_{n+1}^i \wedge p_{n+2}^{k_m}$ reduces $D$ below $k_{m-1}^+, \ldots$ and eventually we get $p_{n+1}^i \wedge p_{n+2}^{k_m} \wedge \cdots \wedge p_{n+m+2}^{k_0}$ in $G(\leq i^+)$ meeting $D$.

Now note that in the above we could have chosen our initial $p_0^i \in P(\leq i)$ to reduce every dense $D \subseteq P \cap L_i$ in $\text{Hull}(\gamma^+ \cup \{i\})$ below $\gamma^+$, for any $\gamma < i$. Thus the resulting generic $G(\leq i^+)$ meets every dense $D \subseteq P \cap L_i$ definable over $L_i$. Now let $G = \cup\{G(\leq i^+) \mid i \in I\}$ and we see that $G$ is $P$-generic over $L$.

To obtain total relevance argue as follows. Given $p \in P$, choose indiscernibles $i_0, \ldots, i_n$ from which $p$ is $L$-definable. By the above we can successively construct $G(\leq i_0^+) \subseteq G(\leq i_1^+) \subseteq \cdots \subseteq G(\leq i_n^+)$ where $G(\leq i_k^+)$ is $P(\leq i_k^+)$-generic and contains $p(\leq i_k^+)$. Then similarly obtain a $P$-generic $G$ containing $G(\leq i_n^+)$. □

In the above proof we use thin supports to guarantee that for $i < j$ in $I$, the "pre-generics" $G_0^i, G_0^j$ agree at $i^+$ (indeed they equal $\emptyset$ at $i^+$). A less severe restriction is to require *coherence* on a CUB:

**Definition.** Let $P$ denote Long Easton forcing at Successors and suppose that $p$ belongs to $P(\leq \kappa^+)$, where $\kappa$ is $L$-regular. For any $\xi \in [\kappa, \kappa^+)$ let $f_\xi$ be the $L$-least 1-1 function from $\kappa$ onto $\xi$. For $s \in P(\kappa^+) = \kappa^+$-Cohen forcing and $\alpha < \kappa$ define $s_\alpha$ as follows: If $\xi = \text{length}(s) \leq \kappa$ or $\alpha \neq \kappa \cap f_\xi[\alpha]$ then $s_\alpha = \emptyset$. Otherwise $s_\alpha$ has domain $[\alpha, \bar{\xi})$ where $\bar{\xi} = \text{ordertype } f_\xi[\alpha]$ and $s_\alpha(\delta) = s(f_\xi(\delta))$. We say that $p$ is *coherent at $\kappa$* if $p(\kappa^+)_\alpha, p(\alpha^+)$ are compatible for CUB-many $\alpha < \kappa$. A condition $p$ in $P$ is *coherent* if for each $L$-inaccessible $\kappa$ in the domain of $p$, $p$ is coherent at $\kappa$. *Coherent Easton forcing at Successors* is the forcing whose conditions are the coherent conditions in Long Easton forcing at Successors.

**Theorem 3.18.** *Let $P$ be Coherent Easton forcing at Successors. Then $P$ is totally relevant.*

*Proof.* Follow the proof of the previous theorem. The only new observation is that by virtue of strong coherence at indiscernibles, we again have the compatibility of $G_0^i$, $G_0^j$ for $i < j$ in $I$. □

**Remark.** Thin Easton forcing at Successors and Coherent Easton forcing at Successors serve as prototypes for Jensen coding, introduced in the next chapter. In Jensen coding, conditions are sequences of pairs $(p_\alpha, p_\alpha^*)$ where strong coherence is used on the "coding strings" $p_\alpha$ and thinness is used on the "restraints" $p_\alpha^*$.

Finally we turn to Long Easton forcing at Successors.

**Theorem 3.19.** *Let $P$ denote Long Easton forcing at Successors. Then $P$ is totally relevant.*

*Proof.* Suppose that $p$ belongs to $P$ and $i$ is an indiscernible. We say that $p$ is *coherent at* $i$ if $p(i^+)$, $\pi(p)(i^+)$ are compatible, where $\pi : L \longrightarrow L$ is an elementary embedding with critical point $i$. Equivalently: $p(i^+)_\alpha$, $p(\alpha^+)$ are compatible for all $\alpha$ in a set $X$ belonging to the $L$-ultrafilter derived from the embedding $\pi$. It suffices to show that if $p$ belongs to $P(\leq i^+)$, $p$ is coherent at indiscernibles $\leq i$ and $D \subseteq P(\leq i^+)$, $D \in L$ is $L$-definable from indiscernibles $\geq i$ then $p$ has an extension meeting $D$ which is coherent at indiscernibles $\leq i$. For then, we can repeat the proof of Theorem 3.17, using conditions which are coherent at indiscernibles $\leq i$ to construct $G_0^i$, and therefore again obtain the compatibility of $G_0^i$, $G_0^j$ for $i < j$ in $I$.

Given $p$, $D$ as above, inductively extend $p(\alpha^+)$, $\alpha < i$, $\alpha$ an $L$-limit cardinal to $q(\alpha^+)$ as follows: If $q \restriction \alpha$ has been defined then let $q(\alpha^+)$ be least so that for some least $r_\alpha \in P(< \alpha)$, $r_\alpha \cup \{q(\alpha^+)\}$ extends $q \restriction \alpha \cup \{p(\alpha^+)\}$ and meets $D$. Now choose $X$ in the ultrafilter derived from $\pi$ ($X$ containing all indiscernibles $< i$) such that the $r_\alpha$ cohere for $\alpha$ in $X$ to a condition $r$ in $P(< i)$. Also define $r(i^+)$ to be $\pi(r)(i^+)$. Then $r$ extends $p$, is coherent at indiscernibles $\leq i$ and meets $D$. □

## Indiscernible Preservation

Though we have shown Easton at Successors and Reverse Easton to be totally relevant, we can further ask for a generic class that preserves indiscernibles. This will be important in the next chapter, where Jensen coding is introduced, as we can only code a class by a real (in $L[0^\#]$) if the class preserves (a periodic subclass of the) indiscernibles.

It is too much to ask that every condition $p$ be included in a generic class that preserves indiscernibles, as $p$ itself may not (only $2^{\aleph_0}$ subclasses of $L$ can).

**Definition.** A class $A \subseteq L$ *preserves indiscernibles* if $I$ is a class of indiscernibles for the structure $\langle L[A], A\rangle$.

**Theorem 3.20.** *For each of Easton at Successors, Reverse Easton, Thin Easton at Successors, Coherent Easton at Successors and Long Easton at Successors there is a generic class $G$ that preserves indiscernibles.*

*Proof.* It is easy to see that the generic classes built earlier for Thin Easton at Successors, Coherent Easton at Successors and Long Easton at Successors preserve indiscernibles. We now treat the case of Reverse Easton forcing, using the technique of Friedman [85a]. It suffices to build $H \subseteq L_{i_\omega}$ which is $P(< i_\omega)$-generic over $L_{i_\omega}$ and such that $t(j_1, \ldots, j_n) \in H$ iff $t(j_1', \ldots, j_n') \in H$ whenever $j_1 < \cdots < j_n$, $j_1' < \cdots < j_n'$ belong to $I \cap i_\omega$, $i_\omega = \omega^{\text{th}}$ indiscernible. For then define $t(k_1, \ldots, k_n) \in G$ iff $t(i_1, \ldots, i_n) \in H$, $i_1 < \cdots < i_n$ the first $n$ indiscernibles. This is well-defined using the above property of $H$. And $G$ is $P$-generic over $L$: It suffices to consider predense $D \in L$ as $P$ has the $\infty$-chain condition. Now write $D \in L$ as $s(l_1, \ldots, l_m)$, $l_1 < \cdots < l_m$ in $I$, and then $\bar{D} = s(i_1, \ldots, i_m)$ is predense on $P(< i_\omega)$. If $\bar{p} = t(i_1, \ldots, i_n) \in H$ meets $\bar{D}$ then $p = t(l_1, \ldots, l_m, l_{m+1}, \ldots, l_n)$ meets $D$, where $l_m < l_{m+1} < \cdots < l_n$ belong to $I$. Also $p \in G$ by definition of $G$. Finally, note that if $k_1 < \cdots < k_m < l_1 < \cdots < l_m$ and $l_1, \ldots, l_m$ are in $\operatorname{Lim} I$, $k_1, \ldots, k_m$ in $I$ then for any $\varphi$, $\langle L[G], G\rangle \vDash \varphi(k_1, \ldots, k_m) \Longleftrightarrow \varphi(l_1, \ldots, l_m)$ by the Truth Lemma and the fact that $G$ obeys the same invariance property that characterized $H$. So $I$ is a class of indiscernibles for $\langle L[G], G\rangle$.

Now we build $H$. Let $H_2 \subseteq P(< i_2)$ be a $P(< i_2)$-generic in $L[0^\#]$ and $H_1 = H_2 \cap P(< i_1)$. We must now define $H_3 \subseteq P(< i_3)$ to be $P(< i_3)$-generic so that $t(i_1, \vec{j}) \in H_2$ iff $t(i_2, \vec{j}) \in H_3$, where $\vec{j}$ is an increasing sequence from $I - i_\omega$. Note that $H_2(i_1)$, a subset of $i_1$ generic over $L[H_1]$, is a condition in the $i_2$-Cohen forcing defined over $L[H_2]$; choose $H_3(i_2)$ to be a generic for this forcing extending $H_2(i_1)$. Now note that for each $n$ there is $t_n(i_1, \vec{j}_n) = p_n \in H_2$ which reduces all predense $D \subseteq P(< i_2)$ in $\operatorname{Hull}(i_1 \cup \{i_1, k_1, \ldots, k_n\})$ below $i_1$, where $i_\omega \leq k_1 < \cdots < k_n$ belong to $I$, using the $i_1^+$-distributivity of $P(> i_1)^{H_2(\leq i_1)}$ in $L[H_2(\leq i)]$. So if we define $H_3' = \{t_n(i_2, \vec{j}_n) \mid n \in \omega\}$ we have that $H_3'$ reduces all predense $D \subseteq P(< i_3)$, $D \in L$ below $i_2$. So the desired $H_3$ can be defined by $H_3 = \{p \in P(< i_3) \mid p(\leq i_2) \in H_3(\leq i_2)$, $p$ compatible with $H_3'\}$. By construction, $t(i_1, \vec{j}) \in H_2$ iff $t(i_2, \vec{j}) \in H_3$. Note that $H_3$ was uniquely determined by this last condition, once a choice of $H_3(i_2)$ was made.

$H_4$ is uniquely determined by $P(< i_4)$-genericity and the condition $t(i_1, i_2, \vec{j}) \in H_3$ iff $t(i_2, i_3, \vec{j}) \in H_4$, as the forcing to add $H_3(i_2)$ is $i_1^+$-distributive (and the forcing to add $H_3(> i_2)$ is $i_2^+$-distributive). We must check that $t(i_1, i_3, \vec{j}) \in H_4$ iff $t(i_2, i_3, \vec{j}) \in H_4$. Now any condition in $H_4$ is extended by one of the form $p = (p_0, p_1)$ where $p_0 \in H_4(\leq i_3)$ and $p_1 = t(i_3, \vec{j})$, as such $p$ reduce all dense $D \subseteq P(< i_4)$, $D \in L$ below $i_3$. So it suffices to show that $t(i_1, i_3, \vec{j}) \in H_4(\leq i_3)$ iff $t(i_2, i_3, \vec{j}) \in H_4(\leq i_3)$. By definition of $H_4$ we have $t(i_2, i_3, \vec{j}) \in H_4(\leq i_3)$ iff $t(i_1, i_2, \vec{j}) \in H_3(\leq i_2)$. But the latter implies that $t(i_1, i_2, \vec{j}) = t(i_1, i_3, \vec{j})$ and as $H_3(\leq i_2)$ extends $H_2(\leq i_1)$ we have that $H_4(\leq i_3)$ extends $H_3(\leq i_2)$. So $t(i_1, i_2, \vec{j}) \in H_3(\leq i_2)$ iff $t(i_1, i_2, \vec{j}) \in H_4(\leq i_3)$ iff $t(i_1, i_3, \vec{j}) \in H_4(\leq i_3)$.

In general define $H_{m+3}$ by the condition $t(i_m, i_{m+1}, \vec{j}) \in H_{m+2}$ iff $t(i_{m+1}, i_{m+2}, \vec{j}) \in H_{m+3}$. As above we get that $H_{m+3}$ is $P(< i_{m+3})$-generic and $t(i_1, \ldots, i_{m+1}, \vec{j}) \in$

$H_{m+2}$ iff $t(i_1, \ldots, i_m, i_{m+2}, \vec{j}) \in H_{m+3}$. Finally let $H = \cup\{H_m \mid m \in \omega\}$. Then $H$ is $P(< i_\omega)$-generic over $L$ and for any $k_1 < \cdots < k_{l+2} < \vec{j}$ in $I$, $k_{l+2} < i_\omega \leq \vec{j}$ we have $t(k_1, \ldots, k_{l+1}, \vec{j}) \in H$ iff $t(k_1, \ldots, k_l, k_{l+2}, \vec{j}) \in H$. This is enough to imply that $t(\vec{k}_0) \in H$ iff $t(\vec{k}_1) \in H$ whenever $\vec{k}_0, \vec{k}_1$ are increasing sequences from $I \cap i_\omega$. This completes the proof in the case of Reverse Easton forcing.

Easton forcing at Successors can be handled in the same way without need to consider $H(i)$ for $i \in I$, as $H(\alpha)$ is nontrivial only when $\alpha$ is a successor $L$-cardinal. (Indeed, without the latter restriction the construction fails as there is no available choice for $H(i_2)$.) □

*Chapter 4*

# The Coding Theorem

The class forcings discussed in the previous two chapters provide examples of set-theoretic universes which neither contain $0^\#$ nor are obtainable by forcing over $L$ by the traditional method of forcing, with sets of conditions. Notice however that these universes are "locally set-generic" over $L$: Each set belongs to an intermediate set-generic extension of $L$.

Solovay posed three questions the solutions to which require use of a new kind of forcing, where *sets* are produced using a *class* of forcing conditions. Jensen developed this technique to prove his Coding Theorem, which says that any universe can be class-generically extended to one of the form $L[R]$, $R$ a real.

It is the purpose of this chapter to introduce the Solovay questions and to prove the Coding Theorem. In subsequent chapters we will provide the solutions to the Solovay problems, using the coding technique.

The Coding Theorem is easier to prove if we assume that $0^\#$ is not present in the universe to be coded, so that we may take advantage of the Covering Theorem 3.10. In Section 4.2 we treat this case using the idea of "coding delays" introduced in Friedman [97a]; the proof makes no use of Jensen's fine structure theory. The general case is handled in Section 4.3. There we provide a proof, using the fine-structure theory, which is adaptable to the more delicate construction of Chapter 7, where Solovay's third question (concerning the Admissibility Spectrum) is answered. Finally, in Section 4.4 we show that Jensen Coding is relevant in the sense of Chapter 3, and verify that certain large cardinal properties consistent with $V = L$ are preserved by these forcings.

## 4.1 Three Questions of Solovay

Solovay's three problems each demand the existence of a real that neither constructs $0^\#$, nor is in a set-generic extension of $L$.

**Definition.** If $x, y$ are sets of ordinals then we write $x \leq_L y$ for $x \in L[y]$ and $x <_L y$ for $x \leq_L y$, $y \not\leq_L x$.

**Genericity Problem**. Does there exist a real $R <_L 0^\#$ such that $R$ does not belong to a set-generic extension of $L$?

It was to affirmatively answer this question that Jensen proved his Coding Theorem. Roughly speaking he showed that if $G$ is generic for Easton forcing at Successors and

$G$ preserves indiscernibles (see Theorem 3.20) then there is a real $R <_L 0^\#$, obtained by class forcing over $\langle L[G], G\rangle$, such that $L[G] \subseteq L[R]$ and $G$ is definable over $L[R]$. Then $R$ does not belong to a set-generic extension of $L$ as $L[G]$ is not included in any set-generic extension of $L$.

Solovay's second problem concerns definability of reals.

**Definition.** $R$ is an *Absolute Singleton* if for some formula $\varphi$, $R$ is the unique solution to $\varphi$ in every inner model containing $R$.

Shoenfield's Absoluteness Theorem (Shoenfield [61]) states that if $\varphi$ is $\Pi^1_2$ (i.e., of the form $\forall R \exists S \psi$, $\psi$ arithmetical) then $\varphi(R) \iff M \vDash \varphi(R)$ where $M$ is any inner model containing $R$. Thus any $\Pi^1_2$-Singleton (i.e., the unique solution to a $\Pi^1_2$ formula) is an Absolute Singleton; $0^\#$ is an example. Also 0 is trivially an example. Solovay asked if there are any examples lying strictly between these two.

**$\Pi^1_2$-Singleton Problem**. Does there exists a real $R$, $0 <_L R <_L 0^\#$ such that $R$ is a $\Pi^1_2$-Singleton?

Note that it follows from the Covering Theorem (relative to $R$) that if $R <_L 0^\#$ then $R^\# \in L[0^\#]$ where $R^\#$ is defined relative to $L[R]$ the way we defined $0^\#$ relative to $L$. In particular $L_{\aleph_1}[R]$ is elementary in $L[R]$ and therefore if $R$ is in a $P$-generic extension of $L$, $P \in L$ then there is such a $P$ in $L_{\aleph_1}$. As $\aleph_1$ is inaccessible in $L$, there are only countably-many subsets of $P$ in $L$ and therefore we can build a $P$-generic containing any condition in $P$. So we conclude that if $R$ is a nonconstructible real in a $P$-generic extension of $L$ then $R$ cannot be a $\Pi^1_2$-Singleton, as there must be other $P$-generic extensions with reals $R' \neq R$ satisfying any $\Pi^1_2$ formula satisfied by $R$. This is why the $\Pi^1_2$-Singleton Problem requires Jensen's method: An affirmative answer to the $\Pi^1_2$-Singleton Problem implies an affirmative answer to the Genericity Problem.

Solovay's third problem concerns Admissibility Spectra. Let $T$ be a subtheory of $ZF$ and $R$ a real. The *$T$-spectrum* of $R$, $\Lambda_T(R)$, is the class of all ordinals $\alpha$ such that $L_\alpha[R] \vDash T$. A general problem is to characterize the possible $T$-spectra of reals for various theories $T$. An important special case is when $T = T_0 =$ ($ZF$ without the Power Set Axiom and with Replacement restricted to $\Sigma_1$ formulas). We may refer to $T_0$ as "admissibility theory," as an ordinal $\alpha$ is $R$-admissible (see Barwise [75]) if and only if it is $\omega$ or belongs to the $T_0$-spectrum of $R$. We refer to the $T_0$- spectrum of $R$ as the *admissibility spectrum* of $R$ and denote it by $\Lambda(R)$.

There are some basic facts which limit the possibilities for $\Lambda(R)$ (see Friedman [85b]): First, if $R$ belongs to a set-generic extension of $L$ then $\Lambda(R)$ contains $\Lambda - \beta$ for some ordinal $\beta$, where $\Lambda = \Lambda(0)$. This is because if $\alpha \in \Lambda$, $P \in L_\alpha$ then $L_\alpha[G] \vDash T_0$ for $P$-generic $G$. Second, if $0^\# \leq_L R$ then $\Lambda(R) - \beta \subseteq L$-inaccessibles for some $\beta$. This is because if $0^\# \in L_\beta[R]$ then every $\alpha$ in $\Lambda(R) - \beta$ is in $\Lambda(0^\#)$ and hence is a Silver indiscernible.

Thus to get a nontrivial admissibility spectrum for $R$ without $0^\#$ we need Jensen's methods. An ordinal is *recursively inaccessible* if it is admissible and also the limit of admissibles.

**Admissibility Spectrum Problem.** Does there exist a real $R \leq_L 0^\#$ such that $\Lambda(R) =$ the recursively inaccessible ordinals?

Of course we must in fact have $R <_L 0^\#$ as otherwise $\Lambda(R)$ is too small. In Chapter 7 we will give an affirmative answer to this question using an extension of Jensen's methods called Strong Coding.

## 4.2 The Coding Theorem without $0^\#$

In this section we use the Covering Theorem to prove the following result of Jensen.

**Theorem 4.1.** *Suppose that $A \subseteq$ ORD and $\langle L[A], A\rangle$ is a model of* ZFC + GCH $+0^\#$ *does not exist. Then there is an $\langle L[A], A\rangle$-definable class forcing $P$ such that if $G \subseteq P$ is $P$-generic over $\langle L[A], A\rangle$:*

(a) *$\langle L[A, G], A, G\rangle$ is a model of* ZFC + GCH.

(b) *$L[A, G] = L[R]$ for some real $R$ and $A, G$ are definable over $L[R]$ from the parameter $R$.*

(c) *$L[A]$ and $L[R]$ have the same cofinalities.*

The proof makes use of the following consequence of the Covering Theorem, proved as Theorem 3.13(a):

**Fact.** Assume that $0^\#$ does not exist. If $j : L_\alpha \longrightarrow L_\beta$ is $\Sigma_1$-elementary, $\alpha \geq \omega_2$ and $\kappa =$ critical point of $j$ then $\alpha < (\kappa^+)^L$.

Note the following consequence of Theorem 4.1:

**Corollary 4.2.** *Suppose $\langle M, A\rangle$ is a model of* ZFC $+0^\#$ *does not exist. Then there is a $\langle M, A\rangle$-definable class forcing $P$ such that if $G \subseteq P$ is $P$-generic over $\langle M, A\rangle$:*

(a) *$\langle M[G], A, G\rangle$ is a model of* ZFC.

(b) *$\langle M[G], A, G\rangle \vDash$ For some real $R$, $V = L[R]$ and $A, G$ are definable from $R$.*

*Proof.* We may assume that $\langle M, A\rangle$ is a ground model, by modifying $A$ if necessary. It suffices to show that there is a tame forcing $P^*$ defined in $\langle M, A\rangle$ such that if $G^* \subseteq P^*$ is $P^*$-generic over $\langle M, A\rangle$ then for some $B \subseteq$ ORD$(M)$, $B$ is definable over

$\langle M[G^*], A, G^*\rangle$, and $\langle M[G^*], A, G^*\rangle \vDash V = L[B] + \mathrm{GCH} + 0^\#$ does not exist and $G^*$, $A$ are definable relative to $B$. For then we can apply Theorem 4.1 to $\langle L[B], B\rangle$ via a forcing $P^{**}$ and let $P = P^* * P^{**}$.

$P^*$ is itself a 2-step iteration (actually a product $P_0^* \times P_1^*$). First force over $\langle M, A\rangle$ with $P_0^*$ = the amenable forcing of Chapter 2: conditions are $p : \alpha \longrightarrow 2$, $p \in M$ ordered by $p \leq q$ iff $p$ extends $q$. If $G_0^*$ is $P_0^*$-generic over $\langle M, A\rangle$ then $M[G_0^*] = M$ and $\langle M, A, G_0^*\rangle$ is a model of ZFC, $M = L[G_0^*]$ by genericity (and $AC$ in $M$) and we can identify $G_0^*$ (and therefore, $A$) with a class of ordinals.

Now over the ground model $\langle M, A, G_0^*\rangle$ we define a forcing $P_1^*$ to make the GCH true. For $\kappa$ a cardinal of $M$ let $P(\kappa)$ be the partial-ordering defined in $M$ consisting of all $p : \alpha \longrightarrow 2^\kappa, \alpha < \kappa^+$ ordered by $p \leq q$ iff $p$ extends $q$. If $G(\kappa)$ is $P(\kappa)$-generic over $M$ then in $M[G(\kappa)]$, $(\kappa^+)^M$, $((2^\kappa)^+)^M$ are adjacent cardinals and $2^\kappa = \kappa^+$. We take $P_1^*$ to be the Easton product of the $P(\kappa)$'s: A condition is $p : \alpha(p) \longrightarrow M, p \in M$ such that for $\beta < \alpha(p)$, $p(\beta) = \emptyset$ unless $\beta$ is an infinite $M$-cardinal in which case $p(\beta) \in P(\beta)$; also require that $\{\gamma < \beta \mid p(\gamma) \neq \emptyset\}$ is bounded in $\beta$ for inaccessible $\beta \leq \alpha$.

Now the proof of tameness, cardinal-preservation for the basic example of Easton forcing in Chapter 2 shows that $P_1^*$ is tame and preserves all $M$-cardinals that it does not intentionally collapse: The infinite cardinals of $M[G_1^*]$, when $G_1^*$ is $P_1^*$-generic over $\langle M, A, G_0^*\rangle$ are the regular $M$-cardinals $\kappa$ with the property: $\bar{\kappa}^+ < \kappa, \bar{\kappa}$ an infinite $M$-cardinal $\Longrightarrow \kappa > 2^{\bar{\kappa}}$, together with their limits. And for such $\kappa$ we have that $2^\kappa$ in $M[G_1^*]$ is at most $(2^\kappa)^M$, and hence equals $\kappa^+$ in $M[G_1^*]$. So we get the GCH in $M[G_1^*]$. To see that $M[G_1^*] \vDash 0^\#$ does not exist, just observe that there are singular cardinals $\kappa$ in $M[G_1^*]$ such that $\kappa^+$ in $M[G_1^*] = \kappa^+$ in $M = \kappa^+$ in $L^M$. □

We are ready to discuss the proof of Theorem 4.1. We make the following assumption about the predicate $A$: If $H_\alpha$, $\alpha$ an infinite $L[A]$-cardinal, denotes $\{x \in L[A] \mid$ transitive closure $(x)$ has $L[A]$-cardinality $< \alpha\}$ then we assume that $H_\alpha = L_\alpha[A]$. This is easily arranged using the fact that the GCH holds in $L[A]$.

The basic idea of the proof is simple. Let Card denote all infinite $L[A]$-cardinals. Also $\mathrm{Card}^+ = \{\alpha^+ \mid \alpha \in \mathrm{Card}\}$ and $\mathrm{Card}'$ = all uncountable limit cardinals. If $a \subseteq \alpha^{++}, \alpha \in \mathrm{Card}$ we can attempt to "code" $a$ by $b \subseteq \alpha^+$ as follows. We associate a subset $b_\xi$ of $\alpha^+$ to each $\xi < \alpha^{++}$ and design $b$ so that $\xi \in a$ iff $b, b_\xi$ are almost disjoint, i.e. have intersection bounded in $\alpha^+$. There is a natural forcing $R^a$ for doing this, invented by Solovay (see Jensen–Solovay [70]). A condition in $R^a$ is a pair $(p, \bar{p})$ where $p$ is a bounded subset of $\alpha^+$ and $\bar{p}$ consists of at most $\alpha$-many $b_\xi$'s with $\xi \in a$. When extending $(p, \bar{p})$ to $(q, \bar{q})$, $q$ must end-extend $p$, $\bar{q}$ must contain $\bar{p}$ and $q - p$ must be disjoint from all $b_\xi$ in $\bar{p}$.

Of course the forcing $R^a$ does not really code $a$ by a subset of $\alpha^+$ without some assumptions about the $b_\xi$'s. For example each $b_\xi$ should be almost disjoint from the union of $\alpha$-many other $b_\xi$'s; this is easy to arrange. More seriously, we need to know how to find $b_\xi$ in $L[a \cap \xi]$ in a uniform way, so that $a$ can be inductively recovered from our generic $b \subset \alpha^+$. The latter is possible only if $\xi < \alpha^{++} \Longrightarrow L[a \cap \xi] \vDash \mathrm{card}(\xi) \leq \alpha^+$.

If this fails then we must first "reshape" $a$ to make it true, by forcing with bounded subsets of $\alpha^{++}$ which do have this property up to their supremum.

It is not clear that the forcing for the purpose of reshaping $a$ is cardinal-preserving unless we can apply it in $L[c]$, where $c$ is an already-reshaped subset of $\alpha^{+++}$. Jensen's solution to this problem is to both reshape $A \cap \alpha^+$ and code $A \cap \alpha^+$ into a subset of $\alpha$, for all $\alpha$ simultaneously. Then in effect, the forcing to reshape $A \cap \alpha^+$ takes place in $L[c]$ where $c$ is a reshaped subset of $\alpha^{++}$ that codes $L[A]$.

As suggested in the previous paragraph there is a forcing analogous to $R^a$ for coding a reshaped $a \subseteq \alpha^+$ into a subset of $\alpha$, for $\alpha$ a limit cardinal. Thus if we combine all of these forcings we obtain a single forcing $P$ for coding $A$ by a real. A condition is of the form $p = \langle (p_\alpha, p^*_\alpha) \mid \alpha \in \text{Card}, \alpha \le \alpha(p) \rangle$ where $p_\alpha$ is a (reshaped) bounded subset of $\alpha^+$, $p^*_\alpha$ is the "restraint" imposed on $p_\alpha$ to ensure that $p_{\alpha^+}$ is coded, and where for $\alpha \in \text{Card}'$ we require that $\langle p_{\bar\alpha} \mid \bar\alpha \in \text{Card} \cap \alpha \rangle$ code $p_\alpha$.

We make two more comments before beginning the formal proof. First, there is a serious difficulty in showing that conditions can be "arbitrarily extended" due to potential conflicts between the codings at different limit cardinals. Jensen resolved these conflicts by coding differently at singulars than at inaccessibles, which led to heavy uses of the fine structure theory. The proof presented here instead uses "coding delays" which enables us to avoid any use of fine structure. Second, it is in the proof of cardinal-preservation that we make use of the $\sim 0^\#$ hypothesis. An "imperfection" arises if one attempts to carry out the natural argument and Jensen handles this using the *Fact* mentioned above. In the next section we present a different forcing, defined *using* fine structure, which enables the natural argument to succeed without the $\sim 0^\#$ hypothesis.

*Proof of Theorem* 4.1. Let $\alpha$ belong to Card.

**Definition** (Strings). $S_\alpha$ consists of all $s : [\alpha, |s|) \longrightarrow 2$, $\alpha \le |s| < \alpha^+$ such that $|s|$ is a multiple of $\alpha$ and for all $\eta \le |s|$, $L_\delta[A \cap \alpha, s \restriction \eta] \vDash \text{card}(\eta) \le \alpha$ for some $\delta < (\eta^+)^L \cup \omega_2$.

Thus for $\alpha$ equal to $\omega$ or $\omega_1$, elements of $S_\alpha$ are "reshaped" in the natural sense, but for $\alpha \ge \omega_2$ we insist that $s \in S_\alpha$ be "quickly reshaped" in that $\eta \le |s|$ be collapsed relative to $A \cap \alpha, s \restriction \eta$ before the next $L$-cardinal. This will be important when we use $\sim 0^\#$ to establish cardinal-preservation, via the above-mentioned Fact. The requirement that $|s|$ be a multiple of $\alpha$ is a technical convenience. Elements of $S_\alpha$ are called "strings". Note that we allow the empty string $\emptyset_\alpha \in S_\alpha$, where $|\emptyset_\alpha| = \alpha$. For $s, t$ in $S_\alpha$ we write $s < t$ for $s \le t, s \ne t$.

**Definition** (Coding Structures). For $s \in S_\alpha$ define $\mu^{<s}, \mu^s$ inductively by: $\mu^{<\emptyset_\alpha} = \alpha$, $\mu^{<s} = \bigcup\{\mu^t \mid t < s\}$ for $s \ne \emptyset_\alpha$ and $\mu^s$ = least limit of multiples of $\alpha$ $\mu > \mu^{<s}$ such that $L_\mu[A \cap \alpha, s] \vDash s \in S_\alpha$. And $\mathcal{A}^s = L_{\mu^s}[A \cap \alpha, s]$.

Thus by definition, when $\alpha \ge \omega_2$ there is $\delta < \mu^s$ such that $L_\delta[A \cap \alpha, s] \vDash \text{card}(|s|) \le \alpha$ and $L_{\mu^s} \vDash \text{card}(\delta) \le |s|$. The requirement "limit of multiples of $\alpha$" on $\mu^s$ is a technical convenience.

**Definition** (Coding Apparatus). For $\alpha > \omega$, $s \in S_\alpha$, $i < \alpha$ let $H^s(i) = \Sigma_1$ Skolem hull of $i \cup \{A \cap \alpha, s\}$ in $\mathcal{A}^s$ and $f^s(i) =$ ordertype $(H^s(i) \cap \text{ORD})$. For $\alpha \in \text{Card}^+$, $b^s = \text{Range}(f^s \restriction B^s)$ where $B^s =$ the successor elements of $\{i < \alpha \mid i = H^s(i) \cap \alpha\}$.

Note that if $s < t$ belong to $S_\alpha$ then Range $f^s$, Range $f^t$ are almost disjoint in the sense that their intersection is bounded in $\alpha$. The choice of $f^s \restriction B^s$ rather than $f^s$ is a technical convenience.

Using the above we will construct a tame, cofinality-preserving forcing $P$ for coding $\langle L[A], A\rangle$ by a subset $G_\omega$ of $\omega_1$ which is reshaped in the sense that proper initial segments of (the characteristic function of) $G_\omega$ belong to $S_\omega$. Then as $G_\omega$ can be coded into a real by a ccc forcing of size $\omega_1$ by the Solovay technique mentioned earlier, Theorem 4.1 follows.

**Definition** (A Partition of the Ordinals). Let $B, C, D, E$ denote the classes of ordinals congruent to 0, 1, 2, 3 mod 4, respectively. For any ordinal $\alpha$, $\alpha^B$ denotes the $\alpha^{\text{th}}$ element of $B$, when $B$ is listed in increasing order and for any set of ordinals $X$, $X^B$ denotes $\{\alpha^B \mid \alpha \in X\}$. Similarly for $C, D, E$.

**Definition** (The Successor Coding). Suppose $\alpha \in$ Card and $s \in S_{\alpha^+}$. A *condition in* $R^s$ is a pair $(t, t^*)$ where $t \in S_\alpha$, $t^* \subseteq \{b^{s\restriction\eta} \mid \eta \in [\alpha^+, |s|)\} \cup |t|$, $\text{card}(t^*) \leq \alpha$. Extension of conditions is defined by: $(t_0, t_0^*) \leq (t_1, t_1^*)$ iff $t_0 \subseteq t_1$, $t_1^* \subseteq t_0^*$ and:

(1) $|t_1| \leq \gamma^B < |t_0|$, $\gamma \in b^{s\restriction\eta} \in t_1^* \Longrightarrow t_0(\gamma^B) = 0$ or $s(\eta)$.

(2) $|t_1| \leq \gamma^C < |t_0|$, $\gamma = \langle\gamma_0, \gamma_1\rangle$, $\gamma_0 \in A \cap t_1^* \Longrightarrow t_0(\gamma^C) = 0$.

In (2) above, $\langle\cdot,\cdot\rangle$ is an $L$-definable pairing function on ORD so that $\text{card}(\langle\gamma_0, \gamma_1\rangle) = \text{card}\,\gamma_0 + \text{card}\,\gamma_1$ in $L$ for infinite $\gamma_0, \gamma_1$. An $R^s$-generic over $\mathcal{A}^s$ is determined by a function $T : \alpha^+ \longrightarrow 2$ such that $s(\eta) = 0$ iff $T(\gamma^B) = 0$ for sufficiently large $\gamma \in b^{s\restriction\eta}$ and such that for $\gamma_0 < \alpha^+ : \gamma_0 \in A$ iff $T(\langle\gamma_0, \gamma_1\rangle^C) = 0$ for sufficiently large $\gamma_1 < \alpha^+$.

Now we come to the definition of the limit coding, which incorporates the idea of "coding delays." Suppose $s \in S_\alpha$, $\alpha \in \text{Card}'$ and $p = \langle(p_\beta, p_\beta^*) \mid \beta \in \text{Card} \cap \alpha\rangle$ where $p_\beta \in S_\beta$ for each $\beta \in \text{Card} \cap \alpha$. A natural definition of "$p$ codes $s$" would be: For $\eta < |s|$, $p_\beta(f^{s\restriction\eta}(\beta)) = s(\eta)$ for sufficiently large $\beta \in \text{Card} \cap \alpha$. There are a number of problems with this definition however. First, to avoid conflict with the Successor Coding we should use $f^{s\restriction\eta}(\beta)^D$ instead of $f^{s\restriction\eta}(\beta)$. Second, to lessen conflict with codings at $\beta \in \text{Card}' \cap \alpha$ we only require the above for $\beta \in \text{Card}^+ \cap \alpha$. However there are still difficulties in making sure that the coding of $s$ is consistent with the coding of $p_\beta$ by $p \restriction \beta$ for $\beta \in \text{Card}' \cap \alpha$.

We introduce coding delays to facilitate extendibility of conditions. The rough idea is to code not using $f^{s\restriction\eta}(\beta)^D$, but instead just after the least ordinal $\geq f^{s\restriction\eta}(\beta)^D$ where $p_\beta$ takes the value 1. In addition, we "precode" $s$ by a subset of $\alpha$, which is then coded with delays by $\langle p_\beta \mid \beta \in \text{Card} \cap \alpha\rangle$; this "indirect" coding further facilitates extendibility of conditions.

**Definition.** Suppose $\alpha \in \text{Card}$, $X \subseteq \alpha$, $s \in S_\alpha$. Let $\tilde{\mu}^s$ be defined just as we defined $\mu^s$ but with the requirement "limit of multiples of $\alpha$" replaced by the weaker condition "multiple of $\alpha$". Then note that $\tilde{\mathcal{A}}^s = L_{\tilde{\mu}^s}[A \cap \alpha, s]$ belongs to $\mathcal{A}^s$, contains $s$ and $\Sigma_1$ Hull$(\alpha \cup \{A \cap \alpha, s\})$ in $\tilde{\mathcal{A}}^s = \tilde{\mathcal{A}}^s$. Now $X$ *precedes* $s$ if $X$ is the $\Sigma_1$ theory of $\tilde{\mathcal{A}}^s$ with parameters from $\alpha \cup \{A \cap \alpha, s\}$, viewed as a subset of $\alpha$.

**Definition** (Limit Coding). Suppose $s \in S_\alpha$, $\alpha \in \text{Card}'$ and $p = \langle (p_\beta, p^*_\beta) \mid \beta \in \text{Card} \cap \alpha \rangle$ where $p_\beta \in S_\beta$ for each $\beta \in \text{Card} \cap \alpha$. We wish to define "$p$ codes $s$". First we define a sequence $\langle s_\gamma \mid \gamma \leq \gamma_0 \rangle$ of elements of $S_\alpha$: Let $s_0 = \emptyset_\alpha$. For limit $\gamma \leq \gamma_0$, $s_\gamma = \bigcup\{s_\delta \mid \delta < \gamma\}$. Now suppose $s_\gamma$ is defined and let $f_p^{s_\gamma}(\beta) =$ least $\delta \geq f^{s_\gamma}(\beta)$ such that $p_\beta(\delta^D) = 1$, if such a $\delta$ exists. If for cofinally many $\beta \in \text{Card}^+ \cap \alpha$, $f_p^{s_\gamma}(\beta)$ is undefined, then set $\gamma_0 = \gamma$. Otherwise define $X \subseteq \alpha$ by: $\delta \in X$ iff $p_\beta((f_p^{s_\gamma}(\beta) + 1 + \delta)^D) = 1$ for sufficiently large $\beta \in \text{Card}^+ \cap \alpha$. If Even $(X)$ precedes an element $t$ of $S_\alpha$ extending $s_\gamma$ such that $\mathcal{A}^t$ contains $X$ and $f_p^{s_\gamma}$, then set $s_{\gamma+1} = t$. Otherwise let $s_{\gamma+1}$ be $s_\gamma * X^E$, if $f_p^{s_\gamma}$ belongs to $\mathcal{A}^{s_\gamma * X^E}$; if not, then again $\gamma_0 = \gamma$. Now $p$ *exactly codes* $s$ if $s = s_\gamma$ for some $\gamma \leq \gamma_0$ and $p$ *codes* $s$ if $s \leq s_\gamma$ for some $\gamma \leq \gamma_0$.

Note that the Successor Coding only restrains $p_\beta$ from taking certain nonzero values, so there is no conflict between the Successor Coding and these delays. The advantage of delays is that they give us more control over *where* the limit coding takes place, thereby enabling us to avoid conflict between the limit codings at different cardinals.

**Definition** (The Conditions). A *condition in* $P$ is a sequence $p = \langle (p_\alpha, p^*_\alpha) \mid \alpha \in \text{Card}, \alpha \leq \alpha(p) \rangle$ where $\alpha(p) \in \text{Card}$ and:

(1) $p_{\alpha(p)}$ belongs to $S_{\alpha(p)}$ and $p^*_{\alpha(p)} = \emptyset$.

(2) For $\alpha \in \text{Card} \cap \alpha(p)$, $(p_\alpha, p^*_\alpha)$ belongs to $R^{p_{\alpha^+}}$.

(3) For $\alpha \in \text{Card}'$, $\alpha \leq \alpha(p)$, $p \restriction \alpha$ belongs to $\mathcal{A}^{p_\alpha}$ and exactly codes $p_\alpha$.

(4) For $\alpha \in \text{Card}'$, $\alpha \leq \alpha(p)$ if $\alpha$ is inaccessible in $\mathcal{A}^{p_\alpha}$ then there exists a CUB $C \subseteq \alpha$, $C \in \mathcal{A}^{p_\alpha}$ such that $p^*_\beta = \emptyset$ for $\beta \in C$.

For $\alpha \in \text{Card}$, $P^{<\alpha}$ denotes the set of all conditions $p$ such that $\alpha(p) < \alpha$. Conditions are ordered by: $p \leq q$ iff $\alpha(p) \geq \alpha(q)$, $p(\alpha) \leq q(\alpha)$ in $R^{p_{\alpha^+}}$ for $\alpha \in \text{Card} \cap \alpha(p) \cap (\alpha(q) + 1)$ and $p_{\alpha(p)}$ extends $q_{\alpha(p)}$ if $\alpha(q) = \alpha(p)$. Also for $s \in S_\alpha$, $\omega < \alpha \in \text{Card}$, $P^s$ denotes $P^{<\alpha}$ together with all $p \restriction \alpha$ for conditions $p$ such that $\alpha(p) = \alpha$, $p_{\alpha(p)} \leq s$. To order conditions in $P^s$, first define $p^+$ for $p$ in $P^s$ as follows: $p^+ = p$ for $p \in P^{<\alpha}$; for $p \in P^s - P^{<\alpha}$, $p^+ \restriction \alpha = p$ and $p^+(\alpha) = (s \restriction \eta, \emptyset)$ where $\eta$ is least such that $p \in P^{s \restriction \eta}$. Now $p \leq q$ in $P^s$ iff $p^+ \leq q^+$ in $P$. Finally, $P^{<s} = \bigcup\{P^{s \restriction \eta} \mid \eta < |s|\} \cup P^{<\alpha}$.

It is worth noting that (3) above implies that $f^{p_\alpha}$ dominates the coding of $p_\alpha$ by $p \restriction \alpha$, in the sense that $f^{p_\alpha}$ strictly dominates each $f_{p \restriction \alpha}^{p_\alpha \restriction \eta}$, $\eta < |p_\alpha|$ on a tail of

$\mathrm{Card}^+ \cap \alpha$. The purpose of (4) is to guarantee that extendibility of conditions at (local) inaccessibles is not hindered by the Successor Coding (see the proof of Extendibility below).

We now embark on a series of lemmas which together show that $P$ preserves cofinalities and that if $G$ is $P$-generic over $\langle L[A], A\rangle$ then for some reshaped $X \subseteq \omega_1$, $L[A, G] = L[X]$ and $A$ is $L[X]$-definable from the parameter $X$.

**Lemma 4.3** (Distributivity for $R^s$). *Suppose $\alpha \in \mathrm{Card}$, $s \in S_{\alpha^+}$. Then $R^s$ is $\alpha^+$-distributive in $\mathcal{A}^s$: If $\langle D_i \mid i < \alpha\rangle \in \mathcal{A}^s$ is a sequence of dense subsets of $R^s$ and $p \in R^s$ then there is $q \leq p$ such that $q$ meets each $D_i$.*

*Proof.* Choose $\mu < \mu^s$ to be a large enough limit ordinal such that $p$, $\langle D_i \mid i < \alpha\rangle$ and $\mathcal{A}^{<s}$ belong to $\mathcal{A} = L_\mu[A \cap \alpha^+, s]$. Let $\langle \alpha_i \mid i < \alpha\rangle$ enumerate the first $\alpha$ elements of $\{\beta < \alpha^+ \mid \beta = \alpha^+ \cap \Sigma_1 \text{ Hull of } (\beta \cup \{p, \langle D_i \mid i < \alpha\rangle, \mathcal{A}^{<s}\}) \text{ in } \mathcal{A}\}$.

Now write $p$ as $(t_0, t_0^*)$ and successively extend $p$ to $(t_i, t_i^*)$ for $i \leq \alpha$ as follows: $(t_{i+1}, t_{i+1}^*)$ is the least extension of $(t_i, t_i^*)$ meeting $D_i$ such that: (a) $t_{i+1}^*$ contains $\{b^{s\restriction\eta} \mid \eta \in H_i \cap [\alpha^+, |s|)\}$ where $H_i = \Sigma_1$ Hull of $\alpha_i \cup \{p, \langle D_i \mid i < \alpha\rangle, \mathcal{A}^{<s}\}$ in $\mathcal{A}$. (b) If $b^{s\restriction\eta} \in t_i^*$, $s(\eta) = 1$ then $t_{i+1}(\gamma^\beta) = 1$ for some $\gamma \in b^{s\restriction\eta}$, $\gamma > |t_i|$. (c) If $\gamma_0 \notin A$, $\gamma_0 < |t_i|$ then $t_{i+1}(\langle\gamma_0, \gamma_1\rangle^C) = 1$ for some $\gamma_1 > |t_i|$.

The lemma reduces to:

**Claim.** $(t_\lambda, t_\lambda^*) =$ greatest lower bound to $\langle (t_i, t_i^*) \mid i < \lambda\rangle$ exists for limit $\lambda \leq \alpha$.

*Proof of Claim.* We must show that $t_\lambda = \bigcup\{t_i \mid i < \lambda\}$ belongs to $S_\alpha$. Note that $\langle t_i \mid i < \lambda\rangle$ is definable over $\overline{H}_\lambda =$ transitive collapse of $H_\lambda$ and by construction, $t_\lambda$ codes $\overline{H}_\lambda$ definably over $L_{\bar\mu_\lambda}[t_\lambda]$, where $\bar\mu_\lambda =$ height of $\overline{H}_\lambda$. So $t_\lambda$ is reshaped, as $|t_\lambda|$ is definably singular over $L_{\bar\mu_\lambda}[t_\lambda]$. By the *Fact*, $\bar\mu_\lambda < (|t_\lambda|^+)^L$ if $\alpha \geq \omega_2$. So $t_\lambda$ belongs to $S_\alpha$. □ (Claim)

□

The next lemma illustrates the use of coding delays.

**Lemma 4.4** (Extendibility for $P^s$). *Suppose that $\alpha$ is a limit cardinal, $s$ belongs to $S_\alpha$, and $p \in P^s$. Suppose also that $X \subseteq \alpha$ belongs to $\mathcal{A}^s$. Then there exists $q \leq p$ in $P^s$ such that $X \cap \beta \in \mathcal{A}^{q_\beta}$ for each $\beta \in \mathrm{Card} \cap \alpha$.*

*Proof.* By induction on $\alpha$. Let $Y \subseteq \alpha$ be chosen so that $\mathrm{Even}(Y)$ precodes $s$ and $\mathrm{Odd}(Y)$ is the $\Sigma_1$ theory of $\mathcal{A}$ with parameters from $\alpha \cup \{A \cap \alpha, s\}$, where $\mathcal{A}$ is an initial segment of $\mathcal{A}^s$ of limit height large enough to extend $\widetilde{\mathcal{A}}^s$ and contain $X$, $p$. For $\beta \in \mathrm{Card} \cap \alpha$ let $\overline{\mathcal{A}}_\beta$ be the transitive collapse of $H_\beta = \Sigma_1 \mathrm{Hull}(\beta \cup \{A \cap \alpha, s\})$ in $\mathcal{A}$ and suppose that $\beta$ is large enough so that $H_\beta$ contains $p$. If $H_\beta \cap \alpha = \beta$ then Even $(Y \cap \beta)$ precodes $s_\beta \in S_\beta$ where $s_\beta$ is the pre-image of $s$ under the natural embedding $\overline{\mathcal{A}}_\beta \longrightarrow \mathcal{A}$. If $H_\beta \cap \alpha \neq \beta$ then $|p_\beta| < (\beta^+)^{\overline{\mathcal{A}}_\beta}$, in which case $f^{p_\beta}$ is dominated by the function $g(\gamma) = (\gamma^+)^{\overline{\mathcal{A}}_\gamma}$ on a final segment of $\mathrm{Card}^+ \cap \beta$.

Now define $q$ as follows: If Even$(Y \cap \beta)$ precodes $s_\beta \in S_\beta$, then $q_\beta = s_\beta$. For other $\beta \in \text{Card}' \cap \alpha$, $q_\beta = p_\beta * (Y \cap \beta)^E$. For $\beta \in \text{Card}^+ \cap \alpha$, $q_\beta = p_\beta * \vec{0} * 1 * (Y \cap \beta)^D$ where $\vec{0}$ has length $g(\beta)$.

As $g \restriction \beta$ and $Y \cap \beta$ are definable over $\overline{\mathcal{A}}_\beta$ for $\beta \in \text{Card}' \cap \alpha$ we get $g \restriction \beta$, $Y \cap \beta \in \mathcal{A}^{s_\beta}$ when Even $(Y \cap \beta)$ precodes $s_\beta \in S_\beta$. Also $g \restriction \beta$, $Y \cap \beta \in \mathcal{A}^{q_\beta}$ for other $\beta \in \text{Card}' \cap \alpha$ as Odd $(Y \cap \beta)$ codes $\overline{\mathcal{A}}_\beta$. And note that for sufficiently large $\beta \in \text{Card}' \cap \alpha$, $g \restriction \beta$ dominates $f^{p_\beta}$ on a final segment of $\text{Card}^+ \cap \beta$ (and hence $q \restriction \beta$ exactly codes $q_\beta$), unless Even $(Y \cap \beta)$ precodes $s_\beta$ and $s_\beta = p_\beta$, in which case $q \restriction \beta$ exactly codes $q_\beta = s_\beta$ because $p \restriction \beta$ does.

So we conclude that for sufficiently large $\beta \in \text{Card}' \cap \alpha$, $q \restriction \beta$ exactly codes $q_\beta$ and $X \cap \beta \in \mathcal{A}^{q_\beta}$. Apply induction on $\alpha$ to obtain this for all $\beta \in \text{Card}' \cap \alpha$. Finally, note that the only problem in verifying $q \leq p$ is that the restraint $p^*_\beta$ may prevent us from making the extension $q_\beta$ of $p_\beta$ when $q_\beta = s_\beta$ and Even $(Y \cap \beta)$ precodes $s_\beta$. But property (d) in the definition of condition guarantees that $p^*_\beta = \emptyset$ for $\beta$ in a CUB $C \subseteq \alpha$, $C \in \mathcal{A}^s$. We may assume that $C \in \mathcal{A}$ and hence for sufficiently large $\beta$ as above we get $\beta \in C$ and hence $p^*_\beta = \emptyset$. So $q \leq p$ on a final segment of $\text{Card} \cap \alpha$, and we may again apply induction to get $q \leq p$ everywhere. □

We come now to the verification of distributivity for $P^s$. Before we can state and prove this property we need some preliminary definitions.

**Definition.** Suppose $i < \beta \in \text{Card}$ and $D \subseteq P^s$, $s \in S_{\beta^+}$. $D$ is *$i^+$-predense* on $P^s$ if $\forall p \in P^s \exists q \in P^s (q \leq p, q$ meets $D$ and $q \restriction i^+ = p \restriction i^+)$. $X \subseteq \text{Card} \cap \beta^+$ is *thin* if for each inaccessible $\gamma \leq \beta$, $X \cap \gamma$ is not stationary in $\gamma$. A function $f : \text{Card} \cap \beta^+ \longrightarrow V$ is *small* if for each $\gamma \in \text{Card} \cap \beta^+$, $\text{card}(f(\gamma)) \leq \gamma$ and $\text{Support}(f) = \{\gamma \in \text{Card} \cap \beta^+ \mid f(\gamma) \neq \emptyset\}$ is thin. If $D \subseteq P^s$ is predense and $p \in P^s$, $\gamma \in \text{Card} \cap \beta^+$ we say that *$p$ reduces $D$ below $\gamma$* if for some $\delta \leq \gamma$ in $\text{Card}^+$, every $q \leq p$ can be extended to $r \leq q$ such that $r$ meets $D$ and $r \restriction [\delta, \beta] = q \restriction [\delta, \beta]$. Finally, for $p \in P^s$, $f$ small, $f$ in $\mathcal{A}^s$ we define $\Sigma^p_f$ to consist of all $q \leq p$ in $P^s$ such that whenever $\gamma \in \text{Card} \cap \beta^+$, $D \in f(\gamma)$, and $D$ is predense on $P^{p_{\gamma^+}}$, we have that $q$ reduces $D$ below $\gamma$.

**Lemma 4.5** (Distributivity for $P^s$). *Suppose $s \in S_{\beta^+}$, $\beta \in \text{Card}$.*

(a) *If $\langle D_i \mid i < \beta \rangle \in \mathcal{A}^s$, $D_i$ $i^+$-dense on $P^s$ for each $i < \beta$ and $p \in P^s$ then there is $q \leq p$ such that $q$ meets each $D_i$.*

(b) *If $p \in P^s$, $f$ small, $f$ in $\mathcal{A}^s$ then there exists $q \leq p$, $q \in \Sigma^p_f$.*

*Proof.* We demonstrate (a) and (b) by a simultaneous induction on $\beta$. If $\beta = \omega$ or belongs to $\text{Card}^+$ then by induction, (a) and (b) reduce to the following: If $S$ is a collection of $\beta$-many predense subsets of $P^s$, $S \in \mathcal{A}^s$ then $\{q \in P^s \mid q$ reduces each $D \in S$ below $\beta\}$ is dense on $P^s$. The latter follows from Lemma 4.3, since $P^s$ factors as $R^s * Q$ where $R^s \Vdash Q$ is $\beta^+$-cc, and hence any $p \in P^s$ can be extended to $q \in P^s$ such that $D^q = \{r \mid r \cup q(\beta)$ meets $D\}$ is predense $\leq q \restriction \beta$ for each $D \in S$.

Now suppose that $\beta$ is inaccessible. We first show that (b) holds for $f$, provided $f(\beta) = \emptyset$. First select a CUB $C \subseteq \beta$ in $\mathcal{A}^s$ such that $\gamma \in C \Longrightarrow f(\gamma) = \emptyset$ and extend $p$ so that $f \upharpoonright \gamma$, $C \cap \gamma$ belong to $\mathcal{A}^{p_\gamma}$ for each $\gamma \in \mathrm{Card} \cap \beta^+$. Then we can successively extend $p$ on $[\beta_i^+, \beta_{i+1}]$ in the $L[A]$-least way so as to meet $\Sigma_f^p$ on $[\beta_i^+, \beta_{i+1}]$, where $\langle \beta_i \mid i < \beta \rangle$ is the increasing enumeration of $C$. At limit stages $\lambda$, we still have a condition, as the sequence of first $\lambda$ extensions belongs to $\mathcal{A}^{p_{\beta_\lambda}}$. The final condition, after $\beta$ steps, is an extension of $p$ in $\Sigma_f^p$.

Now we prove (a) in this case. Suppose $p \in P^s$ and $\langle D_i \mid i < \beta \rangle \in \mathcal{A}^s$ and $D_i$ is $i^+$-dense on $P^s$ for each $i < \beta$. Let $\mu_0 < \mu^s$ be a large enough limit ordinal so that $\langle D_i \mid i < \beta \rangle$, $p$ and $\tilde{\mu}^s$ belong to $L_{\mu_0}[A \cap \beta^+, s]$. For $i < \beta$, $\mu_i$ denotes $\mu_0 + \omega \cdot i < \mu^s$. For any $\gamma$ we let $H_i(\gamma)$ denote $\Sigma_1$ Hull$(\gamma \cup \{\langle D_i \mid i < \beta \rangle, p, \tilde{\mu}^s, s, A \cap \beta^+\})$ in $L_{\mu_i}[A \cap \beta^+, s]$.

Let $f_i : \mathrm{Card} \cap \beta \longrightarrow V$ be defined by: $f_i(\gamma) = H_i(\gamma)$ if $i < \gamma \in H_i(\gamma)$ and $f_i(\gamma) = \emptyset$ otherwise. Then each $f_i$ is small in $\mathcal{A}^s$ and we inductively define $p = p^0 \geq p^1 \geq \cdots$ in $P^s$ as follows: $p^{i+1} = L[A]$-least $q \leq p^i$ such that:

(1) $q(\beta)$ meets all predense $D \subseteq R^s$, $D \in H_i(\beta)$,

(2) $q$ meets $\Sigma_{f_i}^{p^i}$ and $D_i$,

(3) $q \upharpoonright i^+ = p^i \upharpoonright i^+$.

For limit $\lambda \leq \beta$ we take $p^\lambda$ to be the greatest lower bound to $\langle p^i \mid i < \lambda \rangle$, whose existence is guaranteed by the following Claim.

**Claim.** $p^\lambda$ is a condition in $P^s$, where $p^\lambda(\gamma) = (\cup\{p_\gamma^i \mid i < \lambda\}, \cup\{p_\gamma^{i\,*} \mid i < \lambda\})$ for each $\gamma \in \mathrm{Card} \cap \beta^+$.

Suppose that $\gamma$ belongs to $H_\lambda(\gamma) \cap \beta$. First we verify that $p_\gamma^\lambda = \cup\{p_\gamma^i \mid i < \lambda\}$ belongs to $S_\gamma$. Let $\bar{H}_\lambda(\gamma)$ be the transitive collapse of $H_\lambda(\gamma)$ and write $\bar{H}_\lambda(\gamma)$ as $L_{\bar{\mu}}[\bar{A}, \bar{s}]$, $\bar{P}$ = image of $P^s \cap H_\lambda(\gamma)$ under transitive collapse, $\bar{\beta}$ = image of $\beta$ under collapse. Also write $\bar{P}$ as $\bar{R}^{\bar{s}} * P^{\bar{G}_{\bar{\beta}}}$ where $\bar{G}$ denotes an $\bar{R}^{\bar{s}}$-generic (just as $P^s$ factors as $R^s * P^{G_\beta}$, $G_\beta$ denoting an $R^s$-generic).

Now the construction of the $p^i$'s (see conditions (1), (2)) was designed to guarantee: (i) $\bar{G}_{\bar{\beta}} = \{\bar{p} \in R^{\bar{s}} \mid \bar{p}$ is extended by some $\bar{p}^i(\bar{\beta}), i < \lambda\}$ is $R^{\bar{s}}$-generic over $\bar{H}_\lambda(\gamma)$, where $\bar{p}^i$ = image of $p^i$ under collapse, and (ii) For each $\bar{\delta}$ in (Card$^+$ of $\bar{H}_\lambda(\gamma)$), $\gamma < \bar{\delta} < \bar{\beta}$, $\{\bar{p} \mid \bar{p}$ is extended by some $\bar{p}^i \upharpoonright [\gamma, \bar{\delta})$ in $\bar{P}_\gamma^{\bar{p}_{\bar{\delta}}^i}\}$ is $\bar{P}_\gamma^{\bar{G}_{\bar{\delta}}}$-generic over $\mathcal{A}^{\bar{G}_{\bar{\delta}}} = \cup\{\mathcal{A}^{\bar{p}_{\bar{\delta}}^i} \mid i < \lambda\}$, where $\bar{P}_\gamma^{\bar{G}_{\bar{\delta}}} = \cup\{\bar{P}_\gamma^{\bar{p}_{\bar{\delta}}^i} \mid i < \lambda\}$ and $\bar{P}_\gamma^{\bar{p}_{\bar{\delta}}^i}$ denotes the image under collapse of $P_\gamma^{p_\delta^i} = \{q \upharpoonright [\gamma, \delta) \mid q \in P^{p_\delta^i}\}$, $\bar{\delta}$ = image of $\delta$ under collapse.

**Note.** We do *not* necessarily have property (ii) above for $\bar{\delta} = \bar{\beta}$, and this is the source of our need for $\sim 0^\#$ in this proof.

By induction, we have the distributivity of $P^t$ for $t \in S_\delta$, $\delta \in \text{Card}^+ \cap \beta$, and hence that of $\bar{P}^{\bar{t}}$ for $\bar{t} \in \bar{S}_{\bar{\delta}}$, $\bar{\delta} \in$ (Card$^+$ of $\bar{H}_\lambda(\gamma)$), $\bar{\delta} < \bar{\beta}$. So the "weak" genericity of the preceding paragraph implies that:

(4) $L_{\bar{\mu}}[A \cap \gamma, p_\gamma^\lambda] \models |p_\gamma^\lambda|$ is $\Sigma_1$-singular.

Also:

(5) $L_{\bar{\beta}}[A \cap \gamma, p_\gamma^\lambda] \models |p_\gamma^\lambda|$ is a cardinal.

Thus $p_\gamma^\lambda \in S_\gamma$ (by (4)) provided we can show that when $\gamma \geq \omega_2$, $\bar{\mu} < (|p_\gamma^\lambda|^+)^L$. But $\bar{H}_\lambda(\gamma) \xrightarrow{\sim} H_\lambda(\gamma)$ gives a $\Sigma_1$-elementary embedding with critical point $|p_\gamma^\lambda|$, so by the *Fact*, this is true.

The key point is that we also get $p^\lambda \restriction \gamma \in \mathcal{A}^{p_\gamma^\lambda}$, since $p^\lambda \restriction \gamma$ is definable over $\bar{H}_\lambda(\gamma)$ and we defined $\mathcal{A}^{p_\gamma^\lambda}$ to be large enough to contain $\bar{H}_\lambda(\gamma)$, since $L_{\bar{\beta}} \models |p_\gamma^\lambda|$ is a cardinal by (5) and $\bar{\beta}$ is a cardinal of $L_{\bar{\mu}}$.

The previous argument applies also if $\gamma = \beta$, using the distributivity of $R^s$, or if $\gamma = \beta \cap H_\lambda(\gamma)$, using the fact that $p_\beta^\lambda$ collapses to $p_\gamma^\lambda$. If $\gamma < \gamma^* = \min(H_\lambda(\gamma) \cap [\gamma, \beta))$ then we can apply the first argument to get the result for $\gamma^*$, and then the second argument to get the result for $\gamma$.

Finally, to prove the Claim we must verify the restraint condition (4) in the definition of $P$. Suppose $\gamma$ is inaccessible and for $i < \lambda$ let $C^i$ be the least CUB subset of $\gamma$ in $\mathcal{A}^{p_\gamma^i}$ disjoint from $\{\bar{\gamma} < \gamma \mid p_{\bar{\gamma}}^{i\,*} \neq \emptyset\}$. If $\lambda < \gamma$ then $\cap\{C^i \mid i < \lambda\}$ witnesses the restraint condition for $p^\lambda$ at $\gamma$. If $\gamma < \lambda$ then the restraint condition for $p^\lambda$ at $\gamma$ follows by induction on $\lambda$. And if $\gamma = \lambda$ then $\Delta\{C^i \mid i < \lambda\}$ witnesses the restraint condition for $p^\lambda$ at $\gamma$, where $\Delta$ denotes diagonal intersection.

Thus the Claim and therefore (a) is proved in case $\beta$ is inaccessible. To verify (b) in this case, note that as we have already proved (b) when $f(\beta) = \emptyset$, it suffices to show: If $\langle D_i \mid i < \beta \rangle \in \mathcal{A}^s$ is a sequence of dense subsets of $P^s$ then every $p \in P^s$ cna be extended to $q \in P^s$ that reduces each $D_i$ below $\beta$. But using (a) we see that $D_i^* = \{q \mid q$ reduces $D_i$ below $i^+\}$ is $i^+$-dense for each $i < \beta$, so again by (a) there is $q \leq p$ reducing $D_i$ below $i^+$ for each $i$.

We are now left with the case where $\beta$ is singular. The proof of (a) can be handled using the ideas from the inaccessible case as follows. Choose $\langle \beta_i \mid i < \lambda_0 \rangle$ to be a continuous and cofinal sequence of cardinals $< \beta$, $\lambda_0 < \beta_0$. As before, we first we argue that $p \in P^s$ can be extended to meet $\Sigma_f^p$ for any small $f$ in $\mathcal{A}^s$ provided $f(\beta) = \emptyset$: Extend $p$ if necessary so that for each $\gamma \in \text{Card} \cap \beta^+$, $f \restriction \gamma$ and $\{\beta_i \mid \beta_i < \gamma\}$ belong to $\mathcal{A}^{p_\gamma}$. Now perform a construction like the one used in the inaccessible case, successively extending $p$ this time on $[\beta_0, {\beta_i}^+]$ so as to meet $\Sigma_f^p$ on $[\beta_0, {\beta_i}^+]$ as well as $\Sigma_{f_i}^{p^i}$'s defined on $[\beta_0, {\beta_i}^+]$, to guarantee that $p^\lambda$ is a condition for limit $\lambda \leq \lambda_0$. Note that each extension is made on a bounded initial segment of $[\beta_0, \beta)$ and therefore by induction $\Sigma_f^p$, $\Sigma_{f_i}^{p^i}$ can be met on these intervals. The result is that $p$ can be extended to meet $\Sigma_f^p$ on a final segment of Card $\cap \beta$ and therefore by induction can be extended to meet $\Sigma_f^p$. Second, use the density of $\Sigma_f^p$ when $f(\beta) = \emptyset$ to carry out the proof of (a) as we did in the inaccessible case. And again, the general case of (b) follows from (a). This completes the proof of Lemma 4.5. $\square$

The argument of the previous lemma also shows:

**Lemma 4.6.** *$P$ is $\Delta$-distributive at $\kappa$ for all regular $\kappa$.*

Thus by Lemma 2.23, $P$ is tame and preserves cofinalities. As $L[A, G] = L[X]$ where $X \subseteq \omega_1$, we also have GCH-preservation. This completes the proof of Theorem 4.1. $\square$

## 4.3 The Coding Theorem in the General Case

In this section we prove Theorem 4.1 without the assumption that $L[A] \vDash 0^\#$ does not exist. The fundamental obstacle to overcome is the lack of "full genericity over the collapsed hull" in the proof of distributivity (see the *Note* in the proof of Lemma 4.5). The solution presented here is to refine our class of forcing conditions so as to guarantee full genericity over collapsed hulls. This refinement is reminiscent of Jensen's construction of a $\kappa^+$-Souslin tree in $L$ for limit cardinals $\kappa$ (see Jensen [72]) and therefore makes use of versions of the combinatorial principles $\diamondsuit$ and $\square$. In particular we use the fine structure theory, as it is needed to establish $\square$. It is also used in the proof of extendibility and distributivity, which are now established via a "fine-structural" induction.

We may assume that $V = L[A]$ where $H_\alpha = L_\alpha[A]$ for each infinite cardinal $\alpha$. Define Card =all infinite cardinals, $\text{Card}^+ = \{\alpha^+ \mid \alpha \in \text{Card}\}$ and $\text{Card}' =$ all uncountable limit cardinals.

**Definition** (Strings). For $\alpha \in$ Card, $S_\alpha$ consists of all $s : [\alpha, |s|) \longrightarrow 2, \alpha \leq |s| < \alpha^+$ such that for all $\eta \leq |s|$, $L[A \cap \alpha, s \restriction \eta] \vDash \text{Card}(\eta) \leq \alpha$.

Thus we allow the empty string $\emptyset_\alpha$, $|\emptyset_\alpha| = \alpha$. For $s, t \in S_\alpha$ we write $s < t$ for $s \subseteq t, s \neq t$.

**Definition** (Coding Structures). For $s \in S_\alpha$ define $\mu^{<s}, \mu^s$ inductively by: $\mu^{<\emptyset_\alpha} = \alpha$, $\mu^{<s} = \bigcup\{\mu^t \mid t < s\}$ for $s \neq \emptyset_\alpha$ and $\mu^s =$ least $\mu > \mu^{<s}$ such that $\mu'\mu = \mu$ for $0 < \mu' < \mu$ and $L_\mu[A \cap \alpha, s] \vDash \text{Card}(|s|) \leq \alpha$. And $\mathcal{A}^s = L_{\mu^s}[A \cap \alpha, s]$, $\mathcal{A}^{<s} = \langle L_{\mu^{<s}}[A \cap \alpha, \hat{s}], A \cap \alpha, \hat{s}\rangle$ where $\hat{s} = \{\mu^{<s \restriction \eta} \mid s(\eta) = 1\}$. Also define $\hat{\mu}^s$ to be the largest $\mu > \mu^{<s}$ such that $\mu'\mu = \mu$ for $0 < \mu' < \mu$ and $L_\mu[A \cap \alpha, s] \vDash |s|$ is a cardinal, if such a $\mu$ exists; $\hat{\mu}^s = \mu^{<s}$ otherwise. And $\hat{\mathcal{A}}^s = \langle L_{\hat{\mu}^s}[A \cap \alpha, \hat{s}], A \cap \alpha, \hat{s}\rangle$.

Thus $\mu^s$ is the least $\mu > \hat{\mu}^s$ such that $\hat{\mu}^s\mu = \mu$, and therefore equals $(\hat{\mu}^s)^\omega$.

**Definition** (Coding Apparatus). For $\alpha \in \text{Card}$, $\alpha > \omega$, $s \in S_\alpha$ and $i < \alpha$ let $H^s(i) = \Sigma_1$ Hull of $i \cup \{A \cap \alpha, s\}$ in $\mathcal{A}^s$ and $f^s(i) = \text{ordertype}\,(H^s(i) \cap \text{ORD})$. For $\alpha \in \text{Card}^+$, $b^s = \text{Range}(f^s \restriction B^s)$ where $B^s =$ the successor elements of $\{i < \alpha \mid i = \alpha \cap H^s(i)\}$.

We take $B^s$ to be *successor* elements of $\{i < \alpha \mid i = \alpha \cap H^s(i)\}$ to facilitate the proof of persistence for the forcing $R^s$ (see Lemma 4.8 below).

**Definition** (A Partition of the Ordinals). Let $B, C, D, E$ denote the classes of ordinals congruent to 0, 1, 2, 3 mod 4, respectively. For any ordinal $\alpha$, $\alpha^B$ denotes the $\alpha^{\text{th}}$ element of $B$, when $B$ is listed in increasing order and for any set of ordinals $X$, $X^B$ denotes $\{\alpha^B \mid \alpha \in X\}$. Similarly for $C, D, E$.

**Definition** (The Successor Coding). Suppose $\alpha \in \text{Card}$, $s \in S_{\alpha^+}$. A condition in $R^s$ is a pair $(t, t^*)$ where $t \in S_\alpha$, $t^* \subseteq \{b^{s\restriction\eta} \mid \eta \in [\alpha, |s|)\} \cup |t|$ and $\text{Card}(t^*) \leq \alpha$. Extension of conditions is defined by: $(t_0, t_0^*) \leq (t_1, t_1^*)$ iff $t_1 \subseteq t_0$, $t_1^* \subseteq t_0^*$ and:

(1) $|t_1| \leq \gamma^B < |t_0|, \gamma \in b^{s\restriction\eta} \in t_1^* \Longrightarrow t_0(\gamma^B) = 0 \text{ or } s(\eta)$.

(2) $|t_1| \leq \gamma^C < |t_0|, \gamma = \langle\gamma_0, \gamma_1\rangle, \gamma_0 \in A \cap t_1^* \Longrightarrow t_0(\gamma^C) = 0$.

In the above, $\langle\cdot,\cdot\rangle$ is an $L$-definable pairing function on ORD so that $\text{Card}(\langle\gamma_0, \gamma_1\rangle) = \text{Card}\,\gamma_0 + \text{Card}\,\gamma_1$, for infinite $\gamma_0, \gamma_1$. An $R^s$-generic is determined by a function $T : \alpha^+ \longrightarrow 2$ such that $s(\eta) = 0$ iff $T(\gamma^B) = 0$ for sufficiently large $\gamma \in b^{s\restriction\eta}$ and such that for $\gamma_0 < \alpha^+ : \gamma_0 \in A$ iff $T(\gamma^C) = 0$ for sufficiently large $\gamma$ of the form $\langle\gamma_0, \gamma_1\rangle < \alpha^+$.

We no longer use Coding Delays at limit cardinals $\alpha$ but instead the natural coding given by the $\langle f^s \mid s \in S_\alpha\rangle$.

**Definition** (The Limit Coding). Suppose $\alpha \in \text{Card}'$, $s \in S_\alpha$ and $p = \langle(p_\beta, p_\beta^*) \mid \beta \in \text{Card} \cap \alpha\rangle$ where $p_\beta \in S_\beta$ for all $\beta \in \text{Card} \cap \alpha$. We say that *$p$ codes $s$* if for each $\eta$, $\alpha \leq \eta < |s|$: $s(\eta) = p_\beta(f^{s\restriction\eta}(\beta)^D)$ for sufficiently large $\beta \in \text{Card}^+ \cap \alpha$.

We now come to the definition of condition. The crucial definition is that of $P^s$, $s \in S_\alpha$, $\alpha \in \text{Card}'$. Roughly speaking, we take $P^s$ to be $P^{<\alpha}$ (defined inductively) together with all $p = \langle(p_\beta, p_\beta^*) \mid \beta \in \text{Card} \cap \alpha\rangle \in \mathcal{A}^s$ such that $p$ codes $s$, $(p_\beta, p_\beta^*) \in R^{p_{\beta^+}}$ for $\beta \in \text{Card} \cap \alpha$ and $p \restriction \beta \in P^{p_\beta} - P^{<p_\beta}$ for $\beta \in \text{Card}' \cap \alpha$ (defined inductively). However, for the sake of the proofs of extendibility and distributivity, we add a number of extra requirements to this definition. To state these requirements we must use the definitions of $J$-model, collapsibility and Relativized $\diamondsuit$, given in Chapter 1, as well as introduce the notions of projectibility and good approximations. For the convenience of the reader, we reproduce here the needed definitions from Chapter 1.

A *$J$-model* is an amenable structure $\langle\mathcal{A}, C\rangle$ of the form $\langle\widetilde{J}_\mu[B], B, C\rangle$. We define $\mathcal{A}^+$ to be $\langle\widetilde{J}_{\mu^*}[B], B\rangle$ where $\mu^* \geq \mu$ is the least limit ordinal such that $\widetilde{J}_{\mu^*+\omega}[B] \models \mu$ is not a cardinal (if it exists). Now $\langle\mathcal{A}, C\rangle$ is *collapsible* if $\mathcal{A}^+$ exists and whenever $\pi : \langle\bar{\mathcal{A}}, \bar{C}\rangle \longrightarrow \langle\mathcal{A}, C\rangle$ is $\Sigma_1$ elementary then $\bar{\mathcal{A}}^+$ exists, $\bar{C}$ is definable over $\bar{\mathcal{A}}^+$ and $\pi$ lifts to a $\Sigma_1$ elementary $\pi^+ : \bar{\mathcal{A}}^+ \longrightarrow \mathcal{A}^+$.

**Relativized $\diamondsuit$.** Suppose $\alpha$ is an uncountable limit cardinal. Then there exists $\langle D^s \mid s \in S_\alpha \rangle$ such that:

(a) $D^s \subseteq \mathcal{A}^{<s}$, $\langle D^t \mid t$ an initial segment of $s\rangle \in \mathcal{A}^s$.

(b) If $D \subseteq \mathcal{A}^{<s}$, $D \in \hat{\mathcal{A}}^s \neq \mathcal{A}^{<s}$ then $\{\eta < |s| \mid D^{s \restriction \eta} = D \cap \mathcal{A}^{<s \restriction \eta}\}$ is stationary in $\hat{\mathcal{A}}^s$.

(c) If $\mu^{<s \restriction \eta} \in \operatorname{Lim} C^s$, $\eta < |s|$ then $D^{s \restriction \eta} = \emptyset$. If $\hat{\mathcal{A}}^s \models |s|^{++}$ exists then $D^s = \emptyset$. And if $\pi : \langle \mathcal{A}^{<\bar{s}}, \bar{C} \rangle \longrightarrow \langle \mathcal{A}^{<s}, C^s \rangle$ is $\Sigma_1$-elementary, $\pi(\bar{\alpha}) = \alpha$ where $\bar{s} \in S_{\bar{\alpha}}$ then $D^{\bar{s}} = \pi^{-1}[D^s]$.

**Definition** (Projectibility). A $J$-model $\langle \mathcal{A}, C \rangle$ is *$k$-projectible to $\alpha$*, where $\alpha \leq \mu = \operatorname{ORD}(\mathcal{A})$ and $0 < k \in \omega$, if $\langle \mathcal{A}, C \rangle = \Sigma_k$ Hull in $\langle \mathcal{A}, C \rangle$ of $\alpha \cup \{x\}$ for some parameter $x \in \mathcal{A}$.

**Definition** (Approximations). Suppose $s \in S_\alpha$, $\alpha \in \text{Card}'$. An *approximation* to $\mathcal{A}^s$ is a $J$-model $\langle \mathcal{A}, C \rangle \in \mathcal{A}^s$ of height $\mu$ such that $A \cap \alpha$ is $\Delta_1 \langle \mathcal{A}, C \rangle$, $\mu^{<s} \leq \mu < \mu^s$ and $\mathcal{A}$ is $\Delta_1$-definable over $\mathcal{A}^s \restriction \mu = \langle \tilde{J}_\mu[A \cap \alpha, \hat{s}], A \cap \alpha, \hat{s} \rangle$. $\langle \mathcal{A}, C \rangle$ is a *good approximation* to $\mathcal{A}^s$ if in addition it is collapsible and $k$-projectible to $\alpha$ for some $k$. For such $k$, we define $g = g_k^{\langle \mathcal{A}, C \rangle}$, the $k^{th}$ *canonical small function* (from $\text{Card} \cap \alpha$ into $V$) *associated to* $\langle \mathcal{A}, C \rangle$, as follows: $g(\beta) = \emptyset$ unless $\beta \in \Sigma_k \operatorname{Hull}(\beta \cup \{x\})$ in $\langle \mathcal{A}, C \rangle$, in which case $g(\beta) = \Sigma_k \operatorname{Hull}(\beta \cup \{x\})$ in $\langle \mathcal{A}, C \rangle$, where $x =$ least parameter in $\mathcal{A}$ such that $\langle \mathcal{A}, C \rangle = \Sigma_k \operatorname{Hull}(\alpha \cup \{x\})$ in $\langle \mathcal{A}, C \rangle$.

**Remark.** Good approximations arise in our proof of extendibility. We introduce them now so that we can give a complete definition of the class of forcing conditions. However the motivation for this notion will not become apparent until much later in this section.

With one more definition we are prepared to define our class of forcing conditions.

**Definition** (Predensity Reduction). Suppose $\alpha \in \text{Card}'$, $s \in S_\alpha$. Assume that $P^{<s}$ has been defined and consists of conditions of the form $p$: $\text{Card} \cap \alpha' \longrightarrow V$ where $\alpha' \in \text{Card} \cap \alpha^+$. Now for any $\beta \in \text{Card}^+ \cap \alpha$ define $p \leq_\beta q$ iff $p \leq q$, $p \restriction \beta = q \restriction \beta$. If $D \subseteq P^{<s}$ then $D$ is *$\beta$-predense* on $P^{<s}$ if for each $p \in P^{<s}$ there is $q \leq_\beta p$ such that $q$ meets $D$. And $p$ *$\beta$-reduces $D$* if every $q \leq p$ can be extended to $r$ such that $r$ meets $D$ and $r \restriction [\beta, \alpha) = q \restriction [\beta, \alpha)$. For arbitrary $\gamma \in \text{Card} \cap \alpha$, *$p$ reduces $D$ below $\gamma$* if $p$ $\beta$-reduces $D$ for some $\beta \in \text{Card}^+$, $\beta \leq \gamma$.

**Definition** (Definition of $P^s$). Suppose that $s \in S_\alpha$, $\alpha \in \text{Card}'$. Then $P^s$ consists of $P^{<s} = \cup\{P^t \mid t < s\} \cup P^{<\alpha}$, together with all $p$: $\text{Card} \cap \alpha \longrightarrow V$ such that:

(1) (Basics) $p(\beta) \in R^{p_{\beta^+}}$ for all $\beta \in \text{Card} \cap \alpha$, $p$ codes $s$ and $p \restriction \beta \in P^{p_\beta} - P^{<p_\beta}$ for all $\beta \in \text{Card}' \cap \alpha$.

(2) (Predensity Reduction)

(i) If $D^s$ is contained in $P^{<s}$ and is $\beta$-predense for all $\beta \in \text{Card}^+ \cap \alpha$ then $p$ meets $D^s$ (where $D^s$ comes from the Relativized $\diamondsuit$ Principle).

(ii) If $|s|$ is a successor ordinal, $D \subseteq P^{<s}$, $D$ predense and $D \in \mathcal{A}^{<s}$ then $p$ $\beta$-reduces $D$ for some $\beta \in \text{Card}^+ \cap \alpha$.

(3) (Restriction) $\eta < |s| \Longrightarrow$ There exists $q \in P^{s\restriction\eta} - P^{<s\restriction\eta}$ such that $p \leq q$.

(4) (Nonstationary Restraint) If $\alpha$ is inaccessible in $\mathcal{A}^s$ then there exists a CUB $C^* \subseteq \alpha$, $C^* \in \mathcal{A}^s$ such that $\beta \in C^* \Longrightarrow p^*_\beta = \emptyset$.

(5) (Growth Requirement) There is a good approximation $\langle \mathcal{A}, C\rangle$ to $\mathcal{A}^s$ such that for some $k$: $p$ is $\Delta_1\langle L_\alpha[A], g_k^{\langle \mathcal{A}, C\rangle}\rangle$ and $g_k^{\langle \mathcal{A}, C\rangle} \restriction \beta \in \mathcal{A}^{p_\beta}$ for sufficiently large $\beta \in \text{Card} \cap \alpha$.

If $s \in S_{\alpha^+}, \alpha \in \text{Card}$ then $P^s$ consists of $P^{<\alpha}$ together with all $p : \text{Card} \cap \alpha^+ \longrightarrow V$ such that $p(\alpha) \in R^s$, $p \restriction \alpha \in P^{p_\alpha}$ and $\alpha \in \text{Card}' \Longrightarrow p \restriction \alpha \in P^{p_\alpha} - P^{<p_\alpha}$.

**Definition** (The Conditions). $P$ consists of all $p$ : $\text{Card} \cap \alpha(p)^+ \longrightarrow V$ where $\alpha(p) \in \text{Card}$, such that $p(\alpha(p)) = (p_{\alpha(p)}, \emptyset)$, $p_{\alpha(p)} \in S_{\alpha(p)}$, $p \restriction \alpha(p) \in P^{p_{\alpha(p)}}$ and, if $\alpha(p) \in \text{Card}'$, $p \restriction \alpha(p) \notin P^{<p_{\alpha(p)}}$. For $\alpha \in \text{Card}$, $P^{<\alpha} = \{p \in P \mid \alpha(p) < \alpha\}$. Order conditions in $P$ by: $p \leq q$ iff $\alpha(p) \geq \alpha(q)$, $p(\alpha) \leq q(\alpha)$ in $R^{p_{\alpha^+}}$ for $\alpha \in \text{Card} \cap \alpha(p) \cap \alpha(q)^+$ and $q_{\alpha(p)} \leq p_{\alpha(p)}$ in $S_{\alpha(p)}$ if $\alpha(p) = \alpha(q)$.

To order conditions in $P^s$, $s \in S_\alpha$, first define $p^+ \in P$ for $p \in P^s$ by: $p^+ = p$ if $p \in P^{<\alpha}$, and otherwise $p^+ \restriction \alpha = p$, $p^+(\alpha) = (s \restriction \eta, \emptyset)$ where $\eta$ is least so that $p \in P^{s\restriction\eta}$. Then $p \leq q$ in $P^s$ iff $p^+ \leq q^+$ in $P$.

Extendibility and distributivity for $P^s$ are established in conjunction with a more fundamental property, called "$\Sigma_f$-density," which we describe next.

**Definition.** Suppose $\alpha \in \text{Card}$, $s \in S_\alpha$ and $X \subseteq \alpha$. Then $X$ is *thin in* $\mathcal{A}^s$ if $X \in \mathcal{A}^s$ and for each $\mathcal{A}^s$-inaccessible $\beta \leq \alpha$, $\mathcal{A}^s \models X \cap \beta$ is not stationary in $\beta$. A function $f : \text{Card} \cap \alpha \longrightarrow \mathcal{A}^s$ in $\mathcal{A}^s$ is *small in* $\mathcal{A}^s$ if for each $\beta \in \text{Card} \cap \alpha$, $\text{Card}(f(\beta)) \leq \beta$ in $\mathcal{A}^s$ and $\text{Support}(f) = \{\beta \in \text{Card} \cap \alpha \mid f(\beta) \neq \emptyset\}$ is thin in $\mathcal{A}^s$. For $p \in P^s$, $f$ small in $\mathcal{A}^s$ we define $\Sigma_f^p$ to consist of all $q \leq p$ in $P^s$ such that whenever $\beta \in \text{Card} \cap \alpha$, $D \in f(\beta)$, $D$ is predense on $P^{p_{\beta^+}}$ and $D \in \mathcal{A}^{p_{\beta^+}}$, we have that $q$ reduces $D$ below $\beta$.

Now fix $\alpha \in \text{Card}$, $s \in S_\alpha$.

**$\Sigma_f$-Density for $P^s$.** For $p \in P^s$, $f$ small in $\mathcal{A}^s$ there exists $q \leq p$, $q \in \Sigma_f^p$.

**Extendibility for $P^s$.** If $\alpha \in \text{Card}'$ and $p \in P^s$, there exists $q \leq p$, $q \in P^s - P^{<s}$.

**Distributivity for $P^{<s}$.** If $\langle D_i \mid i < \alpha\rangle \in \hat{\mathcal{A}}^s$, each $D_i$ is $i^+$-predense on $P^{<s}$ and $p \in P^{<s}$ then there exists $q \leq p$ in $P^{<s}$ such that $q$ meets each $D_i$, $i < \alpha$.

We now embark on a series of lemmas, leading to a proof of the above three properties.

**Lemma 4.7** (Distributivity for $R^s$). *Suppose $\alpha \in$ Card, $s \in S_{\alpha^+}$. Then $R^s$ is $\alpha^+$-distributive in $\mathcal{A}^s$: If $\langle D_i \mid i < \alpha\rangle \in \mathcal{A}^s$ is a sequence of predense subsets of $R^s$ and $p \in R^s$ then there is $q \leq p$ such that $q$ meets each $D_i$, $i < \alpha$.*

*Proof.* Choose $\mu < \mu^s$ to be a large enough limit ordinal so that $p$, $\langle D_i \mid i < \alpha\rangle$, $\mathcal{A}^{<s} \in L_\mu[A \cap \alpha^+, s] = \mathcal{A}$. Let $\langle \alpha_i \mid i < \alpha\rangle$ enumerate the first $\alpha$ elements of $\{\beta < \alpha^+ \mid \beta = \alpha^+ \cap \Sigma_1 \text{ Hull}(\beta \cup \{p, \langle D_i \mid i < \alpha\rangle, \mathcal{A}^{<s}\}) \text{ in } \mathcal{A}\}$.

Now write $p$ as $(t_0, t_0^*)$ and successively extend $p$ to $(t_i, t_i^*)$, $i \leq \alpha$ as follows: $(t_{i+1}, t_{i+1}^*)$ is the $L[A]$-least extension of $(t_i, t_i^*)$ meeting $D_i$ such that $t_{i+1}^*$ contains $\{b^{s\restriction\eta} \mid \eta \in H_i \cap |s|\}$ where $H_i = \Sigma_1 \text{ Hull}(\alpha_i \cup \{p, \langle D_i \mid i < \alpha\rangle, \mathcal{A}^{<s}\})$ in $\mathcal{A}$, and:

(1) If $b^{s\restriction\eta} \in t_i^*$, $s(\eta) = 1$ then $t_{i+1}(\gamma^B) = 1$ for some $\gamma \in b^{s\restriction\eta}$, $\gamma > |t_i|$.

(2) If $\gamma_0 \notin A$, $\gamma_0 < |t_i|$ then $t_{i+1}(\langle\gamma_0, \gamma_1\rangle^C) = 1$ for some $\gamma_1 > |t_i|$.

**Claim.** $(t_\lambda, t_\lambda^*) =$ greatest lower bound to $\langle (t_i, t_i^*) \mid i < \lambda\rangle$ exists for limit $\lambda \leq \alpha$.

*Proof of Claim.* We must show that $t_\lambda = \bigcup\{t_i \mid i < \lambda\}$ belongs to $S_\alpha$. Note that $\langle t_i \mid i < \lambda\rangle$ is definable over $\bar{H}_\lambda =$ transitive collapse of $H_\lambda$, and by construction, $t_\lambda$ codes $\bar{H}_\lambda$ definably over $L_{\bar{\mu}_\lambda}[t_\lambda]$, where $\bar{\mu}_\lambda =$ height of $\bar{H}_\lambda$. So $t_\lambda \in S_\alpha$, as $|t_\lambda|$ is definably singular over $L_{\bar{\mu}_\lambda}[t_\lambda]$. □ (Claim)

Finally let $q = (t_\alpha, t_\alpha^*)$. Then $q \leq p$ and $q$ meets each of the $D_i$'s. □

**Lemma 4.8** (Persistence for $R^s$). *Suppose $s \subseteq t$ in $S_{\alpha^+}$, $D \subseteq R^s$ is predense and $D \in \mathcal{A}^s$. Then $D$ is predense on $R^t$.*

*Proof.* We must show that each $(u, u^*) \in R^t$ is compatible with an element of $D$. We may assume that $s \neq t$ and $|u| = \beta = \alpha^+ \cap H$ where $H = \Sigma_1 \text{ Hull}(\beta \cup A \cap \alpha^+, s)$ in $\mathcal{A}^s$ and that $D$ belongs to $H$. We may also assume that $u^* = \{b^{t\restriction\eta} \mid \eta \in H \cap \mid s \mid\}$. Let $u_0^* = u^* \cap \mathcal{A}^{<s}$, $u_1^* = u^* - \mathcal{A}^{<s}$ and note that $(u, u_0^*) \in R^s$ as $[\mathcal{A}^{<s}]^\alpha \subseteq \mathcal{A}^s$. Let $(v, v_0^*)$ be the $L[A]$-least extension of $(u, u_0^*)$ in $R^s$ meeting $D$. Then $\beta \in \text{Lim } B^s$ and $|v| < \min(B^s - \beta) \leq \min(B^{t\restriction\eta} - \beta)$ for all $\eta \in [|s|, |t|) \cap H$.

**Claim.** $(v, v^*) = (v, v_0^* \cup u_1^*)$ is an extension of $(u, u^*)$.

*Proof of Claim.* We must check that in extending $u$ to $v$, we have not violated any of the restraint imposed by $u_1^*$. But all of this restraint involves ordinals outside the interval $[\beta, |v|)$ by the remark just before the Claim. □ (Claim)

Thus $(v, v^*)$ is an extension of $(u, u^*)$ meeting $D$ and hence $(u, u^*)$ is compatible with an element of $D$ □

**Corollary 4.9.** *For $s \in S_{\alpha^+}$ let $R^{<s} = \bigcup\{R^t \mid t < s\}$. Then $R^{<s}$ is $\alpha^+$-distributive in $\hat{\mathcal{A}}^s$ and if $D \subseteq R^{<s}$, $D$ is predense on $R^{<s}$ and belongs to $\hat{\mathcal{A}}^s$, then $D$ is predense on $R^t$ for all $t$ extending $s$.*

*Proof.* Note that $R^{<s}$ is $\alpha^{++}$-cc in $\hat{\mathcal{A}}^s$ and hence we may replace $\hat{\mathcal{A}}^s$ by $\mathcal{A}^{<s}$. Then both conclusions follow immediately by induction on $|s|$, using Lemmas 4.7, 4.8. □

**Lemma 4.10.** *Suppose $s \in S_\alpha$, $\alpha \in \mathrm{Card}'$ and Distributivity for $P^{<s}$ holds. Then:*

(a) *$P^{<s}$ has the $\alpha^+$-cc in $\hat{\mathcal{A}}^s$ (if $\hat{\mathcal{A}}^s \neq \mathcal{A}^{<s}$).*

(b) *If $D$ is predense on $P^{<s}$, $D$ belongs to $\hat{\mathcal{A}}^s$, $s \subseteq t$ and $p \in P^t - P^{<s}$ then $p$ $\beta$-reduces $D$ for some $\beta \in \mathrm{Card}^+ \cap \alpha$.*

(c) *If $D$ is predense on $P^{<s}$, $D$ belongs to $\hat{\mathcal{A}}^s$ and $s \subseteq t$ then $D$ is predense on $P^t$.*

*Proof.* (a) If $D \subseteq P^{<s}$, $D$ belongs to $\hat{\mathcal{A}}^s \neq \mathcal{A}^{<s}$ and $D$ is predense then consider $D^* = \{p \in P^{<s} \mid p\ \beta\text{-reduces } D \text{ for some } \beta < \alpha\}$. Then $D^* \in \hat{\mathcal{A}}^s$ and by Distributivity for $P^{<s}$, $D^*$ is $\beta$-predense for all $\beta \in \mathrm{Card}^+ \cap \alpha$. Apply Relativized $\diamondsuit$ (b) to get $\eta < |s|$ such that $D^{s\restriction\eta} = D^* \cap \mathcal{A}^{<s\restriction\eta}$ and $D^{s\restriction\eta}$ is $\beta$-predense for all $\beta \in \mathrm{Card}^+ \cap \alpha$. By Predensity Reduction and Restriction in the definition of $P^s$, $D^* \cap \mathcal{A}^{<s\restriction\eta}$ is predense on $P^{<s}$ and therefore so is $D \cap \mathcal{A}^{<s\restriction\eta}$, a subset of $D$ of $\hat{\mathcal{A}}^s$-cardinality $\leq \alpha$.

(b) By (a) we may assume that $D \in \mathcal{A}^{<s}$ and therefore by induction we may assume that $|s|$ is a successor ordinal. By Restriction in the definition of $P$, we need only show: If $p \in P^s - P^{<s}$, $D \subseteq P^{<s}$ is predense and $D$ belongs to $\mathcal{A}^{<s}$ then $p$ $\beta$-reduces $D$ for some $\beta \in \mathrm{Card}^+ \cap \alpha$. But this is part (2)(ii) of Predensity Reduction in the definition of $P^s$.

(c) follows from (b). □

**Lemma 4.11.** *Suppose that $\Sigma_f$-Density for $P^{s_0}$, Extendibility for $P^{s_0}$, Distributivity for $P^{<s_0}$ hold for all $s_0 < s$. Then Distributivity holds for $P^{<s}$.*

*Proof.* By Lemma 4.7 and induction we may assume that $\alpha$ is a limit cardinal. And vacuously we may assume that $|s| > \alpha$. Now additionally assume that $\alpha$ is regular in $\hat{\mathcal{A}}^s$ and that $|s|$ is limit ordinal. Let $\langle D_i \mid i < \alpha\rangle \in \hat{\mathcal{A}}^s$, each $D_i$ $i^+$-predense on $P^{<s}$ and $p \in P^{<s}$. We will show that there exists $q \leq p$ meeting each $D_i$. If $\hat{\mathcal{A}}^s = \mathcal{A}^{<s}$ then $\langle D_i \mid i < \alpha\rangle \in \mathcal{A}^{s\restriction\eta}$ for some $\eta < |s|$ so by induction we can find the desired $q$. So we may assume that $\hat{\mathcal{A}}^s \neq \mathcal{A}^{<s}$ and then by reflection we may choose $\eta < |s|$ such that cofinality $(\eta) = \alpha$, $p \in P^{s\restriction\eta}$, $D_i \cap P^{<s\restriction\eta}$ is $i^+$-dense on $P^{<s\restriction\eta}$ for each $i$ and $\langle D_i \cap P^{<s\restriction\eta} \mid i < \alpha\rangle \in \tilde{J}_\mu[A \cap \alpha, \widehat{s \restriction \eta}]$ for some limit ordinal $\mu$ such that $\tilde{J}_\mu[A \cap \alpha, \widehat{s \restriction \eta}] \vDash \eta$ is a cardinal. It follows then from Relativized □ (d) that $\langle D_i \cap P^{<s\restriction\eta} \mid i < \alpha\rangle$ is $\Delta_1\langle \mathcal{A}^{<s\restriction\eta}, C^{s\restriction\eta}\rangle$. We shall by induction on $i$ define $p^i, \eta_i, f_j$ for $j < i \leq \alpha$, where $p = p^0 \geq p^1 \geq \cdots$; the desired $q$ will be $p^\alpha$. We may assume that $p$ meets $D^{s\restriction\eta}$ (as required by Predensity Reduction (2)(i)), if the latter is a subset of $P^{<s\restriction\eta}$ which is $\beta$-predense for all $\beta \in \mathrm{Card}^+ \cap \alpha$. Let $x \in \mathcal{A}^{<s\restriction\eta}$ be a parameter that defines $\langle D_i \cap P^{<s\restriction\eta} \mid i < \alpha\rangle$ in a $\Delta_1$ way over $\langle \mathcal{A}^{<s\restriction\eta}, C^{s\restriction\eta}\rangle$.

Let $p^0$ be $p$ and let $\eta_0$ be least so that $p$ and $x$ belong to $\mathcal{A}^{<s\restriction\eta_0}$. Now suppose that $p^i, \eta_i$ have been defined. For any $\beta \in \mathrm{Card} \cap \alpha$ let $H_i(\beta) = \Sigma_1$ Hull of $\beta \cup \{p, x, \alpha\}$ in $\langle \mathcal{A}^{<s\restriction\eta_i}, C^{s\restriction\eta_i}\rangle$ and define $f_i(\beta) = H_i(\beta)$ if $i < \beta \in H_i(\beta)$, $f_i(\beta) = \emptyset$ otherwise. Then $p^{i+1}$ is the least $q \leq p^i$ such that $q \restriction i^+ = p^i \restriction i^+$, $q \notin P^{<s\restriction\eta_i}$ and $q$ meets both $D_i$ and $\Sigma^{p^i}_{f_i}$. This is possible by $\Sigma_f$-density and Extendibility for $P^{s\restriction\eta_i}$. And $\eta_{i+1}$ is

least so that $\mu^{<s\restriction\eta_{i+1}} \in \mathrm{Lim}\, C^{s\restriction\eta}$ and $\langle \mathcal{A}^{<s\restriction\eta_{i+1}}, C^{s\restriction\eta_{i+1}}\rangle \vDash p^{i+1}$ meets $D_i$. For limit $\lambda$, $\eta_\lambda = \bigcup\{\eta_i \mid i < \lambda\}$ and $p^\lambda =$ greatest lower bound to $\langle p^i \mid i < \lambda\rangle$, if defined.

**Claim.** $p^\lambda$ is defined for limit $\lambda \le \alpha$.

*Proof of Claim.* Of course we must have $p^\lambda(\beta) = (p^\lambda_\beta, p^{\lambda *}_\beta)$ where $p^\lambda_\beta = \bigcup\{p^i_\beta \mid i < \lambda\}$, $p^{\lambda *}_\beta = \bigcup\{p^{i *}_\beta \mid i < \lambda\}$. The first thing to show is that $p^\lambda_\beta \in S_\beta$ for each $\beta \in \mathrm{Card} \cap \alpha$.

Let $\bar{H}_\lambda(\beta) = \langle \bar{\mathcal{A}}, \bar{C}\rangle = \langle L_{\bar{\mu}}[\bar{A}, \hat{\bar{s}}], \bar{A}, \hat{\bar{s}}, \bar{C}\rangle$ be the transitive collapse of $H_\lambda(\beta)$ and first assume that $\beta \in H_\lambda(\beta)$. Let $P^{<s\restriction\eta_\lambda}_\beta = \{q \restriction [\beta, \alpha) \mid q \in P^{<s\restriction\eta_\lambda}\}$ and $\bar{P} =$ image of $P^{<s\restriction\eta_\lambda}_\beta$ under collapse. If $D \in \mathcal{A}^{<s\restriction\eta_\lambda}$ is predense on $P^{<s\restriction\eta_\lambda}$ then by Distributivity for $P^{<s\restriction\eta'}$, $\eta' \le \eta_\lambda$ and Lemma 4.10 (b), $p^i$ $\beta'$-reduces $D$ for some $\beta' \in \mathrm{Card}^+ \cap \alpha$ and for some $i < \lambda$.

Now suppose that in addition $D \in H_\lambda(\beta)$. Then by virtue of the fact that $p^{i+1}$ meets $\Sigma^{p^i}_{f_i}$ we get that some $p^i$, $i < \lambda$ reduces $D$ below $\beta$. It follows that if $D \in H_\lambda(\beta)$, $D$ predense on $P^{<s\restriction\eta_\lambda}_\beta$ then some $p^i$, $i < \lambda$ meets $D$. Thus $\bar{G} = \{\bar{p} \in \bar{P} \mid \bar{p}^i \le \bar{p}$ for some $i < \lambda\}$ is $\bar{P}$-generic over $\bar{H}_\lambda(\beta)$ (for predense $\bar{D} \in \bar{H}_\lambda(\beta)$), where $\bar{p}^i =$ image of $p^i$ under collapse. Note that $L_{\bar{\mu}}[\bar{A}, \hat{\bar{s}}, \bar{G}] = L_{\bar{\mu}}[A \cap \beta, p^\lambda_\beta]$ and $\bar{A}, \hat{\bar{s}}, \bar{G}$ are definable over this model.

Now we see that $p^\lambda_\beta \in S_\beta$: Indeed, $\langle p^i_\beta \mid i < \lambda\rangle$ is definable over $\langle \bar{\mathcal{A}}, \bar{C}\rangle$ and by collapsibility of $\langle \mathcal{A}^{<s\restriction\eta}, C^{<s\restriction\eta}\rangle$, $\bar{C}$ is definable over $\bar{\mathcal{A}}^+ \in L[\bar{A}, \hat{\bar{s}}] \subseteq L[A \cap \beta, p^\lambda_\beta]$. So $|p^\lambda_\beta|$ is collapsed in $L[A \cap \beta, p^\lambda_\beta]$, as desired.

However, to verify that $p^\lambda$ is a condition we will need to know that $\bar{C}$ belongs to $\mathcal{A}^{p^\lambda_\beta}$. This will follow if we can show that $\bar{\mu}$ is a cardinal in $L_{\hat{\bar{\mu}}}[A \cap \beta, p^\lambda_\beta]$, where. $\hat{\bar{\mu}}$ is the largest $\bar{\nu}$ such that $0 < \nu' < \bar{\nu} \Longrightarrow \nu'\bar{\nu} = \bar{\nu}$ and $L_{\bar{\nu}}[\bar{A}, \hat{\bar{s}}] \vDash \bar{\mu}$ is a cardinal). For then, $\mu^{p^\lambda_\beta}$ is greater than $\hat{\bar{\mu}}$, hence greater than $\bar{\mu}^+ =$ height of $\bar{\mathcal{A}}^+$ so $\bar{C}$ is definable over $L_{\bar{\mu}^+}[\bar{A}, \hat{\bar{s}}] \in L_{\mu^{p^\lambda_\beta}}[A \cap \beta, p^\lambda_\beta] = \mathcal{A}^{p^\lambda_\beta}$.

To see that $\bar{\mu}$ is a cardinal in $L_{\hat{\bar{\mu}}}[A \cap \beta, p^\lambda_\beta]$, it will suffice to show that not only is $\bar{G}$ $\bar{P}$-generic over $L_{\bar{\mu}}[\bar{A}, \hat{\bar{s}}]$ but that it is $\bar{P}$-generic over $L_{\hat{\bar{\mu}}}[\bar{A}, \hat{\bar{s}}]$, and that $\bar{P}$ preserves cardinals over $L_{\hat{\bar{\mu}}}[\bar{A}, \hat{\bar{s}}]$. Now the collapsibility of $\langle \mathcal{A}^{<s\restriction\eta_\lambda}, C^{s\restriction\eta_\lambda}\rangle$ implies that the embedding $L_{\bar{\mu}}[\bar{A}, \hat{\bar{s}}] \longrightarrow \mathcal{A}^{<s\restriction\eta_\lambda}$ lifts to a $\Sigma_1$-elementary embedding of $L_{\hat{\bar{\mu}}}[\bar{A}, \hat{\bar{s}}] \longrightarrow L_{\hat{\mu}}[A \cap \alpha, \widehat{s \restriction \eta_\lambda}]$ where $0 < \mu' < \hat{\mu} \Longrightarrow \mu'\hat{\mu} = \hat{\mu}$ and $\hat{\mu} \le \hat{\mu}^{s\restriction\eta_\lambda}$. But the distributivity of $P^{s\restriction\eta'}$, $\eta' \le \eta_\lambda$ and Lemma 4.10 (a) give that $P^{s\restriction\eta_\lambda}$ is $\alpha^+$-cc in $\hat{\mathcal{A}}^{s\restriction\eta_\lambda}$ and hence $\bar{P}$ is $\bar{\alpha}^+$-cc in $L_{\hat{\bar{\mu}}}[\bar{A}, \hat{\bar{s}}]$. So we get genericity and cardinal-preservation for $\bar{G}$ over $L_{\hat{\bar{\mu}}}[\bar{A}, \hat{\bar{s}}]$.

Thus $\bar{C}$ does belong to $\mathcal{A}^{p^\lambda_\beta}$ in the case $\beta \in H_\lambda(\beta)$. If $\beta \notin H_\lambda(\beta)$ then argue as follows: Let $\beta^* = \min(H_\lambda(\beta) - \beta)$ and if $\beta^* < \alpha$ then instead of $\bar{P} =$ collapse of $P^{<s\restriction\eta_\lambda}_\beta$ use $\bar{P} =$ collapse of $P^{<s\restriction\eta_\lambda}_{\beta^*}$. Then the argument is the same as above. If

$\beta^* = \alpha$ then there is no need for $\bar{P}$ as $p^\lambda_\beta = \bar{s} =$ collapse of $s \restriction \eta_\lambda$, $\bar{\mu} = |p^\lambda_\beta|$ and hence the fact that $\bar{C}$ belongs to $\mathcal{A}^{p^\lambda_\beta}$ follows from the obvious fact that $(\mathcal{A}^{<\bar{s}})^+ \in \mathcal{A}^{\bar{s}}$.

Now we are ready to verify that $p^\lambda$ is indeed a condition. As $p^\lambda \restriction \beta$ is definable over $\langle \bar{\mathcal{A}}, \bar{C}\rangle$ (notation as before) and $\langle \bar{\mathcal{A}}, \bar{C}\rangle \in \mathcal{A}^{p^\lambda_\beta}$ we get that $p^\lambda \restriction \beta$ codes $p^\lambda_\beta$ and belongs to $\mathcal{A}^{p^\lambda_\beta}$. Predensity Reduction (2)(i) holds at $\alpha$ when $\lambda = \alpha$ by choice of $p$ and when $\lambda < \alpha$ by the first part of Relativized $\diamondsuit$ (c). For $\beta$ such that $\beta = \alpha \cap H_\lambda(\beta)$ it holds at $\beta$ by the last part of Relativized $\diamondsuit$ (c). For $\beta \neq \alpha \cap H_\lambda(\beta)$ it holds by the second part of Relativized $\diamondsuit$ (c). Predensity Reduction (2)(ii) does not apply. Restriction is clear by induction. Nonstationary Restraint holds at $\beta \neq \lambda$ by induction and otherwise by diagonal intersection. Finally for the Growth Requirement at $\beta$ use the good approximation $\langle \bar{\mathcal{A}}, \bar{C}\rangle =$ collapse of $H_\lambda(\beta)$ and set $k = 1$. □ (Claim)

Thus the lemma holds when $|s|$ is a limit ordinal and $\alpha$ is regular in $\hat{\mathcal{A}}^s$. Now suppose that $|s|$ is a limit ordinal but $\alpha$ is singular in $\hat{\mathcal{A}}^s$, and hence singular in $\mathcal{A}^{<s}$. Let $\lambda$ be the cofinality of $\alpha$. It suffices to show: If $\langle D_i \mid i < \lambda\rangle \in \hat{\mathcal{A}}^s$ are $\lambda^+$-dense on $P^{<s}$ and $p \in P^{<s}$ then there exists $q \leq p$ such that $q$ meets each $D_i$. For, if $\langle D'_i \mid i < \alpha\rangle$ are as in the hypothesis of Distributivity for $P^{<s}$ and $\langle \alpha_i \mid i < \lambda\rangle \in \hat{\mathcal{A}}^s$ is cofinal in $\alpha$ then $D_i = \{q \mid q$ meets $D_j$ for each $j \in [\lambda^+, \alpha_i)\}$ is $\lambda^+$-dense for each $i < \lambda$ and $\langle D_i \mid i < \lambda\rangle \in \hat{\mathcal{A}}^s$; so by the above, any $p$ can be extended to $q$ which meets $D_i$ for each $i \geq \lambda^+$. Finally, by induction (on $\alpha$) we can extend $q$ further to meet all the $D_i$'s.

The proof that any $p \in P^{<s}$ can be extended to meet each $D_i$, $i < \lambda$ where $\langle D_i \mid i < \lambda\rangle \in \hat{\mathcal{A}}^s$, $D_i$ $\lambda^+$-dense for each $i$, is very similar to the first case of this proof, where we assumed $\alpha$ to be regular in $\hat{\mathcal{A}}^s$. Again we may assume $\hat{\mathcal{A}}^s \neq \mathcal{A}^{<s}$ and use reflection, this time to obtain $\eta < |s|$ such that cofinality $(\eta) \geq \lambda$, $p$ belongs to $P^{<s\restriction\eta}$, $D_i \cap P^{<s\restriction\eta}$ is $\lambda^+$-dense on $P^{<s\restriction\eta}$ for each $i$ and $\langle D_i \cap P^{<s\restriction\eta} \mid i < \lambda\rangle$ is $\Delta_1\langle \mathcal{A}^{<s\restriction\eta}, C^{s\restriction\eta}\rangle$ in some parameter $x$. Now inductively define: $p^0 = p$, $\eta_0$ least so that $x, p \in P^{<s\restriction\eta_0}$; $H_i(\beta) = \Sigma_1$ Hull of $\beta \cup \{p, x, \alpha\}$ in $\langle \mathcal{A}^{<s\restriction\eta_i}, C^{s\restriction\eta_i}\rangle$, $f_i(\beta) = H_i(\beta)$ if $\lambda^+ \leq \beta \in H_i(\beta)$, $f_i(\beta) = \emptyset$ otherwise. Then $p^{i+1}$ is defined to be the least $q \leq p^i$ such that $q \restriction \lambda^+ = p^i \restriction \lambda^+$, $q \notin P^{<s\restriction\eta_i}$ and $q$ meets both $D_i$ and $\Sigma^{p^i}_{f_i}$. And $\eta_{i+1}$ is least so that $\mu^{<s\restriction\eta_{i+1}} \in \operatorname{Lim} C^{s\restriction\eta}$ and $\langle \mathcal{A}^{<s\restriction\eta_{i+1}}, C^{s\restriction\eta_{i+1}}\rangle \models p^{i+1}$ meets $D_i$. For limit $\lambda' \leq \lambda$, $\eta_{\lambda'} = \bigcup\{\eta_i \mid i < \lambda'\}$ and $p^{\lambda'} =$ greatest lower bound to $\langle p^i \mid i < \lambda'\rangle$, if defined. The verification that indeed $p^\lambda$ is defined is as before.

Finally, suppose that $|s|$ is a successor ordinal. Let $s$ be the $S_\alpha$-successor to $s_0$ and note that by Extendibility for $P^{s_0}$ we can extend any given $p \in P^{<s} = P^{s_0}$ into $P^{s_0} - P^{<s_0}$. Now we imitate the earlier constructions as follows. Choose $\langle \mu_i \mid i < \lambda\rangle \in \mathcal{A}^{s_0}$, $\lambda = \mathcal{A}^{s_0}$-cofinality $(\alpha)$ such that the given $p$, $\langle D_i \mid i < \lambda\rangle$ belong to $\mathcal{A}^{s_0} \restriction \mu_0 = \bar{J}_\mu[A \cap \alpha, s_0]$, $\alpha$ is the largest cardinal of $\mathcal{A}^{s_0} \restriction \mu_0$ and $\mu_i = \mu_0 + \omega \cdot i$ for $i < \lambda$. Then proceed as before, using the $\mu_i$'s instead of $C^{s\restriction\eta}$ and with $H_i(\beta) = \Sigma_1$ Hull of $\beta \cup \{A \cap \alpha, s_0\}$ in $\mathcal{A}^{s_0} \restriction \mu_i$. Note that each $\langle \mathcal{A}^{s_0} \restriction \mu_i, \emptyset\rangle$ is a good approximation to $\mathcal{A}^{s_0}$. As there is no $\bar{C}$ to be concerned about, the proof of the Claim is now straightforward.

This completes the proof of Lemma 4.11. □

**Remark.** The argument used to prove Lemma 4.11 also shows: If $\beta \in \text{Card} \cap \alpha$, $\alpha$ inaccessible, $\langle D_i \mid i < \beta \rangle$ $\Sigma_n$ over $\langle L_\alpha[A], A \cap \alpha \rangle$ with $D_i$ $i^+$dense on $P^{<\alpha}$ for each $i < \beta$ then any $p \in P^{<\alpha}$ can be extended to meet each $D_i$. This time, inductively choose $p^{i+1}$ to extend $p^i$ and to meet $D_i$, $\Sigma_{f_i}^{p^i}$ where $f_i$ is defined using $H_i(\beta) = \Sigma_1$ Hull of $\beta \cup \{p, x\}$ in $\langle L_{\alpha_{i+1}}[A], A \cap \alpha_{i+1} \rangle$, where $\langle L_{\alpha_{i+1}}[A], A \cap \alpha_{i+1} \rangle$ is $\Sigma_n$-elementary in $\langle L_\alpha[A], A \cap \alpha \rangle$, $p^i \in P^{<\alpha_{i+1}}$ and $\langle D_i \mid i < \beta \rangle$ is $\Sigma_n$-definable over $\langle L_\alpha[A], A \cap \alpha \rangle$ with parameter $x$. For this argument, one does not need that $\alpha$ is inaccessible but only $\Sigma_{n+1} \langle L_\alpha[A], A \cap \alpha \rangle$ regular.

Using an argument similar to that used in the proof of the previous lemma we now show:

**Lemma 4.12.** *Suppose $\Sigma_f$-Density, Extendibility hold for $P^{s_0}$, $s_0 < s$ and Distributivity holds for $P^{<s_0}$, $s_0 \leq s$. Then Extendibility holds for $P^s$.*

*Proof.* We may assume that $|s| > \alpha$. First additionally assume that $\text{Lim}\, C^s$ is unbounded in $\mu^{<s}$ and let $\langle \eta_i \mid i < \lambda \rangle$ be the final segment of $\text{Lim}\, C^s$ where $\eta_0$ is least such that $p \in P^{s \upharpoonright \eta_0}$ and $\mu^{<s \upharpoonright \eta_0} \in \text{Lim}\, C^s$. Now as in the previous lemma, define $p^i$, $f_j$ for $j < i$ by induction on $i$: $p^0 = p$ and for $\beta \in \text{Card} \cap \alpha$, $H_i(\beta) = \Sigma_1$ Hull of $\beta \cup \{p, \alpha\}$ in $\langle \mathcal{A}^{<s \upharpoonright \eta_i}, C^{s \upharpoonright \eta_i} \rangle$, $f_i(\beta) = H_i(\beta)$ if $i < \beta \in H_i(\beta)$ and $f_i(\beta) = \emptyset$ otherwise. Then $p^{i+1}$ is the least $q \leq p^i$ such that $q \upharpoonright i^+ = p^i \upharpoonright i^+$, $q \in P^{s \upharpoonright \eta_{i+1}} - P^{s \upharpoonright \eta_i}$, and $q$ meets $\Sigma_{f_i}^{p^i}$. For limit $\lambda' \leq \lambda$, $p^{\lambda'}$ is the greatest lower bound to $\langle p^i \mid i < \lambda' \rangle$, if it exists.

The proof that $p^{\lambda'}$ exists for limit $\lambda' < \lambda$ is just as in the previous lemma. When $\lambda' = \lambda$ the proof is also the same, except one must use the Distributivity of $P^{<s}$ to obtain the $\alpha^+$-cc for $P^{<s}$ in $\hat{\mathcal{A}}^s$.

Now suppose that $\text{Lim}\, C^s$ is bounded in $\mu^{<s}$ and $|s|$ is a limit ordinal. Then we choose $\langle \eta_i \mid i < \omega \rangle$ to be cofinal in $|s|$ and $\Delta_1 \langle \mathcal{A}^{<s}, C^s \rangle$ with $\eta_0$ large enough so that $p \in P^{s \upharpoonright \eta_0}$ and $\mathcal{A}^{<s \upharpoonright \eta_0}$ contains a parameter $x$ for the definition of $\langle \eta_i \mid i < \omega \rangle$. Using $H_i(\beta) = \Sigma_1$ Hull of $\beta \cup \{p, x, \alpha, C^s \cap \mu^{<s \upharpoonright \eta_i}\}$ in $\mathcal{A}^{<s \upharpoonright \eta_i}$, we can argue as in the previous case.

Finally, suppose that $|s|$ is a successor ordinal. Let $s_0$ be the $S_\alpha$-predecessor to $s$ and this time use $H_i(\beta) = \Sigma_1$ Hull of $\beta \cup \{p, x, \alpha, C^s \cap \mu_i\}$ in $\mathcal{A}^{s_0} \upharpoonright \mu_i$, where $\langle \mu_i \mid i \in \omega \rangle$ is cofinal in $\mu^{s_0}$ and is $\Delta_1 \langle \mathcal{A}^{<s}, C^s \rangle$ in the parameter $x \in \mathcal{A}^{s_0} \upharpoonright \mu_0 = \tilde{J}_{\mu_0}[A \cap \alpha, s_0]$. Then define $p^i$, $i \in \omega$ as before, getting $p^\omega = q$ obeying all the requirements for being a condition in $P^s - P^{s_0}$, except for the fact that $s(|s_0|)$ may not be coded by $q$ along $f^{s_0}$ on a final segment of $\text{Card}^+ \cap \alpha$. (The $\Sigma_{f_i}^{p^i}$'s guarantee that Predensity Reduction (2)(ii) is satisfied.) To repair this problem, extend $q_\beta$ be assigning the value $s(|s_0|)$ at $|q_\beta|$ for each $\beta$ such that $\beta = \alpha \cap H_\omega(\beta)$, where $H_\omega(\beta) = \bigcup\{H_i(\beta) \mid i \in \omega\}$, and for $\beta \in \text{Card}^+ \cap \alpha$ extend $q_\beta$ to have value $s(|s_0|)$ at $(f^{s_0}(\beta))^D$ when $q_\beta$ is not already defined there. Note that $f^{s_0} \upharpoonright \beta$ belongs to $\mathcal{A}^{q_\beta}$ for $\beta \neq \alpha \cap H_\omega(\beta)$ so we get $q^* \upharpoonright \beta \in \mathcal{A}^{q_\beta^*}$ for the resulting $q^* \leq q$, for all $\beta \in \text{Card} \cap \alpha$. Also if $\alpha$ is regular in $\mathcal{A}^{<s}$ then the restraint $p_\beta^*$ is empty on a CUB in $\mathcal{A}^{<s}$, and hence for sufficiently large $\beta$ such that $\beta = \alpha \cap H_\omega(\beta)$; so the same is true for $q$ and $q^*$ and there is no problem in

defining $q^*_\beta(|q_\beta|)$ as we did, for sufficiently large such $\beta$. And these extensions do not affect the Growth Requirement. So $q^*$ is a condition. □

Our final main lemma establishes $\Sigma_f$-density for $P^s$, using a "fine-structural" induction. This is the key idea of this proof. The technique used here is related to the Fine Scale Principle of Chapter 1.

**Lemma 4.13.** *Suppose Extendibility holds for $P^s$ and Distributivity holds for $P^{<s}$. Then $\Sigma_f$-density holds for $P^s$.*

*Proof.* It is here that the Growth Requirement on conditions is used. By Lemma 4.7 and induction we may assume that $\alpha \in \mathrm{Card}'$. By Extendibility for $P^s$ we may assume that the $p$ given in the hypothesis of $\Sigma_f$-density belongs to $P^s - P^{<s}$ and therefore there is a good approximation $\langle \mathcal{A}, C\rangle$ to $\mathcal{A}^s$ and $k_0$ such that $p$ is $\Delta_1\langle L_\alpha[A], g_{k_0}^{\langle \mathcal{A},C\rangle}\rangle$ and $g_{k_0}^{\langle \mathcal{A},C\rangle} \restriction \beta \in \mathcal{A}^{p_\beta}$ for sufficiently large $\beta \in \mathrm{Card} \cap \alpha$.

Now note that $L_{\mu^s}[\langle \mathcal{A}, C\rangle] = L_{\mu^s}[A \cap \alpha, s] = \mathcal{A}^s$, because $A \cap \alpha$ is $\Delta_1\langle \mathcal{A}, C\rangle$, $p$ is $\Delta_1\langle L_\alpha[A], g_{k_0}^{\langle \mathcal{A},C\rangle}\rangle$ and $p$ codes $s$. Thus if $f$ is small in $\mathcal{A}^s$ then $f$ belongs to $L_{\mu^s}[\langle \mathcal{A}, C\rangle]$.

Let $\mu_0 = \mathrm{ORD}(\mathcal{A})$. For each limit $\mu$, $\mu_0 \le \mu < \mu^s$ let $\mathcal{A}_\mu$ be the structure $\langle \tilde{J}_\mu[B, C], B, C\rangle$ where $\mathcal{A} = \langle \tilde{J}_{\mu_0}[B], B\rangle$. For each such $\mu$ and each positive $k \in \omega$ (such that $k \ge k_0$ when $\mu = \mu_0$) define a canonical $g_{\mu,k}$ small in $\mathcal{A}^s$ by: $g_{\mu,k}(\beta) = \Sigma_k\,\mathrm{Hull}(\beta \cup \{x\})$ in $\mathcal{A}_\mu$ when $\beta \in \Sigma_k\,\mathrm{Hull}(\beta \cup \{x\})$ and $g_{\mu,k}(\beta) = \emptyset$ otherwise, where $x =$ least parameter in $\mathcal{A}$ such that $\langle \mathcal{A}, C\rangle = \Sigma_{k_0}\,\mathrm{Hull}(\alpha \cup \{x\})$ in $\langle \mathcal{A}, C\rangle$. Thus, $g_{\mu_0,k_0} = g_{k_0}^{\langle \mathcal{A},C\rangle}$ and it suffices to show that for each $\mu, k$ as above, $p$ can be extended to a condition $q$ meeting $\Sigma^p_{g_{\mu,k}}$ such that $g_{\mu,k} \restriction \beta \in \mathcal{A}^{q_\beta}$ for sufficiently large $\beta \in \mathrm{Card} \cap \alpha$.

We prove the latter statement by induction on $\mu$, and for fixed $\mu$ by induction on $k$. The base case $(\mu, k) = (\mu_0, k_0)$ is proved as follows: Let $g$ denote $g_{\mu_0,k_0}$. If $\alpha$ is $\Sigma_1\langle L_\alpha[A], g\rangle$-singular then we may choose $\langle \alpha_i \mid i < \lambda_0\rangle$ to be a continuous $\Sigma_1\langle L_\alpha[A], g\rangle$ sequence cofinal in $\alpha$ of ordertype $\lambda_0 < \alpha_0$ such that for limit $\lambda \le \lambda_0$, $p \restriction \alpha_\lambda$ is $\Sigma_1\langle L_{\alpha_\lambda}[A], g \restriction \alpha_\lambda\rangle$. Now successively extend $p$ on $[\lambda_0^+, \alpha_i]$ to meet $\Sigma^p_g$ on these intervals (and afterwards, extend below $\lambda_0^+$). There is no difficulty in verifying that one obtains a condition at limit stages of the construction since we may assume that $g \restriction \alpha_\lambda$ belongs to $\mathcal{A}^{p_{\alpha_\lambda}}$ for all limit $\lambda \le \lambda_0$. If $\alpha$ is $\Sigma_1\langle L_\alpha[A], g\rangle$-regular then we can choose $C$ CUB in $\alpha$, with $p \restriction \beta$ and $C \cap \beta$ $\Delta_1\langle L_\beta[A], g \restriction \beta\rangle$ for all $\beta \in C \cup \{\alpha\}$ such that $\beta \in C \implies g(\beta) = \emptyset$. Now successively extend $p$ to meet $\Sigma^p_g$ on the intervals $[\alpha_i^+, \alpha_{i+1})$, where $\langle \alpha_i \mid i < \alpha\rangle$ is the increasing enumeration of $C$. Again there is no problem at limit stages since we may assume that $g \restriction \alpha_\lambda$ belongs to $\mathcal{A}^{p_{\alpha_\lambda}}$ for all limit $\lambda \le \alpha$.

Next suppose that $\mu > \mu_0$ and we wish to prove the result for $(\mu, 1)$. Let $g = g_{\mu,1}$. Note that $\mathcal{A}_\mu$ has $\Sigma_1$-cofinality $\le \alpha$ and fix $\langle \mu_i \mid i < \lambda_0\rangle$ to be $\Sigma_1(\mathcal{A}_\mu)$ of ordertype $\lambda_0 \le \alpha$ such that for limit $\lambda \le \lambda_0$, $\langle \mu_i \mid i < \lambda\rangle$ is $\Sigma_1(\mathcal{A}_{\mu_\lambda})$ in a parameter belonging to $\mathcal{A}_{\mu_0}$. As in the proof of Lemma 4.12 we can successively extend $p$ to $p^i$, $i \le \lambda_0$ so as to meet $\Sigma^{p^i}_{f_i}$ where $f_i = g_{\mu_i,1}$, as by induction these $\Sigma^{p^i}_{f_i}$'s are dense below $p^i$. The resulting $p^{\lambda_0}$ is an extension of $p$ meeting $\Sigma^p_g$.

Finally we come to the key case, where $k > 1$ (and $k > k_0$ if $\mu = \mu_0$). By induction we may assume that $p$ meets $\Sigma^p_{g_0}$ where $g_0 = g_{\mu,k-1}$, $p$ is $\Delta_1\langle L_\alpha[A], g_0\rangle$ and $g_0 \restriction \beta \in \mathcal{A}^{p_\beta}$ for all $\beta \in \text{Card} \cap \alpha$. If $\alpha$ is $\Sigma_1\langle L_\alpha[A], g_0\rangle$-singular then choose $\langle \alpha_i \mid i < \lambda_0\rangle$ continuous and cofinal in $\alpha$ so that $\lambda_0 < \alpha$ and for limit $\lambda \le \lambda_0$, $p \restriction \alpha_\lambda$ is $\Delta_1\langle L_{\alpha_\lambda}[A], g_0 \restriction \alpha_\lambda\rangle$. Now approximate $g = g_{\mu,k}$ in $\lambda_0$ steps by the functions $g^i : \text{Card} \cap \alpha_i \longrightarrow V$ defined by $g^i(\beta) = \Sigma_k \,\text{Hull}(\beta \cup \{x\})$ in $(\Sigma_{k-1} \,\text{Hull}(\beta \cup \{x\})$ in $\mathcal{A}_\mu)$, where $x$ is the parameter used to define the $g_{\mu,k}$'s. Then $g(\beta) = \cup\{g^i(\beta) \mid i < \lambda_0\}$ and $\langle g^i \mid i < \lambda\rangle \in \mathcal{A}^{p_{\alpha_\lambda}}$ for limit $\lambda \le \lambda_0$. So we may successively extend $p$ to meet $\Sigma^p_{g^i}$, $i < \lambda_0$ on the intervals $[\lambda_0^+, \alpha_i]$ and verify that one has a condition at limit stages, as before. The desired extension of $p$ is obtained at stage $\lambda_0$ after extending below $\lambda_0^+$. If $C = \{\beta < \alpha \mid \beta = \alpha \cap \Sigma_k \,\text{Hull}(\beta \cup \{x\}) \text{ in } \mathcal{A}_\mu\}$ is unbounded in $\alpha$ then we can successively extend $p$ to meet $\Sigma^p_g$ on the intervals $[\alpha_i^+, \alpha_{i+1})$, where $\langle \alpha_i \mid i < \lambda_0\rangle$ is the increasing enumeration of $C$, noting that $\beta \in C \Longrightarrow g(\beta) = \emptyset$. Otherwise $\alpha$ has $\Sigma_1\langle L_\alpha[A], g\rangle$ cofinality $\omega$ and we can successively extend $p$ on $(\alpha_i, \alpha_{i+1}]$ to meet $\Sigma^p_g$, where $\alpha_0 = 0$ and $\langle \alpha_i \mid i \in \omega\rangle$ is $\Sigma_1\langle L_\alpha[A], g\rangle$ and cofinal in $\alpha$. □

This completes the verification of $\Sigma_f$-density, Extendibility and Distributivity for the forcings $P^s$, $P^{<s}$.

Now note that by the remark immediately following the proof of Lemma 4.11, $P$ is now seen to be $\Delta$-distributive at $\kappa$ and hence is tame and preserves cofinalities. By Extendibility, if $G \subseteq P$ is $P$-generic over $\langle L[A], A\rangle$ then $A, G$ are $L[X]$-definable, $L[A, G] = L[X]$ where $X$ is a reshaped subset of $\omega_1$. Then $X$ can be coded by a real $R$ via a ccc forcing, as in Jensen–Solovay [70]. This completes the proof of the Coding Theorem.

## 4.4 Large Cardinal Preservation and Relevance

In this section we show that the forcings of the previous two sections preserve the large cardinal properties Mahlo and $\alpha$-Erdös, $\alpha < \omega_1$. In addition they can be modified to also preserve $\Sigma^n_m$-indescribability for all, $n, m \in \omega$. And we verify that these forcings are relevant when applied to the ground model $\langle L, \emptyset\rangle$.

**Definition.** $\kappa$ is *Mahlo* if $\kappa$ is inaccessible and $\{\alpha < \kappa \mid \alpha \text{ is inaccessible}\}$ is stationary in $\kappa$. For $\alpha \le \kappa$, $\alpha$ a limit ordinal, $\kappa$ is *$\alpha$-Erdös* if whenever $C$ is CUB in $\kappa$, $f : [C]^{<\omega} \longrightarrow \kappa$ is regressive (i.e., $f(a) < 1 + \min(a)$ for each $a \in [C]^{<\omega}$ = the set of finite subsets of $C$) then there is $H \subseteq C$ of ordertype $\alpha$ such that for each $n$, $f$ is constant on $[H]^n$ = the set of size $n$ subsets of $H$. $H$ is said to be *homogeneous* for $f$.

Let $P$ denote one of the forcings of the previous two sections, used to establish Theorem 4.1. Thus $P$ is defined over a model $\langle L[A], A\rangle$ of ZFC + GCH and adds a real $R$ such that $A$ is definable over $L[R]$. Cofinalities are preserved.

To preserve Mahloness we only need the following.

**Proposition 4.14.** *If $C \subseteq \kappa$ is* CUB*, $C \in L[R]$, $\kappa$ inaccessible in $L[A]$ then there exists a* CUB *$D \subseteq C$, $D \in L[A]$.*

*Proof.* Let $\sigma$ be a name for $C$. If $p \Vdash \sigma$ is CUB then by $\Delta$-distributivity there is $q \leq p$ such that $q$ reduces $\Delta_i = \{r \mid$ For some $\alpha < \kappa, r \Vdash i^{\text{th}}$ element of $\sigma = \hat{\alpha}\}$ below $i^+$ for all $i < \kappa$. But then $D = \{\gamma < \kappa \mid$ Whenever $i < \gamma$ and $r \Vdash i^{\text{th}}$ element of $\sigma = \hat{\alpha}$ for some $r \leq q$ then $\alpha < \gamma\}$ is CUB in $\kappa$ and $q \Vdash D \subseteq \sigma$. □

The previous proposition is also useful for showing that the $\alpha$-Erdös property is preserved, which we consider next.

**Proposition 4.15.** *If $\kappa$ is $\alpha$-Erdös in $L[A]$, $\alpha < \omega_1^{L[A]}$ then $\kappa$ is $\alpha$-Erdös in $L[R]$.*

*Proof.* Suppose that $p \in P$ forces that $f : [C]^{<\omega} \longrightarrow \kappa$ is regressive, $C$ CUB in $\kappa$ (where we identify $f, C$ with names representing them). By Proposition 4.14 we may assume that $C \in L[A]$, for the purpose of showing that there is a homogeneous $H \subseteq C$ of ordertype $\alpha$. Note that by Lemma 4.8 the $P$-generic $R$ is also $P^{<\kappa^+}$-generic over $\mathcal{A}^{\emptyset}$ where $\emptyset$ denotes $\emptyset_{\kappa^+}$ and $f$ has a $P^{<\kappa^+}$-name in $\mathcal{A}^{\emptyset}$. So we assume that $p$ belongs to $P^{<\kappa^+}$ and our goal is to find $q \leq p$ in $P^{<\kappa^+}$ such that $q$ forces the existence of the desired $H$.

Well-order $[C]^{<\omega}$ by maximum difference: $a < b$ iff $\alpha \in b$ where $\alpha = \max((a - b) \cup (b - a))$. And in this case define $\gamma(a, b) = \text{card}(\max(a \cap \alpha))$ for the $\alpha$ so defined. We define $p_a \in P^{<\kappa^+}$ by induction on this ordering so that $p_\emptyset = p$ and $p_b \leq q_b = \cup\{p_a \restriction [\gamma(a, b), \kappa] \mid a < b\}$.

Suppose that $p_b$ is defined for $b < a$ and $q_a$ as defined above belongs to $P^{<\kappa^+}$. For $\gamma \in \text{Card} \cap [\omega, \kappa]$ define $X_a^\gamma$ = least $X \prec \mathcal{A}^{\emptyset}$ such that $\gamma \cup \{a, C, \text{name for } f\} \subseteq X$ and for each $b < a$ such that $\gamma(b, a) \leq \gamma$, $X$ contains $b$, $p_b \restriction [\gamma(b, a), \kappa]$ and $X_b^\gamma$. And define $f_a(\gamma) = X_a^\gamma \cap \mathcal{A}^{q_a}{}_{\gamma^+}$ if $\gamma \in X_a^\gamma \cap \kappa$, $f_a(\gamma) = \emptyset$ otherwise. Then $p_a$ is defined to be the $\mathcal{A}^{\emptyset}$-least $p \in P^{<\kappa^+}$ such that $p \leq q_a$ and:

(1) $p \in \Sigma_{f_a}^{q_a}$.

(2) $p(\kappa)$ meets all predense $D \subseteq R^{\emptyset}$ in $X_a^\kappa$ ($\emptyset$ denotes $\emptyset_{\kappa^+}$).

(3) $p \Vdash f(a) = \hat{\alpha}$, for some $\alpha < 1 + \min(a)$.

The inductive proof that the resulting $q_a$'s are indeed conditions in $P^{<\kappa^+}$ is just like the distributivity arguments of Lemma 4.5(a), Lemma 4.11.

Now form the structure $\mathcal{S} = \langle L_{\kappa^+}[A], A \cap \kappa^+, E\rangle$ where $E = \{\langle a, p_a, \alpha\rangle \mid a \in [C]^{<\omega}, p_a \Vdash f(a) = \hat{\alpha}\}$. The fact that $\kappa$ is $\alpha$-Erdös in $L[A]$ implies that there is $I \subseteq C$ of ordertype $\alpha$ such that for each $\gamma < \kappa$, $I - \gamma$ is a set of indiscernibles for $\mathcal{S}$, including names for each ordinal less than $\gamma$. Let $I$ be the $\mathcal{A}^{\emptyset}$-least such set and note that for $a, b \in [I]^n$ we have $p_a \Vdash f(a) = \hat{\beta}$ iff $p_b \Vdash f(b) = \hat{\beta}$.

So we will be done if we can show that $q = \bigcup\{p_a \mid a \in [I]^{<\omega}\}$ is a condition in $P^{<\kappa^+}$. It suffices to produce $\langle b_i \mid i \in \omega\rangle$ in $[I]^{<\omega}$ such that $q = \bigcup\{p_{b_i} \mid i \in \omega\}$ and $p_{b_0} \geq p_{b_1} \geq \cdots$ is definable in $\mathcal{A}^{\emptyset}$ (from $C$, name for $f$ and $p$); for then we can apply the usual distributivity argument to conclude that $q$ is condition.

To obtain the $b_i$'s, let $\langle \beta_i \mid i \in \omega\rangle$ be the $\mathcal{A}^{\emptyset}$-least increasing cofinal sequence in $I$, let $I^* = \{\beta_i \mid i \in \omega\}$, $\langle a_i \mid i \in \omega\rangle =$ the increasing enumeration of $[I^*]^{<\omega}$ under the "maximum difference" order defined earlier. Then define $b_i$ inductively by: $b_0 = a_0$; if $b_i$ has been defined to be $a_k$ then $b_{i+1} = a_{k+1}$ if $\text{card}(a_{k+1}) = \text{card}(a_k) + 1$ and otherwise $b_{i+1} = a_j$ where $j$ is least such that $j > k$ and $\text{card}(a_j) = \text{card}(a_k)$.

**Claim.** $i \leq j \Longrightarrow p_{b_j} \leq p_{b_i}$.

*Proof.* If $\text{card}(a_{k+1}) = \text{card}(a_k) + 1$ then it must be that $a_{k+1} = \{\beta_0\} \cup a_k$ and so $\gamma(a_k, a_{k+1}) = 0$. Thus by definition $p_{a_{k+1}} \leq p_{a_k}$. If $j$ is least so that $j > k$ and $\text{card}(a_j) = \text{card}(a_k)$ then we must have $a_k = \{\gamma_0, \ldots, \beta_\ell, \ldots, \gamma_n\}$, $a_j = \{\gamma_0, \ldots, \beta_{\ell+1}, \ldots, \gamma_n\}$ and $\gamma(a_k, a_j) = \beta_\ell$. So by definition $p_{a_j} \leq p_{a_k} \restriction [\beta_\ell, \kappa]$. But $p_{a_k} \restriction \beta_\ell = p_{a_j} \restriction \beta_\ell$ by indiscernibility and the fact that $a_k$, $a_j$ have the same cardinality and agree below $\beta_\ell$. □

**Claim.** For $a \in [I]^{<\omega}$ there is $i \in \omega$ such that $p_{b_i} \leq p_a$.

*Proof.* There exists $a^* \in [I^*]^{<\omega}$ such that $\max(a) < \min(a^*)$ and $\text{card}(a) = \text{card}(a^*)$. We claim that $p_{a^*} \leq p_a$: Write $a = \{\nu_1, \ldots, \nu_n\}$ and $a^* = \{\nu_1^*, \ldots, \nu_n^*\}$. Then $p_a$, $p_{a^*}$ agree below $\nu_1$ by indiscernibility. If $a_1^* = \{\nu_1, \nu_2^*, \ldots, \nu_n^*\}$ then $p_a$, $p_{a_1^*}$ agree below $\nu_2$ by indiscernibility and $p_{a^*} \leq p_{a_1^*} \restriction [\nu_1, \kappa]$ by definition. Continue using $a_2^* = \{\nu_1, \nu_2, \nu_3^*, \ldots, \nu_n^*\}, \ldots$ and conclude in $n$ steps that $p_{a^*} \leq p_a$. So we may assume that $a$ belongs to $[I^*]^{<\omega}$.

Suppose that $a = a_\ell \in [I^*]^{<\omega}$ and choose $i \leq \ell < j$ such that for some $k$, $a_i = b_k$ and $a_j = b_{k+1}$. We show that there is $m$, with $\ell < m \leq j$, such that $p_{a_m} \leq p_{a_\ell}$ (this suffices). If $\text{card}(a_{\ell+1}) = \text{card}(a_\ell)$ or $\text{card}(a_\ell) + 1$ then as before $p_{a_{\ell+1}} \leq p_{a_\ell}$. So suppose $\text{card}(a_{\ell+1}) < \text{card}(a_\ell)$. We can assume that $i < \ell < j$ and $\text{card}(a_i) = \text{card}(a_j) > \text{card}(a_\ell)$, by definition of $b_{k+1}$. So there is $m$, with $\ell + 1 < m < j$, such that $\text{card}(a_m) = \text{card}(a_\ell)$ and if we take the least such $m$ we have $p_{a_m} \leq p_{a_\ell}$. □

This completes the proof of $\alpha$-Erdös preservation. □

**Definition.** Let $0 < m, n < \omega$. $\kappa$ is $\Sigma_m^n$*-indescribable* iff whenever $B \subseteq \kappa$ and $\varphi$ is a $\Sigma_m$ formula, $V_{\kappa+n} \vDash \varphi(B)$ then there exists $\bar{\kappa} < \kappa$ such that $V_{\bar{\kappa}+n} \vDash \varphi(B \cap \bar{\kappa})$.

In the preceding definition, $V_{\kappa+n}$ can be replaced by $H_{\kappa^{(+n)}}$ where $\kappa^{(+n)}$ is the $n^{\text{th}}$ cardinal after $\kappa$ and $H_\lambda = \{x \mid \text{card (transitive closure of } x) < \lambda\}$. We show that the forcings of the previous two sections used to prove Theorem 4.1 can be modified so as to also preserve $\Sigma_m^n$-indescribability. First we need to modify the predicate $A$, to preserve $\Sigma_m^n$ indescribability in the sense of the next definition.

**Definition.** $\kappa$ is *$\Sigma^n_m$-indescribable relative to $A^* \subseteq$ ORD* iff whenever $B \subseteq \kappa$, $\varphi$ a formula, $\langle H_{\kappa^{(+n)}}, A^* \cap \kappa^{(+n)}\rangle \models \varphi(B)$ then there exists $\bar{\kappa} < \kappa$ such that $\langle H_{\bar{\kappa}^{(+n)}}, A^* \cap \bar{\kappa}^{(+n)}\rangle \models \varphi(B \cap \bar{\kappa})$.

**Proposition 4.16.** *Suppose $\langle L[A], A\rangle \models$ ZFC + GCH, $A \subseteq$ ORD. Then there is an $\langle L[A], A\rangle$-definable tame forcing $P^*$ for adding a class $A^* \subseteq$ ORD such that A is definable over $\langle L[A^*], A^*\rangle$ and if $\kappa$ is $\Sigma^n_m$-indescribable in $L[A]$ then $\kappa$ is $\Sigma^n_m$-indescribable relative to $A^*$ in $L[A^*]$.*

*Proof.* For any $\kappa^+$, $\kappa^+$-Cohen forcing is the forcing adding a subset of $\kappa^+$ whose conditions are $q : \alpha \longrightarrow 2$, $\alpha < \kappa^+$, ordered by extension. Now let $P^*$ consist of all $p : \text{Card} \cap \kappa(p)^+ \longrightarrow V$, $\kappa(p) \in$ Card such that for all $\kappa \in \text{Card} \cap \kappa(p)^+$:

(1) $p(\kappa) \in \kappa^+$-Cohen forcing.

(2) $\kappa \in \text{Card}' \Longrightarrow (p(\kappa)(\gamma) = 1$ iff $\gamma \in A$, for $\gamma < \kappa, \gamma \in \text{Dom}(p(\kappa)))$.

(3) $\kappa$ inaccessible $\Longrightarrow \{\bar{\kappa} \in \text{Card} \cap \kappa \mid p(\bar{\kappa}) \neq \emptyset\}$ is bounded in $\kappa$.

A $P^*$-generic over $\langle L[A], A\rangle$ can be identified with a class $A^* \subseteq$ ORD and $A$ is definable over $\langle L[A^*], A^*\rangle$ by (b) above. We claim that if $\kappa$ is $\Sigma^n_m$-indescribable in $L[A]$ then $\kappa$ is $\Sigma^n_m$-indescribable relative to $A^*$ in $L[A^*]$. Indeed, suppose $\langle H_{\kappa^{(+n)}}, A^* \cap \kappa^{(+n)}\rangle \models \varphi(B)$ in $L[A^*]$, where $B \subseteq \kappa$, $B \in L[A^*]$ and $\varphi$ is $\Sigma_m$. Then this is forced by some condition $p$ in $P^*(< \kappa^{(+n)})$. By $\Sigma^n_m$-reflection in $L[A]$, there are unboundedly many $\bar{\kappa} < \kappa$ such that $p \restriction \bar{\kappa}^+ \cup q$ forces $\langle H_{\bar{\kappa}^{(+n)}}, A^* \cap \bar{\kappa}^{(+n)}\rangle \models \varphi(B \cap \bar{\kappa})$ for some $q \in P^*[\bar{\kappa}^+, \bar{\kappa}^{(+n)}) = \{p \restriction [\bar{\kappa}, \bar{\kappa}^{(+n)}) \mid p \in P^*\}$. (Note that $B \subseteq \kappa$ has a name of size $\kappa$ by the inaccessibility of $\kappa$ and (c) above, and hence can be reflected to $B \cap \bar{\kappa}$.) By genericity, there is such a $\bar{\kappa}$ and $q$ for which $p \restriction \bar{\kappa}^+ \cup q$ belongs to the $P^*$-generic associated to $A^*$, and hence we get $\langle H_{\bar{\kappa}^{(+n)}}, A^* \cap \bar{\kappa}^{(+n)}\rangle \models \varphi(B \cap \bar{\kappa})$ for some $\bar{\kappa} < \kappa$. □

We can now get the Coding Theorem with $\Sigma^n_m$-indescribable preservation, via the next result.

**Proposition 4.17.** *Suppose $\langle L[A], A\rangle$, $A \subseteq$ ORD is a model of* ZFC + GCH. *Then the conclusion of Theorem* 4.1 *holds with the added clause:*

(d) *If $\kappa$ is $\Sigma^n_m$-indescribable relative to A then $\kappa$ is $\Sigma^n_m$-indescribable relative to R.*

*Proof.* Let $P$ be one of the forcings of the previous two sections to code $A$ by a real and by induction on limit cardinals $\kappa$ we define a forcing $P^*(\leq \kappa)$ contained in $P^{<\kappa^+}$; the desired forcing $P^*$ consists of all $p$ such that $p \restriction \kappa^+ \in P^*(\leq \kappa)$ for each $\kappa \in \text{Card}'$. If $\kappa = \aleph_\omega$ then $P^*(\leq \kappa) = P^{<\kappa^+}$. If $\kappa$ is a limit of limit cardinals then $p$ belongs to $P^*(\leq \kappa)$ iff $p \in P^{<\kappa^+}$ and for all $\bar{\kappa} \in \text{Card}' \cap \kappa$, $p \restriction \bar{\kappa}^+ \in P^*(\leq \bar{\kappa})$.

Now suppose that $P^*(\leq \lambda)$, where $\lambda = \kappa^{(+\omega)}$. First let $P^*_0(\leq \lambda)$ consist of all $p \in P^{<\lambda^+}$ such that $p \restriction \kappa^+$ belongs to $P^*(\leq \kappa)$. Let $p_0 = \emptyset$-condition of $P^*_0(\leq \kappa^+)$. We can successively extend $p_0$ to $p_1 \geq p_2 \geq \cdots$ in $P^*_0(\leq \kappa^+)$ so that $p_i(\alpha) = (\emptyset, \emptyset)$ for $\alpha \in \text{Card} \cap \kappa^+$ and for each $\Sigma_1$ $\varphi$, $B \subseteq \kappa$, $q \in P^*(\leq \kappa)$: If there is some $p$

compatible with each $p_i$ such that $p \restriction \kappa^+ = q$ and $p \Vdash \varphi(B)$ is true in $H_{\kappa^{++}}$, then $q \cup \{\langle \kappa^+, p_i(\kappa^+)\rangle\}$ is such a $p$ for some $i$. Moreover, the sequence of $p_i$'s can be taken to be $\Sigma_1$-definable over $\langle L_{\kappa^{++}}[A], A \cap \kappa^{++}\rangle$ and of length less than $\kappa^{++}$. Let $q$ be the greatest lower bound of the $p_i$'s and successively extend $q$ in the same way, but for $\varphi$ $\Sigma_2$ instead of $\Sigma_1$. Continue through all the $\Sigma_n$'s and let the result be the condition $r$ in $P_0^*(\leq \kappa^{++})$ and repeat the above construction below $r$, with $H_{\kappa^{++}}$, $\langle L_{\kappa^{++}}[A], A \cap \kappa^{++}\rangle$ replaced by $H_{\kappa^{+++}}$, $\langle L_{\kappa^{+++}}[A], A \cap \kappa^{+++}\rangle$. Then handle $\Sigma_2, \Sigma_3, \ldots$ and continue in this way for $H_{\kappa^{+++}}, H_{\kappa^{++++}}, \ldots$ until one finally obtains the condition $r_\kappa \in P_0^*(\leq \lambda)$, defined by taking $r_\kappa \restriction \lambda$ as the greatest lower bound of the conditions produced by the above constructions, with $r_\kappa(\lambda) = (\emptyset, \emptyset)$.

Now define $P^*(\leq \lambda)$ to consist of all $p \in P_0^*(\leq \lambda)$ which are compatible with $r_\kappa$. And define $P^*$ as specified above. Now the conclusion of Theorem 4.1 holds for the forcing $P^*$ and we get clause (d) as follows: Suppose $\kappa$ is $\Sigma_m^n$-indescribable relative to $A$ and in $L[R]$, $R$ a $P^*$-generic real, $H_{\kappa^{(+n)}} \vDash \varphi(B)$ where $\varphi$ is $\Sigma_m$ and $B \subseteq \kappa$. By the way we constructed $P^*(\leq \lambda)$, $\lambda = \kappa^{(+\omega)}$, there is a condition $p \in P^*(\leq \kappa) \subseteq P^*$ that forces this, $p$ belonging to the generic determined by $R$. Now apply $\Sigma_m^n$-reflection relative to $A$ to conclude that $p \restriction \bar{\kappa}^+$ forces $H_{\bar{\kappa}^{(+n)}} \vDash \varphi(B \cap \bar{\kappa})$ for some $\bar{\kappa} \in \mathrm{Card}' \cap \kappa$. So clause (d) holds. □

**Remarks.** (1) One could also consider $\Pi_m^n$-indescribability. When $n = 1$ there is no difficulty as $\Pi_m^1$-indescribability is the same as $\Sigma_{m+1}^1$-indescribability. Whereas Proposition 4.17 holds for $\Pi_m^n$ (and $\Sigma_m^n$, simultaneously) the author does not know how to establish Proposition 4.16 for $\Pi_m^n$-indescribability.

(2) The forcings of Propositions 4.16, 4.17 preserve the Mahlo and $\alpha$-Erdös properties. Thus Theorem 4.1 holds with the extra requirement of simultaneous Mahlo, $\alpha$-Erdös and $\Sigma_m^n$-indescribable preservation.

## Relevance

**Theorem 4.18.** *Let $P$ be one of the forcings of the two previous sections used to establish Theorem* 4.1, *when $A = \emptyset$. Then there is a class $G$ which is $P$-generic over $L$, which is definable in $L[0^\#]$ and which preserves indiscernibles.*

*Proof.* This is similar to the proof of Theorem 3.17. For any indiscernible $i$ let $j_n$ be the first $n$ indiscernibles $\geq i$. Then define $s_n \in S^{i^+}$ and $p^n \in P^{s_n}$ inductively, meeting the following conditions: $s_0 = \emptyset$ and $p^n$ is the trivial condition. $s_{n+1} = \pi_i(p^n)_{i^+}$ where $\pi_i : L \longrightarrow L$ is an elementary embedding with critical point $i$, and $p^{n+1}$ is the least $q \leq p^n$ in $P^{s_n}$ meeting $\Sigma_{f_n}^{p^n}$ where for $\beta \in \mathrm{Card} \cap i^+$, $f_n(\beta) = \mathrm{Hull}(\beta \cup j_n)$ if $\beta \in \mathrm{Hull}(\beta \cup j_n)$ and $f_n(\beta) = \emptyset$ otherwise. (When $\beta = i$ we take $p_{\beta^+}^n$ to be $s_n$.) Let $G_0^i = \{p \mid p \text{ is extended by some } p^n\}$.

$G_0^i$ need *not* be $P^{s_n}$-generic over $\mathscr{A}^{s_n}$ as all conditions in $G_0^i$ have empty restraint at indiscernibles $< i$. But notice that for $i_0 < i_1 < \cdots < i_n \leq i$ in $I$, $G_0^{i_0} \cup \cdots \cup G_0^{i_n}$

is a compatible set of conditions. We take $G^i$ to be $\{p \mid p$ is extended by $q_0 \wedge \cdots \wedge q_n$ for some choice of $q_l \in G_0^{i_l}$, $i_0 < \cdots < i_n \leq i$ in $I\}$. Now we claim that $G^i$ is $P^{s_n}$-generic over $\mathcal{A}^{s_n}$ for each $n$. Indeed, if $D$ is predense on $P^{s_n}$ and belongs to $\mathcal{A}^{s_n}$, $D \in \mathrm{Hull}(\{k_0, \ldots, k_m\} \cup j_n)$ with $k_0 < \cdots < k_m < i$ in $I$ then $p^{n+1}$ reduces $D$ below $k_m^+$, $p^{n+2}$ reduces $D$ below $k_{m-1}^+, \ldots$ and eventually we get $p^{n+m+2}$ in $G^i$ meeting $D$.

It follows that $G^i(< i) = G^i \cap P^i$ is generic over $L_i$ (for $L_i$-definable dense sets) and hence $G$ is $P$-generic over $L$ where $G = \cup\{G^i(< i) \mid i \in I\}$. Clearly $G$ preserves indiscernibles. □

**Corollary 4.19.** *If $A \subseteq \mathrm{ORD}$ preserves indiscernibles then there is a real $R \in L[A, 0^\#]$ such that $R$ preserves indiscernibles and $A$ is definable in $L[R]$. If $L[A] \models \mathrm{GCH}$ then $L[A]$, $L[R]$ have the same cofinalities.*

*Proof.* If $L[A] \models\sim \mathrm{GCH}$ then use the forcing $P^*$ from the proof of Corollary 4.2 to obtain the GCH. The proofs of Theorems 3.15, 3.20 show that we may do this so as to preserve indiscernibles. Then apply the technique of Theorem 4.18. □

In fact, it is possible to characterize those $A \subseteq \mathrm{ORD}$ which are coded by reals $R$ such that $0^\# \not\leq_L R$:

**Definition.** For $\alpha, \beta < \omega_1$, $\beta \neq 0$ let $I_{\alpha,\beta} = \{i_{\alpha+\beta\cdot\gamma} \mid \gamma \in \mathrm{ORD}\}$ where $\langle i_\alpha \mid \alpha \in \mathrm{ORD}\rangle$ is the increasing enumeration of $I$.

**Corollary 4.20.** *If $A \subseteq \mathrm{ORD}$ and for some $\alpha, \beta < \omega_1$ the class $I_{\alpha,\beta}$ forms a generating class of indiscernibles for $\langle L[A], A\rangle$ then $A$ is definable in $L[R]$ for some real $R$ such that $0^\# \notin L[R]$.*

In the next chapter we will use the preceding corollary to show that $A \subseteq \mathrm{ORD}$ is definable in $L[R]$ for some real $R$, $0^\# \notin L[R]$ iff $I_{\alpha,\beta}$ forms a class of indiscernibles for $\langle L[A], A\rangle$ for some $\alpha, \beta < \omega_1$, $\beta \neq 0$. Moreover there are reals $R$ such that $I^R = I_{\alpha,\beta}$, for any $\alpha, \beta < \omega_1$, $\beta \neq 0$ (where $I^R$ denotes the Silver indiscernibles for $L[R]$).

Note that the forcings $P$ for coding over $\langle L, \emptyset\rangle$ cannot be *totally* relevant, as $P$ has arbitrarily large antichains, yet there can be at most $2^{\aleph_0}$ $P$-generics.

Using the techniques used to prove the relevance of Easton forcing and Jensen Coding over $L$, it is possible to show that if $A$ obeys the hypothesis of Corollary 4.20 then the forcings of Propositions 4.16, 4.17 have generics in $L[A, 0^\#]$. Thus in the conclusion of Corollary 4.20 we may add the statement that $R$ preserve the Mahlo, $\alpha$-Erdös and $\Sigma_m^n$-indescribability properties over $L[A]$.

*Chapter 5*

# The Genericity Problem

In this chapter we use the method of Jensen coding to discuss the first of the three Solovay questions, concerning genericity over $L$. It follows readily from the Coding Theorem that it is consistent for a real to neither be set-generic over $L$ nor to construct $0^\#$; the relevance of Jensen coding over $L$ (see Theorem 4.18) implies that such a real can exist in $L[0^\#]$. Thus Solovay's version of the Genericity Problem has a negative answer.

This led Beller–Jensen–Welch to propose a new version of the problem, asking if a real $R <_L 0^\#$ must necessarily be *class-generic* over $L$. In this chapter we also provide a negative answer to this version of the Genericity Problem.

Then we turn to the study of generic classes that do *not* preserve indiscernibles. There are three uses of these techniques: First, to present the characterizations (promised in Chapter 4) of those classes of ordinals definable from reals not constructing $0^\#$, as well as of the possible patterns of indiscernibles relative to such reals. Second, to show that unlike for set forcing, it is possible for $L[R]$, $R \subseteq \omega$ to not be a class-generic extension of $L$, even though $R$ does belong to such an extension. And third, we present a result relevant to the open question as to whether $0^\#$ itself is class-generic over some model not containing it.

We close this chapter with a study of generic saturation: We show that if a class forcing over $L$ has a generic then it has one definable in a set-generic extension of $L[0^\#]$, assuming that ORD is $\omega + \omega$-Erdös.

## 5.1 Jensen's Example and a New Conjecture

**Theorem 5.1** (Jensen). *There is a real $R <_L 0^\#$ that is not set-generic over $L$.*

*Proof.* Take $R \in L[0^\#]$ to result from applying the proof of the Coding Theorem to the ground model $\langle L, \emptyset\rangle$, obtaining a generic $G$ coded by $R$. Note that in $L[G] = L[R]$ there are $P(\kappa^+)$-generic sets for each infinite successor $L$-cardinal $\kappa^+$, where $P(\kappa^+) = \kappa^+$-Colen forcing. In a $P$-generic extension of $L$, where $P \in L$, there can be no $\kappa^+$-Cohen set where $\kappa = L$-cardinality $(P)$. So $L[R]$ is not a set-generic extension of $L$. □

Note also that $R$ as in Theorem 5.1 can be chosen to preserve both $L$-cofinalities and indiscernibles.

The natural next question takes the form of a conjecture from Beller–Jensen–Welch [85]:

**Conjecture** (Beller–Jensen–Welch). If $R <_L 0^\#$ then $R$ is class-generic over $L$.

Our next goal is to prove the following.

**Theorem 5.2.** *There is a real $R <_L 0^\#$ such that:*

(a) *$L, L[R]$ have the same cofinalities.*

(b) *$I = I^R$ (where $I^R$ denotes the Silver indiscernibles for $L[R]$).*

(c) *$R$ is not an element of $L[G]$ for any $G$ which is generic for a tame class forcing over a ground model of the form $\langle L, A\rangle$.*

The above Conjecture is refuted by this result. We abbreviate (c) above by simply saying that "$R$ is not generic over $L$." The source for this property lies with Tarski's undefinability of Truth:

**Definition.** For an amenable $\langle L, A\rangle$ let $\text{Sat}\langle L, A\rangle$ denote the canonical satisfaction relation for $\langle L, A\rangle$, viewed as a predicate on $L$. We say that $R \subseteq \omega$ *codes the* Sat *operator* if $\text{Sat}\langle L, A\rangle$ is definable over $\langle L[R], A\rangle$, for each amenable $\langle L, A\rangle$.

**Proposition 5.3.** *If $R$ codes the Sat operator then $R$ is not generic over $L$.*

*Proof.* Suppose $R \in L[G]$ where $G$ is $P$-generic over $\langle L, A\rangle$, $P$ is tame and $P$ is $\langle L, A\rangle$-definable. As $R$ codes the Sat operator we have that $\text{Sat}\langle L, A\rangle$ is definable over $\langle L[R], A\rangle$. The Truth Property for $P$-forcing gives us $\langle L[G], G, A\rangle \models \varphi$ iff $\exists p \in G(p \Vdash \varphi)$ and therefore by the Definability Property for $P$-forcing, $\text{Sat}\langle L[G], G, A\rangle$ is definable over $\langle L[G], G, \text{Sat}\langle L, A\rangle\rangle$. As $\text{Sat}\langle L, A\rangle$ is definable over $\langle L[R], A\rangle$ and hence over $\langle L[G], A\rangle$, we conclude that $\text{Sat}\langle L[G], G, A\rangle$ is definable over $\langle L[G], G, A\rangle$, in contradiction to Tarski's well-known result. $\square$

Thus Theorem 5.2 reduces to showing that there is a cofinality and indiscernible preserving real $R <_L 0^\#$ which in addition codes the Sat operator. In fact it will suffice to produce a class $G$ with these properties (such that $G$ preserves indiscernibles and is definable over $L[0^\#]$) for then we can code $G$ by the desired real $R$.

Of course $G$ cannot be generic over $L$ in the usual sense. (The whole point is to avoid class-genericity!) We now describe a "hyperclass forcing" designed to produce the desired $G$.

Let $\infty$ denote an $L$-regular cardinal and $\infty^+$ its $L$-cardinal successor. We in fact think of $\infty$ as denoting ORD and $\infty^+$ as denoting $\text{ORD}^+$, the least "$L$-cardinal past ORD"; our purpose is to create $G$ to be a subclass of $L$, generic for a forcing $P$ of size $\text{ORD}^+$. More accurately, we obtain $G$ as the union of $G^\infty \subseteq L_\infty$ generic for $P^\infty \subseteq L_{\infty^+}$, as $\infty$ ranges over Silver indiscernibles.

To describe the nature of $P^\infty$ we need some preliminary definitions. For $\alpha \leq \infty$ let $2^\alpha$ denote all $f : \alpha \longrightarrow 2$ in $L$ (or amenable to $L$ if $\infty = \text{ORD}$). And $2^{<\alpha} = \bigcup\{2^\beta \mid \beta < \alpha\}$. For $S \in 2^\infty$, $\mu(S)$ denotes the least limit $\mu > \infty$ such that $S \in L_\mu$ and $C_S = \{\alpha < \infty \mid \alpha = \infty \cap \text{Skolem Hull of } \alpha \text{ in } L_{\mu(S)}\}$. Thus $C_S$ is CUB in $\infty$

and $\langle L_\alpha, S \restriction \alpha\rangle$ is elementary in $\langle L_\infty, S\rangle$ for sufficiently large $\alpha \in C_S$. It follows that $\mathrm{Sat}\langle L_\infty, S\rangle$ is definable over $\langle L_\infty, S, C_S\rangle$.

The purpose of $P^\infty$ is to produce a generic function $f_G : 2^{<\infty} \longrightarrow \infty$ such that for $S \in 2^\infty$, $\{f_G(S \restriction \alpha) \mid \alpha < \infty\}$ is a good approximation to the complement of $C_S$. This approximation will be good enough to conclude that $C_S$, and hence $\langle L_\infty, S\rangle$, is definable over $\langle L_\infty[f_G], f_G, S\rangle$.

For the precise definition of $P^\infty$ we need a few more definitions. An *Easton set of ordinals* is a set of ordinals $X$ such that $X \cap \alpha$ is bounded in $\alpha$ for regular $\alpha \le \infty$. An *Easton set of strings* is a set $D \subseteq 2^{<\infty}$ such that $D \cap 2^{<\alpha}$ has cardinality less than $\alpha$ for every regular $\alpha$. For any $X \subseteq \infty$ let $\mathrm{Seq}(X) = \bigcup\{2^\alpha \mid \alpha \in X\}$.

**Definition.** A *condition* in $P^\infty$ is $(X, F, D, f)$ where:

(1) $X \subseteq \infty$ is an Easton set of ordinals.

(2) $F : X \longrightarrow P(2^\infty) =$ Power Set of $2^\infty$, such that for $\alpha \in X$, $F(\alpha)$ has cardinality $\le \alpha$.

(3) $D \subseteq \mathrm{Seq}(X)$ is an Easton set of strings.

(4) $f : D \longrightarrow \infty$, $f(s) > \mathrm{length}(s)$ for $s \in D$.

Extension of conditions is defined by: $(Y, G, E, g) \le (X, F, D, f)$ iff:

(5) $Y \supseteq X$, $E \supseteq D$, $G(\alpha) \supseteq F(\alpha)$ for $\alpha \in X$, $g$ extends $f$.

(6) If $s \in E - D$ then the interval $(\mathrm{length}(s) + 1, g(s)]$ contains no element of $X$, and if for some $\alpha \le \mathrm{length}(s)$ in $X$, $s \subseteq S \in F(\alpha)$ then $g(s) \notin C_S$.

**Lemma 5.4.** *If $p = (X_p, F_p, D_p, f_p)$ belongs to $P^\infty$ and $\alpha < \infty$, $S \in 2^\infty$, $s \in 2^{<\infty}$ then $p$ has an extension $q$ such that $\alpha \in X_q$, $S \in F_q(\alpha)$ and $s \in D_q$.*

*Proof.* Easy, given the fact that if $s \notin D_p$ then we can add $s$ by defining $f_q(s) = \mathrm{length}(s) + 1$. □

**Lemma 5.5.** *$P^\infty$ has the $\infty^+$-chain condition.*

*Proof.* Any two conditions of the form $(X, F, D, f)$, $(X, G, D, f)$ are compatible and there are only $\infty$-many $(X, D, f)$'s □

**Lemma 5.6.** *For $P^\infty$-generic $G$ let $f_G = \bigcup\{f \mid f = f_p$ for some $p \in G\}$. If $S \in 2^\infty$ then $\{\alpha < \infty \mid f_G(S \restriction \alpha) \in C_S\}$ is an Easton set of ordinals.*

*Proof.* $G$ contains a condition $p$ such that $0 \in X_p$ and $S \in F_p(0)$; then if $s \subseteq S$, $s \notin D_p$ it must be that $f_G(s) \notin C_S$ by the definition of extension. Also $D_p$ is Easton, so the result follows. □

**Lemma 5.7.** *For $P^\infty$-generic $G$, if $\alpha < \infty$ is regular, $S \in 2^\infty$ and $\alpha \notin \mathrm{Lim}\, C_S$ then $\{f_G(S \restriction \beta) \mid \beta < \alpha\}$ intersects every constructible unbounded subset of $\alpha$.*

*Proof.* Given $A \subseteq \alpha$, constructible and unbounded, and a condition $p$ we show that $p$ can be extended to $q$ such that $f_q(S \upharpoonright \delta) \in A$ for some $\delta$. Choose $\delta < \alpha$ large enough so that $\cup(X_p \cap \alpha) < \delta$ and $S \upharpoonright \delta$ is not an initial segment of any $T \neq S$ in $\cup\{F_p(\beta) \mid \beta \in X_p \cap \alpha\}$. This is possible since there are fewer than $\alpha$-many such $T$'s. Then let $X_q = X \cup \{\delta\}$, $F_q(\delta) = \emptyset$, $D_q = D_p \cup \{S \upharpoonright \delta\}$ and $f_q(S \upharpoonright \delta) = \beta \in A - C_S - \delta$. Such a $\beta$ exists since $A$ is unbounded in $\alpha$ and $C_S \cap \alpha$ is not. □

**Corollary 5.8.** *For $P^\infty$-generic $G$, regular $\alpha < \infty$ and $S \in 2^\infty$: $\alpha \notin \operatorname{Lim} C_S$ iff $\{f_G(S \upharpoonright \beta) \mid \beta < \alpha\}$ intersects every constructible unbounded subset of $\alpha$. Thus $\operatorname{Sat}\langle L_\infty, S\rangle$ is definable over $\langle L_\infty[f_G], f_G, S\rangle$.*

*Proof.* The "only if" is Lemma 5.7. Conversely, if $\alpha \in \operatorname{Lim} C_S$ then $\{f_G(S \upharpoonright \beta) \mid \beta < \alpha\}$ is disjoint from a final segment of $C_S$, a constructible unbounded subset of $\alpha$, by Lemma 5.6. The final statement follows as $\alpha \in \operatorname{Lim} C_S \Longrightarrow \langle L_\infty, S \upharpoonright \alpha\rangle$ is elementary in $\langle L_\infty, S\rangle$. □

**Lemma 5.9.** *For regular $\kappa < \infty$ and $p \in P^\infty$ let $(p)_\kappa$ denote $(X_p - \kappa, F_p \upharpoonright X_p - \kappa, D_p \cap \operatorname{Seq}(\infty - \kappa), f_p \upharpoonright D_p \cap \operatorname{Seq}(\infty - \kappa))$ and $P^\infty_\kappa = \{(p)_\kappa \mid p \in P\}$. Also let $(p)^\kappa$ denote $(X_p \cap \kappa, F_p \upharpoonright X_p \cap \kappa, D_p \cap \operatorname{Seq}(\kappa), f_p \upharpoonright D_p \cap \operatorname{Seq}(\kappa))$ and $P^{\infty,\kappa} = \{p \in P^\infty \mid X_p \subseteq \kappa$ and $\operatorname{Range}(f_p) \subseteq \kappa\}$. Then for any $L$-definable predense $\Delta \subseteq P^\infty$ and $p \in P^\infty$ there is $q \leq p$ such that $(q)^\kappa = (p)^\kappa$ and $\Delta^q$ is predense below $q$ on $P^\infty$, where $\Delta^q = \{r \in P^{\infty,\kappa} \mid r$ is compatible with $q$, $F_r(\alpha) \subseteq F_q(\kappa)$ for all $\alpha \in X_r$, and the greatest lower bound of $r, q$ meets $\Delta\}$.*

*Proof.* Let $r \wedge q$ denote the greatest lower bound of $r, q$. First extend $p$ to ensure $\kappa \in X_p$. Then note that $q \leq p$, $s \in D_q - D_p$, $s \in 2^{<\kappa} \Longrightarrow f_q(s) < \kappa$, by the definition of extension. Thus $\mathcal{A} = \{(X_q \cap \kappa, D_q \cap \operatorname{Seq}(\kappa), f_q \upharpoonright D_q \cap \operatorname{Seq}(\kappa)) \mid q \leq p\}$ has cardinality $\kappa$. Now successively extend $p = q_0 \geq q_1 \geq \cdots$ in $\kappa$ steps so that $(q_i)^\kappa = (p)^\kappa$ and for all $(X, D, f) \in \mathcal{A}$, if the resulting $q = q_\kappa$ has an extension $r$ such that $r$ meets $\Delta$ and $(X, D, f) = (X_r \cap \kappa, D_r \cap \operatorname{Seq}(\kappa), f_r \upharpoonright D_r \cap \operatorname{Seq}(\kappa))$, then for some $r$, $(q)_\kappa \leq (r)_\kappa$, $r$ meets $\Delta$, $(X, D, f) = (X_r \cap \kappa, D_r \cap \operatorname{Seq}(\kappa), f_r \upharpoonright D_r \cap \operatorname{Seq}(\kappa))$ and $F_r(\alpha) \subseteq F_r(\kappa)$ for all $\alpha \in X_r \cap \kappa$. This is easily accomplished by the $\kappa^+$-closure of $P^\infty_\kappa$.

We claim that $\Delta^q$ is predense below $q$ on $P^\infty$. If $r \leq q$ then of course $(X, D, f) = (X_r \cap \kappa, D_r \cap \operatorname{Seq}(\kappa), f_r \upharpoonright D_r \cap \operatorname{Seq}(\kappa))$ belongs to $\mathcal{A}$, and if $r$ meets $\Delta$ then for some $r'$, $(q)_\kappa \leq (r')_\kappa$, $r'$ meets $\Delta$, $(X_{r'} \cap \kappa, D_{r'} \cap \operatorname{Seq}(\kappa), f_{r'} \upharpoonright D_{r'} \cap \operatorname{Seq}(\kappa)) = (X, D, f)$, and $F_{r'}(\alpha) \subseteq F_{r'}(\kappa)$ for all $\alpha \in X_{r'} \cap \kappa$. Then $r', r$ are compatible: For, $(r)_\kappa \leq (r')_\kappa$ and $(r)^\kappa = (r')^\kappa$ except possibly $F_r(\alpha)$ may differ from $F_{r'}(\alpha)$ for $\alpha \in X_r \cap \kappa$. But $F_{r'}(\alpha) \subseteq F_{r'}(\kappa)$ for $\alpha \in X_{r'} \cap \kappa$ so the extension $(r)_\kappa \leq (r')_\kappa$ obeys all restraint imposed by $F_{r'}(\alpha)$, $\alpha \in X_{r'} \cap \kappa$. So we obtain a common extension of $r, r'$ by taking $F(\alpha)$ to be $F_r(\alpha) \cup F_{r'}(\alpha)$ for $\alpha \in X_r \cap \kappa$.

Now define $r_0 = (X_{r'} \cap \kappa, F_{r'} \upharpoonright X_{r'} \cap \kappa, D_{r'} \cap \operatorname{Seq}(\kappa) - D_q, f_{r'} \upharpoonright D_{r'} \cap \operatorname{Seq}(\kappa) - D_q)$. Then $r_0$ belongs to $P^{\infty,\kappa}$ and is compatible with $r$. Also, $r_0 \wedge q \leq r'$, using the fact that $F_{r'}(\alpha) \subseteq F_{r'}(\kappa)$ for $\alpha \in X_{r'} \cap \kappa$, and therefore $r_0 \wedge q$ meets $\Delta$. So $r_0$ belongs to $\Delta^q$ and is compatible with $r$, as desired. □

**Corollary 5.10.** *$P^\infty$ is tame and preserves both cofinalities and the GCH.*

*Proof.* By Lemma 5.9 and the $\kappa^+$-closure of $P^\infty_\kappa$ we can, given any $L$-definable sequence $\langle \Delta_i \mid i < \kappa \rangle$ of predense classes and $p \in P^\infty$, extend $q$ and select $\langle d_i \mid i < \kappa \rangle$ such that $\mathrm{card}(d_i) \le \kappa$, $d_i \subseteq \Delta_i$ and $d_i$ is predense $\le q$. This gives pretameness and cofinality-preservation. For the GCH, it suffices to note that by Lemma 5.9 each subset of $\kappa$ in the generic extension belongs to $L[f_G \restriction \mathrm{Seq}(\kappa^+)]$ and therefore there can be only $\kappa^+$ of them. □

The main part of the proof consists in showing that for $\infty$ a Silver indiscernible, a $P^\infty$-generic can be obtained in $L[0^\#]$. For this we need to show that $P^i$-generics for countable $i \in I$ can be selected in a coherent way.

**Lemma 5.11.** *Suppose $i < j$ are adjacent countable Silver indiscernibles. Let $\pi = \pi_{ij}$ denote the elementary embedding $L \longrightarrow L$ which shifts indiscernibles $\ge i$ to the next indiscernible and leaves indiscernibles $< i$ fixed. Then there is a $P^j_i$-generic $G^j_i$ such that if $p \in G^j_i$ and $S \subseteq i$, $S \in L$ then $f_p(\pi(S) \restriction \alpha) \notin C_{\pi(S)}$ for all $\pi(S) \restriction \alpha \in D_p$.*

*Proof.* For any $k \in \omega$ let $\ell_1 < \cdots < \ell_k$ be the first $k$ indiscernibles greater than $j$ and $j_k = j^+ \cap \Sigma_1$ Hull of $j \cup \{j, \ell_1, \dots, \ell_k\}$ in $L$, $i_k = i^+ \cap \Sigma_1$ Hull of $i \cup \{i, \ell_1, \dots, \ell_k\}$ in $L$. Let $j^*_k$ = least limit ordinal $\lambda > j_k$ such that $L_\lambda \models \mathrm{card}(j_k) = j$. And define $C_k = \{\gamma < j \mid \gamma = j \cap \Sigma_1 \text{ Hull}(\gamma \cup \{j, \ell_1, \dots, \ell_k\})$ in $L$, a CUB subset of $j$.

Note that if $S \subseteq i$, $S \in L - L_{i_k}$ then $C_{\pi(S)} \subseteq C_k \cup \gamma$ for some $\gamma < i$: For, as $\pi(S) \notin L_{j_k}$, we have $\mu_{\pi(S)} \ge j^*_k$ and hence $C_{\pi(S)} \subseteq C_k \cup \gamma$ for some $\gamma < j$, as $C_k$ is an element of $L_{j^*_k}$ and for any CUB $C \in L_{\mu_{\pi(S)}}$, $C_{\pi(S)} \subseteq C \cup \gamma$ for some $\gamma < j$. But the least such $\gamma$ is $L$-definable from parameters in $i \cup \{\text{indiscernibles} \ge j\}$, as $\pi(S)$ and $C_k$ are, and hence must be less than $i$.

Thus we may conclude: The set $\{\pi(S) \mid S \subseteq i, S \in L\}$ is the union of $\{\pi(S) \mid S \subseteq i, S \in L_{i_k}\}$, an element of $L$ of cardinality $i$, and $\{\pi(S) \mid S \subseteq i, S \in L - L_{i_k}\}$, whose elements $\pi(S)$ satisfy the property that $C_{\pi(S)}$ is disjoint from the interval $(i, \gamma_k)$, where $\gamma_k$ is the least element of $C_k$ greater than $i$.

Using these observations we now build $G^j_i$ as $\{p \in P^j_i \mid p_k \le p \text{ for some } k \in \omega\}$, where $p_0 \ge p_1 \ge \cdots$ is defined as follows: Let $\langle \Delta_k \mid k \in \omega \rangle$ be a list of all constructible dense subsets of $P^j_i$, so that for all $k \in \omega$, $\Delta_k$ belongs to the $\Sigma_1$ Hull of $i \cup \{i, j, \ell_1, \dots, \ell_k\}$ in $L$. This is possible because any constructible subset of $P^j_i$ belongs to $L_{j^{++}}$ and hence to the $\Sigma_1$ hull in $L$ of $i \cup \{i, j, \ell_1, \dots, \ell_k\}$ for some $k$. Let $p_0$ be the empty condition $(\emptyset, \emptyset, \emptyset, \emptyset)$. If $p_k$ is defined then $p_{k+1}$ is obtained by first defining $F(i)$ to include $\{\pi(S) \mid S \subseteq i, S \in L_{i_k}\}$ and then extending in the $L$-least way to meet $\Delta_k$.

Clearly the resulting $G^j_i$ is $P^j_i$-generic; now we verify that $f_{p_k}(\pi(S) \restriction \alpha) \notin C_{\pi(S)}$ whenever $\pi(S) \restriction \alpha \in D_{p_k}$, $S$ a constructible subset of $i$. We prove this by induction on $k$. Given the result for $k$, we have it for $k+1$ if $S \in L_{i_k}$, as we initially chose $F_{p_{k+1}}(i)$ to include $\pi(S)$. Now also note that $\{\pi(S) \mid S \subseteq i, S \in L_{i_k}\}$ belongs to $\Sigma_1 \text{ Hull}(i^+ \cup \{j, \ell_1, \dots, \ell_k\})$ in $L$, as it is the range of a partial function $h$

on $i$ which is $\Sigma_1(L)$ in parameters $\{i, j, \ell_1, \dots, \ell_k\}$, and hence is $\Sigma_1(L)$-definable from $\{x, i, j, \ell_1, \dots, \ell_k\}$ where $x = \text{Dom}(h) \in L_{i^+}$. It follows that $p_{k+1}$ also belongs to $\Sigma_1$ Hull$(i^+ \cup \{j, \ell_1, \dots, \ell_k\})$) in $L$ and hence Range$(f_{p_{k+1}}) \subseteq \gamma_k$. But then Range$(f_{p_{k+1}})$ is disjoint from $C_{\pi(S)}$ when $S \notin L_{i_k}$. This proves the result for $k+1$ and therefore the lemma. □

**Lemma 5.12.** *Let $i < j$, $\pi$ and $G_i^j$ be as in the previous lemma and let $G^i$ be $P^i$-generic. Then there exists a $P^j$-generic $G^j$ such that $G_i^j = \{(p)_i \mid p \in G^j\}$ and $p \in G^i$ iff $\pi(p) \in G^j$.*

*Proof.* Recall that $P^{j,i}$ denotes $\{p \in P^j \mid X_p \subseteq i$ and Range$(f_p) \subseteq i\}$. For any $p \in P^j$ let $P^{p,i}$ denote $\{q \in P^{j,i} \mid q, p$ are compatible and $F_q(\alpha) \subseteq F_p(i)$ for each $\alpha \in X_q\}$. Define a modification $\bar{p}$ of $p$ for each $p \in P^{j,i}$ as follows: $p$ and $\bar{p}$ agree except with respect to $F_p$, $F_{\bar{p}}$; $F_{\bar{p}}(\alpha) = \{\bar{S} \mid S \in F_p(\alpha)\}$ where $\bar{S} = \pi(S \restriction i)$ if $i \in C_S$ and otherwise $\bar{S} = \pi(T)$ where $T \subseteq i$ is least so that $(T, C_T)$, $(S, C_S)$ agree through $\sup(C_S \cap i)$. Note that $F_{\bar{p}}(\alpha)$ is in $L$ as $\pi$ is constructible when restricted to a set of $L$-cardinality $\leq i$.

Now define $G^j = \{p^* \mid p^*$ has an extension $p \in P^j$ such that $(p)_i \in G_i^j$, $(p)^i \in P^{p,i}$, $\overline{(p)^i} \in \pi[G^i]\}$. Any two conditions in $G^j$ are compatible by the compatibility of $G_i^j$, the compatibility of $G^i$ and the fact that the restraint of $(p)^i \in P^{p,i}$ is included both in $F_p(i)$ and in the restraint of $\overline{(p)^i}$. Clearly $G_i^j = \{(p)_i \mid p \in G^j\}$ and the property $p \in G^i$ iff $\pi(p) \in G^j$ follows from the fact that $G_i^j$ obeys the conclusion of the previous lemma. So it only remains to show that $G^j$ is $P^j$-generic.

By Lemma 5.9 it is enough to show that for $p \in G^j$, $G^j \cap P^{p,i}$ meets all constructible $\Delta$ which are contained in $P^{p,i}$ and are predense below $p$ on $P^j$. For such a $\Delta$ let $\Delta_0 = \{p_0 \in P^i \mid \pi(p_0) = \bar{q}$ for some $q$ meeting $\Delta\}$. Note that $\Delta_0$ is constructible since $\Delta$ has $L$-cardinality $\leq i$. Also $\Delta_0$ is predense on $P^i$: If $p_0 \in P^i$ then $\pi(p_0)$ can be extended to some $q$ meeting $\Delta$, and $\bar{q} = \pi(p_1)$ for some $p_1 \leq p_0$, $p_1$ meeting $\Delta_0$. Now apply the genericity of $G^i$ to get $p_0 \in G^i \cap \Delta_0$; then $\pi(p_0) = \bar{q}$ for some $q$ meeting $\Delta$ and by definition, $q \in G^j$. □

**Lemma 5.13.** *Let $i_1 < i_2 < \cdots$ be the first $\omega$ Silver indiscernibles and let $i_\omega$ be their supremum. Then there exists $\langle G^{i_n} \mid n \geq 1\rangle$ such that $G^{i_n}$ is $P^{i_n}$-generic and whenever $\pi : L \longrightarrow L$ is elementary, $\pi(i_\omega) = i_\omega$ we have $p \in G^{i_n}$ iff $\pi(p) \in G^{\pi(i_n)}$.*

*Proof.* Note that any $\pi$ as in the statement of the lemma restricts to an increasing map from $\{i_n \mid n \geq 1\}$ to itself, so $G^{\pi(i_n)}$ makes sense. We define $G^{i_n}$ by induction on $n \geq 1$. Select $G^{i_1}$ to be the $L[0^\#]$-least $P^{i_1}$-generic. Select $G_{i_1}^{i_2}$ as in Lemma 5.11 and we use Lemma 5.12 to define $G^{i_2}$ from $G^{i_1}$, $G_{i_1}^{i_2}$. Now suppose $G^{i_n}$ has been defined, $n \geq 2$. First define $G_{i_n}^{i_{n+1}}$ to be $\{p \in P_{i_n}^{i_{n+1}} \mid \pi(q) \leq p$ for some $q \in G_{i_1}^{i_2}\}$ where $\pi : L \longrightarrow L$ satisfies $\pi(i_k) = i_{k+n-1}$ for $k \geq 1$. Then $G_{i_n}^{i_{n+1}}$ is $P_{i_n}^{i_{n+1}}$-generic, using the $i_1^+$-closure of $P_{i_1}^{i_2}$ and the fact that the collection of constructible dense subsets of $P_{i_n}^{i_{n+1}}$ is the union of sets of the form $\pi_n(A)$, $A$ of $L$-cardinality $i_1$. And $G^{i_{n+1}}_{i_n}$ obeys the property stated in Lemma 5.11 since $G^{i_2}_{i_1}$ does and $\pi$ is elementary. Now define $G^{i_{n+1}}$ from $G^{i_n}$, $G^{i_{n+1}}_{i_n}$ using Lemma 5.12.

To prove $p \in G^{i_n}$ iff $\pi(p) \in G^{\pi(i_n)}$ for $\pi : L \longrightarrow L$, $\pi(i_\omega) = i_\omega$ first note that it suffices to handle the case $\pi = \pi_m$ for some $m$, where $\pi_m(i_n) = i_{n+1}$ for $n \geq m$, $\pi_m(i_n) = i_n$ for $n < m$: For, given $p$ and $n$, the statement depends only on $\pi \upharpoonright L_{i_k}$ for some finite $k$, and $\pi$ agrees on $L_{i_k}$ with the finite composition of maps of the form $\pi_m$. Now we prove $p \in G^{i_n}$ iff $\pi_m(p) \in G^{\pi_m(i_n)}$ by induction on $n$ (for fixed $m$). This is trivial for $n < m$. For $n = m$ this follows from the fact that $G^{i_{n+1}}$ was defined from $G^{i_n}$, $G^{i_{n+1}}_{i_n}$ via Lemma 5.12. Suppose that the result holds for $n$ where $n \geq m$ and we wish to show it for $n + 1$. But $G^{i_{n+2}}$ is defined from $G^{i_{n+1}}$, $G^{i_{n+2}}_{i_{n+1}}$ just as $G^{i_{n+1}}$ is defined from $G^{i_n}$, $G^{i_{n+1}}_{i_n}$ and by induction $p \in G^{i_n}$ iff $\pi_m(p) \in G^{i_{n+1}}$. By definition it is clear that $q \in G^{i_{n+1}}_{i_n}$ iff $\pi_m(q) \in G^{i_{n+2}}_{i_{n+1}}$. So we may conclude that $p \in G^{i_{n+1}}$ iff $\pi_m(p) \in G^{i_{n+2}}$. □

**Lemma 5.14.** *There exists $\langle G^i \mid i \in I\rangle$ such that $G^i$ is $P^i$-generic and whenever $\pi : L \longrightarrow L$ is elementary, $p \in G^i$ iff $\pi(p) \in G^{\pi(i)}$.*

*Proof.* Let $t$ denote a Skolem term for $L$; thus $L = \{t(j_1, \ldots, j_n) \mid t$ a $n$-ary Skolem term, $j_1 < \cdots < j_n$ in $I\}$. Define $t(j_1, \ldots, j_n) \in G^i$ iff $t(\sigma(j_1), \ldots, \sigma(j_1)) \in G^{\sigma(i)}$ where $\sigma$ is the unique order-preserving map from $\{i, j_1, \ldots, j_n\}$ onto an initial segment of $I$, and where $G^{i_m}$ for finite $m$ is defined as in Lemma 5.13. Clearly the $G^i$'s obey the property stated in the lemma, provided we show that they are well-defined. If $t_1(j_1, \ldots, j_n) = t_2(k_1, \ldots, k_m)$ then let $\sigma^*$ be the unique order-preserving map from $\{i, j_1, \ldots, j_n, k_1, \ldots, k_m\}$ onto an initial segment of $I$. Then $t_1(\sigma^*(j_1), \ldots, \sigma^*(j_n)) = t_2(\sigma^*(k_1), \ldots, \sigma^*(k_m))$.

But by the property stated in Lemma 5.13, $t_1(\sigma^*(j_1), \ldots, \sigma^*(j_n)) \in G^{\sigma^*(i)}$ iff $t_1(\sigma_1(j_1), \ldots, \sigma_1(j_n)) \in G^{\sigma_1(i)}$ where $\sigma_1$ is the unique order-preserving map from $\{i, j_1, \ldots, j_n\}$ onto an initial segment of $I$. The analogous holds for $t_2$, so the $G^i$'s are well-defined. □

For $i \in I$ let $f^i = \cup\{f_p \mid p \in G^i\}$. Thus $f^i : 2^{<i} \longrightarrow i$ and if $f = \cup\{f^i \mid i \in I\}$ then $f : 2^{<\infty} \longrightarrow \infty$, where $\infty$ now denotes the real $\infty$, that is $\infty = \mathrm{ORD}$.

**Lemma 5.15.** (a) *For $L$-amenable $A \subseteq \mathrm{ORD}$, $\mathrm{Sat}\langle L, A\rangle$ is definable over $\langle L[f], f, A\rangle$.*

(b) *$I$ is a class of indiscernibles for $\langle L[f], f\rangle$.*

(c) *$L[f] \models \mathrm{GCH}$ and $L$, $L[f]$ have the same cofinalities.*

*Proof.* (a) follows from Corollary 5.8 as $f$ obeys the property stated there for $f_G$. (b) follows by Lemma 5.14 and (c) by Corollary 5.10. □

*Proof of Theorem* 5.2. Apply Corollary 4.19 to code $f$ by a real $R$ so that $R$ preserves indiscernibles and $L[f]$-cofinalities and $R$ belongs to $L[f, 0^\#] = L[0^\#]$. Then $R$ obeys the theorem, for by Lemma 5.15(a), $R$ codes the Sat operator. □

The above work is not special to the Sat operator. We now present a more general result, first for operators on $P_L(\omega_1) = \{X \subseteq \omega_1 \mid X \in L\}$ and then for $P_L(\infty) =$ all $L$-amenable classes.

**Theorem 5.16.** *Suppose $F : P_L(\omega_1) \longrightarrow P_L(\omega_1)$ is constructible. Then there is a real $R <_L 0^\#$ such that $F(A)$ is definable over $\langle L_{\omega_1}[R], A\rangle$, uniformly for $A \in P_L(\omega_1)$.*

*Proof.* For any $A \in P_L(\omega_1)$, let $C_A$ denote $\{\alpha < \omega_1 \mid \alpha = \omega_1 \cap$ Skolem Hull of $\alpha$ in $L_{\mu(A)}\}$ where $\mu(A) =$ least limit $\mu > \omega_1$ such that $A \in L_\mu$. (This is the same definition as before with $\infty = \omega_1$.) Then note that we may assume that $F(A) = \operatorname{Lim} F'(A)$ where $F'(A)$ is a CUB subset of $C_A$, as $C_A$ is definable over $\langle L_{\omega_1}, A, B\rangle$ for any unbounded $B \subseteq C_A$. Now choose $\alpha < \omega_1$ so that $F'$ is definable in $L$ from parameters in $\alpha \cup (I - \omega_1)$ and for $\alpha \leq i \leq \omega_1$ define $P^i$ as we did before but using $F'(A)$ in place of $C_A$. Then as before we can produce a generic $f : 2^{<\omega_1} \longrightarrow \omega_1$ such that for $A \in P_L(\omega_1)$, $F(A) = \operatorname{Lim} F'(A)$ is definable over $\langle L_{\omega_1}[f], f, A\rangle$. Finally code $f$ by a real $R <_L 0^\#$, using Corollary 4.19 and the fact that $\alpha$ is countable. □

To deal with operators on $P_L(\infty)$ we need to keep track of parameters.

**Definition.** Suppose $i < j$ belong to $I$ and $F_i : P_L(i) \longrightarrow P_L(i)$ is constructible. Then $F_i^j : P_L(j) \longrightarrow P_L(j)$ is defined as follows: Write $F_i = t(\alpha, i, \vec{k})$ where $t$ is a Skolem term for $L$, $\alpha < i$ and $\vec{k}$ are indiscernibles greater that $j$. Then $F_i^j = t(\alpha, j, \vec{k})$. Also define $F_i^\infty$: Given $A \in P_L(\infty)$ choose $t$ and $\alpha$ so that for $j \in I - \alpha$, $A \cap j = t(\alpha, j, \vec{k})$ where $\vec{k}$ is a sequence of indiscernibles greater than $j$. Then $F_i^\infty(A) = \cup\{F_i^j(A \cap j) \mid \alpha < j \in I\}$. An operator $F : P_L(\infty) \longrightarrow P_L(\infty)$ is *countably constructible* if it is of the form $F_{\omega_1}^\infty$ where $F_{\omega_1} : P_L(\omega_1) \longrightarrow P_L(\omega_1)$ is constructible.

**Theorem 5.17.** *If $F : P_L(\infty) \longrightarrow P_L(\infty)$ is countably constructible then there exists $R <_L 0^\#$ such that $F(A)$ is definable over $\langle L[R], A\rangle$, uniformly for $A \in P_L(\infty)$.*

*Proof.* Apply Theorem 5.16 to $F_{\omega_1}$, where $F = F_{\omega_1}^\infty$. The resulting real $R$ obeys the conclusion of the present theorem. □

## 5.2 Perturbing the Indiscernibles

In this section we consider generic classes which explicitly do *not* preserve indiscernibles. The indiscernibles relative to a class which is generic over $L$ can be very chaotic and we shall exploit this fact to show how $0^\#$ can be approximated by such classes. However the indiscernibles relative to a *real* generic over $L$ are "eventually periodic" in $I$ (in the sense of Corollary 4.20), by a result of Paris [74]. We first show that the Paris result is optimal. A further application of the method is to show that an inner model $L[R]$, $R \subseteq \omega$ of a class-generic extension of $L$ need not itself be such an extension.

Let $\langle i_\alpha \mid \alpha \in \text{ORD}\rangle$ be the increasing enumeration of $I$.

**Theorem 5.18.** *Suppose $\alpha, \beta$ are countable ordinals, $\beta$ nonzero. Then there exists a* CUB $C \subseteq$ ORD *such that $C \cap I = I_{\alpha,\beta} = \{i_{\alpha+\beta\cdot\gamma} \mid \gamma \in \text{ORD}\}$ and $I_{\alpha,\beta}$ is a class of indiscernibles for $\langle L[C], C\rangle$.*

*Proof.* Consider the reverse Easton iteration $P$ with Easton support, where $P(\leq \alpha) = P(< \alpha) * P(\alpha)$, $P(< \alpha) \Vdash P(\alpha)$ is the forcing to add a CUB subset of $\alpha$ if $\alpha$ is regular, $P(\alpha)$ is the trivial forcing, otherwise. In the nontrivial case, $P(\alpha)$ consists of bounded closed subsets of $\alpha$, ordered by $p \leq q$ iff $p$ end extends $q$. As with the basic example of Reverse Easton forcing, $P$ preserves cofinalities and the GCH.

We define a $P(< i_{\alpha+\beta\cdot\omega})$-generic $G(< i_{\alpha+\beta\cdot\omega})$ so that $n < m \implies G(i_{\alpha+\beta\cdot n})$ is an initial segment of $G(i_{\alpha+\beta\cdot m})$: $G(< i_\alpha)$ is an arbitrarily chosen $P(< i_\alpha)$-generic, which exists by the countability of $(i_\alpha^+)^L$. Then $G(i_\alpha)$ is chosen to be $P(i_\alpha)$-generic (relative to the ground model $L[G(< i_\alpha)]$) and disjoint from $I$. Such a $G(i_\alpha)$ exists, by the following argument: Our property is automatic if $\alpha = 0$. If $\alpha = \beta + 1$ then choose $G(i_\alpha)$ to contain the condition $\{i_\beta + 1\}$. Otherwise we can list the relevant dense sets $D_0, D_1, \ldots$ and choose $\delta_0 < \delta_1 < \cdots$ cofinal in $i_\alpha$ so that $D_n \cap L_{\delta_n}[G(< i_\alpha)]$ is dense in $P(i_\alpha) \cap L_{\delta_n}[G(< i_\alpha)]$ for each $n$ and no $\delta_n$ belongs to $I$; then choose $G(i_\alpha)$ to be the limit of $p_0 \geq p_1 \geq \cdots$ where $p_n$ meets $D_n$ and $p_n - p_{n-1}$ is contained in the interval $(\sup(I \cap \delta_n), \delta_n)$. Then $G(i_\alpha) \cap I = \emptyset$ and $G(i_\alpha)$ is $P(i_\alpha)$-generic (relative to $L[G(< i_\alpha)]$).

Next choose $G(< i_{\alpha+\beta})$ to be an arbitrary $P(< i_{\alpha+\beta})$-generic extending $G(\leq i_\alpha)$. Now, similarly to the preceding paragraph, select $G(i_{\alpha+\beta})$ to be $P(i_{\alpha+\beta})$-generic (relative to $L[G(< i_{\alpha+\beta})]$), extending the condition $G(i_\alpha) \cup \{i_\alpha\}$ and disjoint from $I \cap (i_\alpha, i_{\alpha+\beta})$. Note that $G(i_{\alpha+\beta})$ need only meet dense $D$ definable in $\langle L[G(< i_{\alpha+\beta})], G(< i_{\alpha+\beta})\rangle$ from elements of $I - i_\alpha$, by the $i_\alpha^+$-closure of $P(i_{\alpha+\beta})$. Hence $G(i_{\alpha+\beta})$ can be selected as the limit of $p_0 \geq p_1 \geq \cdots$ where we have fixed a decomposition $\bigcup_n F_n$ of $\beta$ into finite subsets and $p_n$ is definable from $\{i_{\alpha+\beta'} \mid \beta' \in F_n\} \cup \{i_{\alpha+\beta}, \ldots, i_{\alpha+\beta+n}\}$.

The rest of $G(< i_{\alpha+\beta\cdot\omega})$ is now determined by the requirements of the theorem, much as in the proof of Lemma 5.13: For each $n$ let $\pi_n : L_{i_{\alpha+\beta\cdot\omega}} \longrightarrow L_{i_{\alpha+\beta\cdot\omega}}$ have critical point $i_{\alpha+\beta\cdot n}$ and be defined by sending $i_{\alpha+\beta\cdot k+\beta'}$ to $i_{\alpha+\beta\cdot k+\beta+\beta'}$ for $k \geq n$ and $\beta' < \beta$. Then $G(\leq i_{\alpha+\beta}), \pi_0[G(\leq i_{\alpha+\beta})]$ determine a $P(\leq i_{\alpha+\beta+\beta})$-generic $G(\leq i_{\alpha+\beta+\beta}) = \{p \in P(\leq i_{\alpha+\beta+\beta}) \mid p$ is extended by the greatest lower bound of $q, \pi_0(q)$ for some $q \in G(\leq i_{\alpha+\beta})\}$, as $\pi_0$ maps $I \cap [i_\alpha, i_{\alpha+\beta})$ onto $I \cap [i_{\alpha+\beta}, i_{\alpha+\beta+\beta})$. And $G(\leq i_{\alpha+\beta+\beta}), \pi_1[G(\leq i_{\alpha+\beta+\beta})]$ similarly determine a $P(\leq i_{\alpha+\beta+\beta+\beta})$-generic; continue in this way to obtain $G(< i_{\alpha+\beta\cdot\omega})$ so as to be preserved by each of the $\pi_n$'s. It follows that $I_{\alpha+\beta} \cap i_{\alpha+\beta\cdot\omega}$ is a set of indiscernibles for $\langle L_{i_{\alpha+\beta\cdot\omega}}[G(< i_{\alpha+\beta\cdot\omega})], G(< i_{\alpha+\beta\cdot\omega})\rangle$. Finally, exactly as in the proof of Lemma 5.14, "stretch" $G(< i_{\alpha+\beta\cdot\omega})$ to a $P$-generic $G$ and let $C = \bigcup\{G(i_{\alpha+\beta\cdot\gamma}) \mid \gamma \in \text{ORD}\}$. □

**Corollary 5.19.** (a) *Suppose that $\alpha$ and $\beta$ are countable ordinals, $\beta$ nonzero. Then there is a real $R$ such that $I^R = I_{\alpha,\beta}$.*

(b) *If $\alpha$ and $\beta$ are countable ordinals, $\beta$ nonzero, and $I_{\alpha,\beta}$ a class of indiscernibles for $\langle L[A], A\rangle$ then there is a real $R$ such that $I^R = I_{\alpha,\beta}$ and $A$ is definable in $L[R]$.*

*Proof.* Let $C$ be as in the previous theorem, and now add a class $A$, generically over $\langle L[C], C\rangle$ such that $C$ is definable over $\langle L[A], A\rangle$ and whenever $\alpha < \beta$ are adjacent in $C$, then in $L[A]$, the cardinality of $\beta$ is at most $(\alpha^+)^L$. This can be arranged using another reverse Easton iteration, preserving the indiscernibility of $I_{\alpha,\beta}$.

Note that $I_{\alpha,\beta}$ is a *generating* class of indiscernibles for $\langle L[A], A\rangle$. Indeed, if not then the least ordinal $\delta$ not in the Hull of $I_{\alpha,\beta}$ in $\langle L[A], A\rangle$ belongs both to $I$ and to $C$, contradicting the defining property of $C$.

Now apply Corollary 4.20. (b) is obtained by relativizing to $A$. □

Question: Are there reals as in the previous corollary that preserve cofinalities?

The converse is due to Paris [74].

**Theorem 5.20.** *If $0^\# \notin L[R]$ then $I^R = I_{\alpha,\beta}$ for some countable $\alpha$, $\beta$.*

*Proof.* Let $I^R = \{i^R_\alpha \mid \alpha \in \mathrm{ORD}\}$ and let $\pi_{mn} : L[R] \longrightarrow L[R]$ be the elementary embedding determined by $\pi_{mn} \restriction i^R_m = \text{identity}$, $\pi_{mn}(i^R_{\alpha+m}) = i^R_{\alpha+n}$, for $m \le n$. We claim that $\pi_{01}$ maps $I \cap (i^R_0, i^R_1)$ onto $I \cap (i^R_1, i^R_2)$. Now $i \in I \iff \pi_{01}(i) \in I$; so suppose that $\gamma = t^R(i^R_0, i^R_1, \infty) \in I \cap (i^R_1, i^R_2)$, $\gamma \notin \mathrm{Range}\,\pi_{01}$. This means that $\gamma$ is not of the form $s^R(\beta, i^R_1, \infty)$ for any term $s^R$, $\beta < i^R_0$.

Let $\delta_n = t^R(i^R_n, i^R_\omega, \infty) \in (i^R_\omega, i^R_{\omega+1})$, and $\delta = \cup_n \delta_n$. Then the $\delta_n$ are distinct as for each $n$, $\delta_n$ is not of the form $t^R(\beta, i^R_\omega, \infty)$ for any $\beta < i^R_n$. So as $\delta$ is a limit point of $\{t^R(\beta, i^R_\omega, \infty) \mid \beta < i^R_\omega\}$ we conclude that $L[R] \models \mathrm{cof}(\delta) \le i^R_\omega$. Also as each $\delta_n$ belongs to $I$, so does $\delta$, so $\delta$ is regular in $L$.

However by the Covering Theorem, $\mathrm{cof}^L(\delta) < (i^R_\omega)^+$ of $L$. For a contradiction we claim that $(i^R_\omega)^+$ of $L = (i^R_\omega)^+$ of $L[R]$: Indeed suppose that $\delta = (i^R_\omega)^+$ of $L$ has cardinality $i^R_\omega$ in $L[R]$. Let $\pi : L[R] \longrightarrow L[R]$ be an elementary embedding that fixes ordinals $< i^R_\omega$ and sends $i^R_\omega$ to $i^R_{\omega+1}$. Then $\pi \restriction L_\delta : L_\delta \longrightarrow L_{\pi(\delta)}$ belongs to $L[R]$ and hence we get $0^\# \in L[R]$.

So $\pi_{01}$ maps $I \cap (i^R_0, i^R_1)$ onto $I \cap (i^R_1, i^R_2)$ and the same argument works for $\pi_{n,n+1}$, any $n > 0$. Now by induction on $n$ it follows that $\pi_{0,n}$ maps $I \cap (i^R_0, i^R_1)$ onto $I \cap (i^R_n, i^R_{n+1})$. So we get that $I^R = I_{\alpha,\beta}$ where $i^R_0 = i_\alpha$ and $\beta =$ ordertype of $I \cap [i^R_0, i^R_1)$. □

## Strict Genericity

The principal result of this subsection is that an inner model of a generic extension of $L$ need not itself be such an extension, even when it is of the form $L[R]$, $R \subseteq \omega$.

**Definition.** Let $\langle M, A\rangle$ be a ground model. $G$ is *literally generic* over $\langle M, A\rangle$ if for some tame $P$ defined over $\langle M, A\rangle$, $G$ is $P$-generic over $\langle M, A\rangle$. $S$ is *generic over $M$* if for some ground model $\langle M, A\rangle$ and $G$ literally generic over $\langle M, A\rangle$, $S$ is definable over $\langle M[G], A, G\rangle$. $S$ is *strictly generic over $M$* if we also require that $G$ be definable over $\langle M[S], A, S\rangle$.

The next result states that generic implies strictly generic for set forcing.

**Proposition 5.21.** *If $G$ is $P$-generic over $\langle M, A\rangle$ where $P$ is an element of $M$ and $S$ is definable over $\langle M[G], A\rangle$ then $S$ is strictly generic over $M$.*

*Proof.* We shall make use of the Boolean-valued approach to forcing; for details see Jech [78]. First note that we may assume that $S \subseteq \alpha$ for some $\alpha \in \mathrm{ORD}(M)$ as the fact that $P$ belongs to $M$ implies that $S$ is definable over $\langle M[S \cap V_\alpha], A\rangle$ for some $\alpha \in \mathrm{ORD}(M)$ and $M$ satisfies choice. We may also assume that $P$ is a complete Boolean algebra. Let $\sigma$ be a name such that $\sigma^G = S$ and consider the complete subalgebra $P_0$ of $P$ generated by the Boolean values of the sentences "$\hat{\beta} \in \sigma$", for $\beta < \alpha$. Then $G_0 = G \cap P_0$ is $P_0$-generic over $M$ and witnesses that $S$ is strictly generic over $M$. □

Now note the following necessary condition for strict genericity.

**Proposition 5.22.** *If $S$ is strictly generic over $M$ then for some $A$, $\langle M, A\rangle$ is amenable and* $\mathrm{Sat}\langle M[S], S, A\rangle$ *is definable over* $\langle M[S, \mathrm{Sat}\langle M, A\rangle], S, \mathrm{Sat}\langle M, A\rangle\rangle$, *where* Sat *denotes the satisfaction relation.*

*Proof.* Suppose that $A$, $G$ witness that $S$ is strictly generic over $M$, via the tame forcing $P$ defined over $\langle M, A\rangle$. Then we have: $\langle M[S], S, A\rangle \models \varphi$ iff $\exists p \in G(p \Vdash \varphi^*)$, where the operation $\varphi \longmapsto \varphi^*$ is $\langle M, A\rangle$-definable. As $G$ is definable over $\langle M[S], S, A\rangle$ and $\Vdash$ is definable over $\langle M[\mathrm{Sat}\langle M, A\rangle], \mathrm{Sat}\langle M, A\rangle\rangle$ the result follows. □

**Corollary 5.23.** *If $S$ is strictly generic over $L$ then for some $L$-amenable $A$,* $\mathrm{Sat}\langle L[S], S\rangle$ *is definable over $\langle L[S], S, A\rangle$.*

*Proof.* Note that if $A$ is $L$-amenable then so is $\mathrm{Sat}\langle L, A\rangle$, as we have assumed that $0^\#$ exists. The result now follows directly from Proposition 5.22. □

Our strategy for producing $S$ which is generic but not strictly generic over $L$ is to violate the conclusion of Corollary 5.23. First we accomplish this with a class $S$ and afterwards with a real.

**Theorem 5.24.** *There is a class $S$, definable in $L[0^\#]$, which is generic but not strictly generic over $L$.*

*Proof.* We produce $S$ via the Reverse Easton iteration defined as follows. $P(0)$ is the trivial forcing and for limit $\lambda \leq \infty$, $P(<\lambda) = \mathrm{Direct\,Lim}\langle P(<\alpha) \mid \alpha < \lambda\rangle$ for $\lambda$ regular, $P(<\lambda) = \mathrm{Inverse\,Lim}\langle P(<\alpha) \mid \alpha < \lambda\rangle$ for $\lambda$ singular. For $\alpha > 0$, $P(\leq \alpha) = P(<\alpha) * Q(\alpha)$ where $Q(\alpha)$ is defined as follows. If $\alpha$ is not inaccessible then $Q(\alpha)$ is the trivial forcing and otherwise a condition in $Q(\alpha)$ is an $\omega$-sequence $p = \langle p(0), p(1), \ldots \rangle$, where for some $\alpha(p) < \alpha$, each $p(n)$ is either a function $p(n) : \alpha(p) \longrightarrow 2$ or is empty. To define extension of conditions in $Q(\alpha)$ first let $\langle \cdot, \cdot \rangle$ be a canonical pairing function on ORD and for $\gamma \in$ ORD, $b_\gamma = \{2\langle \gamma, \delta \rangle + 1 \mid \delta \in \text{ORD}\}$. Then $p \leq q$ iff $\alpha(p) \geq \alpha(q)$, $p(n)$ extends $q(n)$ for each $n$ and: $q(n+1)(\gamma) = 1$, $\beta \in b_\gamma \cap [\alpha(q), \alpha(p)) \Longrightarrow p(n)(\beta) = 0$. Thus if $G$ is $Q(\alpha)$-generic and we let $G(n) = \cup\{p(n) \mid p \in G\}$ we get: $G(n+1)(\gamma) = 1$ iff $G(n)(\beta) = 0$ for sufficiently large $\beta \in b_\gamma$.

We now build a particular $P = P(\leq \infty)$-generic $G(\leq \infty)$ and the desired $S$ will be $G(\infty)(0) = \cup\{p(0) \mid p \in G(\infty)\}$. $G(\leq i_\alpha)$ is defined by induction on $\alpha \leq \infty$ where $\langle i_\alpha \mid \alpha \leq \infty \rangle$ is the increasing enumeration of $I \cup \{\infty\}$. $G(\leq i_0)$ is the $L[0^\#]$-least $P(\leq i_0)$-generic and for limit $\lambda \leq \infty$ : $G(< i_\lambda) = \cup\{G(< i_\alpha) \mid \alpha < \lambda\}$, $G(i_\lambda)(n) = \cup\{G(i_{2\alpha})(n) \mid \alpha < \lambda\}$ for each $n$. (We shall define $G(i_{2\alpha})$ so that the $G(i_{2\alpha})(n)$ for $\alpha < \lambda$ are compatible, for each $n$.)

Suppose that $G(\leq i_{\lambda+n})$ is defined, $\lambda$ limit or 0, $n \in \omega$ and we wish to define $G(\leq i_{\lambda+n+1})$. First suppose that $n$ is even. $G(< i_{\lambda+n+1})$ is the $L[0^\#]$-least $P(< i_{\lambda+n+1})$-generic extending $G(\leq i_{\lambda+n})$. To define $G(i_{\lambda+n+1})$ first form the condition $p \in Q(i_{\lambda+n+1})$ (of $L[G(< i_{\lambda+n+1})]$) defined by: $\alpha(p) = i_{\lambda+n} + 1$, $p(m) \restriction i_{\lambda+n} = G(i_{\lambda+n})(m)$ for all $m$, $p(m)(i_{\lambda+n}) = 1$ iff $m > n$. Then $G(i_{\lambda+n+1})$ is the $L[0^\#]$-least $Q(i_{\lambda+n+1})$-generic containing the condition $p$. If $n$ is odd, then again $G(< i_{\lambda+n+1})$ is the $L[0^\#]$-least $P(< i_{\lambda+n+1})$-generic extending $G(\leq i_{\lambda+n})$. To define $G(i_{\lambda+n+1})$, first form the condition $p \in Q(i_{\lambda+n+1})$ (of $L[G(< i_{\lambda+n+1})]$) defined by: $\alpha(p) = i_{\lambda+n}$, $p(m)(\gamma) = G(i_{\lambda+n})(m)(\gamma)$ for $\gamma \neq i_{\lambda+n-1}$ and $p(m)(i_{\lambda+n-1}) = 0$ for all $m$. Then $G(i_{\lambda+n+1})$ is the $L[0^\#]$-least $Q(i_{\lambda+n+1})$-generic extending $p$. This complete the definition of $G(\leq \infty)$.

For $i \in I \cup \{\infty\}$ and $n \in \omega$ let $S_n(i) = \cup\{p(n) \mid p \in G(i)\}$ and $S(i) = S_0(i)$, $S = S(\infty)$. Note that $S_m(i_{\lambda+n}) = S_m(\infty) \cap i_{\lambda+n}$ for $m \leq n$ and this holds for all $m$ if $n$ is even ($\lambda$ limit or 0). We now proceed to show that $S$ is not strictly generic over $L$.

**Definition.** For $X \subseteq$ ORD, $\alpha \in$ ORD and $n \in \omega$ we say that $\alpha$ is $X$-$\Sigma_n$ *stable* if $\langle L_\alpha[X], X \cap \alpha \rangle$ is $\Sigma_n$-elementary in $\langle L[X], X \rangle$. $\alpha$ is $X$-*stable* if $\alpha$ is $X$-$\Sigma_n$ stable for all $n$.

**Lemma 5.25.** *For $\lambda$ limit or 0, $n$ even, $i_{\lambda+n+1}$ is not $S$-stable.*

*Proof of Lemma* 5.25. Let $i = i_{\lambda+n}$ and $j = i_{\lambda+n+1}$. Note that for each $m$, $S_m(j)$ is defined from $S_0(j) = S \cap j$ exactly as $S_m(\infty)$ is defined from $S$. But for $m > n$, $i \in S_m(j) - S_m(\infty)$, so $j$ is not $S$-stable □ (Lemma 5.25)

**Lemma 5.26.** *For $L$-amenable $A \subseteq$ ORD, $i_{\lambda+n+1}$ is $(S, A)$-$\Sigma_n$ stable for all $n$, for sufficiently large limit $\lambda$.*

*Proof of Lemma* 5.26. Let $i = i_{\lambda+n+1}$ where $\lambda$ is large enough to guarantee that $i$ is $A$-stable. (This is possible as for some $\vec{j}_0$ from $I$ and term $t$, $A \cap j = t(\vec{j}_0, j, \vec{j}_1)$ for all $\vec{j}_0 < j < \vec{j}_1$ in $I$ of the right length; then $j$ is $A$-stable for all $j \in I$, $j > \vec{j}_0$.)

For $p \in P(\leq i) = P(< i) * Q(i)$ and $m \in \omega$, let $(p)_m$ be obtained from $p$ by redefining $p(i)(m')$ to be $\emptyset$ for $m' > m$, and otherwise leaving $p$ unchanged.

**Claim.** Suppose $\varphi$ is $\Pi_m$ with parameters relative to $S(i)$, $B$, where $B \subseteq i$ is constructible. If $p \in P(\leq i)$ and $p \Vdash \varphi$ then $(p)_m \Vdash \varphi$.

*Proof of Claim.* By induction on $m \geq 1$. For $m = 1$ write $\varphi$ as $\forall x \psi$, $\psi$ $\Sigma_0$; if the conclusion failed then we could choose $q \leq (p)_1, q(< i) \Vdash (q(i) \Vdash\sim \psi(x)$ for some $x \in L_{\alpha(q)}[G(< i)])$; then $(q)_0 \Vdash\sim \psi(x)$ for some $x$ and $(q)_0$ is compatible with $p$. This contradicts the hypothesis that $p \Vdash \varphi$. Given the result for $m$, write a given $\Pi_{m+1}$ $\varphi$ as $\forall x \psi$, $\psi$ $\Sigma_m$ and then if the conclusion failed we could choose $q \leq (p)_{m+1}$ such that $q(< i) \Vdash (q(i) \Vdash\sim \psi(x)$ for some $x \in L_{\alpha(q)}[G(< i)])$. By induction we get $(q)_m \Vdash\sim \psi(x)$ for some $x$, in contradiction to the facts that $(q)_m, p$ are compatible and $p \Vdash \varphi$. □ (Claim)

Now we prove the lemma. Suppose $\varphi$ is $\Pi_n$ and true of $(S(i), A \cap i)$. Choose $p \in G(\leq i), p \Vdash \varphi$. By the Claim, $(p)_n \Vdash \varphi$ (in $P(\leq i)$). Now we may also canonically view $(p)_n$ as a condition in $P(\leq \infty)$ and in fact by construction, $(p)_n \in G(\leq \infty)$ in the sense that $(p)_n(< i) \in G(< i) \subseteq G(< \infty)$ and $(p)_n(i) \in G(\infty)$. By the $A$-stability of $i$, $(p)_n \Vdash \varphi$ in $P(\leq \infty)$ so $\varphi$ is true of $(S, A)$. □ (Lemma 5.26)

Now to complete the proof of Theorem 5.24, we argue as follows. If $S$ were strictly generic over $L$ then for some $L$-amenable $A$ we would have $\text{Sat}\langle L[S], S\rangle$ definable over $\langle L[S], S, A\rangle$. But then for some $n$, all sufficiently large $(S, A)$-$\Sigma_n$ stables would necessarily be $S$-stable. This contradicts the previous two lemmas. □

To obtain a generic, not strictly generic real we must first refine the construction of Theorem 5.24.

**Theorem 5.27.** *Let $\langle A(i) \mid i \in I\rangle$ be a sequence such that $A(i)$ is a constructible subset of $i$ for each $i \in I$. Then there exists $S$ obeying Lemmas* 5.25, 5.26 *such that in addition, $A(i)$ is definable over $\langle L_i[S], S \cap i\rangle$ for $i \in \text{Odd}(I) = \{i_{\lambda+n} \mid \lambda$ limit or* 0, *$n$* odd$\}$.

*Proof.* We use a slightly different Reverse Easton iteration: $Q(\alpha)$ specifies $n(\alpha) \leq \omega$ and if $n(\alpha) < \omega$, it also specifies a constructible $A(\alpha) \subseteq \alpha$; then conditions and extension are as before, except we now require that if $n(\alpha) < \omega$ then for $p$ to extend $q$, we must have $p(n(\alpha))(2\beta + 2) = 1$ iff $\beta \in A(\alpha)$, for $2\beta + 2 \in [\alpha(q), \alpha(p))$. Thus if $n(\alpha) < \omega$, the $Q(\alpha)$-generic will code $A(\alpha)$ definably (though the complexity of the definition depends upon $n(\alpha) < \omega$).

Now in the construction of $G(\leq i_\alpha)$, $\alpha \leq \infty$ we proceed as before, with the following additional specifications: $n(i_{\lambda+n}) = n$ for odd $n$ and $n(i_{\lambda+n}) = \omega$ for even $n$ ($\lambda$ limit or 0). And for odd $n$ we specify $A(i_{\lambda+n})$ to be the $A(i), i = i_{\lambda+n}$ as given in the hypothesis of the theorem.

Lemma 5.25 holds as before; we need a new argument for Lemma 5.26. Note that for $i \in \text{Odd}(I)$ it is no longer the case that $P(< i) \Vdash Q(i) = Q(\infty) \cap L_i[G(< i)]$. Let

$Q^*(i)$ denote $Q(\infty) \cap L_i[G(< i)]$, i.e., the forcing $Q(i)$ where $n(i)$ has been specified as $\omega$. Define $(p)_m$ as before for $p \in P(\le i)$.

**Claim.** Suppose $m \le n+1$, $n$ is even, $i = i_{\lambda+n+1}$ ($\lambda$ limit or 0) and $\varphi$ is $\Pi_m$ with parameters relative to $S(i)$, $B$, where $B \subseteq i$ is constructible. If $p \in P(\le i)$ (where $n(i) = n+1$) then $p \Vdash \varphi$ in $P(\le i)$ iff $(p)_m \Vdash \varphi$ in $P^*(\le i) = P(< i) * Q^*(i)$ iff $p \Vdash \varphi$ in $P^*(\le i)$.

*Proof of Claim.* This is as in the proof of the corresponding Claim in the proof of Lemma 5.26. If $m = 1$ and $p \Vdash \varphi$ in $P(\le i)$, then if the conclusion failed, we could choose $q \le (p)_1$ in $P^*(\le i)$, $q \Vdash\sim \varphi$; then (we can assume) $(q)_0 \Vdash\sim \varphi$ in $P(\le i)$, but $(q)_0$ and $p$ are compatible. The other implications are clear, as $P(\le i) \subseteq P^*(\le i)$. Given the result for $m \le n$, $\varphi$ $\Pi_{m+1}$ and $p \Vdash \varphi$ in $P(\le i)$, if the conclusion failed we could choose $q \le (p)_{m+1}$ in $P^*(\le i)$, $q \Vdash\sim \varphi$ (indeed, $q \Vdash\sim \psi(x)$ some $x$, where $\varphi = \forall x \psi$, $\psi$ $\Sigma_m$); then $q \Vdash\sim \varphi$ in $P(\le i)$, $(q)_m \Vdash\sim \varphi$ in $P^*(\le i)$, $(q)_m \Vdash\sim \varphi$ in $P(\le i)$ by induction. But this is a contradiction, as $(q)_m$, $p$ are compatible in $P(\le i)$, using the fact that $m \le n$ and $q \le (p)_{m+1}$. And again the other implications follow, as $P(\le i) \subseteq P^*(\le i)$. □ (Claim)

Now we can apply the proof of Lemma 5.26, using the new version of the Claim. □

The choice of $\langle A(i) \mid i \in I \rangle$ that we have in mind comes from the next proposition.

**Proposition 5.28.** *For each n let $A_n = \{\alpha \mid$ For $i < j_1 < \cdots < j_n$ in $I$ with $\alpha < i$, the sequences $(\alpha, j_1, \ldots, j_n)$ and $(i, j_1, \ldots, j_n)$ satisfy the same formulas in $L$ with parameters $< \alpha\}$. Then any $L$-amenable $A$ is $\Delta_1$-definable over $\langle L, A_n \rangle$ for some $n$.*

*Proof.* For some $\vec{\jmath}_0$ from $K$ and term $t$, $A \cap L_\jmath = t(\vec{\jmath}_0, \jmath, \vec{\jmath}_1)$ for all $\vec{\jmath}_0 < \jmath < \vec{\jmath}_1$ in $I$ of the right length. Let $n$ be the length of these $\vec{\jmath}_1$; we show that $A$ is $\Delta_1$-definable over $\langle L, A_{n+1} \rangle$. It suffices to show that for $\vec{\imath} < \vec{\jmath}$ increasing sequences from $A_{n+1}$ of length $n+1$, $\vec{\imath}$ and $\vec{\jmath}$ satisfy the same formulas in $L$ with parameters $< \min(\vec{\imath})$. But by definition, for $\alpha < \min(\vec{\imath})$ and $\vec{\imath} = \{i_0, \ldots, i_n\}$, $\vec{\jmath} = \{j_0, \ldots, j_n\}$ we get: $L \models \varphi(\alpha, j_0, \ldots, j_n) \iff \varphi(\alpha, i_0, j_1, \ldots, j_n) \iff \varphi(\alpha, i_0, i_1, j_2, \ldots, j_n) \iff \cdots \iff \varphi(\alpha, i_0, \ldots, i_n)$. □

Now for $i \in I$ write $i = i_{\lambda+n}$, $\lambda$ limit or 0, $n \in \omega$ and let $A(i) = A_n \cap i$. Thus by Theorem 5.27 there is $S$ obeying Lemmas 5.25, 5.26 such that $A_n \cap i$ is definable over $\langle L_i[S], S \cap i \rangle$ for $i = i_{\lambda+n+1}$, $n$ even.

**Theorem 5.29.** *There is a real $R \in L[0^\#]$ which is generic but not strictly generic over $L$.*

*Proof.* First observe that we may build $G(\le \infty)$ to satisfy Theorem 5.27 for the preceding choice of $\langle A(i) \mid i \in I \rangle$ and in addition preserve the indiscernibility of (the generating indiscernibles) Lim $I$. Then we may code $(G(< \infty), S)$ by a real $R$, where $S = G_0(\infty)$. The resulting $R$ obeys Lemma 5.25 because $S$ is definable from $R$; to obtain Lemma 5.26 for $R$ we must modify the coding of $(G(< \infty), S)$ by $R$ in the following way: For inaccessible $\kappa$ we require that any coding condition with $\kappa$ in its domain reduce any dense $D \subseteq P^{<\kappa} = \{q \mid \alpha(q) < \kappa\}$ strictly below $\kappa$, when $D$ is definable over $\langle L_\kappa[G(<\infty), S], G(<\kappa), S \cap \kappa\rangle$. This extra requirement does not interfere with the proofs of extendibility and distributivity for the coding conditions.

Now to obtain Lemma 5.26 for $R$ argue as follows: Given $L$-amenable $A$, choose $n$ and $\lambda$ large enough so that $A$ is $\Delta_1$-definable from $A_n$ with parameters $< i_\lambda$. Then $i_{\lambda+n+1}$ is $(G(<\infty), S, A)$-$\Sigma_n$ stable. And also $A \cap i_{\lambda+n+1}$ is definable over $\langle L_i[G(<i), S\cap i], G(<i), S\cap i\rangle$ where $i = i_{\lambda+n+1}$. Thus if $\varphi$ is $\Pi_n$ and true of $G(<i), S\cap i, A\cap i$ then $\varphi$ is forced by some coding condition $p \in P^{<i}$ ($p$ in the generic determined by $R$) and hence by the $(G(<\infty), S, A)$-$\Sigma_n$ stability of $i$, we get that $\varphi$ is true of $G(<\infty), S, A$. □

We built $R$ as in Theorem 5.29 by perturbing the indiscernibles. However with extra care we can in fact obtain indiscernible preservation.

**Theorem 5.30.** *There is a real $R \in L[0^\#]$ such that $R$ is generic but not strictly generic over $L$, $R$ preserves $L$-cofinalities and $I^R = I$.*

*Proof.* Instead of using the $i_{\lambda+n}$, $n \in \omega$ ($\lambda$ limit or 0) use the $i^n_\alpha$, $n \in \omega$ where $i^n_\alpha =$ least element of $A_n$ greater than $i_\alpha$. Thus $\bigcup\{i^n_\alpha \mid n \in \omega\} = i_{\alpha+1}$ and as above we can construct $S$ to preserve indiscernibles and $L$-cofinalities and to satisfy: (a) No $i^n_\alpha$, $n$ odd is $S$-stable; (b) For any $L$-amenable $A$, $i^{n+1}_\alpha$ is $(S, A)$-$\Sigma_n$ stable for large enough $\alpha, n$; and (c) $A_n \cap i^{n+1}_\alpha$ is definable over $\langle L_i[S], S \cap i\rangle$ for $i = i^{n+1}_\alpha$, $n$ even. Then code $(G(<\infty), S)$ by a real, preserving indiscernibles and cofinalities, requiring as before that for inaccessible $\kappa$, any coding condition with $\kappa$ in its domain reduces dense $D \subseteq P^{<\kappa}$ strictly below $\kappa$, when $D$ is definable over $\langle L_\kappa[G(<\kappa), S\cap\kappa], G(<\kappa), S\cap\kappa\rangle$. Then for any $L$-amenable $A$, $i^{n+1}_\alpha$ will be $(R, A)$-$\Sigma_n$ stable for sufficiently large $\alpha, n$. This implies as before that $R$ is not strictly generic. □

**Remark.** A similar argument shows: For any $n \in \omega$ there is a real $R \in L[0^\#]$ which is strictly generic over $L$, yet $G$ is not $\Sigma_n\langle L[R], R, A\rangle$ whenever $R \in L[G]$ and $G$ is literally generic over $\langle L, A\rangle$. Thus there is a strict hierarchy within strict genericity, given by the level of definability of the literally generic $G$ from the strictly generic real.

## Is $0^\#$ Generic?

An interesting open question is whether $0^\#$ is generic over some inner model $M$, $0^\# \notin M$. In this subsection we describe a definability condition sufficient to imply an affirmative answer to this question, and study it in the context of the minimal model of ZFC $+0^\#$ exists. We show that in some sense $0^\#$ is "close" to being generic over a proper inner model of $L[0^\#]$ and in addition there is a predicate $A \subseteq$ ORD not constructing $0^\#$ which "minimizes" the universe in the sense that no $\langle L_\alpha[A], A \cap \alpha\rangle$ is elementary in $\langle L[A], A\rangle$.

**Definition.** Suppose $\alpha < \beta$, $A \subseteq \mathrm{ORD}$, $n \in \omega$. We say that $\alpha$ is *$A, \beta$-$\Sigma_n$ stable*, written $\alpha <_n^A \beta$, if $\langle L_\alpha[A], A \cap \alpha\rangle$ is $\Sigma_n$-elementary in $\langle L_\beta[A], A \cap \beta\rangle$. Also let $A_\beta^{(n)}$ denote the $\Sigma_n$-satisfaction predicate for $\langle L_\beta[A], A \cap \beta\rangle$.

**Proposition 5.31.** *Suppose $A \subseteq \mathrm{ORD}$ and there is an $\langle L[A], A\rangle$-definable relation $S(n, \alpha, \beta)$ such that:*

(a) *Whenever $i <_n^{0^\#} j$ and $i, j$ are $<_n^{0^\#}$-*limits *then $S(n, i, j)$.*

(b) *Whenever $i < \beta$, $i$ belongs to $I$, $\beta$ $L$-regular and $S(n+1, i, \beta)$, then $i <_1^{0^\#, A_\beta^{(n)}} \beta$.*

*Then $0^\#$ is strictly generic over $L[A]$.*

*Proof.* In the language having names for elements of $L[A]$ as well as a name $\hat{R}$ for $0^\#$, let $T$ consist of all $\Delta_0$ sentences of the form:

$$\text{“}\alpha\ \hat{R}\text{-admissible} \Longrightarrow \alpha <_1^{\hat{R}, A_\beta^{(n)}} \beta\text{”}$$

whenever $\alpha < \beta$ are $L$-regular and $S(n+1, \alpha, \beta)$. A $\Delta_0$ sentence $\varphi$ is consistent with $T$ if for each $\alpha$ there is a solution to $(T \cap L_\alpha[A]) \cup \{\varphi\}$ in a set-generic extension. Let $P$ be all equivalence classes $[\varphi]$ of $\Delta_0$ sentences $\varphi$ consistent with $T$, under the equivalence relation $\varphi_0 \sim \varphi_1$ iff $\varphi_0 \not\Longleftrightarrow \varphi_1$ is inconsistent with $T$. Order $P$ by $[\varphi_0] \le [\varphi_1]$ iff $(\varphi_0 \wedge \sim \varphi_1)$ is inconsistent with $T$.

Let $G = \{[\varphi] \mid \varphi \text{ is true when } \hat{R} = 0^\#\}$. Then $G$ is $P$-generic over $\langle L[A], A\rangle$: If not then choose an $\langle L[A], A\rangle$-definable predense $\Delta \subseteq P$, $\Delta \cap G \neq \emptyset$. Suppose that $\Delta^* = \{\varphi \mid [\varphi] \in \Delta\}$ is $\Sigma_n\langle L[A], A\rangle$ and choose $i <_{n+2}^{A,0^\#} < j <_{n+2}^{A,0^\#} < \infty$, where all parameters in the definition of $\Delta$ are $< i$. Let $\varphi = \wedge\{\sim \psi \mid \psi \in \Delta^* \cap L_i[A]\}$, where $\wedge$ denotes conjunction. Then using the axioms in $T$ and $S(n+1, i, j)$, we have that $T \cup \{\varphi\} \cup \{\psi\}$ is inconsistent for $\psi \in L_j[A] \cap \Delta^*$. As $j$ can be chosen arbitrarily large we get that $[\varphi]$ is incompatible with each condition in $\Delta$, in contradiction to the predensity of $\Delta$. Note that $\varphi$ is consistent with $T$ since $T \cup \{\varphi\} \subseteq G^* = \{\gamma \mid [\gamma] \in G\} = \{\gamma \mid \gamma$ is true when $\hat{R} = 0^\#\}$. □

Now let us focus on the minimal model of ZFC $+0^\#$ exists. First note the following.

**Proposition 5.32.** *Suppose that there is no transitive set model of* ZFC $+0^\#$ *exists and $0^\#$ is generic via a forcing definable over $\langle L[A], A\rangle$. Then for sufficiently large $\alpha$, $\langle L_\alpha[A], A \cap \alpha\rangle$ is not elementary in $\langle L[A], A\rangle$.*

*Proof.* Otherwise choose $\langle L_\alpha[A], A \cap \alpha\rangle$ to be elementary in $\langle L[A], A\rangle$ and to contain the parameter needed to define $P$ such that $0^\# \in L[G]$, $G$ $P$-generic over $\langle L[A], A\rangle$. Then we have that $P \cap L_\alpha[A]$ forces ZFC $+0^\#$ exists over $\langle L_\alpha[A], A \cap \alpha\rangle$. By taking a countable elementary submodel and collapsing we in fact get a countable transitive model of ZFC $+0^\#$ exists, in contradiction to our minimality assumption. □

Our next result says that indeed the conclusion of Proposition 5.32 is possible, with $0^\# \notin L[A]$.

**Theorem 5.33.** *Suppose that for no $\alpha$ is $L_\alpha[0^\#]$ a model of* ZFC. *Then there is $A \subseteq$ ORD definable over $L[0^\#]$ such that $0^\# \notin L[A]$ and for no $\alpha$ is $\langle L_\alpha[A], A \cap \alpha\rangle$ elementary in $\langle L[A], A\rangle$.*

To prove this theorem, we introduce *stability systems.*

**Definition.** A *stability system $p$* consists of a successor ordinal $|p| = \alpha(p) + 1$ and functions $f_k = f_k^p$, $k > 0$ such that:

(1) Dom $f_1 =$ Lim $\cap\, |p|$, $f_1(\alpha) \leq \alpha$ for $\alpha \in$ Dom $f_1$, $f_1(\alpha) = \underline{\lim}\langle f_1(\bar\alpha) \mid \bar\alpha \in$ Lim $\cap\, \alpha\rangle$ for $\alpha \in \text{Lim}^2 \cap |p|$. ($\underline{\lim}$ denotes lim inf and $\lim^2 =$ limit of limits.) Define $\alpha <_1 \beta$ iff $\alpha < \beta$ and $\alpha < \gamma \leq \beta$, $\gamma \in$ Dom $f_1 \Longrightarrow f_1(\gamma) \geq \alpha$. Then $\alpha \in$ Dom $f_1 \Longrightarrow f_1(\alpha) \leq_1 \alpha$.

(2) Dom $f_{k+1} = \{\alpha < |p| \mid \alpha <_k$ -limit$\}$, $f_{k+1}(\alpha) \leq \alpha$ for $\alpha \in$ Dom $f_{k+1}$, $f_{k+1}(\alpha) = \underline{\lim}\langle f_{k+1}(\bar\alpha) \mid \bar\alpha <_k \alpha, \bar\alpha \in$ Dom $f_{k+1}\rangle$ for $\alpha < |p|$, $\alpha \in <_k$-$\lim^2$. Define $\alpha <_{k+1} \beta$ iff $\alpha <_k \beta$ and $\alpha < \gamma \leq_k \beta, \gamma \in$ Dom $f_{k+1} \Longrightarrow f_{k+1}(\gamma) \geq \alpha$. Then $\alpha \in$ Dom $f_{k+1} \Longrightarrow f_{k+1}(\alpha) \leq_{k+1} \alpha$.

The intuitive idea is for $f_k(\alpha)$ to formally represent the supremum of the ordinals which are "$\Sigma_k$ stable in $\alpha$."

**Definition.** Suppose $\kappa$ is regular, $\ell > 0$, $\gamma < \kappa$. The forcing $P(\kappa, \ell, \gamma)$ consists of all stability systems $p$ such that $\gamma \leq_\ell^p \alpha(p) < \kappa$. Extension of conditions is defined by: $q \leq p$ iff $f_k^q \supseteq f_k^p$ for all $k$ and $\alpha(p) \leq_{\ell-1}^q \alpha(q)$, where $\leq_0^q = \leq$. We will see that $\leq$ is transitive.

**Lemma 5.34.** *For any stability system $p$ and any $k$, $\leq_k^p = \leq_k$ is a tree ordering and $\alpha \leq \beta \leq_k \gamma, \alpha \leq_{k+1} \gamma \Longrightarrow \alpha \leq_{k+1} \beta$. Also $\{\alpha \mid \alpha <_k \beta\}$ is closed in $\beta$.*

*Proof.* We prove that $\leq_k$ is a tree ordering, by induction on $k$. For $k = 0$, $\leq_0 = \leq$ so the result is obvious. Suppose it holds for $k$ and we first show that $\leq_{k+1}$ is transitive. Suppose $\alpha \leq_{k+1} \beta \leq_{k+1} \gamma$. Then $\alpha \leq_k \gamma$ by induction. Let $\delta \in$ Dom $f_{k+1}$, $\alpha < \delta \leq_k \gamma$. If $\delta > \beta$ then since $\beta \leq_{k+1} \gamma$ we get $f_{k+1}(\delta) \geq \beta \geq \alpha$. If $\delta \leq \beta$ then $\delta \leq_k \beta$ since $\leq_k$ is a tree ordering and so since $\alpha \leq_{k+1} \beta$ we have $f_{k+1}(\delta) \geq \alpha$. So $\alpha \leq_{k+1} \gamma$ and we have shown that $\leq_{k+1}$ is transitive. To see that $\leq_{k+1}$ is a tree, suppose that $\alpha \leq \beta$ are both $\leq_{k+1} \gamma$. By induction $\alpha \leq_k \beta$. If $\alpha < \delta \leq_k \beta$, $\delta \in$ Dom $f_{k+1}$ then $\delta \leq_k \gamma$ so $f_{k+1}(\delta) \geq \alpha$ since $\alpha \leq_{k+1} \gamma$. So we have shown that $\alpha \leq_{k+1} \beta$ and so $\leq_{k+1}$ is a tree.

If $\alpha \leq \beta \leq_k \gamma$ and $\alpha \leq_{k+1} \gamma$ then we first get $\alpha \leq_k \beta$ as $\leq_k$ is a tree ordering. If $\alpha < \delta \leq_k \beta$ and $\delta \in$ Dom $f_{k+1}$ then $\delta \leq_k \gamma$ so $f_{k+1}(\delta) \geq \alpha$ since $\alpha \leq_{k+1} \gamma$. So $\alpha \leq_{k+1} \beta$.

Finally we show that $\{\alpha \mid \alpha <_k \beta\}$ is closed in $\beta$, by induction on $k$. This is obvious for $k = 0$. Suppose it holds for $k$ and $\bar\alpha$ is a limit of $\{\alpha \mid \alpha <_{k+1} \beta\}$, $\bar\alpha < \beta$. Then $\bar\alpha <_k \beta$ by induction. Suppose $\bar\alpha < \gamma \leq_k \beta$, $\gamma \in$ Dom $f_{k+1}$. Then $f_{k+1}(\gamma) \geq \alpha$ for all $\alpha <_{k+1} \beta$, $\alpha < \gamma$. So $f_{k+1}(\gamma) \geq \bar\alpha$ since $\bar\alpha$ is a limit of such $\alpha$. Thus $\bar\alpha <_{k+1} \beta$ and we have shown that $\{\alpha \mid \alpha <_{k+1} \beta\}$ is closed in $\beta$. □

**Lemma 5.35.** *Let $f_k$, $<_k$ come from a stability system. If $\alpha \in \mathrm{Dom}\, f_k$, $f_k(\alpha) < \alpha$, then $f_k(\alpha) = \text{largest}\, \bar{\alpha} <_k \alpha$. If $\alpha \in <_k\text{-lim}^2$, $f_{k+1}(\alpha) = \alpha$ then $\{\bar{\alpha} \mid \bar{\alpha} <_{k+1} \alpha\}$ is unbounded in $\alpha$.*

*Proof.* By definition of stability system, $f_k(\alpha) \leq_k \alpha$. If $f_k(\alpha) < \beta < \alpha$ then $\beta \not<_k \alpha$ as $f_k(\alpha) < \beta$. So the first statement of the lemma holds.

Suppose $\alpha \in <_k\text{-lim}^2$, $f_{k+1}(\alpha) = \alpha$. Then $\alpha = \lim\{f_{k+1}(\bar{\alpha}) \mid \bar{\alpha} <_k \alpha, \bar{\alpha} <_k\text{-lim}\}$. There are unboundedly many $\bar{\alpha} <_k \alpha$, $\bar{\alpha} <_k$-lim such that $f_{k+1}(\bar{\bar{\alpha}}) \geq f_{k+1}(\bar{\alpha})$ for all similar $\bar{\bar{\alpha}} \geq \bar{\alpha}$; the resulting $f_{k+1}(\bar{\alpha})$'s are unbounded in $\alpha$ and $<_k \alpha$. We claim that in fact these $f_{k+1}(\bar{\alpha})$'s are $<_{k+1} \alpha$: Suppose $f_{k+1}(\bar{\alpha}) < \beta \leq_k \alpha$ and $\beta \in \mathrm{Dom}\, f_{k+1}$. If $\beta \leq \bar{\alpha}$ then $\beta \leq_k \bar{\alpha}$ so $f_{k+1}(\beta) \geq f_{k+1}(\bar{\alpha})$ since $f_{k+1}(\bar{\alpha}) <_{k+1} \bar{\alpha}$. If $\beta > \bar{\alpha}$ then by choice of $\bar{\alpha}$, $f_{k+1}(\beta) \geq f_{k+1}(\bar{\alpha})$. So indeed $f_{k+1}(\bar{\alpha}) <_{k+1} \alpha$ and so $\{\beta \mid \beta <_{k+1} \alpha\}$ is unbounded in $\alpha$. □

**Lemma 5.36.** *$p \leq q \leq r$ in $P(\kappa, \ell, \gamma) \Longrightarrow p \leq r$.*

*Proof.* We need to check that $\alpha(r) \leq_{\ell-1} \alpha(p)$. But $\leq_{\ell-1}$ is transitive and we have $\alpha(r) \leq_{\ell-1} \alpha(q) \leq_{\ell-1} \alpha(p)$. □

**Lemma 5.37.** *Suppose $p \in P(\kappa, \ell, \gamma)$ and $\alpha(p) \leq \alpha < \kappa$. Then there exists $q \leq p$, $\alpha(q) = \alpha$.*

*Proof.* For a limit $\lambda \in (\alpha(p), \alpha]$ define $f_k^q(\lambda) = \lambda$, on the appropriate domains. □

**Lemma 5.38.** *Suppose $p_0 \geq p_1 \geq \cdots$ is a sequence of conditions in $P(\kappa, \ell, \gamma)$ of the length $< \kappa$. Then there is a greatest lower bound $p$ to the $p_i$'s satisfying $\alpha(p) = \bigcup_i \alpha(p_i)$.*

*Proof.* Let $\alpha = \bigcup_i \alpha(p_i)$, which we may assume to be greater than each $\alpha(p_i)$. We define $f_k^p(\alpha) = f_k$ by induction on $k$. If $\alpha \in \mathrm{Lim}^2$ then $f_1(\alpha) = \underline{\lim}\langle f_1(\bar{\alpha}) \mid \bar{\alpha} < \alpha$, $\bar{\alpha}$ limit$\rangle$. Otherwise $f_1(\alpha) = \alpha$. Assuming $f_k(\alpha)$ is defined it makes sense to ask if $\alpha$ is a $<_k$-limit; if not then $f_{k+1}(\alpha)$ is undefined. If $\alpha$ is a $<_k$- $\lim^2$ then $f_{k+1}(\alpha) = \underline{\lim}\langle f_{k+1}(\bar{\alpha}) \mid \bar{\alpha} <_k \alpha, \bar{\alpha} <_k\text{-limit}\rangle$. Otherwise $f_{k+1}(\alpha) = \alpha$. It is clear that $p$ as defined in this way is a condition, as any failure of $f_k(\alpha) \leq_k \alpha$ would arise from a failure of $f_k(\bar{\alpha}) \leq_k \bar{\alpha}$ for some $\bar{\alpha} <_{k-1} \alpha$. Similarly, $\alpha(p_i) <_k \alpha$ for each $i$ follows by induction on $k \leq \ell - 1$. So $p$ extends each $p_i$. □

To prove Theorem 5.33, we iterate the forcings $P(\kappa, \ell, \gamma)$. That is, $P$ is the reverse Easton iteration where at regular $\kappa$, $Q(\kappa)$ first selects $(\ell_\kappa, \gamma_\kappa)$ with $0 < \ell_\kappa \in \omega$, $\gamma_\kappa < \kappa$ and then applies the forcing $P(\kappa, \ell_\kappa, \gamma_\kappa)$. Our goal is to build $\langle G(\leq i) \mid i \in I\rangle$ and select ordinals $\alpha_i \in [i, i^*)$, $i < i^*$ adjacent Silver indiscernibles, such that $G(\leq i)$ is $P(\leq i)$-generic and:

(1) $i \leq j$ in $I \Longrightarrow G(j)$ extends $G(i)$.

(2) For $i \in I$, $\ell_i$ is the least $\ell$ such that the $L[0^\sharp]$-cardinals which are $0^\sharp$-$\Sigma_\ell$ stable in $i$ are bounded in $i$. Also $\gamma_i = \alpha_j$ where $j = \bigcup\{L[0^\sharp]$-cardinals which are $0^\sharp$-$\Sigma_{\ell_i}$ stable in $i\} \geq 0$. (By convention, $\alpha_0 = 0$.)

(3) Let $f_k = \bigcup\{f_k^p \mid p \in G(i)$ for some $i \in I\}$. Then for $i \in I$, $f_k(i) = i$ for $k < \ell_i$ and $f_{\ell_i}(i) = \gamma_i$.

(4) For $i \in I$, $i \leq_{\ell_i} \alpha_i$ and $f_{\ell_i+1}(\alpha_i) = \alpha_j$ where $j = \bigcup\{L[0^\sharp]$-cardinals which are $0^\sharp$-$\Sigma_{\ell_i+1}$ stable in $i\}$.

Suppose that the $G(i)$'s have been constructed to obey (1)–(4) and let $A = \langle f_k \mid k \in \omega\rangle$. We claim that for no $\alpha$ is $\langle L_\alpha[A], A \cap \alpha\rangle$ elementary in $\langle L[A], A\rangle$. Indeed, we now show that for each $k$ there are unboundedly many $\alpha <_k \infty$ (where $<_k$ is defined using $f_\ell$, $\ell \leq k$) yet no $\alpha$ is $<_k \infty$ for all $k$ simultaneously:

**Lemma 5.39.** *Let $i < j$ be indiscernibles, $i$ an $L[0^\sharp]$-cardinal, $i$ $0^\sharp$-$\Sigma_k$ stable in $j$ and either $k \leq 2$ or $j$ a limit of $0^\sharp$-$\Sigma_{k-2}$ stables in $j$. Then $i \leq_k \alpha_i <_k j$.*

*Proof.* By induction on $k$ and for fixed $k$ by induction on $j$. We need only check $\alpha_i <_k j$ as (4) gives us $i \leq_{\ell_i} \alpha_i$ and by (2), $\ell_i \geq k$.

Suppose $k = 1$. If $\alpha_i \not<_1 j$ then choose $\alpha_i < \beta \leq j$ such that $f_1(\beta) < \alpha_i$. Then $\beta < j$ by (2), (3). There are no indiscernibles between $\beta$ and $j$, as otherwise we can apply induction on $j$. So $\bar{j} \leq \beta < j$ where $\bar{j} = I$-predecessor to $j$. In fact $\bar{j} < \beta$ by (2), (3). If $f_1(j) < \beta$ then since $f_1(j) <_1 j$ we have $f_1(\beta) \geq f_1(j) \geq \alpha_i$, contrary to assumption. So $f_1(j) \geq \beta > \bar{j}$ and by (3), $f_1(j) = \alpha_{\bar{j}}$, $\bar{j}$ is an $L[0^\sharp]$-cardinal and $\bar{j}$ is $0^\sharp$- $\Sigma_1$ stable in $j$. And $\bar{j} < \beta \leq \alpha_{\bar{j}}$. But $\bar{j} \leq_1 \alpha_{\bar{j}}$ so $f_1(\beta) \geq \bar{j} > \alpha_i$, contrary to assumption.

Suppose that the lemma holds for $k$ and we prove it for $k+1$. If $\alpha_i \not<_{k+1} j$ then choose $\alpha_i < \beta \leq_k j$ such that $f_{k+1}(\beta) < \alpha_i$. By (2), (3) we have $\beta <_k j$. By induction on $j$ there can be no $\bar{j}$ such that $\beta \leq \bar{j}$ and $\bar{j}$ is an $L[0^\sharp]$-cardinal, $\bar{j}$ $0^\sharp$-$\Sigma_k$ stable in $j$, as otherwise by induction on $k$, $\bar{j} \leq_k \alpha_{\bar{j}} <_k j$ so $\bar{j} <_k j$ and $\beta \leq_k \bar{j}$ by Lemma 5.34. By (2) and (3), $f_k(j)$ is equal to $\alpha_{\bar{j}}$ where $\bar{j} = \bigcup\{L[0^\sharp]$-cardinals which are $0^\sharp$-$\Sigma_k$ stable in $j\}$ and since $\beta <_k j$ we have $\beta \leq \alpha_{\bar{j}}$. And $\bar{j} < \beta \leq \alpha_{\bar{j}}$, by the fourth sentence of this paragraph. Now $\alpha_{\bar{j}} = f_k(j) <_k j$ so since $\beta <_k j$ we have $\beta \leq_k \alpha_{\bar{j}}$ by Lemma 5.34. Now let $\ell = \ell_{\bar{j}}$. Clearly $\ell \geq k$ since $\bar{j}$ is $0^\sharp$-$\Sigma_k$ stable in $j$. But if $\ell > k$ then by (4), $\bar{j} <_{k+1} \alpha_{\bar{j}}$ and this contradicts $f_{k+1}(\beta) < \alpha_i < \bar{j}$. So $\ell = k$, $\bar{j} <_k \alpha_{\bar{j}}$ and by (4), $f_{k+1}(\alpha_{\bar{j}}) = \alpha_{\bar{\bar{j}}}$ where $\bar{\bar{j}} = \bigcup\{0^\sharp$-$\Sigma_{k+1}$ stables in $\bar{j}\}$. But $i$ is $0^\sharp$-$\Sigma_{k+1}$ stable in $j$, $\bar{j}$ is $0^\sharp$-$\Sigma_k$ stable in $j$ and $i < \bar{j}$; so $i$ is $0^\sharp$-$\Sigma_{k+1}$ stable in $\bar{j}$ and we get $i \leq \bar{\bar{j}}$. Thus $\alpha_i \leq f_{k+1}(\alpha_{\bar{j}}) < \beta$. This contradicts $\beta \leq_k \alpha_{\bar{j}}$, $f_{k+1}(\beta) < \alpha_i$ since $f_{k+1}(\alpha_{\bar{j}}) <_{k+1} \alpha_{\bar{j}}$. □

**Corollary 5.40.** *For each $k$ there are cofinally many $\alpha <_k \infty$. No $\alpha$ is $<_k \infty$ for all $k$.*

*Proof.* By the previous lemma, if $i$ is an $L[0^\sharp]$-cardinal, $i$ $0^\sharp$-$\Sigma_k$ stable in $\infty$ then $i <_k \infty$; the class of such $i$ is unbounded in the ordinals. As we have assumed that no $L_\alpha[0^\sharp]$ is a model of ZFC, for any $\alpha$ there is $k$ such that the least $0^\sharp$-$\Sigma_k$ stable $L[0^\sharp]$-cardinal $i$ is greater than $\alpha$. Then $i <_k \infty$ and $f_k(i) = \alpha_0 = 0$. So $\alpha \not<_k \infty$. □

Thus to prove Theorem 5.33 it remains only to show that $\langle G(\leq i) \mid i \in I\rangle$ can be constructed so as to satisfy (1)–(4).

## The Construction

By induction on $i \in I$ we define $G(\leq i)$; also when defining $G(\leq i^*)$, $i^* = I$-successor to $i$, we define $\alpha_i \in [i, i^*)$.

Let $i_0 = \min I$. Then $G(< i_0)$ is the $L[0^\#]$-least generic for $P(< i_0)$. And $G(i_0)$ is the $L[0^\#]$-least generic for the $P(i_0, 1, 0)$ of $L[G(< i_0)]$, $\ell_{i_0} = 1$, $\gamma_{i_0} = 0 = \alpha_0$.

Suppose that $G(\leq i)$ is defined and we wish to define $G(\leq i^*)$, $\alpha_i$. Then $G(< i^*)$ is the $L[0^\#]$-least generic extending $G(\leq i)$. The key is to define $G(i^*)$.

Choose $n \geq 0$ so that the ordertype of the $L[0^\#]$-cardinals which are $0^\#$-$\Sigma_\ell$ stable in $i$ is $\lambda + n$ ($\lambda$ limit or 0), $\ell = \ell_i =$ least $\ell$ such that the $L[0^\#]$-cardinals which are $0^\sharp$-$\Sigma_\ell$ stable in $i$ have supremum $j < i$.

Let $\bar{P}$ be the forcing $P(i^*, \ell + 1, \alpha_j)$ defined in $L[G(< i^*)]$. Let $p_0 \in \bar{P}$ be the condition defined by: $\alpha(p_0) = i$, $f_k^{p_0} \restriction i = \bigcup\{f_k^p \mid p \in G(i)\}$, $f_k^{p_0}(i) = i$ if $k < \ell$, $f_\ell^{p_0}(i) = \gamma_i = \alpha_j$. (We will verify later that $p_0$ is indeed a condition in $\bar{P}$.) Choose $p_1 \leq p_0$ in $\bar{P}$ to meet all dense $D$ in $L[G(< i^*)]$ definable in $L[G(< i^*)]$ from $G(< i^*)$ and parameters in $(i + 1) \cup \{j_1, \ldots, j_n\}$ where $j_1, \ldots, j_n$ are the first $n$ indiscernibles $\geq i^*$. Also arrange that $\alpha(p_1)$ is a $<_\ell^{p_1}$-limit, $f_\ell^{p_1}(\alpha(p_1)) = \alpha_j$. Now set $\alpha_i = \alpha(p_1)$ and $\ell_{i^*} = 1$, $\gamma_{i^*} = \alpha_{j'}$, where $j' = \bigcup\{L[0^\#]\text{-cardinals which are } 0^\#\text{-}\Sigma_1 \text{ stable in } i^*\}$. Choose $G(i^*)$ to be the $L[0^\#]$-least $P(i^*, 1, \alpha_{j'})$-generic containing the condition $p_1$.

Finally for $i \in \operatorname{Lim} I$, $G(< i) = \bigcup\{G(< j) \mid j \in I \cap i\}$ and $\ell_i =$ least $\ell$ such that the $L[0^\#]$-cardinals which are $0^\#$-$\Sigma_\ell$ stable in $i$ have supremum $j < i$ and $\gamma_i = \alpha_j$. Let $f^k \restriction i = \bigcup\{f_k^p \mid p \in G(j)$ from some $j \in I \cap i\}$ and $G(i) = P(i, \ell_i, \gamma_i)$-generic determined by $\langle f_k \restriction i \mid k \in \omega\rangle$. (We will verify that $\langle f_k \restriction i \mid k \in \omega\rangle$ does indeed determine a $P(i, \ell_i, \gamma_i)$-generic.)

This completes the construction of the $G(\leq i)$'s. We must verify properties (1)–(4). Assuming that the two verifications mentioned during the construction can be carried out, these properties follow directly from the construction, with the possible exception of the assertion "$i \leq_{\ell_i} \alpha_i$" in (4). But note that in the construction of $G(\leq i^*)$, $i = \alpha(p_0)$, $\alpha_i = \alpha(p_1)$ where $p_1 \leq p_0$ in $P(i^*, \ell_i + 1, \alpha_j)$; so $i \leq_{\ell_i} \alpha_i$ follows from the definition of extension of conditions.

Finally to make the verifications stated in the construction we need:

**Lemma 5.41.** (a) *$G(i)$ is well-defined and $Q_i = P(i, \ell_i, \gamma_i)$-generic over $L[G(< i)]$.*

(b) *Define $p$ by: $\alpha(p) = i$, $f_k^p \restriction i = \bigcup\{f_k^q \mid q \in G(i)\}$, $f_k^p(i) = i$ for $k < \ell_i$, $f_{\ell_i}^p(i) = \gamma_i$. Then $p$ is a stability system.*

(c) *Lemma* 5.39 *holds for indiscernibles $\leq i$.*

(d) *If $p$ is defined as in* (b) *then $p \in \bar{P}$ where $\bar{P}$ is defined as in the construction of $G(\leq i^*)$.*

*Proof.* By induction on $i$.

(a) This follows by induction unless $\ell_i > 1$ and there is a final segment $i_0 < i_1 < \cdots$ of ordertype $\omega$ of the $L[0^\#]$-cardinals which are $0^\#$-$\Sigma_{\ell_i - 1}$ stable in $i$. We may also assume that $i_0$ is big enough so that $j = \bigcup\{L[0^\#]\text{-cardinals which are } 0^\#\text{-}\Sigma_{\ell_i} \text{ stable in } i\} = \bigcup\{L[0^\#]\text{-cardinals which are } 0^\#\text{-}\Sigma_{\ell_i} \text{ stable in } i_n\}$, for all $n$. Note that $\ell_{i_n} = \ell_i - 1$ and the ordertype of the $L[0^\#]$-cardinals which are $0^\#$-$\Sigma_{\ell_{i_n}}$ stable in $i_n$ has ordertype $\lambda + n'$, $n \le n' < \omega$ and $\lambda$ limit or 0. By construction, if $p_1^n$ is the condition in $G(i_n^*)$ with $\alpha(p_1^n) = \alpha_{i_n}$ then $p_1^n$ meets all $\Delta$ dense on $P(i_n^*, \ell_i, \gamma_i)$ which are defined in $L[G(< i_n^*)]$ from $G(< i_n^*)$ and parameters in $(i_n + 1) \cup \{j_1, \dots, j_n\}$ where $j_1, \dots, j_n$ are the first $n$ indiscernibles $\ge i_n^*$. By an inductive use of (c), $i_n <_{l_i - 1} \alpha_{i_n} <_{\ell_i - 1} i_m$ for $m > n$, relative to $p$ as defined in (b). So $\alpha_{i_0} <_{\ell_i - 1} \alpha_{i_1} <_{\ell_i - 1} \cdots$ and $p_1^0 \ge p_1^1 \ge p_1^2 \ge \cdots$ in $P(i, \ell_i, \gamma_i)$ determine the generic $G(i)$.

(b) The genericity of $G(i)$ implies that $i$ is a $<^p_{\ell_i - 1}$-lim$^2$ and $f_k^p(i)$ is determined correctly by $f_k^p(\gamma)$, $\gamma <^p_{k-1} i$, $\gamma$ a $<^p_{k-1}$-lim, for $k \le \ell_i$. So $p$ obeys the requirements for a stability system.

(c) The proof of Lemma 5.39 for indiscernibles $\le i$ only used the facts that $p$ as defined in (b) is a stability system and (1)–(4) hold $\le i$.

(d) We need only verify that $\alpha_j <^p_{\ell+1} i$ in the definition of $G(i^*)$. This follows from (c). □

This completes the proof of Theorem 5.33. Now the same technique also yields the following.

**Proposition 5.42.** *The hypothesis of Proposition* 5.31 *holds for some* $A$, $0^\# \notin L[A]$ *provided in* (b) *we replace "*$\beta$ *L-regular" by "*$\beta \in I$*."*

*Proof.* Build $G(\le i)$ by induction on $i \in I$, as in the preceding proof, and let $f_k = \bigcup\{f_k^p \mid p \in G(i) \text{ for some } i \in I\}$, $<_k$ defined from the $f_k$'s in the usual way. Define $S(n, \alpha, \beta) \iff \alpha <_n \beta$ and $\alpha, \beta$ are $<_n$-limits. (a) in the statement of Proposition 5.31 follows from Lemma 5.39. By (2), (3), if $S(n+1, \alpha, \beta)$ holds and $\alpha < \beta$ are in $I$ then $\alpha, \beta$ are limits of $0^\#$-$\Sigma_n$ stables and $\alpha$ is $0^\#$-$\Sigma_{n+1}$ stable in $\beta$. So (b) in the statement of Proposition 5.31 follows. □

It is in the sense of Propositions 5.31, 5.42 that $0^\#$ is "close" to being generic over a model not containing it.

## 5.3 Generic Saturation

Suppose that $P$ is an $L$-forcing which has a generic; need it have a generic definable in $L[0^\#]$? Not necessarily, as the forcing $P$ could produce a real $R$ that guarantees the countability of $\omega_1^{L[0^\#]}$, and clearly no such real can exist in $L[0^\#]$. However, we can weaken this slightly to obtain a positive result:

**Definition.** Suppose that $M \subseteq N$ are inner models of ZFC. We say that $N$ is *generically saturated over* $M$ if whenever an $M$-forcing has a generic, then it has one definable in a set-generic extension of $N$.

With a mild assumption, we will show that $L[0^\#]$ is generically saturated over $L$. This assumption involves the concept of an *Erdös cardinal.*

**Definition.** A cardinal $\kappa$ is *$\alpha$-Erdös* if whenever $A \subseteq \kappa$ and $C$ is closed unbounded in $\kappa$, there exists $X \subseteq C$ such that ordertype $X = \alpha$ and $\gamma \in X$ implies $X - \gamma$ is a set of indiscernibles for $\langle L[A], A, \delta\rangle_{\delta<\gamma}$. We say that ORD is *$\alpha$-Erdös* if this holds where $\kappa$ is replaced by ORD and indiscernibility is only required for $\Sigma_1$ formulas.

**Proposition 5.43.** *Let $\alpha$ be a limit ordinal, $\alpha \leq \kappa$. Then $\kappa$ is $\alpha$-Erdös iff whenever $\mathcal{A} = \langle T, \in, \ldots\rangle$ is transitive, in a countable language and $\kappa \subseteq T$, $C$ CUB in $\kappa$ then there exists $X \subseteq C$, ordertype $(X) = \alpha$, $X$ a good set of $\Sigma_1$ indiscernibles for $\mathcal{A}$.*

**Theorem 5.44.** *Suppose* ORD *is $\omega + \omega$-Erdös. Then $L[0^\#]$ is generically saturated over $L$.*

Thus assuming that ORD is $\omega+\omega$-Erdös, if an $L$-forcing $P$ has a generic then it has one definable in a set-generic extension of $L[0^\#]$. If $P$ is definable without parameters over the ground model $\langle L, \emptyset\rangle$ then it in fact has a generic definable in $L[0^\#]$.

Our proof is based on the idea of "periodicity."

**Definition.** Let $I = \langle i_\alpha \mid \alpha \in \mathrm{ORD}\rangle$ be the increasing enumeration of the Silver indiscernibles and for $\beta$ nonzero let $I_{\alpha,\beta}$ denote $\{i_{\alpha+\beta\cdot\gamma} \mid \gamma \in \mathrm{ORD}\}$. A class $B$ is *$\alpha, \beta$-periodic* if $I_{\alpha,\beta}$ is a class of indiscernibles for $\langle L[B], B\rangle$. A class forcing $P$ defined over a ground model $\langle L, A\rangle$ is *$\alpha, \beta$-periodic* if there is a $P$-generic $G$ such that $\langle G, A\rangle$ is $\alpha, \beta$-periodic; $P$ is *almost $\alpha, \beta$-periodic* if $P$ is $\alpha, \beta$-periodic in a set-generic extension of $V$.

Almost periodicity is equivalent to "almost codability" in the sense of the next definition.

**Definition.** $P$ is *codable* if $P$ has a generic $G$ such that $\langle G, A\rangle$ is definable in $L[R]$ for some real $R \in L[0^\#]$, $R$ generic over $L$. $P$ is *almost codable* if we say instead that $R$ belongs to a set-generic extension of $L[0^\#]$ (which may not be included in $V$).

**Theorem 5.45.** (a) *If $A = \emptyset$, $P$ definable over $\langle L, \emptyset\rangle$ without parameters then $P$ is codable iff $P$ is almost $\alpha, \beta$-periodic for some $\alpha, \beta$.*

(b) *For arbitrary $P$, $P$ is almost codable iff $P$ is almost $\alpha, \beta$-periodic for some $\alpha, \beta$.*

*Proof.* (a) The "only if" direction follows from Theorem 5.20. Conversely, if $G$ is $\alpha, \beta$-periodic in a set-generic extension of $V$ for some $P$-generic $G$ then by collapsing $\alpha, \beta$ to $\omega$, we may assume that $\alpha, \beta$ are countable in this set-generic extension. Then by absoluteness there is such a $G$ in $L[0^\#]$, as the existence of such a $G$ is a $\Sigma_1$ property with parameter $0^\#$. Now apply Corollary 5.19. The proof of (b) is the same, using now a Lévy collapse of $\gamma$, where $I - \gamma$ is a class of indiscernibles for $A$ and $P$ is definable over $\langle L[A], A\rangle$ with parameters less than $\gamma$. □

Thus Theorem 5.44 reduces to:

**Theorem 5.46.** *Suppose that* ORD *is* $\omega+\omega$*-Erdös and that* $P$ *is an* $L$*-forcing. Then* $P$ *is almost* $\alpha, \beta$*-periodic for some countable* $\beta$.

*Proof.* Suppose that $G \subseteq P$ is $P$-generic over $\langle L, A\rangle$. We shall construct another $P$-generic $G^*$ (in a set-generic extension of $V$) such that $G^*$ has periodic indiscernibles.

Let $X$ be a set of indiscernibles for $\langle L[0^\#, G], G, A\rangle$ of ordertype $\omega+\omega$ such that $\alpha \in X \Longrightarrow \alpha$ is $\Sigma_1$-stable in $0^\#, G, A$. The latter means that $\langle L_\alpha[0^\#, G \cap L_\alpha], G \cap L_\alpha, A \cap L_\alpha\rangle$ is $\Sigma_1$-elementary in $\langle L[0^\#, G], G, A\rangle$. We can obtain $X$ as $C = \{\alpha \mid \alpha$ is $\Sigma_1$-stable in $0^\#, G, A\}$ is closed unbounded.

Choose $\langle D(\alpha_1, \ldots, \alpha_n) \mid \alpha_1 < \cdots < \alpha_n$ in ORD$\rangle$ such that each $\langle L, A\rangle$-definable open dense $D \subseteq P$ is of the form $D(\alpha_1, \ldots, \alpha_n)$ for some $\alpha_1 < \cdots < \alpha_n$ in $I$. Also assume that this sequence is $\Delta_1\langle L, \text{Sat}\langle L, A\rangle\rangle$. Let $D^*(\alpha_1, \ldots, \alpha_n) = \cap\{D(\vec{\beta}) \mid \vec{\beta}$ a subsequence of $\langle \alpha_1, \ldots, \alpha_n\rangle\}$.

For $j_0 \in X$ choose the least $t_{j_0}(\vec{k}_0(j_0), j_0, \vec{k}_1(j_0))$ in $D(j_0) \cap G$. By the choice of the indiscernibles $X$, we can write this as $t_0(\vec{k}_0, j_0, \vec{k}_1(j_0))$, and in addition $\vec{k}_1(j_0) < j_1$ for $j_0 < j_1$ in $X$.

Next for $j_0 < j_1$ in $X$ choose the least $t_{j_0, j_1}(\vec{k}_0^1(j_0, j_1), j_0, \vec{k}_1^1(j_0, j_1), j_1, \vec{k}_2^1(j_0, j_1))$ in $D^*(\vec{k}_0, j_0, \vec{k}_1(j_0), j_1, \vec{k}_1(j_1)) \cap G$. By the choice of $X$ we can write this as $t_1(\vec{k}_0^1, j_0, \vec{k}_1^1(j_0), j_1, \vec{k}_2^1(j_0, j_1))$, and by $\Sigma_1$-stability this is less than $j_2$ whenever $j_1 < j_2$ in $X$. But we want to argue that in fact $\vec{k}_2^1(j_0, j_1)$ can be chosen *independently of* $j_0$.

Assuming the latter, we have $t_1(\vec{k}_0^1, j_0, \vec{k}_1^1(j_0), j_1, \vec{k}_2^1(j_1)) \in D^*(\vec{k}_0, j_0, \vec{k}_1(j_0), j_1, \vec{k}_1(j_1)) \cap G$ for $j_0 < j_1$ in $X$. By modifying $t_1$ we can guarantee that $\vec{k}_1^1(j_0) = \vec{k}_2^1(j_0)$ for all $j_0 \in X$, $j_0 \neq \min X$. Also we can arrange that $\vec{k}_0 \subseteq \vec{k}_0^1$, $\vec{k}_1(j_0) \subseteq \vec{k}_1^1(j_0)$ for $j_0 \in X$. By indiscernibility, the structure $\langle \vec{k}_1^1(j_0), <\rangle$ with a unary predicate for $\vec{k}_1(j_0)$ has isomorphism type independent of the choice of $j_0 \in X$.

Build $t_2(\vec{k}_0^2, j_0, \vec{k}_1^2(j_0), j_1, \vec{k}_1^2(j_1), j_2, \vec{k}_1^2(j_2)) \in D^*(\vec{k}_0^1, j_0, \vec{k}_1^1(j_0), j_1, \vec{k}_1^1(j_1), j_2, \vec{k}_1^1(j_2)) \cap G$ similarly, so that $\vec{k}_0^1 \subseteq \vec{k}_0^2$ and for $j_0 \in X$, $\vec{k}_1^1(j_0) \subseteq \vec{k}_1^2(j_0)$ with the isomorphism type of $\langle \vec{k}_1^2(j_0), <\rangle$ with unary predicates for $\vec{k}_1(j_0), \vec{k}_1^1(j_0)$ independent of $j_0$. Continue with $t_3, t_4, \ldots$.

Let $i_\alpha$ be the minimum of $X$ and let $\beta$ be the ordertype of $\cup\{\vec{k}_1^n(j_0) \mid n \in \omega\}$, an ordinal independent of the choice of $j_0 \in X$. In a generic extension where $\alpha$ is countable we may also arrange that $\cup\{\vec{k}_0^n \mid n \in \omega\} = I \cap i_\alpha$.

For any indiscernible $i_\gamma$ define $\vec{k}_1^n(i_\gamma) \subseteq I \cap (i_\gamma, i_{\gamma+\beta})$ so that $\langle I \cap (i_\gamma, i_{\gamma+\beta}), <\rangle$ with a predicate for $\vec{k}_1^n(i_\gamma)$ is isomorphic to $\langle \cup\{\vec{k}_1^n(j_0) \mid n \in \omega\}, <\rangle$ with a predicate for $\vec{k}_1^n(j_0)$, for $j_0 \in X$. Define: $G^* = \{p \in P \mid p$ is extended by some $t_n(\vec{k}_0^n, i_{\alpha_1}, \vec{k}_1^n(i_{\alpha_1}), \ldots, i_{\alpha_n}, \vec{k}_1^n(i_{\alpha_n}))$ where $\alpha \leq \alpha_1 < \cdots < \alpha_n$ are of the form $\alpha+\beta\cdot\gamma$ for some $\gamma \in$ ORD$\}$. Using the indiscernibility of $I - i_\alpha$ in $\langle L, A\rangle$, $G^*$ is compatible and meets every $\langle L, A\rangle$-definable open dense subclass of $P$. Thus $G^*$ is $P$-generic and $I_{\alpha,\beta}$ is a class of indiscernibles for $\langle L[G^*], A, G^*\rangle$.

To complete the proof we return to the problem of making $\vec{k}_2^1(j_0, j_1)$ independent of $j_0$. First a lemma:

**Lemma 5.47.** *Let $x < y$ by the maximum difference order on finite sets of ordinals: $x < y$ iff $\alpha \in y$ where $\alpha$ is the greatest element of the symmetric difference of $x$ and $y$. For any $j_0 < j_1$ in $X$ and any open dense $D$ definable in $\langle L, A\rangle$ there exists $t(\vec{\ell}_0, j_0, \vec{\ell}_1, j_1, \vec{\ell}_2, \vec{\ell}) \in L_{\min(\vec{\ell})} \cap D \cap G$ such that $\vec{\ell}_0 < j_0 < \vec{\ell}_1 < j_1 < \vec{\ell}_2 < \vec{\ell}$ belong to $I$ and $\vec{\ell}_0 \cup \vec{\ell}_1 \cup \vec{\ell}_2$ is the $<$-least finite set of* ordinals *(not necessarily indiscernibles) $x$ such that $t(x \cap j_0, j_0, x \cap (j_0, j_1), j_1, x - j_1, \vec{\ell})$ belongs to $L_{\min(\vec{\ell})} \cap D \cap G$.*

*Proof.* Let $x$ be $<$-least such that for some $t$ and indiscernibles $\vec{\ell} > \max(x)$, $t(x \cap j_0, j_0, x \cap (j_0, j_1), j_1, x - j_1, \vec{\ell}) \in L_{\min(\vec{\ell})} \cap D \cap G$. If some $\alpha \in x$ were not in $I$ then there would be a $t^*(x^* \cap j_0, j_0, x^* \cap (j_0, j_1), j_1, x^* - j_1, \vec{\ell}^*) = t(x \cap j_0, j_0, x \cap (j_0, j_1), j_1, x - j_1, \vec{\ell})$ with $\vec{\ell}$ an initial segment of $\vec{\ell}^*$ and $x^* - \alpha = x^* - (\alpha + 1)$, as $\alpha$ is $L$-definable from indiscernibles $< \alpha$ and indiscernibles $> \vec{\ell}$. So let $\vec{\ell}_0, \vec{\ell}_1, \vec{\ell}_2$ be $x \cap j_0, x \cap (j_0, j_1), x - j_1$. □

Now for $j_0 < j_1$ in $X$ choose the least $t_{j_0, j_1}(\vec{k}_0^1(j_0, j_1), j_0, \vec{k}_1^1(j_0, j_1), j_1, \vec{k}_{2,0}^1(j_0, j_1), \vec{k}_{2,1}^1(j_0, j_1))$ to satisfy Lemma 5.47 with $D = D^*(\vec{k}_0, j_0, \vec{k}_1(j_0), j_1, \vec{k}_1(j_1))$, and $\vec{\ell}$ denoted by $\vec{k}_{2,1}^1(j_0, j_1)$. By the choice of $X$ we can write this as $t_1(\vec{k}_0^1, j_0, \vec{k}_1^1(j_0), j_1, \vec{k}_{2,0}^1(j_0, j_1), \vec{\infty})$, where $\vec{\infty}$ denotes an arbitrary sequence of large indiscernibles (of the appropriate length). Note that $\langle \vec{k}_0^1, \vec{k}_1^1(j_0), \vec{k}_{2,0}^1(j_0, j_1)\rangle$ is definable in $\langle L[G], A, G\rangle$ from $\vec{k}_0, j_0, \vec{k}_1(j_0), j_1, \vec{k}_1(j_1), \vec{\infty}$ and therefore $\vec{k}_{2,0}^1(j_0, j_1)$ is definable in $\langle L[G], A, G\rangle$ from $\vec{k}_1(j_1), \vec{\infty}$ and ordinals $\leq j_1$.

**Claim.** $\vec{k}_{2,0}^1(j_0, j_1)$ is independent of $j_0$.

*Proof.* Let $j_0 < j_1 < \cdots < j$ be the first $\omega + 1$ elements of $X$ and for any $n, m$ let $\vec{k}(j_n, j)(m) = m^{\text{th}}$ element of $\vec{k}_{2,0}^1(j_n, j)$. If the Claim fails then for some fixed $m$, $\vec{k}(j_0, j)(m) < \vec{k}(j_1, j)(m) < \cdots$ is an increasing sequence of indiscernibles with supremum $\ell \in I$ (using the fact that $X - j$ has ordertype $>$ length$(\vec{\infty})$). As these ordinals are definable in $\langle L[G], A, G\rangle$ from ordinals in $(j+1) \cup \vec{k}_1(j) \cup \vec{\infty}$ we get that $\ell$ has cofinality $\leq j$ in $L[G]$. But $0^\# \notin L[G]$ (as $G$ is generic over $L$) so by Jensen's Covering Theorem, $\ell$ has $L$-cofinality $< (j^+$ in $L[G])$. As $\ell \in I$, $\ell$ is $L$-regular and hence $j^+$ in $L < j^+$ in $L[G]$.

But then in $L[G]$ there is a closed unbounded $C \subseteq j$ such that if $D$ is a constructible CUB subset of $j$ then $C \subseteq D \cup \alpha$ for some $\alpha < j$. Now $I \cap j$ is the intersection of countably many such $D$'s and therefore as $j$ has uncountable cofinality (in $L[G, 0^\#]$) we get $C \subseteq I \cup \alpha$ for some $\alpha < j$. This yields $0^\# \in L[G]$, contradiction.

This proves the Claim. □

With the Claim we see that there is a $P$-generic $G^*$ (in a set-generic extension of $V$) such that $\langle L[G^*], A, G^*\rangle$ has a periodic class of indiscernibles $I_{\alpha,\beta}$, $\beta$ countable. It now follows by absoluteness that there is such a $G^*$ definable in a set-generic extension of $L[0^\#]$ in which $\alpha$ is also countable. This completes the proof of Theorem 5.46. □

**Remarks.** (1) The previous proof only made use of a weaker hypothesis: define $X$ to be a good set of $\Sigma_1$ *n-indiscernibles* for $\mathcal{A} = \langle T, \in, \ldots\rangle$ if $\gamma \in X \Longrightarrow X - \gamma$ is $\Sigma_1$ indiscernible in $\langle \mathcal{A}, \alpha\rangle_{\alpha\in\gamma}$ for *n-tuples*. Our proof only used the existence of $X_1 \supseteq X_2 \supseteq \cdots$ such that each $X_n$ is a good set of $\Sigma_1$ $n$-indiscernibles for $\langle L[G, 0^\#], G, A\rangle$ of ordertype at least $\omega + \omega$ such that $\alpha \in X_1 \Longrightarrow \alpha$ is $\Sigma_1$-stable in $0^\#$, $G$, $A$.

(2) We will show in Chapter 8 that Theorem 5.46 is optimal in the following sense: If $\alpha$, $\beta$ are ordinals, $\beta$ countable then there is a forcing $P$ over $L$ that has a generic but is not almost $\alpha'$, $\beta'$-periodic for $\alpha' \leq \alpha$ or for $\beta' \leq \beta$.

*Chapter 6*

# The $\Pi^1_2$-Singleton Problem

In this chapter we produce relevant $L$-definable forcings with few generics. Our key application is to the construction of a $\Pi^1_2$-singleton $R$ which is *intermediate* in the sense that $0 <_L R <_L 0^\#$. We obtain $R$ as the unique real coding a $P$-generic class for an appropriate $L$-definable forcing $P$. The uniqueness of $R$ emanates from the uniqueness properties of $0^\#$.

Section 6.1 is devoted to the proof of a slightly weaker result: We construct an intermediate *absolute-singleton* $R$. This means that for some formula $\varphi$, $R$ is the unique real such that $L[R] \vDash \varphi(R)$. The forcing used is a mixture of Jensen coding and Reverse Easton forcing. In Section 6.2 we introduce "David's trick" (see David [82a]) which we present as a general method for turning genericity for $L$-definable forcings into a $\Pi^1_2$ property. The solution to the $\Pi^1_2$-singleton problem quickly follows. We also note that our $\Pi^1_2$-singleton can in addition be chosen to be the *provably* unique solution to a $\Pi^1_2$ formula, in a slight extension of ZFC.

Two related results are considered in Section 6.3. We answer a question of Kechris by constructing a countable infinite $\Pi^1_2$ set containing no $\Pi^1_2$-singleton. And we show that over any inner model not containing $0^\#$ one can force a new $\Sigma^1_3$ parameter-free property.

## 6.1 An Absolute Singleton

In this section we construct an $L$-definable forcing $P$ with a unique $P$-generic, coded by a real $R <_L 0^\#$. Our $P$-generic will be determined by a real $R$ in the following way:

(1) There is a $\Sigma_1$ over $L$ class function $\langle \alpha_1, \ldots, \alpha_n \rangle \longmapsto p(\alpha_1, \ldots, \alpha_n)$ such that the $P$-generic determined by $R$ is $G_R = \{p \in P \mid p \text{ is compatible with } p(i_1, \ldots, i_n)$ for all $i_1 < \cdots < i_n$ in $I\}$ where $I$ denotes the Silver indiscernibles for $L$.

(2) $R$ codes $A \subseteq L$ as in Jensen coding.

(3) $A$ "kills" $(\alpha_1, \ldots, \alpha_n)$ whenever $p(\alpha_1, \ldots, \alpha_n)$ "contradicts" $R$.

We offer some further explanation of (3): Think of $(\alpha_1, \ldots, \alpha_n)$ as a "guess" at an increasing $n$-tuple of Silver indiscernibles. We develop a method of "killing" such guesses, by adding certain CUB sets. The key point of this killing procedure is that although the killing of guesses is cofinality-preserving no guess $(i_1, \ldots, i_n)$ consisting of actual Silver indiscernibles can be killed when $i_1$ is regular, assuming the existence of $0^\#$. Now any guess $(\alpha_1, \ldots, \alpha_n)$ produces a condition $p(\alpha_1, \ldots, \alpha_n) \in P$ which has partial information $p(\alpha_1, \ldots, \alpha_n)_0$ about the real $R$ (as $R$ is part of the generic that it determines). We say that $p(\alpha_1, \ldots, \alpha_n)$ "contradicts" $R$ if this information $p(\alpha_1, \ldots, \alpha_n)_0$ is inconsistent with $R$.

We can now argue that there is a unique $P$-generic, as described in (1): Any $P$-generic $G_{R'}$ with $R' \neq R$ would necessitate the killing of all $(i_1, \ldots, i_n)$ from $I$ such that $p(i_1, \ldots, i_n)$ contradicts $R'$. Our construction will ensure that for $n$ sufficiently large, $p(\alpha_1, \ldots, \alpha_n)_0$ will determine the $k^{\text{th}}$ value of $R$, where $R'$ and $R$ disagree at $k$. Hence we must kill all $(i_1, \ldots, i_n)$ from $I$ for $n$ sufficiently large, an impossibility.

## How to Kill a Guess

For $X \subseteq \mathrm{ORD}$ we use $X^n$ to denote all *increasing* $n$-tuples from $X$. And we use $\alpha^+$ to denote $(\alpha^+)^L$.

For $(\alpha_1, \ldots, \alpha_{n+1}) \in \mathrm{ORD}^{n+1}$, $n > 0$ define $I(\alpha_1, \ldots, \alpha_{n+1}) = \{\alpha < \alpha_1 \mid \alpha, \alpha_1$ satisfy the same $\Sigma_1$ properties in $L_{\alpha_{n+1}}$ with parameters from $\alpha \cup \{\alpha_2, \ldots, \alpha_n\}\}$. An *acceptable guess* is a sequence $(\alpha_1, \ldots, \alpha_{n+1})$ such that:

(1) $\alpha_1$ is $L$-inaccessible, $n > 0$.

(2) $1 \leq k \leq \ell \leq n \Longrightarrow \alpha_k \in I(\alpha_\ell, \ldots, \alpha_{n+1})$.

Note that elements of $I(\alpha_1, \ldots, \alpha_{n+1})$ are $L$-inaccessible. But we do not require that $\alpha_2, \ldots, \alpha_{n+1}$ are $L$-inaccessible. In fact we will be interested in "killing" acceptable guesses $(\alpha_1, \ldots, \alpha_{n+1})$ where $\alpha_{n+1} < \alpha_1^+$, by adding a CUB subset of $\alpha_1 - I(\alpha_1, \ldots, \alpha_{n+1})$. This will be sufficient for our purposes, by the following lemma.

**Lemma 6.1.** *If* $(\alpha_1, \ldots, \alpha_{n+1})$ *is an acceptable guess then there exists* $(\alpha_1, \bar{\alpha}_2, \ldots, \bar{\alpha}_{n+1})$ *such that* $(\alpha_1, \bar{\alpha}_2, \ldots, \bar{\alpha}_{n+1})$ *is also an acceptable guess,* $\bar{\alpha}_{n+1} < \alpha_1^+$ *and* $I(\alpha_1, \bar{\alpha}_2, \ldots, \bar{\alpha}_{n+1}) = I(\alpha_1, \ldots, \alpha_{n+1})$.

*Proof.* Let $\bar{\alpha}_i$, $2 \leq i \leq n+1$ be the image of $\alpha_i$ under the transitive collapse of an elementary submodel of $L_{\alpha_{n+1}^+}$ of $L$-cardinality $\alpha_1$ containing $\alpha_1 \cup \{\alpha_1, \ldots, \alpha_{n+1}\}$ as a subset. □

Acceptable guesses arise from, and give rise to, indiscernibles:

**Lemma 6.2.** (a) *Suppose* $(i_1, \ldots, i_{n+1}) \in I^{n+1}$ *and* $n > 0$. *Then* $(i_1, \ldots, i_{n+1})$ *is an acceptable guess.*

(b) *Let* $(\alpha_1, \ldots, \alpha_{n+1})$ *be an acceptable guess and* $(\beta_1, \ldots, \beta_n) \in I(\alpha_1, \ldots, \alpha_{n+1})^n$. *Then* $(\beta_1, \ldots, \beta_n)$, $(\alpha_1, \ldots, \alpha_n)$ *satisfy the same* $\Sigma_1$ *properties in* $L_{\alpha_{n+1}}$ *with parameters from* $\beta_1$.

*Proof.* (a) Clear, by indiscernibility in $L$ of $I$.

(b) By induction on $n \geq 1$. If $n = 1$ then the result is clear from the definition of $I(\alpha_1, \alpha_2)$. Suppose $n > 1$ and the property holds for smaller positive $n$. As

$\beta_2, \ldots, \beta_n \in I(\alpha_1, \ldots, \alpha_{n+1})$ we certainly have $\beta_2, \ldots, \beta_n \in I(\alpha_1, \alpha_3, \ldots, \alpha_{n+1})$. As $\alpha_1 \in I(\alpha_2, \alpha_3, \ldots, \alpha_{n+1})$ we can conclude that $\beta_2, \ldots, \beta_n \in I(\alpha_2, \alpha_3, \ldots, \alpha_{n+1})$. By induction $(\beta_2, \ldots, \beta_n)$, $(\alpha_2, \ldots, \alpha_n)$ satisfy the same $\Sigma_1$ properties in $L_{\alpha_{n+1}}$ with parameters from $\beta_2$; it follows that $(\beta_1, \ldots, \beta_n)$, $(\beta_1, \alpha_2, \ldots, \alpha_n)$ satisfy the same $\Sigma_1$ properties in $L_{\alpha_{n+1}}$ with parameters from $\beta_1$. But $\beta_1 \in I(\alpha_1, \ldots, \alpha_{n+1})$ so $(\beta_1, \alpha_2, \ldots, \alpha_n)$, $(\alpha_1, \ldots, \alpha_n)$ satisfy the same $\Sigma_1$ properties in $L_{\alpha_{n+1}}$ with parameters from $\beta_1$. The last two assertions give the desired result. □

An acceptable guess $(\alpha_1, \ldots, \alpha_{n+1})$ is *killed* by adding a CUB subset to $\alpha_1 - I(\alpha_1, \ldots, \alpha_{n+1})$. Note that if $(i_1, \ldots, i_{n+1}) \in I^{n+1}$ and $i_1$ is regular then $(i_1, \ldots, i_{n+1})$ is an acceptable guess that *cannot* be killed as $I(i_1, \ldots, i_{n+1})$ contains the CUB set $I \cap i_1$.

As mentioned at the start of this section, we shall define a $\Sigma_1$ over $L$ class function $\langle \alpha_1, \ldots, \alpha_n \rangle \longmapsto p(\alpha_1, \ldots, \alpha_n)$ that associates a condition $p(\alpha_1, \ldots, \alpha_n)$ in our desired forcing $P$ to each $\langle \alpha_1, \ldots, \alpha_n \rangle \in \mathrm{ORD}^n$. We refer to this class function as a *procedure*. We now wish to describe how to kill acceptable guesses that via this procedure produce conditions which contradict desired information about our generic real.

A condition in $P$ will be of the form $p = (\bar{s}, q)$ where $\bar{s}$ provides partial information about our desired generic real $R$, and $q$ is partial information both about the class $A \subseteq L$ needed to kill acceptable guesses contradicting $R$, as well as about how $A$ is coded by $R$. It would be simplest if we could take $\bar{s}$ to be an element of $2^{<\omega} =$ {finite sequences of 0's and 1's}, however for technical reasons connected with the construction of a $P$-generic from $0^\#$ we must work with more complex $\bar{s}$. Elements of $(2^{<\omega})^{<\omega} =$ {finite sequences of finite sequences of 0's and 1's} are called *hyperstrings*. A *hypertree* is a function $\bar{s}$ : Hyperstrings $\longrightarrow$ Perfect subtrees of $2^{<\omega}$ such that for all hyperstrings $\langle s_0, \ldots, s_n \rangle$, $\bar{s}(\langle s_0, \ldots, s_n \rangle) \subseteq \bar{s}(\langle s_0, \ldots, s_n \restriction k \rangle)$ for all $k$. A hyperstring $s = \langle s_0, \ldots, s_n \rangle$ *lies on* a hypertree $\bar{s}$ if $s_0 \in \bar{s}(\emptyset)$ and for $i < n$, $s_{i+1} \in \bar{s}(\langle s_0, \ldots, s_i \rangle)$. Otherwise $s$ is *incompatible with* $\bar{s}$.

If $p = (\bar{s}, q) \in P$ then $p_0$ denotes $\bar{s}$.

**Definition.** Let $s$ be a hyperstring. An acceptable guess $(\alpha_1, \ldots, \alpha_{n+1})$ is *$s$-bad* if $p(\alpha_1, \ldots, \alpha_n) \in L_{\alpha_{n+1}}$ and $s$ is incompatible with $p(\alpha_1, \ldots, \alpha_n)_0$.

The forcing $\mathbb{Q}(s)$ kills all $s$-bad acceptable guesses: $\mathbb{Q}_0(s) = \{0\}$, the trivial forcing. $\mathbb{Q}_{i+1}(s) = \mathbb{Q}_i(s) * \mathbb{R}_i(s)$ where $\mathbb{Q}_i(s) \Vdash \mathbb{R}_i(s)$ is the trivial forcing unless $i$ is inaccessible, in which case $\mathbb{Q}_i(s) \Vdash \mathbb{R}_i(s)$ is the forcing to add a CUB subset to $i - I(i, \alpha_2, \ldots, \alpha_{n+1})$, for each $s$-bad acceptable guess $(i, \alpha_2, \ldots, \alpha_{n+1})$ satisfying $\alpha_{n+1} < i^+$, using support of size $< i$. (Thus a condition specifies a bounded closed subset of $i - I(i, \alpha_2, \ldots, \alpha_{n+1})$ for each of $< i$-many $s$-bad acceptable guesses $(i, \alpha_2, \ldots, \alpha_{n+1})$ satisfying $\alpha_{n+1} < i^+$. If there are no such $(i, \alpha_2, \ldots, \alpha_{n+1})$ then $\mathbb{R}_i(s)$ is the trivial forcing.) For limit $\lambda$, $\mathbb{Q}_\lambda(s) =$ Direct Limit$\langle \mathbb{Q}_i(s) \mid i < \lambda \rangle$ for regular $\lambda$ and $\mathbb{Q}_\lambda(s) =$ Inverse Limit$\langle \mathbb{Q}_i(s) \mid i < \lambda \rangle$ for singular $\lambda$. And $\mathbb{Q}(s) =$ Direct Limit$\langle \mathbb{Q}_i(s) \mid i < \infty \rangle$.

The previous definition takes place in $L$ and yields a Reverse Easton forcing very much like the basic example of Section 2.3. The fact that size $< i$ supports are used in the definition of $\mathbb{R}_i(s)$ will imply that $\mathbb{Q}_{i+1}(s)$ is $i^+$-cc for regular $i$. The fact that for any acceptable guess $(i, \alpha_2, \ldots, \alpha_{n+1})$, any singular cardinal $< i$ belongs to $i - I(i, \alpha_2, \ldots, \alpha_{n+1})$ will imply that $\mathbb{Q}_{i+1}(s) \Vdash \mathbb{Q}_{i+1,\infty}(s)$ has an $i^+$-closed dense subclass, where $\mathbb{Q}(s) \simeq \mathbb{Q}_{i+1}(s) * \mathbb{Q}_{i+1,\infty}(s)$. The next lemmas spell this out in more detail.

**Lemma 6.3.** *For each $i$, $\mathbb{Q}_i(s)$ has a dense subordering which is a set of cardinality $\leq i^+$.*

*Proof.* As in Lemma 2.33. □

**Lemma 6.4.** *For $i$ regular and infinite, $\mathbb{Q}_{i+1}(s)$ is $i^+$-cc.*

*Proof.* Note that $\mathbb{Q}_i(s) = \text{Direct Limit}\langle \mathbb{Q}_j(s) \mid j < i \rangle$ so by Lemma 6.3, $\mathbb{Q}_i(s)$ has a dense subordering of cardinality $\leq i$, and hence is certainly $i^+$-cc. Also $\mathbb{Q}_i(s) \Vdash \mathbb{R}_i(s)$ is $i^+$-cc as if $\Delta \subseteq \mathbb{R}_i(s)$ is predense we can choose $\alpha < i^+$ of cofinality $i$ such that $\Delta \cap \mathbb{R}_i^\alpha(s)$ is predense in $\mathbb{R}_i^\alpha(s) = \{r \in \mathbb{R}_i(s) \mid r$ mentions only guesses $(i, \alpha_2, \ldots, \alpha_{n+1})$ satisfying $\alpha_{n+1} < \alpha\}$; but then $\Delta \cap \mathbb{R}_i^\alpha(s)$ is predense on $\mathbb{R}_i(s)$ as supports are of size $< i$. As $\mathbb{R}_i^\alpha(s)$ has size $i$ we get the $i^+$-cc.

Then as in the proof of Lemma 2.34 we get that $\mathbb{Q}_{i+1}(s)$ is $i^+$-cc. □

For $i \leq j \leq \infty$ we let $\mathbb{Q}_{ij}(s)$ be the iteration of length $j - i$ defined just like $\mathbb{Q}(s)$, except beginning at index $i$ and ending after $j - i$ stages.

**Lemma 6.5.** $\mathbb{Q}_j(s) \simeq \mathbb{Q}_i(s) * \mathbb{Q}_{ij}(s)$.

*Proof.* As in Lemma 2.35. □

**Lemma 6.6.** *For $i$ regular and infinite, $\mathbb{Q}_{i+1}(s) \Vdash \mathbb{Q}_{i+1,\infty}(s)$ has an $i^+$-closed dense definable subclass.*

*Proof.* Let $\Delta \subseteq \mathbb{Q}_{i+1,\infty}(s)$ consist of all conditions all of whose closed sets used to kill $s$-bad guesses are either $\emptyset$ or contain an ordinal greater than $i$. Then $\Delta$ is $i^+$-closed as no ordinal greater than $i$ of cofinality $\leq i$ belongs to $I(j, \alpha_2, \ldots, \alpha_{n+1})$ for any acceptable guess $(j, \alpha_2, \ldots, \alpha_{n+1})$. □

Now as $\mathbb{Q}(s)$ is $\infty$-cc, it is pretame and Lemmas 6.4, 6.6 give us cofinality-preservation. GCH-preservation follows from Lemmas 6.3, 6.6.

The forcing $\mathbb{P}_0(s)$ is the 2-step iteration $\mathbb{Q}(s) * \mathbb{R}(s)$ where if $G(s)$ denotes the $\mathbb{Q}(s)$-generic, $\mathbb{R}(s)$ is the coding of $G(s) \subseteq L$ by $X(s) \subseteq \omega_1^L$, as described in Section 4.2. Then $\mathbb{P}_0(s)$ is a cofinality-preserving forcing, defined in $L$ (uniformly in $s$) which adds no new $\omega$-sequences of ordinals.

The forcing $\mathbb{P}(s)$ is the 3-step iteration $\mathbb{Q}(s) * \mathbb{R}(s) * \mathbb{S}(s)$ where $\mathbb{S}(s)$ is the coding of $X(s) \subseteq \omega_1^L$ by a real, using perfect trees as follows. Inductively define $L$-countable ordinals $\mu_\alpha$, $\alpha < \omega_1^L$ by: $\mu_0 = \omega$ and for $\alpha > 0$, $\mu_\alpha$ is the least $\mu > \bigcup\{\mu_\beta \mid \beta < \alpha\}$ such that $L_\mu[X(s) \cap \alpha] \vDash \mathrm{ZF}^- + \omega$ is the largest cardinal. A real $R$ *codes* $X(s)$ *below* $\alpha$ if for all $\beta < \alpha$, $\beta \in X(s)$ iff $L_{\mu_\beta}[X(s) \cap \beta, R] \vDash \mathrm{ZF}^-$. For $T \subseteq 2^{<\omega}$ a perfect tree, let $|T|$ denote the least $\alpha$ such that $T \in L_{\mu_\alpha}[X(s) \cap \alpha]$. A *condition* in $\mathbb{S}(s)$ is a perfect tree $T$ such that $R$ codes $X[s]$ below $|T|$ whenever $R$ is a branch through $T$. $\mathbb{S}(s)$ is ordered by: $T_0 \leq T_1$ iff $T_0$ is a subtree of $T_1$. This is equivalent to $[T_0] \subseteq [T_1]$ where $[T]$ denotes the set of branches through $T$.

**Lemma 6.7.** (a) $T \in \mathbb{S}(s)$, $\alpha < \omega_i^L \Longrightarrow \exists T^* \leq T(|T^*| \geq \alpha)$.

(b) $\mathbb{S}(s)$ *preserves cofinalities.*

*Proof.* (a) By induction on $\alpha$. We may assume that $|T| < \alpha$. If $\alpha = \beta + 1$ then we may also assume by induction that $|T| = \beta$ and hence $T \in \mathcal{A}_\beta = L_{\mu_\beta}[X(s) \cap \beta]$. If $\beta$ belongs to $X(s)$ then we take $T^* \leq T$ to have the property that $R$ is $P_T$-generic over $\mathcal{A}_\beta$ for $R \in [T^*]$, where $P_T$ is the forcing (isomorphic to Cohen forcing) whose conditions are the elements of $T$, ordered by extension. Note that $T^*$ can be chosen in $\mathcal{A}_\alpha = L_{\mu_\alpha}[X(s) \cap \alpha]$ as $\mathcal{A}_\beta$ is a countable element of $\mathcal{A}_\alpha$. Also $L_{\mu_\beta}[X(s) \cap \beta, R] \models \mathrm{ZF}^-$ for $R \in [T^*]$, by the $P_T$-genericity of $R \in [T^*]$. So $T^*$ is a condition and $|T^*| = \alpha$. If $\beta$ does not belong to $X(s)$ then choose a real $R_0$ coding a well ordering of $\omega$ of ordertype $\mu_\beta$, $R_0 \in \mathcal{A}_\alpha$, and take $T^* \leq T$ to be the tree whose branches are exactly the branches $R$ through $T$ such that for all $n$, $n \in R_0$ iff $R$ goes right at the $2n^{\text{th}}$ splitting level of $T$. Then $T^* \in \mathcal{A}_\alpha$ and for $R \in [T^*]$, $(R, T)$ computes $R_0$ and hence $L_{\mu_\beta}[X(s) \cap \beta, R]$ is *not* a model of $\mathrm{ZF}^-$, since it contains $R_0$ as an element.

If $\alpha$ is a limit ordinal then choose $|T| = \alpha_0 < \alpha_1 < \cdots$ to be an $\omega$-sequence cofinal in $\alpha$ which belongs to $\mathcal{A}_\alpha = L_{\mu_\alpha}[X(s) \cap \alpha]$. Define $T_0 \leq_n T_1$ iff $T_0 \leq T_1$ and $T_0, T_1$ have the same first $n$ splitting levels. Note that if $T_{n+1} \leq_n T_n$ for all $n$ then $\bigcap_n T_n$ is a perfect tree.

Now let $T_0 = T$ and for each $n$ let $T_{n+1} \in \mathbb{S}(s)$ be least in $\mathcal{A}_{\alpha_{n+1}}$ such that $|T_{n+1}| = \alpha_{n+1}$ and $T_{n+1} \leq_n T_n$. Such $T_n$'s exist by induction. If $T^* = \bigcap_n T_n$ then $T^* \leq T$ belongs to $\mathcal{A}_\alpha$ and satisfies the requirement for belonging to $\mathbb{S}(s)$. So $T^* \leq T$, $|T^*| = \alpha$, as desired.

(b) We say that $D \subseteq \mathbb{S}(s)$ is *open* if $T^* \in D$ whenever $T^* \leq T \in D \Longrightarrow T^* \in D$, and is *n-dense* if for all $T \in \mathbb{S}(s)$ there is $T^* \leq_n T$, $T^* \in D$. We show that if for each $n$, $D_n$ is open and $n$-dense then for all $T \in \mathbb{S}(s)$ there exists $T^* \leq T$ such that $T^*$ belongs to $D_n$ for each $n$. It follows that $\mathbb{S}(s)$ preserves "cofinality $> \omega$," for if $\sigma$ is a name for a function from $\omega$ into ORD then for each $n$, $D_n = \{T \in \mathbb{S}(s) \mid$ For some finite $d$, $T \Vdash \sigma(n) \in \hat{d}\}$ is $n$-dense and hence our result implies that the range of $\sigma$ is covered by a set countable in the ground model.

So suppose $T \in \mathbb{S}(s)$ and $D_n$ is open and $n$-dense for each $n$. Let $M$ be a countable elementary submodel of the ground model $L[X(s)]$ containing $T$ and $\langle D_n \mid n \in \omega \rangle$ as elements and let $\alpha = M \cap \omega_1^L$. Also let $\alpha_0 < \alpha_1 < \cdots$ be an $\omega$-sequence cofinal in $\alpha$, belonging to $\mathcal{A}_\alpha$. Now we can choose $T = T_0 \geq_0 T_1 \geq_1 T_2 \geq_2 \cdots$ so that $T_{n+1} \in D_n \cap M$ and $|T_{n+1}| \geq \alpha_{n+1}$. Note that $M$ is isomorphic to $\bar{M} \in \mathcal{A}_\alpha$ as if $\bar{M}$ is the transitive collapse of $M$ we get $\bar{M} = L_\mu[X(s) \cap \alpha] \models \alpha = \omega_1$ whereas $L_{\mu_\alpha}[X(s) \cap \alpha] \models \alpha$ is countable. So in fact the sequence $\langle T_n \mid n \in \omega \rangle$ can be chosen as an element of $\mathcal{A}_\alpha$. Then $T^* = \bigcap_n T_n$ belongs to each $D_n$, $T^* \leq T$ and $T^* \in \mathbb{S}(s)$ as $T^* \in \mathcal{A}_\alpha$. $\square$

Next we consider products of the $\mathbb{Q}(s)$, $\mathbb{R}(s)$, $\mathbb{S}(s)$ forcings.

Let $\mathbb{Q} = \Pi_s\mathbb{Q}(s)$.

**Lemma 6.8.** *$\mathbb{Q}$ is cofinality-preserving.*

*Proof.* For $i$ regular and infinite, we use Lemma 6.5 to write $\mathbb{Q} \simeq \mathbb{Q}_{i+1} * \mathbb{Q}_{i+1,\infty}$ where $\mathbb{Q}_{i+1} = \Pi_s\mathbb{Q}_{i+1}(s)$ and $\mathbb{Q}_{i+1,\infty} = \Pi_s\mathbb{Q}_{i+1,\infty}(s)$. More explicitly, if $G_{i+1}$ denotes the $\mathbb{Q}_{i+1}$-generic then $\mathbb{Q}_{i+1,\infty} = \{f \in L[G_{i+1}] \mid \mathrm{Dom}(f) = \text{All hyperstrings and } \forall s(f(s) \in \mathbb{Q}_{i+1,\infty}(s) \subseteq L[G_{i+1}(s)])\}$, where $G_{i+1}(s)$ denotes the $s$-component of $G_{i+1}$. This factoring is valid because $L[G_{i+1}] \cap L^\omega \subseteq L$ and hence any $f \in \mathbb{Q}_{i+1,\infty}$ is $\mathbb{Q}_{i+1}$-named by a constructible sequence of $\mathbb{Q}_{i+1}(s)$-names, $\langle\sigma(s) \mid s$ is a hyperstring$\rangle$.

As in Lemma 6.4, $\mathbb{Q}_{i+1}$ is $i^+$-cc, using the fact that supports have size $< i = i^{\aleph_0}$. The lemma follows if we show that $\mathbb{Q}_{i+1,\infty}$ is $i^+$-distributive in $L[G_{i+1}]$. Suppose $\langle\Delta_j \mid j < i\rangle$ are open dense on $\mathbb{Q}_{i+1,\infty}$, $\langle\Delta_j \mid j < i\rangle \in L[G_{i+1}]$ and $p$ is a condition in $\mathbb{Q}_{i+1,\infty}$. Of course we can successively extend $p$ to meet the $\Delta_j$'s but the problem is to guarantee that at limit stages $\lambda$, our $\lambda^{\text{th}}$ extension $p_\lambda$ satisfies $p_\lambda(s) \in L[G_{i+1}(s)]$ for each $s$ and not just $p_\lambda(s) \in L[G_{i+1}]$. So instead we build a constructible sequence of *names* for conditions as follows. Let $\sigma_0$ be a $\mathbb{Q}_{i+1}$-name, and $q_0$ an element of $G_{i+1}$ such that $\sigma_0^{G_{i+1}} = p$ and $q_0 \Vdash \sigma_0 \in \mathbb{Q}_{i+1,\infty}$ and $\langle\Delta_j \mid j < i\rangle$ are open dense. Now choose a $\mathbb{Q}_{i+1}$-name $\sigma_1$ and $q_1 \leq q_0$ such that $q_1 \Vdash \sigma_1$ meets $\Delta_0$ and $\mathbb{Q}_{i+1} \Vdash \sigma_1 \leq \sigma_0$ in $\mathbb{Q}_{i+1,\infty}$. Then choose $q_2 \leq q_0$, $q_2$ incompatible with $q_1$ and $\sigma_2$ such that $q_2 \Vdash \sigma_2$ meets $\Delta_0$ and $\mathbb{Q}_{i+1} \Vdash \sigma_2 \leq \sigma_1$. Continue in this way, selecting conditions $q_j \leq q$ incompatible with earlier choices, using Lemma 6.6 to guarantee that for limit $\lambda$, $\sigma_\lambda$ is a $\mathbb{Q}_{i+1}$-name for a condition in $\mathbb{Q}_{i+1,\infty}$. As $\mathbb{Q}_{i+1}$ is $i^+$-cc, there will be a stage $j_0 < i^+$ so that $q_0 \Vdash \sigma_{j_0} \leq \sigma_0$, $\sigma_{j_0}$ meets $\Delta_0$. Now repeat the process for $\Delta_1, \Delta_2, \ldots$ until obtaining a $\sigma$ such that $q_0 \Vdash \sigma \leq \sigma_0$, $\sigma$ meets each $\Delta_j$, $j < i$. Then $p^* = \sigma^{G_{i+1}}$ extends $p$ and meets each $\Delta_j$, $j < i$, proving $i^+$-distributivity of $\mathbb{Q}_{i+1,\infty}$. □

Next let $\mathbb{R} = \Pi_s\mathbb{R}(s)$, a forcing defined in $L[G]$, where $G$ denotes the $\mathbb{Q}$-generic.

**Lemma 6.9.** *$\mathbb{Q} \Vdash \mathbb{R}$ is cofinality-preserving.*

*Proof.* For $\alpha$ regular we factor $\mathbb{R}$ as $\mathbb{R}_\alpha * \mathbb{R}^{H_\alpha}$, where $\mathbb{R}_\alpha = \Pi_s\mathbb{R}_\alpha(s)$, $\mathbb{R}_\alpha(s)$ is the coding of $G(s)$ by a subset $H_\alpha(s)$ of $\alpha^+$, and where $\mathbb{R}^{H_\alpha} = \Pi_s\mathbb{R}^{H_\alpha(s)}$, $\mathbb{R}^{H_\alpha(s)}$ the coding of $H_\alpha(s)$ by a subset of $\omega_1^L$.

Now as in the proof of Lemma 6.8, by building antichains in $\mathbb{Q}_{\alpha+1}$, we can establish the $\alpha^+$-distributivity of $\mathbb{Q}_{\alpha+1,\infty} * \mathbb{R}_\alpha$ in $L[G_{\alpha+1}]$, where $G_{\alpha+1}$ denotes the $\mathbb{Q}_{\alpha+1}$-generic. Similarly, if $\Delta_\beta$ is open and *$\beta$-dense* on $\mathbb{Q}_{\alpha+1,\infty} * \mathbb{R}$ for successor cardinals $\beta < \alpha$ then the intersection of the $\Delta_\beta$'s is dense on $\mathbb{Q}_{\alpha+1,\infty} * \mathbb{R}$ (where $\Delta$ is $\beta$-dense iff every $(q, p)$ can be extended to $(q^*, p^*)$ in $\Delta$ such that $p, p^*$ agree below $\beta$). For successor cardinals $\alpha$, $\mathbb{R}^{H_\alpha}$ is $\alpha^+$-cc in $L[G][H_\alpha]$, as restraints in the coding of $H_\alpha$ by a subset of $\alpha$ have size $< \alpha = \alpha^{\aleph_0}$. It follows that $\mathbb{Q} \Vdash \mathbb{R}$ preserves cofinalities. □

Finally consider $\mathbb{S} = \Pi_s \mathbb{S}(s)$, a forcing defined in $L[G * H]$, where $G * H$ denotes the $\mathbb{Q} * \mathbb{R}$-generic. Let $X(s) \subseteq \omega_1^L$ correspond to $H(s)$, where $H = \Pi_s H(s)$.

**Lemma 6.10.** $\mathbb{Q} * \mathbb{R} \Vdash \mathbb{S}$ *is cofinality-preserving.*

*Proof.* Let $\langle s_n \mid n \in \omega \rangle$ be a constructible enumeration of all hyperstrings. Say that $\Delta \subseteq \mathbb{S}$ is *n-dense* if for all $p \in \mathbb{S}$ there is $p^* \leq_n p$, $p^* \in \Delta$ (where $p^* \leq_n p$ iff for all $k \leq n$, $p^*(s_k) \leq_n p(s_k)$). Note that $\mathbb{Q} * \mathbb{R} \Vdash 2^\omega \subseteq L$, and in fact for stationary many $\alpha < \omega_1^L$, $\langle X(s) \cap \alpha \mid s \text{ a hyperstring} \rangle$ belongs to $L_{\mu_\alpha} = L_{\mu_\alpha}[X(s) \cap \alpha]$ for each $s$, so just as in the proof of Lemma 6.7(b) we can show that if $\langle \Delta_n \mid n \in \omega \rangle \in L[G * H]$, $\Delta_n$ $n$-dense and open for each $n$, then $\bigcap_n \Delta_n$ is dense. It follows that $\mathbb{Q} * \mathbb{R} \Vdash \mathbb{S}$ preserves cofinalities. □

## The Forcing $P$

Let $G * H$ be $\mathbb{Q} * \mathbb{R}$-generic and let $J$ be $\mathbb{S}$-generic over $L[G * H]$. $J$ is determined by $\langle \mathbb{R}(s) \mid s \text{ a hyperstring} \rangle$ where $R(s)$ is the unique real belonging to $\bigcap\{[p(s)] \mid p \in J\}$.

Now let $R = \{\langle u_0, \ldots, u_n \rangle \mid u_0 \subseteq R(\emptyset)$ and for all $i < n, u_{i+1} \subseteq R(\langle u_0, \ldots, u_i \rangle)\}$. Thus $R$ is an element of $L[G * H][\langle R(s) \mid s \text{ a hyperstring} \rangle]$, a cofinality-preserving extension of $L$. The desired forcing $P$ is a forcing that produces the extension $L[R]$ of $L$.

We define $P$. For $q^0 \in \mathbb{Q}$ and $s$ a hyperstring let $q^0[s]$ be defined by $q^0[s](t) = \emptyset$ if $t \neq s$, $q^0[s](s) = q^0(s)$. A condition in $P$ is $p = (\bar{s}, q)$ where $q = (q^0, q^1) \in \mathbb{Q} * \mathbb{R}$, $\bar{s}$ is a hypertree and for all hyperstrings $s$, $q^0[s] \Vdash (q^1(s), \bar{s}(s)) \in \mathbb{R}(s) * \mathbb{S}(s)$. And $(\bar{t}, r) \leq (\bar{s}, q)$ iff for all hyperstrings $s$ lying on $\bar{t}$, $r^0[s] \leq q^0[s]$ in $\mathbb{Q}$, $r^0[s] \Vdash r^1(s) \leq q^1(s)$ in $\mathbb{R}(s)$ and $\bar{t}(s)$ is a subtree of $\bar{s}(s)$.

**Lemma 6.11.** *$P$ preserves cofinalities.*

*Proof.* By Lemma 6.10, $\mathbb{Q} * \mathbb{R} * \mathbb{S}$ preserves cofinalities. So it suffices to prove the following:

**Sublemma 6.12.** *Suppose $G * H * J$ is $\mathbb{Q} * \mathbb{R} * \mathbb{S}$-generic. Let $K_0 = \{(\bar{s}, q) \in P \mid (r^0, r^1, \bar{t}) \in G * H * J$ for some $(r^0, r^1, \bar{t})$ such that $r^0(s) = q^0(s)$, $r^0[s] \Vdash r^1(s) = q^1(s)$, $\bar{t}(s) = \bar{s}(s)$ for all hyperstrings $s$ which lie on $\bar{s}\}$ and $K = \{(\bar{s}, q) \in P \mid (\bar{s}, q)$ is extended by an element of $K_0\}$. Then $K$ is $P$-generic.*

*Proof of Sublemma.* First we show that $K_0$ is compatible: If $(\bar{s}_0, q_0)$, $(\bar{s}_1, q_1)$ belong to $K_0$, as witnessed by $(r_0^0, r_0^1, \bar{t}_0)$, $(r_1^0, r_1^1, \bar{t}_1)$ respectively, then by compatibility of $G * H * J$, we can choose $(r^0, r^1, \bar{t}) \in G * H * J$ extending both $(r_0^0, r_0^1, \bar{t}_0)$, $(r_1^0, r_1^1, \bar{t}_1)$. Define $(\bar{s}, q)$ by: $\bar{s} = \bar{t}$, $q^0 = r^0$, $q^1 =$ a canonical $\mathbb{Q}$-term such that for all $s$, $\mathbb{Q} \Vdash q^1(s)$ is the greatest lower bound to $r_0^1(s)$, $r_1^1(s)$ if the latter are compatible, $q^1(s) = \emptyset$ otherwise. Then for all $s$, $q^0[s] \Vdash (q^1(s) \in \mathbb{R}(s)$ and $q^1(s) \Vdash \bar{s}(s) \in \mathbb{S}(s))$, the latter since $\bar{s}(s)$ is a tree, not just a name for a tree. So $(\bar{s}, q) \in P$. And clearly $(\bar{s}, q)$ extends both $(\bar{s}_0, q_0)$, $(\bar{s}_1, q_1)$ so $K_0$ is compatible.

It follows that $K$ is compatible. By definition $K$ is upward closed. Now we show that $K$ meets each (constructible) open dense $\Delta \subseteq P$: Given such a $\Delta$ consider $\Delta^* = \{(r^0, r^1, \bar{t}) \mid r^0 \in \mathbb{Q}, r^0[s] \Vdash r^1(s) \in \mathbb{R}(s)$ for each hyperstring $s$, $\bar{t}$ is a function from Hyperstrings to Perfect Trees, $(r^0[s], r^1(s)) \Vdash \bar{t}(s) \in \mathbb{S}(s)$ for each hyperstring $s$ and $(\bar{t}, (r^0, r^1)) \in \Delta\}$. We claim that $\Delta^*$ is dense on $\mathbb{Q}*\mathbb{R}*\mathbb{S}$: Given $(r^0, r^1, \tau) \in \mathbb{Q}*\mathbb{R}*\mathbb{S}$, we can extend to meet all but the last condition for membership in $\Delta^*$, by first extending $(r^0, r^1, \tau)$ to force $\tau = \bar{t} \in \mathbb{S}$, and then extending $r^1$ so that $\mathbb{Q} \Vdash r^1 \in \mathbb{R}$. Now notice that the resulting $(\bar{t}, (r^0, r^1))$ belongs to $P$ and therefore can be extended into $\Delta$. As $\Delta$ is open, we can assume that this extension $(\bar{t}_*, (r^0_*, r^1_*))$ agrees with $(\bar{t}, (r^0, r^1))$ at all hyperstrings $s$ *not* lying on $\bar{t}_*$. But then $(r^0_*, r^1_*, \bar{t}_*)$ is an extension of $(r^0, r^1, \tau)$ in $\mathbb{Q} * \mathbb{R} * \mathbb{S}$, belonging to $\Delta^*$. So $\Delta^*$ is dense.

Choose $(r^0, r^1, \bar{t}) \in \Delta^* \cap G * H * J$, by genericity. Then $(\bar{t}, (r^0, r^1)) \in \Delta \cap K_0$. So $K$ is $P$-generic. □ (Sublemma)

□

## *P* is Relevant

Suppose $j_1 < \cdots < j_{n+1}$ belong to $I$ and $s$ is a hyperstring incompatible with $p(j_1, \ldots, j_{n+1})_0$. Then there can be no $\mathbb{Q}(s)$-generic, as such a generic would force us to add a CUB $C \subseteq j_1$ whose intersection with $I(j_1, \ldots, j_{n_1})$ is of cardinality $\leq n$, for *any* $j_1 < \cdots < j_{i+1}$ in $I$, by the indiscernibility of $I$ and Lemma 6.2(b). Setting $j_1 = \omega_1$ gives a contradiction.

In particular there is no $\mathbb{Q}$-generic and hence no $\mathbb{Q} * \mathbb{R} * \mathbb{S}$-generic. Nonetheless we show now that there does exist a $P$-generic, which will quickly give us the desired absolute singleton.

More explicitly, we describe a $\Sigma_1$ over $L$ class function or *procedure* $\langle \alpha_1, \ldots, \alpha_n \rangle \longmapsto p(\alpha_1, \ldots, \alpha_n)$ that associates a condition $p(\alpha_1, \ldots, \alpha_n) \in P$ to each $\langle \alpha_1, \ldots, \alpha_n \rangle \in \mathrm{ORD}^n$. The desired $P$-generic is $\{p \in P \mid p$ is compatible with $p(i_1, \ldots, i_n)$ for each $i_1 \cdots < i_n$ in $I\}$. Now we come to a key point: Our definition of "$s$-bad," and in particular our definition of $\mathbb{Q}$, depended upon knowing a $\Sigma_1$ index for the procedure $\langle \alpha_1, \ldots, \alpha_n \rangle \longmapsto p(\alpha_1, \ldots, \alpha_n)$. But of course to describe our procedure we must know the definition of $P$, in order to know for which forcing we wish to build a generic. The escape from this circularity is to use the Recursion Theorem: Thus, given a guess $e$ at a $\Sigma_1$ index for our procedure, we use this guess to define $P$ and build our procedure, with $\Sigma_1$ index $e^*$. This will be a total procedure, even if $e$ is an index for a partial one. The Recursion Theorem implies that we can assume that $e$ is a *correct* guess, in the sense that $e$ and $e^*$ define the same procedure. In other words, in building our total procedure $\langle \alpha_1, \ldots, \alpha_n \rangle \longmapsto p(\alpha_1, \ldots, \alpha_n)$ we may assume that we have a $\Sigma_1$ index for it, and therefore that we have a definition of our forcing $P$. The correctness of our guess will of course be essential in showing that our construction of a $P$-generic is successful.

Our next lemma, based on the Product Lemma 2.31, will be of great help in the construction of a $P$-generic. We make use of the following definitions.

**Definition.** If $p = (\bar{s}, (q^0, q^1)) \in P$ the we write $\bar{s} = \bar{s}(p), q^0 = q^0(p), q^1 = q^1(p)$. As before, we say that $p \in P$ *meets* $\Delta \subseteq P$ if $p$ extends an element of $\Delta$. For $\Delta \subseteq P$, $\Delta$ is *predense on the* $\mathbb{Q}$*-component* if $p \in P \Longrightarrow \exists p^* \leq p$ ($p^*$ meets $\Delta$ and $q^1(p^*) = q^1(p)$). Similarly, $\Delta$ is *predense on the* $\mathbb{R}$*-component* if $p \in P \Longrightarrow \exists p^* \leq p$ ($p^*$ meets $\Delta$ and $q^0(p^*) = q^0(p)$). $K \subseteq P$ is $P$*-generic on the* $\mathbb{Q}$*-component* if $K$ is compatible, upward closed and intersects each $\Delta$ that is predense on the $\mathbb{Q}$-component; $P$*-generic on the* $\mathbb{R}$*-component* is defined similarly.

**Lemma 6.13.** *Suppose $K \subseteq P$ is $P$-generic on the $\mathbb{Q}$-component and also on the $\mathbb{R}$-component. Then $K$ is $P$-generic.*

*Proof.* Suppose $\Delta \subseteq P$ is predense; we want to show that some condition in $K$ meets $\Delta$. For each hypertree $\bar{s}$ consider $\Delta^1(\bar{s}) = \{p \in P \mid q(p)^0 \Vdash$ There is $\bar{t} \leq \bar{s}$ such that $(\bar{t}, (q^0, q(p)^1))$ meets $\Delta$ for some $q^0 \in G\}$, where $G$ denotes the $\mathbb{Q}$-generic. We claim that $\Delta^1(\bar{s})$ is predense on the $\mathbb{R}$-component, for each $\bar{s}$: Suppose $p \in P$ and let $q^1$ be a $\mathbb{Q}$-term such that $q(p)^0 \Vdash$ For some $\bar{t} \leq \bar{s}$, $(\bar{t}, (q^0, q^1))$ meets $\Delta$ for some $q^0 \in G$. Such a $q^1$ exists as the predensity of $\Delta$ implies that $\{q^0 \mid$ For some $\bar{t} \leq \bar{s}$ and some $q^1$, $(\bar{t}, (q^0, q^1))$ meets $\Delta\}$ is dense on $\mathbb{Q}$ below $q(p)^0$. We may also assume that $\mathbb{Q} \Vdash$ If $q(p)^1$ belongs to $\mathbb{R}$ then $q^1 \leq q(p)^1$. Thus $(\bar{s}(p), (q(p)^0, q^1)) \leq p$ meets $\Delta^1(\bar{s})$, so the latter is predense on the $\mathbb{R}$-component, as claimed.

Now for each hypertree $\bar{s}$ choose $p(\bar{s}) \in K$ meeting $\Delta^1(\bar{s})$, by genericity of $K$ on the $\mathbb{R}$-component. Define $\Delta^0(\bar{s}) = \{p \in P \mid$ For some $\bar{t} \leq \bar{s}$, $(\bar{t}, (q(p)^0, q(p(\bar{s}))^1))$ meets $\Delta\} \cup \{p \in P \mid q(p)^0, q(p(\bar{s}))^0$ are incompatible in $\mathbb{Q}\}$. Then $\Delta^0(\bar{s})$ is predense on the $\mathbb{Q}$-component for each $\bar{s}$. In fact $\Delta^0 = \bigcap\{\Delta^0(\bar{s}) \mid \bar{s}$ a hypertree $\}$ is predense on the $\mathbb{Q}$-component, as $\mathbb{Q}$ is $\omega_2^L$-distributive and there are only $\omega_1^L$-many hypertrees $\bar{s}$.

Choose $p^0 \in K$ meeting $\Delta^0$, by genericity of $K$ on the $\mathbb{Q}$-component. Finally let $\bar{\Delta} = \{p \mid (\bar{s}(p), q(p^0)^0, q(p(\bar{s}(p)))^1)$ meets $\Delta\} \cup \{p \in P \mid q(p)^0, q(p(\bar{s}))^0$ are incompatible for all $\bar{s}\}$. Then $\bar{\Delta}$ is predense on the $\mathbb{Q}$-component (and on the $\mathbb{R}$-component) so we can choose $p \in K$ meeting $\bar{\Delta}$.

Then $p^* = (\bar{s}(p), q(p^0)^0, q(p(\bar{s}(p)))^1)$ belongs to $K$ since $p$, $p^0$ and $p(\bar{s}(p))$ do, and in addition by the compatibility of $K$, $p^*$ meets $\Delta$. □

**Remark.** The above use of the $\omega_2^L$-distributivity of $\mathbb{Q}$ can be eliminated if we redefine $P$ so as to allow $\bar{s}$ to be a $\mathbb{Q}$-term.

We are ready to build a $P$-generic $K$. We guarantee that $K$ is $P$-generic on the $\mathbb{Q}$-component using the technique which demonstrated the relevance of Reverse Easton forcing. We guarantee that $K$ is $P$-generic on the $\mathbb{R}$-component using the technique which established the relevance of Jensen coding. Lemma 6.13 enables us to employ these techniques simultaneously, on the respective components.

Let $\langle k_\alpha \mid \alpha > 0\rangle$ be the increasing enumeration of the class $I$ of Silver indiscernibles. We shall build $K_0 \subseteq P \cap L_{k_\omega} = P_0$ to be $P_0$-generic over $L_{k_\omega}$ and to be preserved by all elementary embeddings $L_{k_\omega} \longrightarrow L_{k_\omega}$ obtained by shifting indiscernibles $\langle k_n \mid 0 < n < \omega\rangle$. Then $K$ is obtained from $K_0$ by "stretching." More explicitly, we define $p_n = p(k_1, \ldots, k_n) \in P_0$ by induction on $n$ with $p_0 \geq p_1 \geq \cdots$ and then

$K_0 = \{p \in P_0 \mid p$ is compatible with each $p(\vec{k})$, $\vec{k} \in (I \cap k_\omega)^{<\omega}\}$, $K = \{p \in P \mid p$ is compatible with each $p(\vec{k})$, $\vec{k} \in I^{<\omega}\}$. (Of course we intend $(I \cap k_\omega)^{<\omega}$, $I^{<\omega}$ to consist of all finite *increasing* sequences from $I \cap k_\omega, I$.) The function $\vec{k} \longmapsto p(\vec{k})$ is the restriction to $I^{<\omega}$ of a $\Sigma_1$ over $L$ procedure $\vec{\alpha} \longmapsto p(\vec{\alpha})$, $\vec{\alpha} \in \mathrm{ORD}^{<\omega}$.

We construct the $p_n$'s. Set $p_0 = p_1 =$ the trivial condition $(\emptyset, (\emptyset, \emptyset))$. Assume that $n \geq 1$ and $p_n = (\bar{s}_n, (q_n^0, q_n^1))$ has been defined; we now define $p_{n+1} = (\bar{s}_{n+1}, (q_{n+1}^0, q_{n+1}^1))$ extending $p_n$. There are two cases.

***n* odd.** We assume inductively that $\mathbb{Q} \Vdash$ For all $s$, $q_n^1(s)$ is a coding condition with domain either $\emptyset$ or $\mathrm{Card} \cap [\omega, k_1^+]$ such that $q_n^1(s)^*_{k_1}$, the restraint at $k_1$, is empty and $q_n^1(s)(k_1^+) = (x, \emptyset)$ where if $p$ is defined from $k_2, \ldots, k_n$ as was $p_{n-1}$ from $k_1, \ldots, k_{n-1}$, we have $q(p)^1(s)_{k_1^+} = x$.

Now using the distributivity of Jensen coding choose $\bar{p}_{n+1} \leq p_n$ so that $q(\bar{p}_{n+1})^0 = q_n^0$ and for each $\gamma < k_1$ in $\mathrm{Card}^+ = \{\omega\} \cup$ Infinite Successor cardinals, $\bar{p}_{n+1}$ reduces each $\Delta \in$ Skolem Hull$(\gamma \cup \{k_1, \ldots, k_n\})$ in $L_{k_1^{++}}$ below $\gamma$ on the $\mathbb{R}$-component, provided that $\Delta$ is predense on the $\mathbb{R}$-component. ("Reduces below $\gamma$ on the $\mathbb{R}$-component" is defined just like "Reduces below $\gamma$" for Jensen coding.) Also, if $n = 1$ require that $\bar{p}_2$ reduces each $\Delta \in L_{k_1+1}$ below some $\gamma \in \mathrm{Card}^+ \cap k_1$ on the $\mathbb{R}$-component, provided $\Delta$ is predense on the $\mathbb{R}$-component for the forcing $P \cap L_{k_1}$. Now modify $\bar{p}_{n+1}$ to obtain $p_{n+1}$, as follows: $\bar{s}(p_{n+1}) = \bar{s}_n$, $q(p_{n+1})^0 = q_n^0$, $q(p_{n+1})^1(s) \restriction k_1^+ = q(\bar{p}_{n+1})^1(s) \restriction k_1^+$ for all $s$, however $q(p_{n+1})^1(s)^*_{k_1} = \emptyset$ and $q(p_{n+1})^1(s)(k_1^+) = (x, \emptyset)$ for all $s$ where if $p$ is defined from $k_2, \ldots, k_{n+1}$ as was $p_n$ from $k_1, \ldots, k_n$ then $q(p)^1(s)_{k_1^+} = x$.

***n* even.** Assume that we have fixed in advance a recursive list $\langle t_n \mid n \in \omega\rangle$ of Skolem terms for $L$ such that $\langle t_n(k_1, \ldots, k_n) \mid n \text{ even}\rangle$ enumerates all elements of $L_{k_2^+}$. If $\Delta = t_n(k_1, \ldots, k_n)$ is not predense on the $\mathbb{Q}$-component then set $p_{n+1} = p_n$. Otherwise choose $\bar{p}_{n+1} \leq p_n$ meeting $\Delta$ such that $q(\bar{p}_{n+1})^1 = q_n^1$ and the support of $q(\bar{p}_{n+1})^0$ is contained in $k_2 + 1$.

We would like to include $\bar{p}_{n+1}$ in our $P$-generic, however doing so may conflict with our desire that our $P$-generic be preserved by embeddings derived from shifting indiscernibles. The problem is that $\mathbb{Q}_{k_2}$ may fail to force $q(\bar{p}_{n+1})^0(s)(k_2)(\alpha_1, \ldots, \alpha_m)$ to extend $G_{k_2}(s)(k_1)(\bar{\alpha}_1, \ldots, \bar{\alpha}_m)$ where $G_{k_2}(s)$ denotes the $\mathbb{Q}_{k_2}(s)$-generic and where $(\bar{\alpha}_1, \ldots, \bar{\alpha}_m)$ is defined from $k_1 \cup \{k_1, \ldots, k_{\ell-1}\}$ (some $\ell$) just as $(\alpha_1, \ldots, \alpha_m)$ is defined from $k_1 \cup \{k_2, \ldots, k_\ell\}$. As we are considering $(\alpha_1, \ldots, \alpha_m) \in \mathrm{Dom}(q(\bar{p}_{n+1})^0(s)(k_2))$, we have that $(\alpha_1, \ldots, \alpha_m)$ is definable from $k_2 \cup \{k_2, \ldots, k_n\}$ and hence from $k_1 \cup \{k_2, \ldots, k_\ell\}$. So we may assume $\ell = n$.

We may assume that $\bar{p}_{n+1}$ *strongly* meets $\Delta$ in the sense that it still meets $\Delta$ even after $q(\bar{p}_{n+1})^0(s)(k_2)(\alpha_1, \ldots, \alpha_m) \restriction k_1$ has been constructibly changed for all hyperstrings $s$ and $(\alpha_1, \ldots, \alpha_m) \in \mathrm{Dom}(q(\bar{p}_{n+1})^0(s)(k_2))$. This is because $\mathbb{Q}_{k_2} \Vdash \mathbb{Q}_{k_2,\infty}$ is $k_1^{++}$-distributive.

Now proceed as follows: If possible, extend $\bar{s}(\bar{p}_{n+1})$ to $\bar{s}$ so that for $s$ lying on $\bar{s}$, a condition results if we alter $q(\bar{p}_{n+1})^0(s)(k_2)(\alpha_1, \ldots, \alpha_m)$ so as to be forced by $\mathbb{Q}_{k_2}$ to extend $G_{k_2}(s)(k_1)(\bar{\alpha}_1, \ldots, \bar{\alpha}_m)$ whenever $(\alpha_1, \ldots, \alpha_m) \in \mathrm{Dom}(q(\bar{p}_{n+1})^0(s)(k_2))$ and $(\bar{\alpha}_1, \ldots, \bar{\alpha}_m)$ is defined from $k_1 \cup \{k_1, \ldots, k_{n-1}\}$ just as $(\alpha_1, \ldots, \alpha_m)$ is defined from $k_1 \cup \{k_2, \ldots, k_n\}$. In this *good subcase* we take $p_{n+1}$ to be the result of these changes, with $\bar{s}(p_{n+1}) = \bar{s}$, $q(p_{n+1})^1 = q_n^1$. Otherwise we are in the *bad subcase* and we obtain $p_{n+1}$ by simply extending each tree $\bar{s}(\bar{p}_{n+1})(s)$ to $T_s$ so that stem $T_s$ has length at least $n$, and such that $x_i$ is not a path through $T_s$ for any $i \leq n$ where $\langle x_i \mid i < \omega \rangle$ is the enumeration of the constructible reals defined by: $x_i = t_i(k_1, \ldots, k_i)$ if $i$ is even and this is a real, $x_i = 0$ otherwise. This defines $\bar{s}(p_{n+1})$; we set $q(p_{n+1})^0 = q_n^0$ and $q(p_{n+1})^1 = q_n^1$.

This completes the construction of the $p_n$'s. We obtain a procedure by defining $p(\alpha_1, \ldots, \alpha_n)$ to be that object defined from $(\alpha_1, \ldots, \alpha_n)$ as was $p_n$ defined above from $k_1, \ldots, k_n$. To see that this procedure is $\Sigma_1$ over $L$, simply note that all quantifiers in the definition of $p_{n+1} = p(k_1, \ldots, k_{n+1})$ can be bounded by $L_{k_{n+1}}$.

Now assume that $P$ was defined using a correct guess $e$ at an index for the above procedure, obtained by the Recursion Theorem. Using the correctness of $e$ we show:

**Lemma 6.14.** *The good subcase always occurs when n is even.*

*Proof.* Suppose not. Then note that the bad subcase ensures that $\langle \bar{s}_n \mid n \in \omega \rangle$ converges to a sequence of nonconstructible reals, in the sense that the unique path $R(s)$ through each $\bar{s}_n(s)$, $n \in \omega$ is nonconstructible (for $s$ lying on each $\bar{s}_n$). Consider $\bar{s}_n(\emptyset)$. We claim that for $u \subseteq R(\emptyset)$, $k_1 \notin I(\alpha_1, \ldots, \alpha_m)$ for $\langle u \rangle$-bad $(\alpha_1, \ldots, \alpha_m)$, $\alpha_1 = k_2$ which are definable from $k_1 \cup \{k_2, \ldots, k_n\}$. For otherwise, define $(\beta_1, \ldots, \beta_m)$ from $k_1 \cup \{k_{m+2}, \ldots, k_{m+n}\}$ just as $(\alpha_1, \ldots, \alpha_m)$ is defined from $k_1 \cup \{k_2, \ldots, k_n\}$. We have that $(\beta_1, \ldots, \beta_m)$ is also $\langle u \rangle$-bad, by indiscernibility, and of course $k_1, \ldots, k_m \in I(\beta_1, \ldots, \beta_m)$. By Lemma 6.2(b), $(k_1, \ldots, k_m)$ is also $\langle u \rangle$-bad. But $\bar{s}(p(k_1, \ldots, k_m)) = \bar{s}_m$ and $\langle u \rangle$ lies on $\bar{s}_m$, contradiction.

Thus $\bar{s}(\emptyset)$ has a nonconstructible path, all of whose finite initial segments $u$ have the property that the changes required in the good subcase can be successfully made for $\langle u \rangle$. As a constructible tree with a nonconstructible path has a constructible perfect subtree, we can thin $\bar{s}_n(\emptyset)$ as required in the good subcase. Similarly, we can thin each $\bar{s}_n(s)$, $s$ lying on $\bar{s}$ as required by the good subcase. If we initially extend $\bar{s}_n$ to be of sufficiently large rank, these thinnings will not change rank and will therefore result in $\bar{s}_{n+1}(s) \in \mathcal{C}(s)$. □

Now we have:

**Lemma 6.15.** *Let $K = \{p \in P \mid p$ is compatible with each $p(i_1, \ldots, i_n)$, $i_1 < \cdots < i_n$ in $I\}$. Then $K$ is $P$-generic.*

*Proof.* First we show that $p(\vec{\imath})$, $p(\vec{\jmath})$ are compatible for $\vec{\imath}, \vec{\jmath}$ increasing sequences from $I$. Without loss of generality, we may assume that $\vec{\imath} = \langle i_1, \ldots, i_n \rangle$, $\vec{\jmath} = \langle j_1, \ldots, j_n \rangle$ have the same length and $i_1 \leq j_i$. Note that $\bar{s}(p(\vec{\imath})) = \bar{s}(p(\vec{\jmath})) = \bar{s}_n$. Now $q(p(\vec{\imath}))^0, q(p(\vec{\jmath}))^0$ have support $\subseteq i_2 + 1, j_2 + 1$, respectively and by indiscernibility $q(p(\vec{\jmath}))^0 \restriction j_1 = q(p(\vec{\imath}))^0 \restriction i_1$, as the supports are Easton. If $j_1 = i_1$

then $q(p(\vec{j}))^0 \restriction j_2 = q(p(\vec{\imath}))^0 \restriction i_2$ and therefore $q(p(\vec{\imath}))^0, q(p(\vec{j}))^0$ are compatible. Also if $\{i_1, i_2\} \cap \{j_1, j_2\} = \emptyset$ then $\text{Support}(q(p(\vec{\imath}))^0) \cap \text{Support}(q(p(\vec{j}))^0) \subseteq i_1$ and hence we get the compatibility of $q(p(\vec{\imath}))^0, q(p(\vec{j}))^0$. Thus we are left with the cases: $i_1 < j_1 < i_2 = j_2$ and $i_2 = j_1$. Now by indiscernibility, $q(p(k_2, k_3, \dots, k_{n+1}))^0$ extends $q(p(k_1, k_3, \dots, k_{n+1}))^0$ and hence in the first of these cases, $q(p(\vec{j}))^0$ extends $q(p(\vec{\imath}))^0$. By construction, $q(p(k_1, \dots, k_n))^0$ and $q(p(k_2, \dots, k_{n+1}))^0$ are compatible and hence in the second of these cases, $q(p(\vec{\imath}))^0, q(p(\vec{j}))^0$ are compatible. Thus for any $\vec{\imath}, \vec{j}$ we have the compatibility of $q(p(\vec{\imath}))^0$ and $q(p(\vec{j}))^0$.

Now consider $q(p(\vec{\imath}))^1, q(p(\vec{j}))^1$. These are equal if $i_1 = j_1$, as they have domain contained in $\text{Card} \cap [\omega, i_1^+]$, $\text{Card} \cap [\omega, j_1^+]$ respectively. By construction, $q(p(k_1, \dots, k_n))^1, q(p(k_1, \dots, k_{n+1}))^1$ agree on $\text{Card} \cap [\omega, k_1^+]$, except possibly the latter may have nonempty restraint at $k_1^+$. But in any event these condition are compatible.

It follows that $p(\vec{\imath}), p(\vec{j})$ are compatible for any $\vec{\imath}, \vec{j}$ and hence $K$ contains all $p(\vec{\imath})$, $\vec{\imath}$ from $I$.

Next we show that $K$ meets all $L$-definable $\Delta \subseteq P$ which are predense on the $\mathbb{Q}$-component. By the $\infty$-cc of $\mathbb{Q}$ we may assume that $\Delta$ is an element of $L$; write $\Delta = t(i_1, \dots, i_n, i_{n+1}, \dots, i_{n+m})$ where $\Delta \in L_{i_{n+1}}$, $m$ odd. Using the $i_n^+$-distributivity of $\mathbb{Q}_{i_n^+,\infty}$ we see that there is a predense on the $\mathbb{Q}$-component $\Delta^* \in L_{i_n^+}$ such that each $p$ in $\Delta^*$ reduces all $\Delta$ of the form $t(\alpha_1, \dots, \alpha_{n-1}, i_n, i_{n+1}, \dots, i_{n+m})$ below some $\gamma \in \text{Card}^+ \cap i_n$, provided $\Delta$ is predense on the $\mathbb{Q}$-component. Now $\Delta^*$ is of the form $t^*(i_n, i_{n+1}, \dots, i_{n+m}) \in L_{i_n^+}$ and hence is met by $p(i_n, i_{n+1}, \dots, i_{n+m})$ since by construction $t^*(k_1, k_2, \dots, k_{m+1}) \in L_{k_2}$ is met by $p(k_1, \dots, k_{m+1})$. Thus if we chose $i_n$ to be least in $I$ such that some $\Delta \in L_{i_n^+}$ predense on the $\mathbb{Q}$-component is not met by $K$, we get a contradiction. So $K$ meets every $L$-definable $\Delta$ which is predense on the $\mathbb{Q}$-component.

Now we consider the $\mathbb{R}$-component. Suppose $\gamma \in \text{Card}$ is least such that some $\Delta \in L_{\gamma^{++}}$ is predense on the $\mathbb{R}$-component but is not met by $K$. If $\gamma \notin I$ then $\Delta \in$ Skolem Hull$(\gamma \cup \{i_1, \dots, i_n\})$ in $L$ for some indiscernibles $i_1 < \dots < i_n$ greater than $\gamma$ and hence by construction, $p(i_1, \dots, i_n)$ reduces $\Delta$ below some $\bar{\gamma} \in \text{Card}^+ \cap \gamma$ (unless $\gamma = \omega$, in which case $\Delta$ is met by $K$ by construction). Otherwise $p(\gamma, i_1, \dots, i_n)$ reduces $\Delta$ below some $\bar{\gamma} \in \text{Card}^+ \cap \gamma$ by construction, provided restraint is added to $q(p(\gamma, i_1 \cdots i_n))^1$ at $\gamma$; adding this restraint to $p(\gamma, i_1, \dots, i_n)$ preserves compatibility with all $p(\vec{j})$, $\vec{j}$ from $I$, and therefore results in a condition in $K$. So again a condition in $K$ reduces $\Delta$ below some $\bar{\gamma} \in \text{Card}^+ \cap \gamma$. It follows that $K$ meets every $\Delta \in L$ which is predense on the $\mathbb{R}$-component. To handle $\Delta$ which are $L$-definable classes appeal to the choice of $\bar{p}_2$ in the construction. So $K$ meets every $L$-definable $\Delta$ which is predense on the $\mathbb{R}$-component.

Finally we show that $K$ is compatible. Indeed any condition in $K$ is extended by a condition $p(\vec{\imath})$, provided restraint is added at finitely many indiscernibles to $q(p(\vec{\imath}))^1$. As restraints can be amalgamated we see that any two conditions in $K$ are compatible. So $K$ is $P$-generic. □

In summary, we have proved:

**Theorem 6.16.** *There is a real $R$ such that:*

(a) $0 <_L R <_L 0^\#$.

(b) *For every ordinal $\alpha$, $L$-cof$(\alpha) = L[R]$-cof$(\alpha)$.*

(c) *For some formula $\varphi$, $R$ is the unique real such that $L[R] \vDash \varphi(R)$.*

*Proof.* Let $K$ be as in Lemma 6.15 and set $R = \{$ hyperstrings $s \mid s$ lies on $\bar{s}(p)$ for each $p \in K\}$. (a), (b) hold for $R$, by Lemma 6.11. For (c), note that $L[R] \vDash \varphi(R)$ where $\varphi(R) \equiv$ Every $s$-bad guess $(\alpha_1, \ldots, \alpha_{n+1})$ has been killed, for all $s \in R$. If $S \neq R$ then choose $s \in S - R$ and $(i_1, \ldots, i_{n+1}) \in I^{n+1}$, $s$ not lying on $\bar{s}(p(i_1, \ldots, i_{n+1}))$, $i_1$ regular. Then $(i_1, \ldots, i_{n+1})$ is $s$-bad, $s \in S$ yet $(i_1, \ldots, i_{n+1})$ cannot be killed as $I(i_1, \ldots, i_{n+1})$ contains a CUB subset of $i_1$. So $L[S] \vDash \sim \varphi(S)$. □

Another consequence of the proof is:

**Theorem 6.17.** *There is an $L$-definable forcing $P$ such that:*

(a) *Every $p \in P$ has incompatible extensions.*

(b) *There is a unique $P$-generic.*

*Moreover $P$ preserves cofinalities and if $G$ is the unique $P$-generic then $L[G] = L[R]$ for some real $R$.*

## 6.2 David's Trick

In David [82a] a technique was presented for creating reals $R$ which not only code classes but in addition are *local* $\Pi^1_2$-singletons. The latter means that $L[R] \vDash R$ is the unique solution to a $\Pi^1_2$ formula. We begin this section by putting David's method in a general form, suitable for obtaining a $\Pi^1_2$-singleton $R$, $0 <_L R <_L 0^\#$ from the construction of Section 6.1.

**Theorem 6.18.** *Suppose $A \subseteq$ ORD, $\langle L[A], A\rangle \vDash$ ZFC $+0^\#$ does not exist, $\varphi$ is a formula and suppose that for every infinite cardinal $\kappa$ of $L[A]$, $H_\kappa^{L[A]} = L_\kappa[A]$ and $\langle L_\kappa[A],\ A \cap \kappa\rangle \vDash \varphi$. Then there exists a $\Pi^1_2$ formula $\psi$ such that:*

(a) *If $R$ is a real satisfying $\psi$ then there is $A$ as above, definable over $L[R]$ from the parameter $R$.*

(b) *For some $\langle L[A], A\rangle$-definable, tame, cofinality-preserving forcing $P$, $P \Vdash \exists R\ \psi(R)$.*

*Moreover if $A$ preserves indiscernibles then $\psi$ has a solution in $L[A, 0^\#]$, preserving indiscernibles.*

*Proof.* Our plan is to create an $\langle L[A], A\rangle$-definable, tame, cofinality-preserving forcing $P$ for adding a real $R$ such that whenever $L_\alpha[R] \vDash \mathrm{ZF}^-$ there is $A_\alpha \subseteq \alpha$, definable over $L_\alpha[R]$ (via a definition independent of $\alpha$), such that $L_\alpha[R] \vDash$ For every cardinal $\kappa$, $H_\kappa = L_\kappa[A_\alpha]$ and $\varphi$ is true in $\langle L_\kappa[A_\alpha], A_\alpha \cap \kappa\rangle$. This property $\psi$ of $R$ is $\Pi^1_2$ and gives us (a), (b).

$P$ is obtained as a modification of the forcing used to prove Theorem 4.1. The following definitions take place inside $L[A]$.

**Definition** (Strings). Let $\alpha$ belong to Card, the class of all infinite cardinals. $S_\alpha$ consists of all $s : [\alpha, |s|) \longrightarrow 2$, $\alpha \leq |s| < \alpha^+$ such that $|s|$ is a multiple of $\alpha$ and:

(1) $\eta \leq |s| \Longrightarrow L_\delta[A \cap \alpha, s \restriction \eta] \vDash \mathrm{card}(\eta) \leq \alpha$ for some $\delta < (\eta^+)^L \cup \omega_2$.

(2) If $\mathcal{A} = \langle L_\beta[A \cap \alpha, s \restriction \eta]\rangle \vDash \mathrm{ZF}^-$ and $\eta = \alpha^+$, then over $\mathcal{A}$, $s \restriction \eta$ codes a predicate $A(s \restriction \eta, \beta) = A^* \subseteq \beta$ such that $A^* \cap \alpha = A \cap \alpha$ and for every infinite cardinal $\kappa$ of $L_\beta[A^*]$, $H_\kappa^{L_\beta[A^*]} = L_\kappa[A^*]$ and $\langle L_\kappa[A^*], A^* \cap \kappa\rangle \vDash \varphi$.

**Remark.** When we say that $s \restriction \eta$ *codes* $A^*$ we refer to the canonical coding (relative to $A \cap \alpha$) described by the proof of Theorem 4.1 of a subset of $\beta$ by a subset of $(\alpha^+)^{\mathcal{A}} = \eta$.

The remainder of the definitions from the proof of Theorem 4.1 remain the same in the present context. We now verify that the proofs of the lemmas in the proof of Theorem 4.1 can successfully accommodate the new restriction (clause (2)) on elements of $S_\alpha$.

**Lemma 6.19** (Distributivity for $R^s$). *Suppose $\alpha \in$ Card, $s \in S_{\alpha^+}$. Then $R^s$ is $\alpha^+$-distributive in $\mathcal{A}^s$.*

*Proof.* Proceed as in the proof of Lemma 4.3. The only new point is to verify that in the proof of the Claim, $t_\lambda$ satisfies clause (2) (of the new definition of $S_\alpha$). The fact that $s$ belongs to $S_{\alpha^+}$ and that $t_\lambda$ codes $\bar{H}_\lambda$ imply that clause (2) holds for $t_\lambda$ whenever $\beta$ is at most $\bar{\mu}_\lambda =$ the height of $\bar{H}_\lambda$. But as $|t_\lambda|$ is definably singular over $L_{\bar{\mu}_\lambda}[t_\lambda]$ these are the only $\beta$'s that concern us. □

**Lemma 6.20** (Extendibility for $P^s$). *Suppose that $\alpha$ is a limit cardinal, $s$ belongs to $S_\alpha$, and $p \in P^s$. Suppose also that $X \subseteq \alpha$ belongs to $\mathcal{A}^s$. Then there exists $q \leq p$ in $P^s$ such that $X \cap \beta \in \mathcal{A}^{q_\beta}$ for each $\beta \in$ Card $\cap\, \alpha$.*

*Proof.* Proceed as in the proof of Lemma 4.4. In the definition of $q$, the only instances of clause (2) to check are for $s_\beta$ when $\mathrm{Even}(Y \cap \beta)$ codes $s_\beta$, $s_\beta$ satisfying clause (1) of the definition of membership in $S_\beta$. But the embedding $\mathcal{A}_\beta \longrightarrow \mathcal{A}$ is $\Sigma_1$-elementary and instances of clause (2) refer to ordinals less than the height of $\mathcal{A}$; so the fact that $s$ belongs to $S_\alpha$ implies that $s_\beta$ belongs to $S_\beta$. □

Finally we have:

**Lemma 6.21** (Distributivity for $P^s$). *Suppose $s \in S_{\beta^+}$, $\beta \in$ Card.*

(a) *If* $\langle D_i \mid i < \beta \rangle \in \mathcal{A}^s$, $D_i$ $i^+$*-dense on* $P^s$ *for each* $i < \beta$ *and* $p \in P^s$ *then there is* $q \leq p$, $q$ *meets each* $D_i$.

(b) *If* $p \in P^s$, $f$ *small,* $f$ *in* $\mathcal{A}^s$ *then there exists* $q \leq p$, $q \in \Sigma_f^p$.

*Proof.* Proceed as in the proof of Lemma 4.5. In the Claim we must verify that $p_\gamma^\lambda$ satisfies clause (b). But once again this is clear by the $\Sigma_1$-elementarity of $\bar{H}_\lambda(\gamma) \longrightarrow H_\lambda(\gamma)$, the fact that $p_\gamma^\lambda$ codes $\bar{H}_\lambda(\gamma)$ and the fact that $L_{\bar{\mu}}[A \cap \gamma, p_\gamma^\lambda] \models |p_\gamma^\lambda|$ is $\Sigma_1$-singular, where $\bar{\mu} =$ height of $\bar{H}_\lambda(\gamma)$. □

The argument of the proof of Lemma 6.21 can also be applied to prove the distributivity of $P$, observing that when building sequences of conditions $\langle p^i \mid i < \lambda \rangle$, $\lambda$ limit to meet an $\langle L[A], A \rangle$-definable sequence of dense classes, one has that $p_\gamma^\lambda$ codes $\bar{H}_\lambda(\gamma)$ of height $\bar{\mu}$, where $L_{\bar{\mu}+1}[A \cap \gamma, p^{\lambda_\gamma}] \models |p_\gamma^\lambda|$ is not a cardinal. Thus there is no additional instance of clause (2) to verify beyond those considered in the proof of Lemma 6.21.

Thus $P$ is tame and cofinality-preserving. The final statement of Theorem 6.18 also follows, using the proof of Theorem 4.18. This completes the proof of Theorem 6.18. □

Using the proof of Theorem 6.18 we can improve Theorem 6.16 so as to provide a solution to the $\Pi_2^1$-Singleton Problem.

**Theorem 6.22** (Solution to the $\Pi_2^1$-Singleton Problem). *There exists a* $\Pi_2^1$*-singleton* $R$ *such that:*

(a) $0 <_L R <_L 0^\#$.

(b) *For every ordinal* $\alpha$, $L\text{-cof}(\alpha) = L[R]\text{-cof}(\alpha)$.

(c) $I^R = I$.

*Proof.* The only modification to the forcing used to prove Theorem 6.16 is in the definition of $\mathbb{R}(s)$. Let $G(s)$ denote the $\mathbb{Q}(s)$-generic and notice that for every $L$-cardinal $\kappa$, $L_\kappa[G(s)] \models$ Each $s$-bad acceptable guess has been killed. Apply the proof of Theorem 6.18 to obtain a new forcing $\mathbb{R}(s)$ for adding $X(s) \subseteq \omega_1$ such that whenever $L_\beta[X(s) \cap (\alpha)] \models \mathrm{ZF}^- + \alpha = \omega_1$ then $L_\beta[X(s) \cap \alpha] \models$ Each $s$-bad acceptable guess has been killed. Moreover we have that $\mathbb{R}(s)$ obeys distributivity as in the coding of Theorem 4.1, and if $G(s)$ is $\mathbb{Q}(s)$-generic and preserves indiscernibles then $\mathbb{R}(s)$ has an indiscernible-preserving generic as well. Thus we may carry out the proof of Theorem 6.16 exactly as before, with the new choices for $\langle \mathbb{R}(s) \mid s$ a hyperstring$\rangle$, $\mathbb{R}$ and $P$. The result is that in the proof of Theorem 6.16, the formula $\varphi(R)$ may be chosen to be: $\varphi(R) \equiv$ For every $\alpha$, if $L_\alpha[R] \models \mathrm{ZF}^-$ then $L_\alpha[R] \models$ Every $s$-bad acceptable guess has been killed, for all $s \in R$. This formula is $\Pi_1$ and hence as a property of reals in $\Pi_2^1$. Thus we have the desired $\Pi_2^1$-singleton. □

## Provable $\Pi^1_2$-Singletons

Suppose $T$ is a theory in the language of set theory. A *$T$-provable* $\Pi^1_2$-singleton is a real $R$ such that for some $\Pi^1_2$ formula $\varphi$, $R$ is the unique solution to $\varphi$ and $T \vdash \varphi$ has at most one solution. The real $0^\#$ is a ZFC-provable $\Pi^1_2$-singleton. An open question is whether there exists a ZFC-provable $\Pi^1_2$-singleton $R$, $0 <_L R <_L 0^\#$.

We describe here a subtheory $T$ of ZFC $+0^\#$ exists such that $T$ is consistent with $V = L$ and there exists a $T$-provable intermediate $\Pi^1_2$-singleton.

**Definition.** An acceptable guess $(\alpha_1, \ldots, \alpha_{n+1})$ is *good* if $\alpha_1$ is regular and $I(\alpha_1, \ldots, \alpha_{n+1})$ is stationary in $\alpha_1$.

Let $T$ be the theory ZFC + these exist arbitrarily long good guesses.

**Theorem 6.23.** *There exists a $T$-provable $\Pi^1_2$-singleton $R$, $0 <_L R <_L 0^\#$.*

*Proof.* Repeat the proof of Theorem 6.22, with one slight modification: When defining the sequence of conditions $p_n = (\bar{s}_n, (q^0_n, q^1_n))$ determining the $P$-generic $K$, ensure that $\bar{s}_{n+2}$ "decides" the $n^{\text{th}}$ hyperstring $s_n$ (relative to a fixed enumeration $\langle s_n \mid n \in \omega \rangle$ of hyperstrings) in the sense that either $s_n$ does not lie on $\bar{s}_{n+2}$ or $s_n$ does lie on all extensions of $\bar{s}_{n+2}$. Thus the procedure $(\alpha_1, \ldots, \alpha_n) \longmapsto p(\alpha_1, \ldots, \alpha_n)$ now has the property that for an acceptable guess $(\alpha_1, \ldots, \alpha_{n+2})$, $\bar{s}(p(\alpha_1, \ldots, \alpha_{n+2}))$ decides $s_n$.

As before build $R$ such that for every $\alpha$, $L_\alpha[R] \models \text{ZF}^- \Longrightarrow L_\alpha[R] \models$ Every $s$-bad acceptable guess has been killed, for each $s \in R$. Suppose $S \neq R$ also had this property and choose $s_n \in S - R$. Let $(\alpha_1, \ldots, \alpha_{n+2})$ be a good guess and suppose that $s_n$ does not lie on $\bar{s}(p(\alpha_1, \ldots, \alpha_{n+2}))$. Then the guess $(\alpha_1, \ldots, \alpha_{n+2})$ is killed in $L[S]$ as it is $s_n$-bad. If $s_n$ does lie on $\bar{s}(p(\alpha_1, \ldots, \alpha_{n+2}))$ then $(\alpha_1, \ldots, \alpha_{n+2})$ is $s_m$-bad for some $s_m \in R$, as $\bar{s}(p(\alpha_1, \ldots, \alpha_{n+2}))$ decides $s_n$, so $\alpha_1, \ldots, \alpha_{n+2}$ is killed in $L[R]$. In either case we have that $I(\alpha_1, \ldots, \alpha_{n+2})$ is nonstationary in $\alpha_1$, contradicting goodness. As $T$ proves the existence of good guesses $(\alpha_1, \ldots, \alpha_{n+2})$ we have that $T \vdash$ There is at most one solution to $\varphi$, where $\varphi$ is the $\Pi^1_2$ formula: $\varphi(R) \equiv$ For all $\alpha$, $L_\alpha[R] \models \text{ZF}^- \Longrightarrow L_\alpha[R] \models$ Every $s$-bad acceptable guess has been killed for each $s \in R$. □

The strength of the theory $T$ is measured by "$n$-ineffability."

**Definition.** A cardinal $\kappa$ is *$n$-ineffable* iff whenever $f : [\kappa]^n \longrightarrow \kappa$, $f(a) < 1 + \min(a)$ for all $a$ there exists a stationary $X \subseteq \kappa$ such that $f$ is constant on $[X]^n$.

**Proposition 6.24.** *Suppose that for all $n$ there exists an $n$-ineffable cardinal. Then for all $n$ there exists a good guess of length $n$.*

*Proof.* We define a regressive $f : [\kappa]^{2n} \longrightarrow \kappa$ as follows: Write $a \in [\kappa]^{2n}$ as $(\alpha_1, \ldots, \alpha_{2n})$ where the $\alpha_i$'s are increasing. If $\alpha_1$ is finite or not closed under the standard pairing function on ordinals then $f(a) = 0$. Otherwise, if $(\alpha_1, \ldots, \alpha_n)$, $(\alpha_{n+1}, \ldots, \alpha_{2n})$ do not satisfy the same formulas in $L_\kappa$ with parameters less than $\alpha_1$ then let $f(a) = \langle 0, \varphi, \vec{x} \rangle$ where $(\varphi, \vec{x})$ is the least formula $\varphi$ with parameters $\vec{x}$ from $\alpha_1$ where $(\alpha_1, \ldots, \alpha_n)$, $(\alpha_{n+1}, \ldots, \alpha_{2n})$ differ. Otherwise, if $\alpha_{n+1}$ is singular let $f(a) = \langle 1, \gamma \rangle$ where $\gamma$ is least such that $f_{\alpha_{n+1}}(\gamma) > \alpha_1$, and where $f_{\alpha_{n+1}} : \mathrm{cof}(\alpha_{n+1}) \longrightarrow \alpha_{n+1}$ has been chosen in advance as a cofinal function from $\mathrm{cof}(\alpha_{n+1})$ into $\alpha_{n+1}$. Otherwise, if $C(\alpha_1, \ldots, \alpha_n)$, $C(\alpha_{n+1}, \ldots, \alpha_{2n}) \cap \alpha_1$ differ, where $C(\alpha_1, \ldots, \alpha_n) \subseteq \alpha_1$ has been selected in advance as a CUB set disjoint from $I(\alpha_1, \ldots, \alpha_n)$, for regular $\alpha_1$, set $f(a) = \langle 2, \delta \rangle$ where $\delta$ is least such that $C(\alpha_1, \ldots, \alpha_n)$ and $C(\alpha_{n+1}, \ldots, \alpha_{2n})$ differ at $\delta$. Otherwise, set $f(a) = 3$.

Suppose that $X \subseteq \kappa$ is stationary and $f$ is constant on $[X]^{2n}$. The constant value of $f$ cannot be 0. Suppose that the constant value of $f$ were $\langle 0, \varphi, \vec{x} \rangle$. But then if $\alpha_1 < \cdots < \alpha_{3n}$ are from $X$, there would have to be 3 distinct truth values for $\varphi(\alpha_1, \ldots, \alpha_n, \vec{x})$, $\varphi(\alpha_{n+1}, \ldots, \alpha_{2n}, \vec{x})$, $\varphi(\alpha_{2n+1}, \ldots, \alpha_{3n}, \vec{x})$ in $L_\kappa$, a contradiction. The constant value of $f$ cannot be of the form $\langle 1, \gamma \rangle$ as we may choose $\alpha_{n+1} < \cdots < \alpha_{2n}$ in $X$, $\alpha_{n+1} \in \mathrm{Lim}\, X$ and then we have the contradiction that $f_{\alpha_{n+1}}(\gamma) > \alpha$ for each $\alpha \in X \cap \alpha_{n+1}$. The constant value of $f$ cannot be of the form $\langle 2, \delta \rangle$ as otherwise the 3 sets $C(\alpha_1, \ldots, \alpha_n)$, $C(\alpha_{n+1}, \ldots, \alpha_{2n}) \cap \alpha_1$, $C(\alpha_{2n+1}, \ldots, \alpha_{3n}) \cap \alpha_1$ all disagree at $\delta$ where $\alpha_1 < \cdots < \alpha_{3n}$ belong to $X$.

Now let $C = \cup \{C(\alpha_1, \ldots, \alpha_n) \mid \alpha_1 < \cdots < \alpha_n \text{ in } X\}$. Then $C$ is CUB in $\kappa$ and we may choose $\alpha_1 < \cdots < \alpha_{2n}$ in $X \cap C$. But then $\alpha_1 \notin I(\alpha_{n+1}, \ldots, \alpha_{2n})$ as $\alpha_1 \in C$ but clearly $\alpha_1 \in I(\alpha_{n+1}, \ldots, \alpha_{2n})$ as $(\alpha_1, \ldots, \alpha_n)$, $(\alpha_{n+1}, \ldots, \alpha_{2n})$ satisfy the same formulas in $L_\kappa$ with parameters $< \alpha_1$, and hence $\alpha_1, \alpha_{n+1}$ satisfy the same formulas in $L_{\alpha_{2n}}$ with parameters from $\alpha_1 \cup \{\alpha_{n+2}, \ldots, \alpha_{2n-1}\}$. We escape this contradiction by concluding that for some $\alpha_1 < \cdots < \alpha_n$ in $X$, $C(\alpha_1, \ldots, \alpha_n)$ cannot be defined and hence $I(\alpha_1, \ldots, \alpha_n)$ is stationary in $\alpha_1$. Clearly elements of $X$ are $L$-inaccessible as $\alpha < \beta$ in $X \Longrightarrow \alpha, \beta$ satisfy the same formulas in $L_\kappa$ with parameters $< \alpha$. So $(\alpha_1, \ldots, \alpha_n)$ is a good guess. □

**Corollary 6.25.** *Let $T^* = \mathrm{ZFC} +$ For each $n$ there exists an $n$-ineffable cardinal. Then there is a $T^*$-provable $\Pi^1_2$-singleton $R$, $0 <_L R <_L 0^\#$.*

The theory $T^*$ in fact has the same consistency strength as $T$ since if $(\alpha_1, \ldots, \alpha_{n+3})$ is a good guess then $\alpha_1$ is $n$-ineffable in $L_{\alpha_{n+2}}$ and $L_{\alpha_{n+2}} \models \mathrm{ZFC}$.

## 6.3 Other Applications

We first consider properties of more general countable $\Pi^1_2$ sets. Assume that $R^\#$ exists for every real $R$. Kechris and Woodin showed that a nonempty countable $\Pi^1_2$ set must have an ordinal-definable element; we show that in a sense their result is optimal. First some definitions.

**Definition.** A set of reals $X$ is *$n$-absolute* if for some formula $\varphi$, $R \in X \Longleftrightarrow L[R] \models \varphi(R, \omega_1, \ldots, \omega_n)$, where $\omega_k$ denotes the $\omega_k$ of $V$. An *$n$-absolute singleton* is a real $R$ such that $\{R\}$ is $n$-absolute. When $n = 0$ we say absolute, absolute singleton.

The following is due to Kechris–Woodin.

**Theorem 6.26.** *Assume that $R^\#$ exists for every real $R$. A nonempty countable $\Pi^1_2$ set contains an $n$-absolute singleton for some $n$.*

*Proof.* Every element of $L[R]$, $R$ a real, is definable in $L[R]$ from $R$ and elements of $I^R$, the Silver indiscernibles for $L[R]$. Let $t_0, t_1, \ldots$ be the sequence of functions defined by: $t_k(R, \alpha_1, \ldots, \alpha_k) = L[R]$-least $x$ such that $L[R] \models \varphi_k(R, x, \alpha_1, \ldots, \alpha_k)$ if exists, 0 otherwise where $\varphi_k$ is the $k^{\text{th}}$ formula in a list of all formulas in the language of set theory. (If $\varphi_k$ has only $k_0+2 < k+2$ free variables then we take $\varphi_k(R, x, \alpha_1, \ldots, \alpha_k)$ to be $\varphi_k(R, x, \alpha_1, \ldots, \alpha_{k_0})$.) Then for any $J \subseteq \text{ORD}$, $\{t_k(R, \alpha_1, \ldots, \alpha_k) \mid \alpha_1 < \cdots < \alpha_k$ from $J\}$ equals the Skolem Hull of $\{R\} \cup J$ in $L[R]$.

Now consider the tree $T$, whose $n^{\text{th}}$ level consists of nodes $\sigma$ which decide $R(0), \ldots, R(n)$ and all equality, membership relations between elements of $\{t_0(R), t_1(R, i_1), \ldots, t_n(R, i_1, \ldots, i_n)\}$, consistently realized by some real $R$ in $L[R]$, for all $i_1 < \cdots < i_n$ in a CUB set $C(\sigma) \subseteq \omega_1$ specified by $\sigma$. An infinite branch through $T$ yields a real $R$ together with a CUB set $C_R \subseteq \omega_1$ of indiscernibles generating an uncountable well-founded model $M_R \models \text{ZFC} + V = L[R]$.

Let $T_\varphi$, for a $\Pi^1_2$ formula $\varphi$, be the same as $T$ but with the added requirement that nodes are realized by reals $R$ satisfying $\varphi(R)$. An infinite branch through $T_\varphi$ produces a real $R$ and an uncountable well-founded model $M_R \models \text{ZFC} + V = L[R] + \varphi(R)$. It follows that $\varphi(R)$ is true for any real $R$ resulting from an infinite branch through $T_\varphi$. Conversely if $\varphi(R)$ is true then there is an infinite branch through $T_\varphi$ producing $R$.

Suppose that $\{R \mid \varphi(R)\}$ is countable and nonempty. An infinite branch $B$ through $T_\varphi$ is *isolated* if for some node $\sigma$ on $B$, every infinite branch through $T_\varphi$ containing $\sigma$ produces the same real $R$ as does the branch $B$. $T_\varphi$ must have isolated branches as otherwise $\{R \mid \varphi(R)\}$ contains a perfect closed subset, in violation to its countability. Let $B$ be an isolated branch through $T_\varphi$ and let $\sigma$ witness this, $\sigma$ on the $n^{\text{th}}$ level of $T_\varphi$. Then $R$ is uniquely determined by $\varphi(R)$ together with the information $R(0), \ldots, R(n)$ and the equality, membership relations between elements of $\{t_0(R), t_1(R, \omega_1), \ldots, t_n(R, \omega_1, \ldots, \omega_n)\}$ as the latter relations are realized inside any CUB subset of $\omega_1$ (which must intersect $I^R \cap \omega_1$). Thus $R$ satisfies $\varphi(R)$ and is an $n$-absolute singleton. □

Our next result demonstrates the optimality of the previous theorem.

**Theorem 6.27.** *For each $n$ there is a countable $\Pi^1_2$ set $X_n$ such that no element of $X_n$ is an $n$-absolute singleton.*

*Proof.* Fix $n$. Modify the forcing used to prove Theorem 6.16 to a forcing $P_n$ defined as follows: We now have a $\Sigma_1(L)$-definable procedure that produces an $\omega$-sequence of conditions $\langle p(m, \alpha_1, \ldots, \alpha_\ell) \mid m \in \omega \rangle$ for each acceptable guess $(\alpha_1, \ldots, \alpha_\ell)$ of length $\ell \geq n$. For each hyperstring $s$ and $m \in \omega$ we say that a guess $(\alpha_1, \ldots, \alpha_\ell)$ is *$(m, s)$-bad* if $s$ does not lie on $\bar{s}(p(m, \alpha_1, \ldots, \alpha_\ell))$. Now $P_n^*$ is the iteration $\mathbb{F} * \mathbb{Q} * \mathbb{R} * \mathbb{S}$ where:

(1) $\mathbb{F} = \mathbb{F}_0 * \mathbb{F}_1$ where $\mathbb{F}_0$ adds a function $F : [\mathrm{ORD}]^{n+1} \longrightarrow \omega$ (using Easton-sized restrictions of such functions) and $\mathbb{F}_1$ codes $F$ by $X^F \subseteq \omega_1^L$ via David's trick, so that whenever $\langle L_\alpha[X^F \cap \gamma], X^F \cap \gamma\rangle$ satisfies $\mathrm{ZF}^- + \gamma = \omega_1$ then it also satisfies: $X^F \cap \gamma$ canonically codes (as in Jensen coding) a function $F : [\mathrm{ORD}]^{n+1} \longrightarrow \omega$.

(2) We have $\mathbb{Q} = \Pi_s \mathbb{Q}(s)$ where $\mathbb{Q}(s)$ is the reverse Easton iteration that kills $(F(\kappa, \alpha_1, \ldots, \alpha_n), s)$-bad guesses $(\kappa, \alpha_1, \ldots, \alpha_\ell)$, $\kappa$ $L$-inaccessible, $\ell \geq n$.

(3) $\mathbb{R} = \Pi_s \mathbb{R}(s)$ where $\mathbb{R}(s)$ codes $X^F$ and the $\mathbb{Q}(s)$-generic $G(s)$ by $X(s) \subseteq \omega_1^L$ via David's trick, so that whenever $\langle L_\alpha[X(s) \cap \gamma], X(s) \cap \gamma\rangle$ satisfies $\mathrm{ZF}^- + \gamma = \omega_1$ then it also satisfies: For every $L$-inaccessible $\kappa$, each $(F(\kappa, \alpha_1, \ldots, \alpha_n), s)$-bad guess $(\kappa, \alpha_1, \ldots, \alpha_\ell)$, $\ell \geq n$ has been killed, where $F : [\mathrm{ORD}]^{n+1} \longrightarrow \omega$ is the canonical function coded by $X^F \cap \gamma$.

(4) $\mathbb{S} = \Pi_s \mathbb{S}(s)$ where $\mathbb{S}(s)$ codes $X(s)$ by a real $s$ using perfect trees, as before.

As before, $P_n^*$ preserves cofinalities and we define $P_n$ to consist of all $p = (\bar{s}, (f, q^0, q^1))$ where $\bar{s}$ is a hypertree, $(f, q^0, q^1) \in \mathbb{F} * \mathbb{Q} * \mathbb{R}$ and for all hyperstrings $s$, $(f, q^0[s]) \Vdash (q^1(s), \bar{s}(s)) \in \mathbb{R}(s) * \mathbb{S}(s)$. And $(\bar{t}, (g, r^0, r^1)) \leq (\bar{s}, (f, q^0, q^1))$ iff for all hyperstrings $s$ lying on $\bar{t}$, $(g, r^0[s]) \leq (f, q^0[s])$ in $\mathbb{F} * \mathbb{Q}$, $(g, r^0[s]) \Vdash r^1(s) \leq q^1(s)$ in $\mathbb{R}(s)$ and $\bar{t}(s)$ is a subtree of $\bar{s}(s)$.

$P_n$ preserves cofinalities, as before. Our $\Sigma_1(L)$ procedure $(\alpha_1, \ldots, \alpha_\ell) \longmapsto \langle p(m, \alpha_1, \ldots, \alpha_\ell) \mid m \in \omega\rangle$, $\ell \geq n$ now results from the construction of *$\omega$-many* distinct $P_n$-generic reals $\langle R(m) \mid m \in \omega\rangle$. Initially choose $p_n = p_n^m$ for all $m$ to be the $L$-least condition in $P_n$ that reduces each sentence of the form $L[R] \models \varphi(R, \omega_1, \ldots, \omega_n)$ below $\omega$, in the sense that for any $p \leq p_n$ and any such sentence $\psi$, $\psi$ is decided by simply extending $\bar{s}(p)$ and otherwise leaving $p$ unchanged. This is possible using the $\omega_1$-distributivity of $\mathbb{F} * \mathbb{Q} * \mathbb{R}$. In fact, using fusion for $\mathbb{S}$, we can assume that the extension of $\bar{s}(p)$ is simply of the form $\bar{s}(p)(\leq s)$ for some $s$ lying on $\bar{s}(p)$, where $t$ lies on $\bar{s}(p)(\leq s)$ iff $t$ lies on $\bar{s}(p)$ and $t$ extends $s$. Now let $\langle s_m \mid m \in \omega\rangle$ enumerate the hyperstrings lying on $\bar{s}(p_n)$ and choose $\langle p_\ell^m \mid n \leq \ell \in \omega\rangle$ to determine a $P_n$-generic real $R(m)$ as in the earlier $P$-generic construction, requiring that $\bar{s}(p_{n+1}^m)$ extend $\bar{s}(p_n)(\leq s_m)$ and that $f(p_{n+1}^m) = (f^0, f^1)$ where $f^0(k_1, \ldots, k_{n+1}) = m$. Set $p(m, k_1, \ldots, k_\ell) = p_\ell^m$ and extend this to a $\Sigma_1(L)$ procedure $(\alpha_1, \ldots, \alpha_\ell) \longmapsto \langle p(m, \alpha_1, \ldots, \alpha_\ell) \mid m \in \omega\rangle$, $\ell \geq n$ as before.

Let $X_n$ be $\{R(m) \mid m \in \omega\}$. Each $R$ in $X_n$ satisfies: If $L_\alpha[R]$ satisfies $\mathrm{ZF}^-$ then it also satisfies that for each guess $(\alpha_1, \ldots, \alpha_{n+1})$, $\alpha_1$ $L$-inaccessible there is $m \in \omega$ such that every $(m, s)$-bad guess $(\alpha_1, \ldots, \alpha_{n+1}, \ldots, \alpha_\ell)$, $\ell \geq n+1$ has been killed, for all $s \in R$. Conversely, if $R$ has this property then choose $m$ such that in $L[R]$ every $(m, s)$-bad guess $(\omega_1, \ldots, \omega_{n+1}, \alpha_{n+2}, \ldots, \alpha_\ell)$ has been killed, for all $s \in R$; then as before $R$ must be $R_m$. So $X_n$ is a countable $\Pi_2^1$ set. By construction, if $R \in X_n$, $L[R] \models \varphi(\omega_1, \ldots, \omega_n)$ then the same holds of other $S \in X_n$ as any such property of $R$ will hold of any $S \in X_n$ containing a particular hyperstring $s_m$ lying on $\bar{s}(p_n)$; there are infinitely many such $S$. So no element of $X_n$ is an $n$-absolute singleton. □

Elements of countable $\Pi_2^1$ sets need not be $n$-absolute singletons, for any $n$:

**Theorem 6.28.** *There exists a countable $\Pi^1_2$ set X and $R \in X$ such that for all n, R is* not *an n-absolute singleton.*

*Proof.* Define $P$ as follows: Again we have a $\Sigma_1(L)$ procedure that produces an $\omega$-sequence of conditions $\langle p(m, \alpha_1, \ldots, \alpha_\ell) \mid m < \omega \rangle$ for each acceptable guess $(\alpha_1, \ldots, \alpha_\ell)$. (There is no longer a restriction on $\ell$.) A guess is defined to be $(m, s)$-bad as before. The iteration $P^*_n$ is now replaced with the iteration $P^* = \mathbb{F} * \mathbb{Q} * \mathbb{R} * \mathbb{S}$ where:

(1) $\mathbb{F} = \mathbb{F}_0 * \mathbb{F}_1$ where $\mathbb{F}_0$ adds a function $F : [\mathrm{ORD}]^{<\omega} \longrightarrow \omega$ in such a way that $F(\alpha_1, \ldots, \alpha_\ell) \neq 0 \Longrightarrow F(\alpha_1, \ldots, \alpha_{\ell+1}) = F(\alpha_1, \ldots, \alpha_\ell)$. $\mathbb{F}_1$ is defined as before.

(2) We have $\mathbb{Q} = \Pi_s \mathbb{Q}(s)$ where $\mathbb{Q}(s)$ kills $(F(\kappa, \alpha_1, \ldots, \alpha_\ell), s)$-bad guesses $(\kappa, \alpha_1, \ldots, \alpha_\ell)$, $\kappa$ $L$-inaccessible if $F(\kappa, \alpha_1, \ldots, \alpha_\ell) \neq 0$; if $F(\kappa, \alpha_1, \ldots, \alpha_\ell) = 0$ and $\ell$ is even then $\mathbb{Q}(s)$ kills $(\kappa, \alpha_1, \ldots, \alpha_\ell)$ if it is $(\ell, s \restriction \ell)$-bad (where $s \restriction \ell = \langle s_0 \restriction \ell, \ldots, s_{\bar{\ell}} \restriction \ell \rangle$ for $s = \langle s_0, \ldots, s_n \rangle$, $\bar{\ell} = \min(\ell, n)$).

(3), (4) as before.

$P$ is defined using $P^*$ as was $P_n$ defined from $P^*_n$. We now describe the construction of the $P$-generic reals $R(m), n \geq 0$. For $m > 0$, we initially choose $p^m_m$ to reduce sentences of the form $L[R] \models \varphi(R, \omega_1, \ldots, \omega_m)$ below $\omega$ as before and we define $p^m_{m+1} \leq p^m_m$ so that $f(p^m_{m+1}) = (f_0, f_1)$ where $f_0(k_1, \ldots, k_{m+1}) = m$, $f_0(k_1, \ldots, k_m) = 0$ (the latter guaranteed by the choice of $p^m_m$). Also choose the $\bar{s}(p^{m^*}_{m^*})$, $m^*$ odd to guarantee that each hyperstring on $\bar{s}(p^m_m)$ lies on some $\bar{s}(p^{m^*}_{m^*})$, $m < m^*$ odd, $p^{m^*}_{m^*} \leq p^m_m$. (This guarantees that no $P$-generic containing $p^m_m, m > 0$ produces an $m$-absolute singleton.) Also guarantee that $\{p \mid p$ is extended by some $p^m_m$, *m even* $\}$ is $P$-generic and that for *even m*, $\bar{s}(p^m_m) = \bar{s}(p^m_m)(\leq s_m)$ where the hyperstrings $s_0 \geq s_1 \geq \cdots$ converge to the resulting $P$-generic real $R(0)$. Of course for $m > 0$, $p^m_{m+2} \geq p^m_{m+3} \geq \cdots$ are defined so as to produce the $P$-generic real $R(m)$.

$R(0)$ is not an $m$-absolute singleton for any $m$ as its corresponding $P$-generic contains the condition $p^m_m$ for even $m$. Let $X = \{R(m) \mid m \geq 0\}$. If $R$ belongs to $X$ then $R$ satisfies:

(*) If $L_\alpha[R]$ satisfies $\mathrm{ZF}^-$ then it also satisfies that $R$ canonically codes $F : [\mathrm{ORD}]^{<\omega} \longrightarrow m$ such that if $F(\alpha_1, \ldots, \alpha_\ell) = m \neq 0$, $\alpha_1$ $L$-inaccessible, $s \in R$, $(\alpha_1, \ldots, \alpha_\ell)$ $(m, s)$-bad then $(\alpha_1, \ldots, \alpha_\ell)$ is killed in addition, if $F(\alpha_1, \ldots, \alpha_\ell) = 0$, $\ell$ even, $(\alpha_1, \ldots, \alpha_\ell)$ $(\ell, R \restriction \ell)$-bad then $(\alpha_1, \ldots, \alpha_\ell)$ is killed.

Suppose $R$ satisfies $(*)$, let $F^R : [\mathrm{ORD}]^{<\omega} \longrightarrow \omega$ be canonically coded by $R$ and consider $m_\ell = F^R(\omega_1, \ldots, \omega_\ell)$. If some $m_\ell$ is not 0, then $R$ must equal $R(m_\ell)$. If $m_\ell = 0$ for all $\ell$, then $R \restriction \ell = R(\ell) \restriction \ell$ for all even $\ell$ so $R = R(0)$. This proves $X$ is $\Pi^1_2$. □

**Remark.** Not every absolute singleton belongs to a countable $\Pi^1_2$ set: If a set is $\Sigma^1_2$ (with a constructible parameter) and contains a non-constructible real then it has a constructibly-coded perfect closed subset (see Mansfield [70]), and a code for this perfect closed set can be computed as a $\Sigma^1_2$ function applied to an index $n \in \omega$ for the given $\Sigma^1_2$ set $X_n$. Moreover $\{n \mid X_n$ has a perfect closed subset $\}$ is $\Sigma^1_2$. It follows that in $L$ there is a perfect closed set $C$, with code recursive in the complete $\Sigma^1_2$ subset of $\omega$, such that $R \in C \implies R$ does not belong to any countable $\Pi^1_2$ set. The set $C$ contains elements which are $\Delta^1_3$ in $L$, and hence which are absolute singletons.

An open problem is to provide a revealing characterization of the reals which belong to a countable $\Pi^1_2$ set.

The following result is proved in Harrington–Kechris [77]: If $X$ is a nonempty $\Pi^1_2$ set then $X$ has an element $R$ such that either $R \leq_L 0^\#$ or $0^\# \leq_L R$. Our next result implies that $0^\#$ has least nonzero $L$-degree among reals with this property, even when $X$ is restricted to have a unique element.

**Theorem 6.29.** *There exists a sequence $\langle (R^n_0, R^n_1) \mid n \in \omega \rangle$ of pairs of reals such that:*

(a) *If $R$ is constructible from both $R^n_0$ and $R^n_1$ then $R$ is constructible.*

(b) $\{\langle R, n, i \rangle \mid R = R^n_i\}$ *is* $\Pi^1_2$.

(c) $n \in 0^\#$ *iff* $n \in R^n_0$ *iff* $n \in R^n_1$.

*Proof.* Given $n$, construct a pair of $\Pi^1_2$-singletons $R^n_0$, $R^n_1$ such that (a) holds, using the $\omega$-enumeration of constructible reduction procedures provided by $0^\#$. Also guarantee that $n \in R^n_i$ iff $n \in 0^\#$, by requiring $R^n_i$ to kill all acceptable guesses $(\alpha_1, \ldots, \alpha_{n+1})$ which violate the equivalence $n \in R^n_i \iff L_{\alpha_{n+1}} \models \varphi_n(\alpha_1, \ldots, \alpha_n)$, viewing $0^\#$ as $\{n \mid L_{i_{n+1}} \models \varphi_n(i_1, \ldots, i_n)$ for $i_1 < \cdots < i_{n+1}$ in $I\}$. This is uniform in $n$, so (b) follows. □

**Corollary 6.30.** *Suppose $R$ is non-constructible real and every $\Pi^1_2$-singleton is $\leq_L$-comparable with $R$. Then $0^\# \leq_L R$.*

*Proof.* If for some $n$, $R$ is constructible from both $R^n_0$ and $R^n_1$ then $R$ is constructible, against our hypothesis. So for each $n$ there is $i$ such that $R \geq_L R^n_i$. But then $n \in 0^\#$ iff $L[R] \models \exists S(S = R^n_i$ for some $i$ and $n \in S)$. So $0^\# \leq_L R$. □

Thus $0^\#$ is the least "canonical" $\Pi^1_2$-singleton.

## New $\Sigma^1_3$ Facts

We close this chapter by answering the following question: If $M$ is an inner model (of a universe containing $0^\#$) and $0^\#$ does not belong to $M$, then of course there is a true $\Sigma^1_3$ sentence not holding in $M$, namely the sentence asserting the existence of $0^\#$; can this effect be achieved by forcing over $M$?

**Theorem 6.31.** *There exists an $\omega$-sequence of $\Sigma^1_3$ sentences $\langle \varphi_n \mid n \in \omega \rangle$ such that if $M$ is an inner model, $0^\# \notin M$:*

(a) *$\varphi_n$ is false in $M$ for some $n$.*

(b) *For each $n$, some generic extension of $M$ satisfies $\varphi_n$.*

*Moreover if $M = L[R]$, $R$ a real then the generic extensions in* (b) *can be taken as inner models of $L[R, 0^\#]$.*

We first prove the following, which may be of independent interest.

**Theorem 6.32.** *There exists an $L$-definable function $n : L$-Singulars $\longrightarrow \omega$ such that if $M$ is an inner model, $0^\# \notin M$:*

(a) *For some $n$, $M \models \{\alpha \mid n(\alpha) \leq n\}$ is stationary.*

(b) *For each $n$ there is a generic extension of $M$ in which $0^\#$ does not exist and $\{\alpha \mid n(\alpha) \leq n\}$ is non-stationary.*

In (a), we intend that whenever $C \subseteq \mathrm{ORD}$ is CUB and $M$-definable then there is $\alpha \in C$, $n(\alpha) \leq n$. In (2) we intend that the generic extension satisfy ZFC and have a definable CUB class $C \subseteq \mathrm{ORD}$ such that $\alpha \in C \Longrightarrow n(\alpha) > n$.

*Proof.* We define $n(\alpha)$. Let $\langle C_\alpha \mid \alpha\ L\text{-singular} \rangle$ be an $L$-definable $\Box$-sequence: $C_\alpha$ is CUB in $\alpha$, ot $C_\alpha$ = ordertype $C_\alpha < \alpha$ and $\bar{\alpha} \in \mathrm{Lim}\, C_\alpha \Longrightarrow C_{\bar{\alpha}} = C_\alpha \cap \bar{\alpha}$. If ot $C_\alpha$ is $L$-regular then $n(\alpha) = 0$. Otherwise $n(\alpha) = n(\mathrm{ot}\, C_\alpha) + 1$.

(a) is clear, as otherwise (assuming the existence of a satisfaction predicate for $M$) there is a CUB $C \subseteq L$-regulars, contradicting the Covering Theorem and the hypothesis that $0^\#$ does not belong to $M$.

Now we prove (b). Fix $n \in \omega$. In $M$, let $P$ consist of closed, bounded $p \subseteq \mathrm{ORD}$ such that $\alpha \in p \Longrightarrow \alpha$ $L$-regular or $n(\alpha) \geq n+1$, ordered by $p \leq q$ iff $p$ end extends $q$.

We claim that $P$ is $\infty$-distributive. Suppose that $p \in P$ and $\langle D_\alpha \mid \alpha < \kappa \rangle$ is an $M$-definable sequence of open dense subclasses of $P$, $\kappa$ regular. We wish to find $q \leq p$, $q \in D_\alpha$ for all $\alpha < \kappa$.

Let $C = \{\beta \mid \beta$ is a strong limit cardinal and for all $\alpha < \kappa$, if $r$ belongs to $V^M_\beta$ then there is $s \leq r$ in $V^M_\beta \cap D_\alpha\}$, a CUB subclass of ORD. It suffices to show that $C \cap \{\beta \mid n(\beta) \geq n+1\}$ has a closed subset of ordertype $\kappa + 1$, for then $p$ can be successively extended $\kappa$ times to conditions with maximum in $\{\beta \mid n(\beta) \geq n+1\}$, meeting $D_\alpha$ at stage $\alpha + 1$; the final condition extends $p$ and meets each $D_\alpha$.

**Lemma 6.33.** *Suppose $m \geq n$, $\alpha$ is regular and $C$ is a closed set of ordertype $\alpha^{+m} + 1$, consisting of ordinals greater than $\alpha^{+m}$ (where $\alpha^{+0} = \alpha$, $\alpha^{+(k+1)} = (\alpha^{+k})^+$). Then $C \cap \{\beta \mid n(\beta) \geq n\}$ has a closed subset of ordertype $\alpha^{+(m-n)} + 1$.*

*Proof of Lemma* 6.33. By induction on $n$. Suppose $n = 0$. Let $\beta = \max C$. Then $\beta$ is singular and hence singular in $L$. So $C_\beta$ is defined and $\mathrm{Lim}(C_\beta \cap C)$ is a closed set of ordertype $\alpha^{+m} + 1$ consisting of $L$-singulars. So $\mathrm{Lim}(C_\beta \cap C) \subseteq C \cap \{\gamma \mid n(\gamma) \geq 0\}$ satisfies the lemma.

Suppose the lemma holds for $n$ and let $m \geq n$, $C$ a closed set of ordertype $\alpha^{+(m+1)}+1$ consisting of ordinals greater than $\alpha^{+(m+1)}$. Let $\beta = \max C$. Then $C_p$ is defined and $D = \mathrm{Lim}(C_\beta \cap C)$ is a closed set of ordertype $\alpha^{+(m+1)}+1$. Let $\bar{\beta} = (\alpha^{+m}+\alpha^{+m}+1)^{\text{st}}$ element of $D$. Then $\bar{D} = \{\mathrm{ot}\, C_\gamma \mid \gamma \in D, (\alpha^{+m} + 1)^{\text{st}}$ element of $D \leq \gamma \leq \bar{\beta}\}$ is a closed set of ordertype $\alpha^{+m} + 1$ consisting of ordinals greater than $\alpha^{+m}$. By induction there is a closed $\bar{D}_0 \subseteq \bar{D} \cap \{\gamma \mid n(\gamma) \geq n\}$ of ordertype $\alpha^{+(m-n)} + 1$. But then $D_0 = \{\gamma \in D \mid \mathrm{ot}\, C_\gamma \in \bar{D}_0\}$ is a closed subset of $C \cap \{\gamma \mid n(\gamma) \geq n + 1\}$ of ordertype $\alpha^{+(m-n)} + 1$. As $\alpha^{+(m-n)} = \alpha^{+((m+1)-(n+1))}$ we are done. □ (Lemma 6.33)

By the lemma, $C \cap \{\beta \mid n(\beta) \geq n\}$ has arbitrarily long closed subsets for any $n$ and any CUB $C \subseteq \mathrm{ORD}$, and hence $P$ is $\infty$-distributive. Now to prove (b), we apply the forcing $P$ to $M$, producing $C$ witnessing the nonstationarity of $\{\alpha \mid n(\alpha) \leq n\}$, and then follow this with the forcing to code $\langle M, C\rangle$ by a real, making $C$ definable. Of course this will not produce $0^\#$ as every successor to a strong limit cardinal is preserved in the coding. □

We also note that in Theorem 6.32(b), the generic extension can be formed in $L[R, 0^\#]$ in the case of $M = L[R]$, $R$ a real, using the relevance of both Amenable forcing and Jensen coding.

*Proof of Theorem* 6.31. We use David's trick. Let $\varphi_n$ be the $\Sigma^1_3$ sentence: $\exists R\ \forall\alpha$ (If $L_\alpha[R] \models \mathrm{ZF}^-$ then $L_\alpha[R] \models$ For every limit cardinal $\beta$, $n(\beta) \geq n$). By Theorem 6.32(b), cardinal collapsing and David's trick, $M$ has a generic extension satisfying $\varphi_n$. But Theorem 6.32(a) implies that for some $n$, $\varphi_n$ fails in $M$. And as remarked above, the generic extensions satisfying the $\varphi_n$ can be taken inside $L[R, 0^\#]$ in case $M$ is of the form $L[R]$, $R$ a real. □

Chapter 7

# The Admissibility Spectrum Problem

Recall that for a real $R$, $\alpha$ is *$R$-admissible* if either $\alpha = \omega$ or $L_\alpha[R]$ is a model of the theory $T_0$, whose axioms are those of ZF $-$Power, with Replacement restricted to $\Sigma_1$ formulas. The *admissibility spectrum* of $R$ is $\Lambda(R) = \{\alpha > \omega \mid \alpha \text{ is } R\text{-admissible}\}$. We observed in Section 4.1 that $\Lambda(R)$ is "trivial" when $R$ is set-generic over $L$. On the other hand, if $0^{\#}$ is constructible from $R$ then $\Lambda(R)$ is "very thin" in the sense that $\Lambda(R) \subseteq I \cup \beta$ for some countable $\beta$, where $I$ is the class of Silver indiscernibles. In this chapter we produce a third possibility, by constructing reals $R <_L 0^{\#}$ with non-trivial admissibility spectra.

Let RI $= \{\alpha \mid \alpha$ is both admissible and the limit of admissible ordinals$\}$, the class of *recursively inaccessible* ordinals. RI contains all uncountable $L$-cardinals. In Section 7.1 we produce $R <_L 0^{\#}$ such that $\Lambda(R) \subseteq$ RI. The proof is closely related to David's trick, as described in Section 6.2. We then improve this result in two ways: RI can be replaced by any $\Sigma_1(L)$-definable class containing all $L$-cardinals. And $\Lambda(R)$ can be replaced by $\Lambda^*(R) = \{\alpha \mid$ Every well ordering in $L_\alpha[R]$ has ordertype less than $\alpha\}$.

Section 7.2 introduces the method of Strong Coding, to produce a real $R$ such that $\Lambda(R) =$ RI. Strong Coding is an extension of the coding method where codings associated to the different admissible ordinals are fit together. For this purpose we must use the fine structure theory, as well as the more refined proof of the Coding Theorem, from Section 4.3.

Section 7.3 discusses some generalizations of the above results.

## 7.1 Killing Admissibles

We prove:

**Theorem 7.1.** *There exists $R <_L 0^{\#}$ such that $\Lambda(R) \subseteq$ RI.*

*Proof.* We wish to create an $L$-definable, tame, cofinality-preserving forcing $P$ for adding a real $R$ such that if $\alpha > \omega$ is $R$-admissible then $\alpha$ is recursively inaccessible. As in the proof of Theorem 6.18, $P$ is obtained as a modification of the forcing used to prove Theorem 4.1, followed by a simple coding of $X \subseteq \omega_1^L$ by $R$. The following definitions take place inside $L$.

**Definition** (Strings). Let $\alpha$ belong to Card, the class of all infinite cardinals. $S_\alpha$ consists of all $s : [\alpha, |s|) \longrightarrow 2$, $\alpha \leq |s| < \alpha^+$, $|s|$ a multiple of $\alpha$ such that:

(∗) If $\eta \leq |s|$ and $\langle L_\beta[s \upharpoonright \eta], s \upharpoonright \eta\rangle = L_\beta^{s\upharpoonright\eta}$ is both admissible and satisfies $\eta = \alpha^+$, then $\beta$ is recursively inaccessible.

In the above we allow the possibility that $\beta$ equals $\eta$ and $\alpha$ is the largest cardinal of $L_\beta[s \upharpoonright \eta]$. To say that $L_\beta^{s\upharpoonright\eta} = \langle L_\beta[s \upharpoonright \eta], s \upharpoonright \eta\rangle$ is admissible is to say that $L_\beta^{s\upharpoonright\eta}$ obeys ZF − Power, with Replacement only for $\Sigma_1$ formulas which mention $s \upharpoonright \eta$ as a predicate.

**Definition** (Coding Structures). For $s \in S_\alpha$ define $\mu^s$, $\mu^{<s}$ inductively by: $\mu^{<\emptyset_\alpha} = \alpha$ if $s$ is $\emptyset_\alpha$, the empty string at $\alpha$, $\mu^{<s} = \bigcup\{\mu^t \mid t \subseteq s, t \neq s\}$ for $s \neq \emptyset_\alpha$ and for all $s$, $\mu^s$ is the least $\mu > \mu^{<s}$ such that $\mu'\mu = \mu$ for $0 < \mu' < \mu$, $s \in L_\mu$ and $L_\mu \models \text{Card}(|s|) \leq \alpha$. Also define $\mathcal{A}^s$ to be the set $L_{\mu^s}$.

The remainder of the definitions from the proof of Theorem 4.1 remain the same in the present context, always taking the predicate $A$ to be empty. We now verify that the lemmas in the proof of Theorem 4.1 can be adapted as well.

**Lemma 7.2** (Distributivity for $R^s$). *Suppose $\alpha \in \text{Card}$, $s \in S_{\alpha^+}$. Then $R^s$ is $\alpha^+$-distributive in $\mathcal{A}^s = L_{\mu^s}$.*

*Proof.* Proceed as in the proof of Lemma 4.3. The only new point is to verify that in the proof of the Claim, $t_\lambda$ satisfies (∗) (of the new definition of $S_\alpha$). The facts that $s \in S_{\alpha^+}$ and that $t_\lambda$ codes $\bar{H}_\lambda$ imply that (∗) holds for $t_\lambda$ whenever $\beta$ is at most $\bar{\mu}_\lambda = \bar{H}_\lambda \cap \text{ORD}$ (note that $\bar{\mu}_\lambda$ is not admissible and that $(\alpha^+)^{\bar{H}_\lambda} \in \text{RI}$). As $|t_\lambda|$ is definably singular over $\bar{H}_\lambda$ and $t_\lambda$ is definable over $\bar{H}_\lambda$, these are the only $\beta$'s that concern us. □

**Lemma 7.3** (Extendibility for $P^s$). *Suppose that $\alpha$ is a limit cardinal, $s$ belongs to $S_\alpha$, and $p \in P^s$. Suppose also that $X \subseteq \alpha$ belongs to $\mathcal{A}^s$. Then there exists $q \leq p$ in $P^s$ such that $X \cap \beta \in \mathcal{A}^{q_\beta}$ for each $\beta \in \text{Card} \cap \alpha$.*

*Proof.* Proceed as in the proof of Lemma 4.4. In the definition of $q$, the only new instances of (∗) to check are for $s_\beta$ when $\text{Even}(Y \cap \beta)$ codes $s_\beta$ with domain $[\beta, |s_\beta|)$, $|s_\beta| < \beta^+$. But the embedding $\bar{\mathcal{A}}_\beta \longrightarrow \mathcal{A}$ is $\Sigma_1$-elementary and instances of (∗) refer to ordinals less than the height of $\mathcal{A}$; so the fact that $s$ belongs to $S_\alpha$ implies that $s_\beta$ belongs to $S_\beta$. □

**Lemma 7.4** (Distributivity for $P^s$). *Suppose $s \in S_{\beta^+}$, $\beta \in \text{Card}$.*

(a) *If $\langle D_i \mid i < \beta\rangle \in \mathcal{A}^s$, $D_i$ $i^+$-dense on $P^s$ for each $i < \beta$ and $p \in P^s$ then there is $q \leq p$ such that $q$ meets each $D_i$.*

(b) *If $p \in P^s$, $f$ small, $f$ in $\mathcal{A}^s$ then there exists $q \leq p$, $q \in \Sigma_f^p$.*

*Proof.* Proceed as in the proof of Lemma 4.5. In the Claim we must verify that $p_\gamma^\lambda$ satisfies (∗). But the fact that $p_\gamma^\lambda$ codes a generic over the transitive collapse of $H_\lambda(\gamma) \cap L_\beta$ implies that (∗) is satisfied at ordinals less than $\bar{\beta}$, the image of $\beta$ under the transitive collapse of $H_\lambda(\gamma)$ (note that cardinals of $L_{\bar{\beta}}$ belong to RI). And as $\bar{H}_\lambda(\gamma) \longrightarrow H_\lambda(\gamma)$

is $\Sigma_1$-elementary and $L_{\bar{\mu}}^{p_\gamma^\lambda} \models |p_\gamma^\lambda|$ is $\Sigma_1$-singular (where $\bar{\mu}$ is the height of $\bar{H}_\lambda(\gamma)$), $p_\gamma^\lambda$ obeys $(*)$ at all ordinals $< \mu^{p_\gamma^\lambda}$. □

The previous argument can also be applied to prove the distributivity of $P$, observing that when building sequences of conditions $\langle p^i \mid i < \lambda\rangle$, $\lambda$ limit to meet an $L$-definable sequence of dense classes, one has that $p_\gamma^\lambda$ codes $\bar{H}_\gamma(\lambda)$ of height $\bar{\mu}$, where $L_{\bar{\mu}+1}^{p_\gamma^\lambda} \models |p_\gamma^\lambda|$ is not a cardinal and $\bar{\mu} \in \mathrm{RI}$. Thus there is no additional instance of $(*)$ to verify beyond those considered in the proof of Lemma 7.4.

Thus $P$ is tame and cofinality-preserving. The $P$-generic $X \subseteq \omega_1^L$ can be coded by a real $R$ using Jensen–Solovay [70]; then $\alpha > \omega$, $\alpha$ $R$-admissible implies that $\alpha$ is $X \cap \eta$-admissible where $\eta = (\omega_1)^{L_\alpha[R]}$, and hence $\alpha \in \mathrm{RI}$. The proof of Theorem 4.18 shows that $R$ may be constructed inside $L[0^\#]$. Of course $0^\# \notin L[R]$ as $R$ preserves cofinalities over $L$. So this completes the proof. □

Very little is used in the preceding argument of the special properties of RI. Indeed, we have the stronger result:

**Theorem 7.5.** *Suppose $\varphi$ is $\Sigma_1$ and $L \models \varphi(\kappa)$ whenever $\kappa$ is an $L$-cardinal. Then there exists a real $R <_L 0^\#$ such that $\Lambda(R) \subseteq \{\alpha \mid L \models \varphi(\alpha)\}$.*

*Proof.* The definitions of the previous proof are modified as follows. Again work in $L$.

**Definition** (Strings). Let $\alpha$ belong to Card, the class of all infinite cardinals. $S_\alpha$ consists of all $s : [\alpha, |s|) \longrightarrow 2$, $\alpha \leq |s| < \alpha^+$, $|s|$ a multiple of $\alpha$ such that:

(∗) If $\eta \leq |s|$ and $L_\beta^{s\restriction\eta}$ is both admissible and satisfies $\eta = \alpha^+$ then $L \models \varphi(\beta)$.

Again we allow the possibility that $\beta$ equals $\eta$ and $\alpha$ is the largest cardinal of $L_\beta^{s\restriction\eta}$.

**Definition** (Coding Structures). For $s \in S_\alpha$ define $\mu^s$, $\mu^{<s}$ inductively by: $\mu^{<\emptyset_\alpha} = \alpha$, $\mu^{<s} = \bigcup\{\mu^t \mid t \subseteq s, t \neq s\}$ for $s \neq \emptyset_\alpha$ and for all $s$, $\mu^s$ is the least $\mu > \mu^{<s}$ such that $\mu'\mu = \mu$ for $0 < \mu' < \mu$, $s \in L_\mu$, $L_\mu \models \mathrm{Card}(|s|) \leq \alpha$ and:

(∗∗) If $\eta \leq |s|$, $L_\beta^{s\restriction\eta}$ is admissible and satisfies $\eta = \alpha^+$ then $L_\mu \models \varphi(\beta)$.

The other definitions of the proof of Theorem 4.1 now carry over to the present context, as do the proofs from Theorem 7.1. The one exception is in showing distributivity for $P$: The difficulty is that when building sequences of conditions $\langle p^i \mid i < \lambda\rangle$, $\lambda$ limit to meet an $L$-definable sequence of dense classes, one has that $p_\gamma^\lambda$ codes $\bar{H}_\gamma(\lambda)$ of height $\bar{\mu}$, where $L_{\bar{\mu}+1}^{p_\gamma^\lambda} \models |p_\gamma^\lambda|$ is not a cardinal; but we have no guarantee that $\bar{\mu}$ belongs to $\{\alpha \mid L \models \varphi(\alpha)\}$. (There would be no problem if we were dealing with a class of the form $\{\alpha \mid L_\alpha \models \psi\}$, $\psi$ a sentence, such as RI.)

However we can argue as follows: If $\langle D_i \mid i < \gamma\rangle$ is an $L$-definable sequence of dense classes on $P_\gamma$ there is an $L$-cardinal $\kappa$ of $L$-cofinality greater than $\gamma$ such that $\langle D_i \cap P_\gamma^{<\kappa} \mid i < \gamma\rangle$ is an $L_\kappa$-definable sequence of sets which are dense on $P_\gamma^{<\kappa}$. Now by the usual distributivity argument applied in $\mathcal{A}^{\emptyset_\kappa}$, where $\varphi(\kappa)$ holds, we see that any $p \in P_\gamma^{<\kappa}$ can be extended to $q$ meeting each $D_i$. The $\gamma^+$-distributivity of $P_\gamma$ thereby follows. The rest of the proof is as in Theorem 7.1. □

The previous results do not require the $R$-admissibility of the ordinals in $\Lambda(R)$, but only a weaker property.

**Definition.** $\alpha$ is *quasi R-admissible* if $\alpha > \omega$ and every wellordering in $L_\alpha[R]$ has ordertype less than $\alpha$.

$R$-admissibility implies quasi $R$-admissibility, but not conversely, as the limit of the first $\omega$ $R$-admissibles is quasi $R$-admissible but not $R$-admissible. Let $\Lambda^*(R)$ denote $\{\alpha > \omega \mid \alpha \text{ is quasi } R\text{-admissible}\}$, a CUB class of ordinals containing $\Lambda(R)$.

**Theorem 7.6.** *Suppose $\varphi$ is $\Sigma_1$ and $L \models \varphi(\kappa)$ whenever $\kappa$ is an $L$-cardinal. Then there is a real $R <_L 0^\#$ such that $\Lambda^*(R) \subseteq \{\alpha \mid L \models \varphi(\alpha)\}$.*

*Proof.* The earlier arguments used only three properties of (relativized) admissibility:

(1) $L$-cardinals are admissible.

(2) Admissibility is first-order: $\alpha$ is $x$-admissible iff $\langle L_\alpha[x], x\rangle \models \varphi$, for some fixed sentence $\varphi$.

(3) If $\alpha$ is less than $\kappa^+$ then there exists $x \subseteq \kappa$ such that no ordinal in $(\kappa, \alpha]$ is $x$-admissible.

These properties also hold for quasi-admissibility. □

**Corollary 7.7** (Beller–Jensen–Welch [85], David [82]). *Suppose $\alpha$ is countable, $L_\alpha \models$ ZF. Then for some real R, $\alpha$ is the* least *ordinal such that $L_\alpha[R] \models$ ZF.*

*Proof.* First force over $L_\alpha$ to add $A \subseteq \alpha$ such that for no cardinal $\kappa$ of $L_\alpha[A]$ do we have $\langle L_\kappa[A], A \cap \kappa\rangle \models$ ZF. This can be accomplished by adding a CUB subclass $C$ to $\{\beta < \alpha \mid L_\beta \text{ is not a model of ZF}\}$ and then collapsing cardinals so that every cardinal belongs to $C \cup \{\beta^+ \mid \beta \in C\}$. Code $A$ by a real $R_0$. Then $L_\alpha[R_0] \models \varphi(\kappa)$ for every $L_\alpha[R_0]$-cardinal $\kappa$, where $\varphi(\beta)$ is the formula "$L_\beta[R_0]$ is not a model of ZF". By the forcing used to prove Theorem 7.6 (relativized to $R_0$), we can produce $R$, generic over $L_\alpha[R_0]$, such that $L_\alpha[R] \models$ ZF, $R_0$ is recursive in $R$ and $\Lambda^*(R) \cap \alpha \subseteq \{\beta < \alpha \mid L_\beta[R_0]$ is not a model of ZF$\}$. So $\beta < \alpha \Longrightarrow L_\beta[R]$ is not a model of ZF. □

**Corollary 7.8.** *There is a real $R <_L 0^\#$ such that $\Lambda^*(R) \subseteq \{\alpha \mid L_\alpha \models \text{ZF} - \text{Power}\}$.*

*Proof.* Choose $C$ to be the $L$-amenable CUB class of ordinals such that $\alpha \in C \Longrightarrow L_\alpha \models$ ZF, defined by: $C = \{\alpha \mid L_\alpha \prec L\}$. By Easton forcing collapse cardinals so that every uncountable cardinal belongs to $C \cup \{\alpha^+ \mid \alpha \in C\}$, and then code by a real $R_0 <_L 0^\#$. Now apply Theorem 7.6, relativized to $R_0$, to obtain $R <_L 0^\#$ such that $\Lambda^*(R) \subseteq \{\alpha \mid L_\alpha \models \text{ZF} - \text{Power}\}$. □

**Remark.** Of course if $\Lambda(R)$ is contained in $\{\alpha \mid L_\alpha \models \text{Power}\}$ then $0^\# \leq_L R$, as the Covering Theorem implies that $0^\# \not\leq_L R \Longrightarrow \omega_\omega^+$ of $L[R] = \omega_\omega^+$ of $L$.

## 7.2 Strong Coding

The goal of this section is to prove:

**Theorem 7.9.** *There exists $R <_L 0^\#$ such that $\Lambda(R) = \mathrm{RI}$.*

We approach this as we did Theorem 7.1, with a version of the forcing used to prove the Coding Theorem. However the requirement that the admissibility of elements of RI be preserved greatly complicates matters. In particular we shall need to define not a single forcing $P$, but forcings $P_\kappa^\beta$ for each $\beta \in \mathrm{Lim} = \{\gamma \mid \gamma > \omega, \gamma \text{ limit}\}$ and each $\kappa \in \{\delta \mid \delta \geq \omega, L_\beta \vDash \delta \text{ a cardinal}\}$. The definition of $P_\kappa^\beta$ is by induction on $\beta$. The idea is that a $P_\kappa^\beta$-generic is a set $G_\kappa^\beta \subseteq L_\beta$, coded by $G_\kappa^\beta \cap L_{(\kappa^+)^{L_\beta}}$, with the property that $\beta$ is $G_\kappa^\beta$-admissible iff $\beta \in \mathrm{RI}$. A key property is that if $\kappa < \bar\beta < \beta$ and $\bar\beta \in \mathrm{Lim}$ then $G_\kappa^\beta \cap L_{\bar\beta} = G_\kappa^{\bar\beta}$ is also $P_\kappa^{\bar\beta}$-generic; this guarantees that in addition $\bar\beta$ is $G_\kappa^\beta \cap L_{\bar\beta}$-admissible iff $\bar\beta \in \mathrm{RI}$, for such $\bar\beta$. The net result is that $P = P_\omega^\infty = \bigcup\{p_\omega^\beta \mid \beta \in \mathrm{Lim}\}$ adds $G \subseteq L$, coded by $G \cap L_{\omega_1} = X$, such that $\beta$ is $X \cap (\omega_1)^{L_\beta}$-admissible iff $\beta \in \mathrm{RI}$. Finally the desired real $R$ is obtained by coding $X$ using a simple almost-disjoint coding.

Due to the delicacy of the construction of the $P_\kappa^\beta$'s we cannot use the coding method of Section 4.2; instead we need the more sophisticated technique used to prove the Coding Theorem in the general case.

**Definition** (Strings). We define $S_\kappa^\beta$ for $\beta \in \mathrm{Lim}$ and $\kappa \in \beta\text{-card} = \{\delta \geq \omega \mid \tilde J_\beta \vDash \delta$ a cardinal$\}$, by induction on $\beta$. If $\beta = \omega + \omega$ then $S_\omega^\beta = \{s \mid s : [\omega, |s|) \longrightarrow 2, \omega \leq |s| < \omega + \omega\}$. If $\beta \in \mathrm{Lim}^2 = \{\gamma \mid \gamma$ is a limit of limit ordinals$\}$ then $S_\kappa^\beta = \bigcup\{S_\kappa^{\bar\beta} \mid \kappa < \bar\beta < \beta, \bar\beta \in \mathrm{Lim}\}$. Given that $S_\kappa^\beta$ is defined we take $S_\kappa^{\beta+\omega}$ to consist of all $s : [\kappa, |s|) \longrightarrow 2$, $\kappa \leq |s| < \beta + \omega$ such that either $s \in S_\kappa^\beta$ or $\beta \leq |s|$ and:

(1) $s \restriction \beta$ is sufficiently $P_\kappa^\beta$-generic.

(2) $\beta$ $s \restriction \beta$-admissible $\Longrightarrow \beta \in \mathrm{RI}$.

(3) Let $\eta = \kappa^+$ of $\tilde J_\beta$. Then $s \restriction \eta$ is *uniformly* $\Delta_1\langle \tilde J_\eta, C_\eta\rangle$: For some parameter $x \in \tilde J_\eta$, $s \restriction \bar\eta$ is $\Delta_1\langle \tilde J_{\bar\eta}, C_{\bar\eta}\rangle$ in $x$ for each $\bar\eta \in \mathrm{Lim}\, C_\eta \cup \{\eta\}$ such that $x \in \tilde J_{\bar\eta}$, via a fixed $\Delta_1$ definition.

(4) $\beta \in \mathcal{O}(\kappa) = \{\gamma \mid \tilde J_{\gamma+\omega} \vDash \mathrm{Card}\, \gamma \leq \kappa\}$.

(5) $s(\xi) = 0$ when $\xi$ is a successor ordinal, $\kappa < \xi < |s|$.

**Remarks.** (1) The notion "sufficiently $P_\kappa^\beta$-generic" will be clarified later.

(2) To say that $\beta$ is $s \restriction \beta$-admissible is to say that $\tilde J_\beta^{s\restriction\beta} = \langle \tilde J_\beta[s \restriction \beta], s \restriction \beta\rangle$ is a model of ZF – Power, with Replacement only for $\Sigma_1$ formulas mentioning $s \restriction \beta$ as a predicate.

(3) Recall that $\tilde{J}_{\omega\gamma} = J_\gamma$, the $\gamma^{\text{th}}$ level of Jensen's $J$-hierarchy. This, as well as the canonical global $\square$-sequence $\langle C_\gamma \mid \gamma \text{ singular} \rangle$ for $L$, were defined in Chapter 1.

(4) The strong requirement "$\beta \in \mathcal{O}(\kappa)$" implies that $s \restriction \eta$ is an element of $\tilde{J}_{\beta+\omega}$, because $C_\eta$ is definable over $\tilde{J}_\beta$. If $\eta = \beta$ (we allow this possibility in clause (c)) then we infer that $s \in \tilde{J}_{\beta+\omega}$; otherwise the $P_\kappa^\beta$-genericity of $s \restriction \beta$ will imply that $s \restriction \beta$ is coded over $\tilde{J}_\beta$ by $s \restriction \eta$, and hence again we have $s \in \tilde{J}_{\beta+\omega}$.

**Definition** (Coding Apparatus). Suppose $s \in S_\kappa^\beta$, $\kappa > \omega$, $|s| = \lambda + k$, $\lambda$ limit, $k \in \omega$. Let $\eta = \kappa^+$ of $\tilde{J}_\lambda$ if $\tilde{J}_\lambda \models \kappa^+$ exists and $\eta = \lambda$ otherwise. We set $\mathcal{A}_0^s = \langle \tilde{J}_\eta, C_\eta \rangle$ if $\kappa < \lambda$ and $\mathcal{A}_0^s = \tilde{J}_\kappa = L_\kappa$ otherwise. For $\alpha < \kappa$, $H_0^s(\alpha) = \Sigma_{k+1}$ Hull of $\alpha$ in $\mathcal{A}_0^s$ and we set $\pi_0^s(\alpha) : \bar{H}_0^s(\alpha) \simeq H_0^s(\alpha)$, $\bar{H}_0^s(\alpha) = \langle \tilde{J}_{\bar{\eta}}, C_{\bar{\eta}} \rangle$. Then $\pi_0^s(\alpha)$ extends canonically to a $\Sigma_1$-elementary $\pi^s(\alpha) : \tilde{J}_{\bar{\lambda}} \longrightarrow \tilde{J}_\lambda$ where $\bar{\lambda}$ is least such that $\tilde{J}_{\bar{\lambda}+\omega} \models \text{Card}\, \bar{\eta} \leq \bar{\kappa}$, where $\pi_0^s(\bar{\kappa}) = \kappa$. We define the "approximation" $s_\alpha$ to $s$ by: $\text{Dom}\, s_\alpha = [\bar{\kappa}, \bar{\lambda} + k)$, $s_\alpha \restriction [\bar{\kappa}, \bar{\lambda}) = s \circ \pi^s(\alpha)$ and $s_\alpha(\bar{\lambda} + \ell) = s(\lambda + \ell)$ for $\ell < k$. We say that $\alpha$ is *$s$-good* if $s \restriction \eta$ is uniformly $\Delta_1(\mathcal{A}_0^s)$ in a parameter belonging to $H_0^s(\alpha)$.

**Definition** (The Successor Coding). Suppose $s \in S_{\kappa^+}^\beta$, where $\kappa^+$ denotes $\kappa^+$ of $\tilde{J}_\beta$ and $\kappa \in \beta$-card. A condition in $R^s$ is a pair $(t, t^*)$ where:

(1) $t \in S_\kappa^\beta$

(2) $t^* = \emptyset$ or $t^* = H_0^{s \restriction \eta}(\alpha) \cup \{\eta\}$ for some limit $\eta \leq |s|$, where $\alpha$ is $s \restriction \eta$-good and $\alpha \leq |t|$.

Extension of conditions in $R^s$ is defined by: $(t_0, t_0^*) \leq (t_1, t_1^*)$ iff $t_0$ extends $t_1$, $t_0^*$ contains $t_1^*$ and if $|t_1| < \bar{\alpha}$, $\bar{\eta} \in t_1^*$ and $\bar{\alpha}$ is $s \restriction \bar{\eta}$-good then $t_0$ and $(s \restriction \bar{\eta})_{\bar{\alpha}}$ are compatible.

**Remark.** The $s$-goodness of $\alpha$ is needed to guarantee that $s_\alpha \restriction \lambda_\alpha$ is uniformly $\Delta_1\langle \tilde{J}_{\lambda_\alpha}, C_{\lambda_\alpha} \rangle$, where $\lambda_\alpha = \text{ordertype}(H^s(\alpha) \cap \lambda)$. The requirement "$\alpha \leq |t|$" in Successor Coding (b) is needed for extendibility of conditions in $R^s$. To say that "$t_0$ and $(s \restriction \bar{\eta})_{\bar{\alpha}}$ are compatible" is to say that $t_0$ and $(s \restriction \bar{\eta})_{\bar{\alpha}}$ agree on the intersection of their domains.

**Definition** (Limit Coding). Suppose $s \in S_\kappa^\beta$, $|s|$ limit, $\kappa \in \beta\text{-card}' = \{\delta > \omega \mid \tilde{J}_\beta \models \delta \text{ a limit cardinal}\}$ and $p = \langle (p_\alpha, p_\alpha^*) \mid \alpha \in \beta\text{-card} \cap [\gamma, \kappa) \rangle$ where $(p_\alpha, p_\alpha^*) \in R^{p_{\alpha^+}}$ and $p_{\alpha^+} \in S_{\alpha^+}^\beta$ for each $\alpha \in \beta\text{-card} \cap [\gamma, \kappa)$ ($\alpha^+$ denoting $\alpha^+$ of $\tilde{J}_\beta$). Then *$p$ codes $s$* if for each $\eta \leq |s|$ such that $s \restriction \eta$ belongs to $S_\kappa^\beta$, $(s \restriction \eta)_\alpha \subseteq p_\alpha$ for sufficiently large $\alpha \in \beta\text{-card} \cap \kappa$.

Of course as in Section 4.3 there will be a number of further requirements to impose on $p$ before it can be used in our desired class $P$ of conditions. Before turning to these requirements we discuss $\Sigma$-genericity.

**Definition.** Fix $\beta \in \mathrm{Lim}$ and $\kappa \in \beta$-card. Then $T \subseteq P_\kappa^\beta \times \gamma, \gamma < \beta$ is *persistent* if whenever $(p, \delta) \in T$ and $q \le p$, then $(q, \delta) \in T$. Let $D(T)$ denote all $p \in P_\kappa^\beta$ such that either:

(1) $T_\delta = \{q \mid (q, \delta) \in T\}$ is dense $\le p$ for all $\delta < \gamma$, or

(2) For some $\delta < \gamma$, $(q, \delta) \notin T$ for all $q \le p$.

Then $G \subseteq P_\kappa^\beta$ is $P_\kappa^\beta$ *$\Sigma$-generic over $\tilde{J}_\beta$* if $G$ is compatible, upward-closed and intersects each $D(T)$, for $T$ persistent and $\Sigma_1(\tilde{J}_\beta)$.

A condition in $P_\kappa^\beta$ will be of the form $p = \langle (p_\alpha, p_\alpha^*) \mid \kappa \le \alpha \le \alpha(p), \alpha \in \bar{\beta}\text{-Card}\rangle$ for some $\bar{\beta} \le \beta$, where $p_\alpha \in S_\alpha^{\bar{\beta}}$ and $(p_\alpha, p_\alpha^*) \in R^{p_{\alpha^+}}$ for $\alpha < \alpha(p)$. If $G \subseteq P_\kappa^\beta$ is $P_\kappa^\beta$-generic (i.e., $G$ is compatible, upward-closed and intersects every predense $D \subseteq P_\kappa^\beta$, $D \in \tilde{J}_\beta$) then $G$ will be determined by $s^G = \cup\{p_\alpha \mid p \in G, \kappa \le \alpha \in \beta\text{-card}\}$, a function from $[\kappa, \beta)$ into 2. We say that $s : [\kappa, \beta) \longrightarrow 2$ is *sufficiently $P_\kappa^\beta$-generic* if $s = s^G$ for some $G \subseteq P_\kappa^\beta$, where $G$ is $P_\kappa^\beta$-generic and where $G$ is also $P_\kappa^\beta$ $\Sigma$-generic if $\beta$ is recursively inaccessible.

**Definition** (Predensity Reduction). Suppose $\beta \in \mathrm{Lim}$, $\kappa \in \beta\text{-card}'$, $s \in S_\kappa^\beta$ and $|s| = \lambda + k$ with $\kappa < \lambda$ limit, $k$ finite. Assume that $P^s$ has been defined to contain $P^{<s\restriction\lambda} = \cup\{P^{<s\restriction\eta} \mid \kappa \le \eta < \lambda, \eta \in \mathcal{O}(\kappa)\}$ and to have an open dense set consisting of conditions of the form $p = \langle (p_\alpha, p_\alpha^*) \mid \alpha \in \beta\text{-card} \cap \kappa\rangle$. If $D \subseteq P^{<s\restriction\lambda}$ is predense on $P^{<s\restriction\lambda}$ and $\gamma \in \beta\text{-card}^+ \cap \kappa = \{\delta < \kappa \mid \tilde{J}_\beta \vDash \delta$ is an infinite successor cardinal$\}$ then $p \in P^s$ *reduces $D$ below $\gamma$* (or *$\gamma$-reduces $D$*) iff $p$ is of the form $p = \langle (p_\alpha, p_\alpha^*) \mid \alpha \in \beta\text{-card} \cap \kappa\rangle$ and $\forall q \le p \exists r \le q\ (r \restriction [\gamma, \kappa) = q \restriction [\gamma, \kappa)$ and $r$ meets $D$). And $p$ is *sufficiently $P^{<s\restriction\lambda}$-quasigeneric* if $p$ reduces (i.e., $\gamma$-reduces for some $\gamma \in \beta\text{-card}^+ \cap \kappa$) each predense $D \subseteq P^{<s\restriction\lambda}$, $D \in \tilde{J}_\lambda$ and also when $\lambda$ is recursively inaccessible, $p$ reduces each $D(T)$ where $T \subseteq P^{<s\restriction\lambda} \times \delta, \delta < \lambda$, $T$ is $\Sigma_1$ over $\langle \tilde{J}_\lambda[s \restriction \lambda], s \restriction \lambda\rangle$ and $T$ is persistent.

We shall require that with the above notation, each $p \in P^s - P^{<s\restriction\lambda}$ be sufficiently $P^{<s\restriction\lambda}$-quasigeneric. As in Section 4.3 this will facilitate the proofs of Distributivity, Persistence and Chain Condition for our $P_\kappa^\beta$ forcings.

If $\kappa$ is not the largest cardinal of $\tilde{J}_\lambda$ then $P^{<s\restriction\lambda}$ will be an element of $\tilde{J}_\lambda[s \restriction \lambda]$ and hence will preserve the admissibility of $\tilde{J}_\lambda^{s\restriction\lambda} = \langle \tilde{J}_\lambda[s \restriction \lambda], s \restriction \lambda\rangle$ when $\lambda$ belongs to RI. Otherwise we must be careful to arrange that the forcing relation for $P^{<s\restriction\lambda}$, a class forcing over $\tilde{J}_\lambda^{s\restriction\lambda}$, is $\Delta_1$-definable over $\tilde{J}_\lambda^{s\restriction\lambda}$ when restricted to $\Delta_0$ sentences. Both this property and the quasigenericity property are facilitated through a strong version of Relativized $\diamondsuit$.

**Definition.** Suppose that $\beta, \kappa, s, \lambda, k$ are as in the previous definition and set $\eta(s) = \kappa^+$ of $\tilde{J}_\lambda$ if $\tilde{J}_\lambda \vDash \kappa^+$ exists and $\eta(s) = \lambda$ otherwise. Also $\lambda(s) = \lambda$, $k(s) = k$. We write $t \in E$ if for some $s$ as above, $t = s \restriction \eta$ where $\kappa < \eta \le \eta(s)$ and $C_\eta$ has ordertype $\omega$.

**Strong Relativized ◇.** There exists $\langle D^t \mid t \in E\rangle$ such that:

(a) If $D$ belongs to $\tilde{J}^{s\restriction\lambda(s)}_{\lambda(s)}$, $D \subseteq \tilde{J}_{\eta(s)}$ and $\eta(s) < \lambda(s)$ then $X^D = \{\eta < \eta(s) \mid s \restriction \eta \in E, D^{s\restriction\eta} = D \cap \tilde{J}_\eta\}$ is stationary in $\tilde{J}^{s\restriction\lambda(s)}_{\lambda(s)}$.

(b) If $D$ is $\Pi_1\langle \tilde{J}^{s\restriction\lambda(s)}_{\lambda(s)}, s \restriction \lambda(s)\rangle$, $D \subseteq \tilde{J}_{\eta(s)}$, $\eta(s) < \lambda(s)$ and $\lambda(s)$ is $s \restriction \lambda(s)$-admissible then $X^D$ is $\Pi_1$-stationary in $\langle \tilde{J}^{s\restriction\lambda(s)}_{\lambda(s)}, s \restriction \lambda(s)\rangle$.

(c) If $D$ is $\Delta_1\langle \tilde{J}_{\eta(s)}, s \restriction \eta(s)\rangle$, $D \subseteq \tilde{J}_{\eta(s)}$, $\kappa < \eta(s)$ and $\eta(s)$ is $s \restriction \eta(s)$-admissible then $X^D$ is $\Delta_1$-stationary in $\langle \tilde{J}_{\eta(s)}, s \restriction \lambda(s)\rangle$.

(d) $D^t$ is $\Delta_1\langle \tilde{J}_{|t|}, C_{|t|}\rangle$, uniformly for $t \in E$.

(e) If $\tilde{J}_{\lambda(s)} \models \kappa^{++}$ exists and $s \restriction \eta(s) \in E$ then $D^{s\restriction\eta(s)} = \emptyset$.

This principle is established via a refinement of the proof of Relativized ◇ in Chapter 1.

We are now ready to define the forcings $P^\beta_\kappa$, $\kappa \in \beta$-card and $P^s_\gamma$, $s \in S^\beta_\kappa$, $\gamma \in \beta$-card $\cap\, \kappa$. A condition in $P^s_\gamma$, where $s$ belongs to $S^\beta_\kappa$ and $\gamma < \kappa$ belongs to $\beta$-card, is either an element of $P^{<s\restriction\lambda(s)}_\gamma = \cup\{P^{s\restriction\eta}_\gamma \mid \eta \in \mathcal{O}(\kappa) \cap \lambda(s)\}$ or is of the form $p = \langle (p_\alpha, p^*_\alpha) \mid \gamma \le \alpha \in \beta\text{-card} \cap \kappa\rangle$ where:

(a) (Basics) For $\gamma < \alpha \in \beta$-card$\,\cap\,\kappa$, $p \restriction \alpha$ belongs to $P^{p_\alpha}_\gamma$. If $\kappa = \alpha^+$ of $\tilde{J}_\beta$ then $p(\alpha) = (p_\alpha, p^*_\alpha)$ belongs to $R^s$. If $\kappa$ is a limit cardinal of $\tilde{J}_\beta$ then $p$ codes $s$.

(b) (Predensity Reduction) Suppose $\kappa$ is a limit cardinal of $\tilde{J}_\beta$.

(b1) If $D^{s\restriction\eta(s)} \subseteq P^{<s\restriction\lambda(s)}$ is $\alpha$-predense for all $\alpha \in \beta$-card$^+ \cap \kappa$ then $p$ meets $D^{s\restriction\eta(s)}$.

(b2) If $\langle \tilde{J}_{\eta(s)}, s \restriction \eta(s)\rangle$ is admissible with $\Sigma_1$ projectum $\kappa$ then $p$ reduces each $D(T)$, $T \subseteq P^{<s\restriction\eta(s)} \times \delta$ ($T$ persistent and $\Sigma_1\langle \tilde{J}_{\eta(s)}, s \restriction \eta(s)\rangle$, $\delta$ regular in $\tilde{J}_{\eta(s)}$) below some $\alpha \in \beta$-card$^+ \cap \kappa$.

(b3) If $\eta(s) = \lambda(s) = \bar{\lambda} + \omega$, $D \subseteq P^{<s\restriction\lambda(s)}$ is predense and $D$ belongs to $\tilde{J}_{\lambda(s)}$ then $p$ $\alpha$-reduces $D$ for some $\alpha \in \beta$-card$^+ \cap \kappa$.

(c) (Restriction) If $\kappa$ is a limit cardinal of $\tilde{J}_\beta$, $\eta \le |s|$ and $\eta \in \mathcal{O}(\kappa)$ then $p \le q$ for some $q \in P^{s\restriction\eta}_\gamma - P^{<s\restriction\eta}_\gamma$ (where $P^{<t}_\gamma = \cup\{P^{t\restriction\xi}_\gamma \mid \xi \in \mathcal{O}(\kappa), \xi < |t|\}$).

(d) (Nonstationary Restraint) If $\kappa$ is a limit cardinal of $\tilde{J}_\beta$ and $\kappa$ is $\Sigma_{k(s)+2}\langle \tilde{J}_{\eta(s)}, C_{\eta(s)}\rangle$-regular (or $\Sigma_{k(s)+2}\,(L_\kappa)$-regular if $\lambda(s) = \kappa$) then there is a CUB $C \subseteq \kappa$ such that $\alpha \in C \implies p^*_\alpha = \emptyset$ and $C$ is $\Delta_{k(s)+2}\langle \tilde{J}_{\eta(s)}, C_{\eta(s)}\rangle$ (or $\Delta_{k(s)+2}(L_\kappa)$ if $\lambda(s) = \kappa$).

(e) (Definability) If $\kappa$ is a limit cardinal of $\tilde{J}_\beta$ then $p$ is $\Delta_{k(s)+2}\langle\tilde{J}_{\eta(s)}, C_{\eta(s)}\rangle$ (or $\Delta_{k(s)+2}(L_\kappa)$ if $\lambda(s) = \kappa$).

We order $P^s_\gamma$ by: $p \leq q$ iff $p(\delta) \leq q(\delta)$ for all $\delta \in [\gamma, \kappa) \cap \beta$-card. This completes the definition of $P^s_\gamma$.

Now define $P^\beta_\gamma$ ($\beta$ limit, $\omega \leq \gamma < \beta$, $\gamma$ a cardinal of $\tilde{J}_\beta$) to consist of all $p$ of the form $p' * \{(s', \emptyset)\}$ where for some $\beta' \leq \beta$, $\kappa' \in \beta'$-card, $\gamma \leq \kappa'$ and some $s' \in S^{\beta'}_{\kappa'}$ we have $p' \in P^{s'}_\gamma$ (and $p' \notin P^{<s'}_\gamma$ if $\kappa' \in \beta'$-card$'$). We allow the possibility $p' = \emptyset$, $\kappa' = \gamma$.

We order $P^\beta_\gamma$ as follows: Suppose $p, q \in P^\beta_\gamma$ and let $\delta = \max(\mathrm{Dom}(p) \cap \mathrm{Dom}(q))$. If $\delta = \max(\mathrm{Dom}(q))$ then $p \leq q$ iff $p \restriction \delta \leq q \restriction \delta$ in $P^{p_\delta}_\gamma$ and $q_\delta \subseteq p_\delta$. Otherwise $p \leq q$ iff $q \in P^{\beta'}_\gamma$ with $\beta' \leq |p_\delta|$, $p \restriction \delta \leq q \restriction \delta$ in $P^{p_\delta}_\gamma$ and in addition $q \restriction [\delta, \max(\mathrm{Dom}(q))]$ is a condition in the $P^{\beta'}_\delta$-generic determined by $p_\delta \restriction [\delta, \beta')$.

Finally: $P^s = P^s_\omega$, $P^\beta = P^\beta_\omega$, $P = \cup_\beta P^\beta$.

**Remarks.** In (b1) above, we say that $D$ is *$\alpha$-predense* if for every $p$ there is $q \leq_\alpha p$ such that $q$ meets (i.e., extends an element of) $D$. The relation $\leq_\alpha$ is defined by: $q \leq_\alpha p$ iff $q \leq p$ and $q \restriction \alpha = p \restriction \alpha$. Notice that the Growth Requirement from Section 4.3 has disappeared. This is because the Definability property (e) together with the requirement that $p$ code $s$ imply that $p$ obeys the proper Growth Requirement.

Our plan is to establish the following 5 properties of $(\beta, \kappa, s)$, where $\beta \in$ Lim, $\kappa \in \beta$-card and $s \in S^\beta_\kappa$, by induction on $\beta$, for fixed $\beta$ by induction on $\kappa$, and for fixed $\beta$ and $\kappa$ by induction on $|s|$:

**Extendibility.** Suppose $\gamma \in \beta$-card $\cap \kappa$, $t \in S^\beta_\gamma$ and $|t| \leq \xi \in \mathcal{O}(\gamma) \cap \beta$. Then there exists $u \in S^\beta_\gamma$ such that $t \subseteq u$ and $|u| = \xi$. For $\kappa \in \beta$-card$^+$: If $(t, t^*_0)$, $(t, t^*_1)$ belong to $R^s$ and $\alpha < \kappa$ then there exists $(u, u^*) \leq (t, t^*_0)$, $(t, t^*_1)$ in $R^s$ such that $\alpha \leq |u|$. For $\kappa \in \beta$-card$'$: If $p$ belongs to $P^s$ and $\alpha \in \beta$-card$^+ \cap \kappa$ then there exists $q \leq_\alpha p$ such that $q \in P^s - P^{<s}$.

**Distributivity.** Suppose $\alpha < \kappa$. If $\langle D_i \mid \alpha \leq i < \kappa\rangle \in \tilde{J}^{s \restriction \lambda(s)}_{\lambda(s)}$ and $D_i$ is $i^+$-predense below $p$ on $P^{<s \restriction \lambda(s)}$ for each $i$ then there exists $q \leq_{\alpha^+} p$ such that $q$ meets each $D_i$.

**Chain Condition.** If $D \subseteq P^{<s \restriction \lambda(s)}$ is predense below $p \in P^{<s \restriction \lambda(s)}$ and $D$ belongs to $\tilde{J}_{\lambda(s)}[s \restriction \lambda(s)]$ then $D$ has a subset $d \in \tilde{J}_{\eta(s)}$ which is also predense below $p$. If $D \subseteq P^{<s \restriction \lambda(s)}$ is $\Sigma_1\langle\tilde{J}_{\lambda(s)}[s \restriction \lambda(s)], s \restriction \lambda(s)\rangle$ and predense below $p \in P^{<s \restriction \lambda(s)}$ and $\lambda(s)$ is $s \restriction \lambda(s)$-admissible then $D$ has a subset $d \in \tilde{J}_{\eta(s)}$ which is also predense below $p$.

**Persistence.** Suppose $t \in S^\beta_\kappa$ and $t \subseteq s$. If $D \subseteq P^{<t\restriction\lambda(t)}$ is predense and $D \in \tilde{J}_{\lambda(t)}[t \restriction \lambda(t)]$ then $D$ is predense on $P^{<s}$. If $T \subseteq P^{<t\restriction\lambda(t)} \times \delta$ is $\Sigma_1\langle\tilde{J}_{\lambda(t)}[t \restriction \lambda(t)], t \restriction \lambda(t)\rangle$, $\lambda(t)$ is $t \restriction \lambda(t)$-admissible, $T$ is persistent and $\delta$ is regular in $\tilde{J}_{\lambda(t)}[t \restriction \lambda(t)]$ then $D(T)$ is predense on $P^{<s}$.

As in Section 4.3, we introduce one more property, to aid in establishing Distributivity and Extendibility. We say that $X \subseteq \kappa$ is *thin in* $\tilde{J}_{\eta(s)}$ if $X \in \tilde{J}_{\eta(s)}$ and for each $\tilde{J}_{\eta(s)}$-inaccessible $\alpha \leq \kappa$, $\tilde{J}_{\eta(s)} \vDash X \cap \alpha$ is not stationary in $\alpha$. A function $f : \beta$-card $\cap\, \kappa \longrightarrow \tilde{J}_{\eta(s)}$ in $\tilde{J}_{\eta(s)}$ is *small in* $\tilde{J}_{\eta(s)}$ if for each $\alpha \in \mathrm{Dom}(f)$, Card $f(\alpha) \leq \alpha$ in $\tilde{J}_{\eta(s)}$ and $\mathrm{Support}(f) = \{\alpha \mid f(\alpha) \neq \emptyset\}$ is thin in $\tilde{J}_{\eta(s)}$. For $p \in P^{<s\restriction\lambda(s)}$ and $f$ small in $\tilde{J}_{\eta(s)}$ define $\Sigma^p_f$ = all $q \leq p$ such that whenever $\alpha \in \mathrm{Dom}(f)$, $D \in f(\alpha)$ and $D$ is predense on $P^{p^+_\alpha}$ then $q$ reduces $D$ below some $\bar{\alpha} \leq \alpha$ in $\beta$-card$^+$.

**$\Sigma_f$-Density.** For $p \in P^{<s\restriction\lambda(s)}$ and $f$ small in $\tilde{J}_{\eta(s)}$ there exists $q \leq p$ in $\Sigma^p_f$.

In contrast to Jensen coding, our most difficult task in strong coding will be to establish Extendibility when $\kappa$ is a successor cardinal of $\tilde{J}_\beta$. This is due to our strong requirement that strings $t \in S^\beta_\gamma$ be sufficiently generic when restricted to any limit ordinal.

We proceed now with a series of lemmas which will lead to an inductive proof of the above five properties. Fix $\beta$, $\kappa \in \beta$-card and $s \in S^\beta_\kappa$. We assume that the above five properties hold for smaller $\beta$, for $\beta$ with smaller $\kappa$ and for $\beta, \kappa$ with smaller $|s|$. (Any further hypotheses to the lemmas below are explicity stated.)

**Lemma 7.10** (Extendibility for $S^\beta_\gamma$). *Assume $\Sigma_f$-density for strings in $\cup\{S^{\beta'}_{\kappa'} \mid \kappa' \in \beta'$-card, $\beta' \leq \beta\}$. Suppose $\gamma \in \beta$-card $\cap\, \kappa$, $t \in S^\beta_\gamma$ and $|t| \leq \xi \in \mathcal{O}(\gamma) \cap \beta$. Then there exists $u \in S^\beta_\gamma$ such that $t \subseteq u$ and $|u| = \xi$.*

*Proof.* By induction on $\xi$. The result is immediate by induction if $\xi$ is a successor ordinal, so assume that $\xi$ is a limit ordinal. We must show that $t$ can be extended to $u$ of length $\xi$ which is sufficiently $P^\xi_\gamma$-generic, which kills the admissibility of $\xi$ if $\xi$ is a successor admissible and which obeys the definability condition: $u \restriction \eta$ is uniformly $\Delta_1\langle\tilde{J}_\eta, C_\eta\rangle$ where $\eta = \gamma^+$ of $\tilde{J}_\xi$ (or $\eta = \xi$ if $\tilde{J}_\xi \vDash \gamma$ is the largest cardinal).

**Case 1.** $\tilde{J}_\xi \vDash \gamma$ is the largest cardinal.

If $\xi$ is not the limit of admissible ordinals then let $\bar{\xi} = \cup\{\alpha \mid \alpha$ admissible, $\alpha < \xi\}$ and by induction choose $t \subseteq u_0$, $u_0 \in S^\beta_\gamma$, $|u_0| = \xi_0$ where $\xi_0$ is the least element of $\mathcal{O}(\gamma) - \bar{\xi}$. Then define $u(\delta) = u_0(\delta)$ for $\delta < \xi_0$, $u(\delta) = 1$ iff $\delta \in C_\xi$ for $\xi_0 \leq \delta < \xi$. Clearly $u$ kills the admissibility of $\xi$ and is uniformly $\Delta_1\langle\tilde{J}_\xi, C_\xi\rangle$. Also note that the forcing $P^\xi_\gamma$ is (equivalent to) $S^\xi_\gamma$ and therefore any $v : [\gamma, \xi) \longrightarrow 2$ whose proper restrictions belong to $S^\xi_\gamma$ is sufficiently $P^\xi_\gamma$-generic. It follows that $u$ belongs to $S^\beta_\gamma$, as desired.

Now suppose that $\xi$ is an inadmissible limit of admissible ordinals. Let $\langle \xi_i \mid i < \lambda \rangle$ be the increasing enumeration of $C_\xi - |t|$ and define $\langle t_i \mid i < \lambda \rangle$ by: $t_{i+1}$ = least $t^*$ in $S^\beta_\gamma$ extending $t_i$ such that $|t^*| \geq \xi_i$, and for limit $i$, $t_i = \cup\{t_j \mid j < i\}$. Note that $\xi_i \in \mathcal{O}(\gamma)$ and $|t_i| = \xi_i$ for limit $i$. Also $\xi_i$ is an inadmissible limit of admissible ordinals for limit $i$ as $\xi_i$ is the supremum of the ordinals in the range of a partial $\Sigma_1(\tilde{J}_{\xi_i})$ map from an ordinal less than $\gamma$ into $\xi_i$, a map whose range is $\Sigma_1$-elementary in $\tilde{J}_{\xi_i}$. It therefore follows that $t_i$ belongs to $S^\beta_\gamma$ for limit $i$ and hence the sequence $\langle t_i \mid i < \lambda \rangle$ is well-defined. In this subcase let $u$ equal $\cup\{t_i \mid i < \lambda\}$.

Next suppose that $\xi$ is recursively inaccessible and $\Sigma_1$-*projectible* (i.e., $\Sigma_1$ projectum $(\xi) < \xi$). Then as in the previous case, $\xi_i$ will be an inadmissible limit of admissible ordinals for limit $i$ and hence $\langle t_i \mid i < \lambda \rangle$ is well-defined. However we must ensure in addition that (the $P^\xi_\gamma$-generic corresponding to) $u = \cup\{t_i \mid i < \lambda\}$ is $P^\xi_\gamma$ $\Sigma$-generic. Note that $\lambda = \gamma$ in this case and we may choose $i_0 < \lambda$ large enough so that $\tilde{J}_{\xi_{i_0}}$ contains the least parameter $x$ such that $\xi$ is $\Sigma_1$-projectible to $\gamma$ with parameter $x$. Then for $i \geq i_0$ modify the earlier choice of $t_{i+1}$ to guarantee that either $t_{i+1}$ meets $D_{i-i_0}$ or no extension of $t_{i+1}$ meets $D_{i-i_0}$, where $D_j$ is the $j^{\text{th}}$ set $\Sigma_1$ over $\tilde{J}_\xi$ in parameters $x, j$. Then $|t_{i+1}| < \xi_{i+1}$ and hence again for limit $i$, $|t_i| = \xi_i$ and $t_i$ belongs to $S^\beta_\gamma$. If we now set $u = \cup\{t_i \mid i < \lambda\}$ we have that for every $D$ $\Sigma_1$ over $\tilde{J}_\xi$, $u$ extends a condition $t_i$ such that either $t_i$ meets $D$ or no extension of $t_i$ meets $D$. This weakened version of $\Sigma$-genericity in fact implies full $\Sigma$-genericity in this case, as $P^\xi_\gamma$ is "$\leq \gamma$ $\Sigma$-distributive": Given a uniformly $\Sigma_1(\tilde{J}_\xi)$ sequence $\langle D_j \mid j < \gamma \rangle$ of subsets of $P^\xi_\gamma$ which are dense below a condition $v \in P^\xi_\gamma$, there exists $v^*$ extending $v$ which meets each $D_j$, $j < \gamma$. The latter is easily proved by making successive least extensions of $v$ which meet the $D_j$'s and have inadmissible length.

Finally consider the subcase where $\xi$ is not $\Sigma_1$-projectible. Then proceed as in the subcase where $\xi$ is an admissible limit of admissible ordinals. For limit $i$, $\xi_i$ is not $\Sigma_1$-projectible and in fact is the limit of $\xi_j$, $j < i$ such that $\tilde{J}_{\xi_j}$ is $\Sigma_1$-elementary in $\tilde{J}_{\xi_i}$. Now the $P^{\xi_i}_\gamma$ $\Sigma$-genericity of $t_i$ for limit $i$ follows by reflection, using the $P^{\xi_j}_\gamma$ $\Sigma$-genericity of $t_i \restriction \xi_j$, $j < i$. Thus $u = \cup\{t_i \mid i < \lambda\}$ is the desired extension of $t$, as before.

**Case 2.** $\tilde{J}_\xi \models \gamma^+$ exists.

Let $\xi = \rho_0 \geq \rho_1 \geq \cdots \geq \rho_k = \gamma$ where $\rho_i$ is the $\Sigma_i$-projectum of $\xi$ and $k$ is least such that $\rho_k = \gamma$. For $i < k$ let $x_i$ be least such that $\langle \tilde{J}_{\rho_i}, A_i \rangle = \Sigma_1 \operatorname{Hull}(\rho_{i+1} \cup \{x_i\})$ in $\langle \tilde{J}_{\rho_i}, A_i \rangle$, where $\langle \tilde{J}_{\rho_i}, A_i \rangle$ is the $i^{\text{th}}$ reduct of $\tilde{J}_\xi$ (relative to $\emptyset$) for $i > 0$ and where $A_i = \emptyset$ for $i = 0$. For $i < k$ let $\mathcal{A}_i$ denote $\langle \tilde{J}_{\rho_i}, A_i \rangle$ and for $j < \rho_i$, let $\mathcal{A}^j_i$ denote $\langle \tilde{J}_j, A_i \cap \tilde{J}_j \rangle$. Also for $\delta < \rho_i$, $H_i(\delta) = \Sigma_1 \operatorname{Hull}(\delta \cup \{x_i\})$ in $\mathcal{A}_i$ and $H^j_i(\delta) = \Sigma_1 \operatorname{Hull}(\delta \cup \{x_i\})$ in $\mathcal{A}^j_i$. (In case the structure $\mathcal{A}^j_i$ is not amenable of limit height, we take $H^j_i(\delta)$ to be the closure of $\delta \cup \{x_i\}$ under $h^j_i$, the canonical $\Sigma_1$ Skolem function for $\mathcal{A}_i$ interpreted in $\mathcal{A}^j_i$.)

Define $f_i^j(\delta)$ to be $H_i^j(\delta)$ if $\delta \in H_i^j(\delta)$ and $\emptyset$ otherwise, when $i < k$, $j < \rho_i$ and $\delta \geq \gamma$ is a cardinal of $\tilde{J}_{\rho_i}$. By $\Sigma_f$-density applied to ordinals $\leq \beta$ (which can be assumed by induction and by hypothesis at $\beta$) $\Sigma_{f_i^j}$ is dense on $P_\gamma^{\rho_i}$.

Now construct $\langle p_0^j \mid j < \rho_0 = \xi\rangle$ by choosing $p_{0_\gamma}^0$ to extend $t$ and $p_0^{j+1}$ to be the least extension of $p_0^j$ meeting $\Sigma_{f_0^j}$. At limit stages $j$, $p_0^j$ is the limit of $\langle p_0^{\bar{j}} \mid \bar{j} < j\rangle$. Let $p_0$ be the limit of $\langle p_0^j \mid j < \rho_0\rangle$. Next choose $\langle p_1^j \mid j < \rho_1\rangle$ similarly to extend $p_0$ and to meet the $\Sigma_{f_1^j}$'s, with limit $p_1$. Continue in this way until arriving at $p_k = p$, which meets each $\Sigma_{f_i^j}$. If we can verify that the above construction is well-defined then $G = \{q \in P_\gamma^\xi \mid p \leq q\}$ is $P_\gamma^\xi$-generic, by Predensity Reduction (b3), Restriction and the fact that $p$ meets each $\Sigma_{f_i^j}$.

For limit $j < \rho_i$ and $\gamma \leq \delta \in \rho_i$-card, we have that $p_{i_\delta}^j$ is $\Delta_1$ over $\bar{H}_i^j(\delta) =$ the transitive collapse of $H_i^j(\delta)$. Modify the above construction so that at limit stages $j < \rho_i$, the structure $\mathcal{A}_i^j$ is amenable. Then for limit $j$, $\bar{H}_i^j(\delta)$ is the collapsing structure for $\eta = |p_{i_\delta}^j|$ and therefore $p_{i_\delta}^j$ is $\Delta_1\langle\tilde{J}_\eta, C_\eta\rangle$. If $\xi$ is not a limit of admissibles then we modify the above construction so that if $\bar{\xi} = \cup\{\alpha < \xi \mid \alpha \text{ admissible}\}$ and $\delta =$ the largest $\tilde{J}_\xi$-cardinal, then $p_{0_\delta}$ agrees with $C_\xi$ on $[|t| \cup \bar{\xi}, \xi)$ and $|p_{0_\delta}^0| = |t| \cup \bar{\xi}$. If $\xi$ is an inadmissible limit of admissibles then we modify the construction of $\langle p_0^j \mid j < \rho_0\rangle$ so that $j$ ranges over $C_\xi - |t|$. If $\xi$ is recursively inaccessible and $\Sigma_1$ projectible then as in Case 1 we modify the construction of $p_0$ so that $p_{0_\delta}$ ($\delta$ the largest $\tilde{J}_\xi$-cardinal) "decides" all $\Sigma_1(\tilde{J}_\xi)$ $D \subseteq P_\delta^\xi$. By Distributivity at $\xi$, this is sufficient to guarantee that $G$ is $P_\gamma^\xi$ $\Sigma$-generic. If $\xi$ is not $\Sigma_1$-projectible then we need only modify the construction of $p_0 =$ limit $\langle p_0^j \mid j < \rho_0\rangle$ so that $j$ ranges over $C_\xi - |t|$. Then the $\Sigma$-genericity of $p_{i_\delta}^j$ follows by reflection. We therefore conclude that $p_{i_\delta}^j$ belongs to $S_\delta^\xi$ for limit $j < \rho_i$. To see that $p_i^j \upharpoonright \delta$ is a condition in $P_\gamma^{\rho_i}$ for each $\delta \in \rho_i$-card is now easy, given the definability of the sequence $\langle p_i^j \mid j < \rho_i\rangle$. So in this case we let $u$ be $\cup\{p_\delta \mid \gamma \in \xi\text{-card}\}$. □

We next attack Extendibility for $R^s$. Until now we have only considered strings $s^* \in S_{\kappa^*}^{\beta^*}$, $\kappa^* \in \beta^*$-card. At this point we must generalize this a bit: For any $\kappa^*$, define $S_{\kappa^*}^{\beta^*}$ as before but in the definition of $S_{\kappa^*}^{\beta^*+\omega}$, do not require that $\kappa^*$ be a cardinal of $\tilde{J}_{\beta^*+\omega}$ (though $\kappa^*$ must either be a cardinal of $\tilde{J}_{\beta^*}$ or equal $\beta^*$). Then define $S_{\kappa^*} = \cup\{S_{\kappa^*}^{\beta^*} \mid \kappa^* < \beta^*, \kappa^* \text{ a cardinal of } \tilde{J}_{\bar{\beta}^*} \text{ for all limit } \bar{\beta}^* \in (\kappa^*, \beta^*)\}$, using the new definition of $S_{\kappa^*}^{\beta^*}$. Similarly define $R^{s^*}$ for $s^* \in S_{\kappa^*}$ when $\tilde{J}_{\kappa^*} \models$ There is a largest cardinal. We shall establish a strong form of Extendibility for these $R^{s^*}$ forcings, $s^* \in S_{\kappa^*}$, for $\kappa^* < \kappa$.

We shall use Morass with Square from Chapter 1: Let $\langle C_\alpha \mid \alpha \text{ singular limit}\rangle$, $\bar{\nu} <_1 \nu$, $\pi_{\bar{\nu}\nu}$, $\alpha(\nu)$, $\bar{\nu} <_1^* \nu$ arise from that construction. Then $\bar{\nu} <_1 \nu$ implies that $\pi_{\bar{\nu}\nu}$ extends to a $\Sigma_1$-elementary $\langle\tilde{J}_{\bar{\nu}}, C_{\bar{\nu}}\rangle \longrightarrow \langle\tilde{J}_\nu, C_\nu\rangle$, provided $\alpha(\bar{\nu})$ is at least some $\alpha_0(\nu) < \alpha(\nu)$. Moreover, $\alpha_0(\nu) = \alpha_0(\bar{\nu})$ provided $\bar{\nu} <_1 \nu$ and $\alpha_0(\nu) < \alpha(\bar{\nu})$.

Now for simplicity of notation, we temporarily use $\kappa$, $\beta$ and $s$ as variables. If $\kappa = \gamma^+$ of $\tilde{J}_\beta$, $s \in S_\alpha$, $\alpha < |s|$ limit and $\tilde{J}_\alpha \models \gamma$ is the largest cardinal define $\bar{s} <_1 s$ iff $\bar{s}$ is a good approximation to $s$. Then as in Morass with Square we have the following, where $\alpha(s)$ denotes the $\alpha$ such that $s \in S_\alpha$, $\pi_{\bar{s}s}$ (for $\bar{s} <_1 s$) denotes $\pi_{|\bar{s}||s|}$ and $\bar{s} <_0 s$ means that $\bar{s}$ is a proper initial segment of $s$:

## Relativized Morass with Square

(a) $<_1$ is a tree and if $\bar{s} <_1 s$ then $\alpha(\bar{s}) < \alpha(s)$ and $\bar{s}$ is $<_0$-minimal, successor, limit iff $s$ is $<_0$-minimal, successor, limit.

(b) If $\bar{s} <_1 s$ then $\pi_{\bar{s}s} = \pi : |\bar{s}| \longrightarrow |s|$ is order-preserving, $\bar{s} \upharpoonright \bar{\eta} \in S_{\alpha(\bar{s})}$ iff $s \upharpoonright \pi(\bar{\eta}) \in S_{\alpha(s)}$ and $\pi(\bar{\eta}+\omega) = \pi(\bar{\eta})+\omega$ when $\pi(\bar{\eta})+\omega < |s|$. If $\bar{s}_0 <_0 \bar{s}$ and $s_0 = s \upharpoonright \pi(|\bar{s}_0|)$ then $s_0 \in S_{\alpha(s)}$, $\bar{s} <_1 s_0$ and $\pi_{\bar{s}_0 s_0} = \pi \upharpoonright |\bar{s}_0|$. If $\bar{s}$ is a $<_0$-limit and $\lambda = \bigcup \operatorname{Range} \pi$ then $\bar{s} <_1 s \upharpoonright \lambda$ and $\pi_{\bar{s}s \upharpoonright \lambda} = \pi$. Also if $\pi$ is cofinal and for each $\bar{s}_0 <_0 \bar{s}$, $\alpha = \alpha(s_0')$ for some $s_0' <_1 s \upharpoonright \pi(|\bar{s}_0|)$, then $\alpha = \alpha(s')$ for some $s' <_1 s$.

(c) $\bar{\bar{s}} <_1 \bar{s} <_1 s \Longrightarrow \pi_{\bar{\bar{s}}s} = \pi_{\bar{s}s}\pi_{\bar{\bar{s}}\bar{s}}$. For any $s$, $\{\alpha(\bar{s}) \mid \bar{s} <_1 s\}$ is closed, and this set is unbounded in $\alpha(s)$ unless $s$ is $<_0$-maximal. If $\{\alpha(\bar{s}) \mid \bar{s} <_1 s\}$ is unbounded in $\alpha(s)$ then $|s| = \bigcup\{\operatorname{Range} \pi_{\bar{s}s} \mid \bar{s} <_1 s\}$.

(d) Suppose $s$ is $<_1$-limit and $<_0$-maximal, with $\alpha(s)$ singular. Then for $\alpha$ in a final segment of $C_{\alpha(s)}$ there exists $s_\alpha <_1 s$ in $S_\alpha$ such that if $s$ is a $<_0$-limit, $\lambda_\alpha = \bigcup \operatorname{Range} \pi_{s_\alpha s}$ then $\lambda_\alpha \in C_{|s|} \cup \{|s|\}$, and $\alpha < \beta \Longrightarrow \lambda_\alpha \in \operatorname{Range} \pi_{s_\beta s} \cup \{|s|\}$. And we can in addition require that $s_\alpha$ is $<_0$-maximal, unless there exists $s <_0 t$, $t$ $<_1$-limit.

(e) Suppose $s$ is $<_1$-minimal, $<_0$-limit. Then for $\alpha$ in a final segment of $C_{\alpha(s)}$ there are $s(\alpha) <_1 s_\alpha$, $s(\alpha) \in S_\alpha$, $s(\alpha)$ $<_1$-minimal and $<_0$-limit, $s_\alpha <_0 s$ with $|s_\alpha| \in C_{|s|}$ such that $\alpha < \beta \Longrightarrow |s_\alpha| \in \operatorname{Range} \pi_{s(\beta)s_\beta}$ and $\beta \in \operatorname{Lim} C_{\alpha(s)} \Longrightarrow |s_\beta| = \bigcup\{|s_\alpha| \mid \alpha \in C_\beta\}$.

(f) Suppose $s$ is a $<_1$-successor and a $<_0$-limit. Let $\bar{s} <_1^* s$ be the $<_1$-predecessor to $s$. Then for a final segment of $\alpha$ in $C_{\alpha(s)}$ there are $s(\alpha) <_1 s_\alpha$, $s(\alpha) \in S_\alpha$, $s(\alpha)$ $<_1$-successor and $<_0$-limit, $s_\alpha <_0 s$ with $|s_\alpha| \in C_{|s|}$, such that the final clause of (5) holds and in addition: $\pi_{\bar{s}s}$ cofinal $\Longrightarrow |s_\alpha| = \pi_{\bar{s}s}(|\bar{s}_\alpha|)$ where $\bar{s}_\alpha <_1^* s(\alpha)$, $\bar{s} <_1^* s(\alpha)$, $\pi_{\bar{s}s_\alpha} = \pi_{\bar{s}s}$ and $\lambda = \bigcup \operatorname{Range} \pi_{\bar{s}s} \in \operatorname{Range} \pi_{s(\alpha)s_\alpha}$ provided $\lambda < |s|$.

**Definition.** The relation $t \ll s$ is the smallest relation satisfying:

(1) If $t_0 <_1 t_1$ and either $t_1 \ll s$ or $t_1 = s$, then $t_0 \ll s$.

(2) If $t_0 <_0 t_1$ and $t_0 \ll s$ then $t_1 \ll s$.

**Definition.** Suppose $s \in S_\alpha$, $|s|$ is a limit ordinal greater than $\alpha$ and $\tilde{J}_\alpha \models \gamma$ is the largest cardinal. An $s$-*restraint* is a finite collection $F$ of pairs $(\bar{s}_i, s_i)$, $i < n$ such that:

(1) $\bar{s}_i <_1 s_i$

(2) $s_i \ll s$ or $s_i \leq_0 s$

We write $|F|$ for $\max\{\alpha(\bar{s}_i) \mid i < n\}$. If $t_0 \subseteq t_1$ belong to $S_\gamma$, $|t_0| < \alpha$ we say that the extension $t_0 \subseteq t_1$ *respects* $F$ if whenever $\beta$ is greater than $|t_0|$, $\eta$ belongs to Range $\pi_{\bar{s}_i s_i}$ and $\beta$ is $s_i \upharpoonright \eta$-good, then $t_1$ and $(s_i \upharpoonright \eta)_\beta$ are compatible.

**Lemma 7.11.** *Assume $\Sigma_f$-density for strings in $\cup\{S^{\beta'}_{\kappa'} \mid \kappa' \in \beta'$-card, $\beta' \leq \beta\}$, that $\gamma \in \beta$-card $\cap\, \kappa$, $\alpha$, $s$, $\gamma$ and $F$ are as above and that $t_0 \in S_\gamma$, $|F| \leq |t_0| < \alpha$, $s$ is $<_0$-maximal. Then there exists an extension $t_0 \subseteq t_1$ in $S_\gamma$ respecting $F$ such that $|t_1| = |s|$ and $t_1 \upharpoonright \alpha$ is $R^{<s}$-generic over $\tilde{J}_{|s|}$ (and $R^{<s}$ $\Sigma$-generic over $\tilde{J}^s_{|s|}$ if $|s|$ is recursively inaccessible), where $R^{<s} = \cup\{R^t \mid t \in S_\alpha$, $|t|$ limit, $t$ a proper initial segment of $s\}$.*

*Proof.* By induction on $\alpha$. We first assume that $|s|$ is not recursively inaccessible.

**Case 1.** $s$ is a $<_1$-limit.
Choose a final segment $\langle \alpha_j \mid j < \lambda \rangle$ of $C_\alpha$ such that $|t_0| < \alpha_0$ and $\alpha_j = \alpha(s^*_j)$ for some $<_0$-maximal $s^*_j <_1 s$. We can also insist that for $(\bar{s}_i, s_i) \in F$ if $s_i \leq_0 s$, then $\pi_{\bar{s}_i s_i}$ factors as $\pi_{s^*_0 s}\pi_{\bar{s}_i s^*_0 \upharpoonright \xi_i}$ for some $\xi_i \leq |s^*_0|$, and if $\bar{F}$ is obtained from $F$ by replacing each such $(\bar{s}_i, s_i)$ by $(\bar{s}_i, s^*_0 \upharpoonright \xi_i)$ then $\bar{F}$ is an $s^*_0$-restraint. By induction we can successively choose $t_0 \subseteq u_0 \subseteq u_1 \subseteq \cdots$ where $u_j$ is $R^{<s^*_j}$-generic over $\mathcal{A}^{s^*_j}_0$, $u_j \subseteq u_{j+1}$ respects $\{(s^*_j, s^*_{j+1})\}$ and $t_0 \subseteq u_0$ respects $\bar{F}$. Then $u_{\lambda'} = \cup\{u_j \mid j < \lambda'\}$ is $R^{<s^*_{\lambda'}}$-generic for limit $\lambda'$ by reflection. Finally let $t_1$ equal $u_\lambda \cup s$.

**Case 2.** $\bar{s} <^*_1 s$, $\pi_{\bar{s}s}$ not cofinal.
Choose a final segment $\langle \alpha_j \mid j < \lambda \rangle$ of $C_\alpha$ such that $|t_0| < \alpha_0$ and there are $s^*(j) <_1 s^*_j$, $s^*(j) \in S_{\alpha_j}$, $s^*_j <_0 s$, $\bar{s} <^*_1 s^*(j)$, $\pi_{\bar{s}s^*_j} = \pi_{\bar{s}s}$, such that $j < k \Longrightarrow |s^*_j| \in$ Range $\pi_{s^*(k)s^*_k}$ and $j$ limit $\Longrightarrow |s^*_j| = \cup\{|s^*_{j_0}| \mid j_0 < j\}$. We can assume that if $(\bar{s}_i, s_i) \in F$ and $s_i \leq_0 s$ then $\pi_{\bar{s}_i s_i}$ factors as $\pi_{s^*(0)s^*_0}\pi_{\bar{s}_i s^*(0)\upharpoonright\xi_i}$ for some $\xi_i$ and that $\bar{F}$ is an $s^*(0)$-restraint where $\bar{F}$ is obtained from $F$ by replacing each such $(\bar{s}_i, s_i)$ by $(\bar{s}_i, s^*(0) \upharpoonright \xi_i)$.
By induction we can successively choose $t_0 \subseteq u_0 \subseteq u_1 \subseteq \cdots$ where $u_j$ is $R^{<s^*(j)}$-generic over $\mathcal{A}^{s^*(j)}_0$, $u_j \subseteq u_{j+1}$ respects $\{(s^*(j), s^*(j+1) \upharpoonright \xi^{j+1}_j)\}$ (where $\pi_{s^*(j+1)s^*_{j+1}}(\xi^{j+1}_j) = |s^*_j|$) and $t \subseteq u_0$ respects $\bar{F}$. Let $t_1$ equal $u_\lambda \cup s$.

**Case 3.** $\bar{s} <^*_1 s$, $\pi_{\bar{s}s}$ cofinal, $s$ a $<_0$-limit.
Choose a final segment $\langle \alpha_j \mid j < \lambda \rangle$ of $C_\alpha$ such that $|t_0| < \alpha_0$ and there are $\bar{s}(j) <^*_1 s^*(j) <_1 s^*_j$, $s^*(j) \in S_{\alpha_j}$, $s^*_j <_0 s$, $\bar{s}(j) <_0 \bar{s}$, $|s^*_j| = \pi_{\bar{s}s}(|\bar{s}(j)|)$, such that $j < k \Longrightarrow |s^*_j| \in$ Range $\pi_{s^*(k)s^*_k}$ and $j$ limit $\Longrightarrow |s^*_j| = \cup\{|s^*_{j_0}| \mid j_0 < j\}$. We may assume that if $s_i <_0 s$ and $(\bar{s}_i, s_i) \in F$ then $\pi_{\bar{s}_i s_i}$ factors as $\pi_{s^*(0)s^*_0}\pi_{\bar{s}_i s^*(0)\upharpoonright\xi_i}$ for some $\xi_i$

and that $\bar{F}$ is an $s^*(0)$-restraint where $\bar{F}$ consists of all resulting pairs $(\bar{s}_i, s^*(0) \restriction \xi_i)$, together with all $(\bar{s}_i, s_i)$, $\alpha(s_i) < \alpha$, with $(\bar{s}(0), s^*(0))$, and with all $(\bar{s}_i, \bar{s})$ for $\bar{s}_i <_1 \bar{s}$, $s_i = s$.

By induction we can successively choose $t_0 \subseteq u_0 \subseteq u_1 \subseteq \cdots$ where $u_j$ is $R^{<s^*(j)}$-generic over $\mathcal{A}_0^{s^*(j)}$, $u_j \subseteq u_{j+1}$ respects $\{(s^*(j), s^*(j+1) \restriction \xi_j^{j+1}), (\bar{s}(j+1), s^*(j+1))\}$ (where $\pi_{s^*(j+1)s^*_{j+1}}(\xi_j^{j+1}) = |s_j^*|$) and $t_0 \subseteq u_0$ respects $\bar{F}$. Let $t_1$ equal $u_\lambda \cup s$.

**Case 4.** $\bar{s} <_1^* s$, $|s|$ not a $<_0$-limit.

First suppose that $s$ is not $<_0$-minimal and let $s' = s \restriction (|s| - \omega)$. Choose a final segment $\langle \alpha_j \mid j < \omega \rangle$ of $C_\alpha$ such that $|t_0| < \alpha_0$ and if $s' \neq \emptyset$ then as in Case 1, $\alpha_j = \alpha(s'(j))$ for some $<_0$-maximal $s'(j) <_1 s'$. We can assume that $\alpha(\bar{s}) \leq |t_0|$. Now build $t_0 \subseteq u_0 \subseteq u_1 \subseteq \cdots$ as in Case 1 so that if $t_1 = u_\omega \cup s$ then $t_0 \subseteq t_1$ respects $F$ and $u_\omega$ is $R^{<s'}$-generic over $\mathcal{A}_0^{s'}$. However we must also ensure that $u_\omega$ is $R^{<s}$-generic over $\mathcal{A}_0^s$. By the construction of Case 1, we see that $R^{<s}$ is $\gamma^+$-distributive in $\mathcal{A}_0^s$, so we can in fact choose $u_{j+1}$ to meet all predense $D \subseteq R^{<s}$ which are definable over $\tilde{J}_{|s'|+j}$ via the $j^{\text{th}}$ formula, using parameters in $\alpha_j \cup \{|s'|\}$; the result is that $u_\omega$ meets all predense $D \in \mathcal{A}_0^s$. Moreover $u_\omega$ is $\Delta_1\langle \tilde{J}_\alpha, C_\alpha\rangle$ as it is $\Delta_1(\tilde{J}_{|s|})$. In case $s$ is $<_0$-minimal, use Lemma 7.11.

**Case 5.** $s$ is $<_1$-minimal.

Proceed as in Case 2 if $s$ is a $<_0$-limit and as in Case 4 otherwise, using part (5) of Relativized Morass with Square.

Note that if $|s|$ is not recursively inaccessible then $|s|$ is $\Sigma_1$-projectible; in Cases 1–3 above it follows that the approximations $s^*$ to $s$ ($s_j^*$ in Case 1, $s^*(j)$ in Cases 2, 3) have lengths which are $\Sigma_1$-projectible. In Cases 2, 3 these approximations $s^*$ satisfy "$\Sigma_1$ projectum $(|s^*|) <$ greatest cardinal $(\tilde{J}_{|s^*|})$," so $|s^*|$ is inadmissible. In Case 1 we may assume that the $s_j^*$ are chosen to be large enough (in the $<_1$-predecessors to $s$) so that $|s_j^*|$ is not recursively inaccessible. It follows that in these cases there is no need to be concerned with $\Sigma$-genericity at limit stages in the construction of $\langle u_j \mid j < \lambda\rangle$. In Case 4 there is no concern with $\Sigma$-genericity as $C_\alpha$ has ordertype $\omega$; Case 5 is handled like Cases 2, 4.

Now suppose that $|s|$ is recursively inaccessible. If $C_\alpha$ has ordertype greater than $\omega$ or $|s|$ is not $\Sigma_1$-projectible then $\tilde{J}_{|s|}$ is the direct limit of a $\Sigma_2$-elementary chain of approximations $\tilde{J}_{|s^*|}$ where $|s^*|$ is also recursively inaccessible; thus if we choose the $u_i$'s for $i < \lambda$ to be $\Sigma$-generic (by induction) it will follow that $u_\lambda$ is also $\Sigma$-generic by reflection (for limit $\lambda$). Thus we can assume that $C_\alpha$ has ordertype $\omega$ and $|s|$ is $\Sigma_1$-projectible. (Then we can choose the $u_j$, $j < \omega$ to guarantee $\Sigma$-genericity for $u_\omega$.) As in the proof of Lemma 7.10, to guarantee the $\Sigma$-genericity of $u_\omega$ it suffices to show: Given $p \in R^{<s}$ and a $\Sigma_1(\tilde{J}_{|s|}^s)$ sequence $\langle T_i \mid i < \gamma\rangle$ there is $q \leq p$ such that $q$ meets each $D_0(T_i) = \{r \mid r \text{ meets } T_i \text{ or } s \leq r \Longrightarrow s \notin T_i\}$. This form of distributivity is proved as follows: Let $\langle \alpha_i \mid i \leq \gamma\rangle$ be the first $\gamma + 1$ ordinals $\beta < \alpha$ such that $\beta \notin M_\beta = \Sigma_1$ Hull of $\beta \cup \{x\}$ in $\tilde{J}_{|s|}^s$, where $x$ is a parameter defining $\langle T_i \mid i < \gamma\rangle$, $p$ and a $\Sigma_1(\tilde{J}_{|s|}^s)$ injection from $\tilde{J}_{|s|}^s$ into $\alpha$. Let $\pi_i : M_{\alpha_i} \simeq \tilde{J}_{|s_i|}^{s_i}$ and successively build

$p = p_0 \geq p_1 \geq \cdots$ such that $p_i = (t_i, t_i^*)$ where $t_i$ is $R^{<s_i}$-generic over $\tilde{J}_{|s_i|}$, $t_i \subseteq t_{i+1}$ respects $\{(s_i, s'_{i+1})\}$ (where $\pi_{i+1}\pi_i^{-1} = \pi_{s_i s'_{i+1}}$, $s'_{i+1} \leq_0 s_{i+1}$) and $t_\lambda = \bigcup\{t_i \mid i < \lambda\}$ for limit $\lambda$. Also require that $p_{i+1}$ meet $D_0(\pi_{i+1}[T_i \cap M_{\alpha_{i+1}}])$. Then $p_\gamma = q$ is as desired in the conclusion of the given form of distributivity. □

Now we return to our fixed choice of $s \in S_\kappa^\beta$ for which we wish to establish the 5 basic properties.

**Corollary 7.12** (Extendibility for $R^s$). *If $\kappa \in \beta$-card$^+$, $(t, t_0^*), (t, t_1^*)$ belong to $R^{s^*}$ where $s \leq_0 s^*$ and $\alpha < \kappa$ then there exists $(u, u^*) \leq (t, t_0^*), (t, t_1^*)$ in $R^s$ such that $\alpha \leq |u|$.*

*Proof.* Apply Lemma 7.11 to $\bar{s}$ and $F$, where $\bar{s} <_1 s^*$, $\bar{s}$ is $<_0$-maximal, $\alpha$ and $|t|$ are less than $\alpha(\bar{s})$, and where if $t_i^*$ corresponds to $\bar{s}_i <_1 s_i \leq_0 s^*$ then $F = \{(\bar{s}_0, s'_0), (\bar{s}_1, s'_1)\}$ where $\pi_{\bar{s}_i s_i}$ factors as $\pi_{\bar{s}s^*} \circ \pi_{\bar{s}_i s'_i}$, $s'_i \leq_0 \bar{s}$. The result is $(u, u^*)$ where $u^*$ corresponds to $\bar{s} <_1 s^*$, $|u| = |\bar{s}|$ and $u$ extends $t$. □

**Corollary 7.13** (Distributivity for $R^s$). *Suppose $\kappa \in \beta$-card$^+$, $\gamma < \kappa$, $\langle D_i \mid i < \gamma\rangle \in \tilde{J}_{|s|+\omega}$ is a sequence of predense subsets of $R^s$ and $p \in R^s$. Then there exists $q \leq p$ in $R^s$ such that $q$ meets each $D_i$, $i < \gamma$.*

*Proof.* Write $p = (t, t^*)$ where $t^*$ corresponds to $\bar{s} <_1 s' \leq_0 s$ and apply Lemma 7.11 to $\bar{s}'$ where $\bar{s}' <_1 \bar{s}'' <_1 s$, $\bar{s}''$ is $\leq_0$-maximal and $F = \{(\bar{s}', \bar{s}'')\}$. The result is $(u, u^*)$ which meets all predense $D \subseteq R^{\bar{s}''}$ in Range $\pi_{\bar{s}'\bar{s}''}$. As we can assume that Range $\pi_{\bar{s}s'} \cup \{D_i \mid i < \gamma\} \subseteq$ Range $\pi_{\bar{s}'s}$, $q = (u, u^*)$ is the desired extension of $p$. □

We next consider Persistence and Chain Condition, as in Section 4.3.

**Lemma 7.14** (Persistence for $R^s$). *If $\kappa \in \beta$-card$^+$, $t_0 \leq_0 t_1 \leq_0 s$ in $S_\kappa^\beta$, $D \subseteq R_0^t$ predense, $D \in \tilde{J}_{|t_0|+\omega}$ then $D$ is predense on $R_1^t$.*

*Proof.* We can assume that $t_0 <_0 t_1$ and that $t_0$ and $t_1$ have limit length. We must show that each $(u, u^*)$ in $R_1^t$ is compatible with an element of $D$. We may assume that $u^*$ corresponds to $\bar{t}_1 <_1 t_1$, $|u| = |\bar{t}_1|$, $\bar{t}_1 \subseteq u$ and $D$, $t_0$ belong to Range $\pi_{\bar{t}_1 t_1}$. Let $\pi_{\bar{t}_1 t_1}(\bar{t}_0) = t_0$ and $u_0^*$ correspond to $\bar{t}_0 <_1 t_0$. Then $(u, u_0^*) \in R_0^t$ and we take $(v, v_0^*)$ to be its least extension in $R_0^t$ meeting $D$. Note that for $|t_0| < \eta \leq |t_1|$, $\eta \in$ Range $\pi_{\bar{t}_1 t_1} \cup \{|t_1|\}$, the extension $u \subseteq v$ trivially respects $\{(\bar{t}_1 \restriction \bar{\eta}, t_1 \restriction \eta)\}$ (where $\bar{t}_1 \restriction \bar{\eta} <_1 t_1 \restriction \eta$) as there are no $t_1 \restriction \eta$-good ordinals in the interval $(\alpha(\bar{t}_1), |v|)$.

It follows that $(v, u^*)$ extends $(u, u^*)$ and therefore $(v, v^*)$, $(u, u^*)$ are compatible by Corollary 7.12. □

**Lemma 7.15** (Chain Condition for $R^{<s}$). *If $D \subseteq R^{<s}$ is predense and belongs to $\tilde{J}_{\lambda(s)}^{s \restriction \lambda(s)}$ then $D$ has a predense subset $D_0 \in \tilde{J}_{\eta(s)}$. If $\lambda(s)$ is $s \restriction \lambda(s)$-admissible then the same holds for $D$ which are $\Sigma_1\langle \tilde{J}_{\lambda(s)}^{s \restriction \lambda(s)}, s \restriction \lambda(s)\rangle$.*

*Proof.* The first statement follows from the fact that conditions $(t, t_0^*)$, $(t, t_1^*)$ are compatible. For the second statement we may assume that $\eta(s)$ is less than $\lambda(s)$ as otherwise we can choose $s_0 <_0 s \restriction \lambda(s)$ and $D_0 \subseteq D$, $D_0$ $\Sigma_1\langle \tilde{J}_{|s_0|}, s_0\rangle$, $D_0$ predense on $R^{<s_0}$ and $R^{<s_0} = R^{s_0}$ (the latter is arranged by choosing $|s_0|$ to have cofinality $\kappa$ in $\tilde{J}_{\lambda(s)}$); then $D_0$ is predense on $R^{<s}$ by Lemma 7.14. Now if $\eta(s)$ is less than $\lambda(s)$, $R^{<s}$ is an element of $\tilde{J}_{\lambda(s)}^{s\restriction\lambda(s)}$, so using admissibility $D$ has a predense subset $D_0 \in \tilde{J}_{\lambda(s)}^{s\restriction\lambda(s)}$ and hence we can apply the first statement of the lemma. □

**Lemma 7.16** (Persistence for $R^{<s}$). *Suppose $s \leq_0 t$ in $S_\kappa^\beta$.*

(a) *If $D \subseteq R^{<s}$ is predense, $D \in \tilde{J}_{\lambda(s)}^{s\restriction\lambda(s)}$ then $D$ is predense on $R^t$.*

(b) *If $T \subseteq R^{<s} \times \delta$, $\delta < \lambda(s)$ is persistent and $\Sigma_1\langle \tilde{J}_{\lambda(s)}^{s\restriction\lambda(s)}, s \restriction \lambda(s)\rangle$, and $\lambda(s)$ is $s \restriction \lambda(s)$-admissible then $D(T)$ is predense on $R^t$.*

*Proof.* (a) By Lemma 7.15 we can assume that $D$ is an element of $\tilde{J}_{\eta(s)}$; then the result follows by induction on $|s|$, using Lemma 7.14 to handle the case where $\eta(s)$ is not a limit of limit ordinals.

(b) If $\eta(s) = \lambda(s)$ is $\Sigma_1$-projectible relative to $s \restriction \lambda(s)$ then $R^{<s} = R^s$ and the result follows by Lemma 7.14. If $\eta(s) = \lambda(s)$ is not $\Sigma_1$-projectible relative to $s \restriction \lambda(s)$ then choose $s_0 <_0 s \restriction \lambda(s)$ such that $\langle \tilde{J}_{|s_0|}, s_0\rangle$ is $\Sigma_1$-projectible and $\Sigma_1$-elementary in $\langle \tilde{J}_{\lambda(s)}, s \restriction \lambda(s)\rangle$ and such that $\kappa$, $\delta$ and the defining parameter for $T$ belong to $\tilde{J}_{|s_0|}$. By the above, $D(T \cap \tilde{J}_{|s_0|})$ (as defined in $\langle \tilde{J}_{|s_0|}, s_0\rangle$) is predense on $R^t$ and by Lemma 7.15, $D(T \cap \tilde{J}_{|s_0|})$ is contained in $D(T)$. Finally, suppose that $\eta(s)$ is less than $\lambda(s)$. Define a function $f$ by: $f(u) =$ least $u^*$ such that for some $i < \delta$, no extension of $(u, u^*)$ belongs to $T_i$, if exists; $f(u) = 0$ otherwise. We shall show that $f$ is an element of $\tilde{J}_{\lambda(s)}^{s\restriction\lambda(s)}$; then using $f$ we see that $D(T)$ has a predense subset in $\tilde{J}_{\lambda(s)}^{s\restriction\lambda(s)}$ and hence $D(T)$ is predense on $R^t$ by (a). To see that $f$ belongs to $\tilde{J}_{\lambda(s)}^{s\restriction\lambda(s)}$, note that by Lemma 7.15 and the fact that $\Sigma_1$-projectum $\langle \tilde{J}_{\lambda(s)}^{s\restriction\lambda(s)}, s \restriction \lambda(s)\rangle$ is greater than $\kappa$, $f$ is $\Pi_1\langle \tilde{J}_{\lambda(s)}^{s\restriction\lambda(s)}, s \restriction \lambda(s)\rangle$. Thus $W = \{(i, j) \mid i \in \mathrm{Dom}(f),\ j < f(i)\}$ is $\Sigma_1\langle \tilde{J}_{\lambda(s)}^{s\restriction\lambda(s)}, s \restriction \lambda(s)\rangle$ (where $<$ is the canonical wellordering of $\tilde{J}_{\eta(s)}$). As $\eta(s)$ is regular in $\tilde{J}_{\lambda(s)}^{s\restriction\lambda(s)}$, the canonical enumeration of $W$ terminates in fewer than $\eta(s)$ steps, and hence $f$ belongs to $\tilde{J}_{\lambda(s)}^{s\restriction\lambda(s)}$. □

**Lemma 7.17** (Distributivity for $R^{<s}$). *If $\langle D_i \mid i < \gamma\rangle \in \tilde{J}_{\lambda(s)}^{s\restriction\lambda(s)}$ is a sequence of subsets of $R^{<s}$ predense below $p \in R^{<s}$ then there exists $q \leq p$ meeting each $D_i$. The same holds if $\langle D_i \mid i < \gamma\rangle$ is uniformly $\Sigma_1\langle \tilde{J}_{\lambda(s)}^{s\restriction\lambda(s)}, s \restriction \lambda(s)\rangle$, when $\lambda(s)$ is $s \restriction \lambda(s)$-admissible.*

*Proof.* By Lemma 7.15 we can assume that $\langle D_i \mid i < \gamma\rangle$ belongs to $\tilde{J}_{\eta(s)}$. But then the result follows by induction on $|s|$, using Corollary 7.13 to handle the case where $\eta(s)$ is not a limit of limit ordinals. □

**Lemma 7.18** (Chain Condition for $P^{<s}$). *Assume Distributivity for s. If $D \subseteq P^{<s\restriction\lambda(s)}$ is predense below $p \in P^{<s\restriction\lambda(s)}$, $D \in \tilde{J}^{s\restriction\lambda(s)}_{\lambda(s)}$ then D has a predense below p subset $D_0 \in \tilde{J}_{\eta(s)}$. The same holds for D $\Sigma_1(\tilde{J}^{s\restriction\lambda(s)}_{\lambda(s)})$ when $\lambda(s)$ is $s \restriction \lambda(s)$-admissible.*

*Proof.* We can assume that $\eta(s)$ is less than $\lambda(s)$ and that $\kappa$ is a limit cardinal of $\tilde{J}_\beta$. By distributivity, $D^* = \{p \in P^{s\restriction\lambda(s)} \mid p$ reduces $D$ below some $\gamma < \kappa\}$ is $\alpha$-predense for each $\alpha \in \beta\text{-card}^+ \cap \kappa$, so by (b1) in the definition of condition and Strong Relativized $\diamondsuit$(a), $D^* \cap P^{<s\restriction\xi}$ is predense on $P^{<s\restriction\lambda(s)}$ for some $\xi < \eta(s)$. It follows that $D \cap P^{<s\restriction\lambda\xi} = D_0$ is as desired.

If $\tilde{J}^{s\restriction\lambda(s)}_{\lambda(s)}$ is admissible then by Strong Relativized $\diamondsuit$(b), $D \cap P^{<s\restriction\xi} = D_0$ is as desired for some $\xi < \eta(s)$ as by Distributivity $D^*$ as above has a $\Delta_1(\tilde{J}^{s\restriction\lambda(s)}_{\lambda(s)})$ subset which is $\alpha$-predense for each $\alpha \in \beta\text{-card}^+ \cap \kappa$. □

**Lemma 7.19** (Persistence for $P^{<s}$). *Assume Distributivity for s. Suppose $s \subseteq t \in S^\beta_\kappa$. If $D \subseteq P^{<s\restriction\lambda(s)}$ is predense and $D \in \tilde{J}^{s\restriction\lambda(s)}_{\lambda(s)}$ then D is predense on $P^{<t\restriction\lambda(s)}$. If $\lambda(s)$ is $s \restriction \lambda(s)$-admissible and $T \subseteq P^{<s\restriction\lambda(s)} \times \delta$, $\delta$ is regular in $\tilde{J}^{s\restriction\lambda(s)}_{\lambda(s)}$ and T is $\Sigma_1(\tilde{J}^{s\restriction\lambda(s)}_{\lambda(s)})$ and persistent then $D(T)$ is predense on $P^{<t\restriction\lambda(t)}$.*

*Proof.* We may assume $\kappa$ is a limit cardinal of $\tilde{J}_\beta$. The first statement follows from the chain condition and (b3), (c) from the definition of condition. For the second statement, Distributivity and Chain Condition imply that if $\eta(s)$ is less than $\lambda(s)$ then $D(T)$ contains a $\Pi_1(\tilde{J}^{s\restriction\lambda(s)}_{\lambda(s)})$ predense subset, and that $D(T)^* = \{p \in P^{<s\restriction\lambda(s)} \mid p$ reduces $D(T)$ below some $\gamma < \kappa\}$ contains a $\Pi_1(\tilde{J}^{s\restriction\lambda(s)}_{\lambda(s)})$ subset which is $\alpha$-predense for all $\alpha \in \beta\text{-card}^+ \cap \kappa$. Thus by Strong Relativized $\diamondsuit$(b) and the $\Pi_1(\tilde{J}^{s\restriction\lambda(s)}_{\lambda(s)})$-regularity of $\eta(s)$, we obtain $\xi < \eta(s)$ such that $D(T) \cap P^{<s\restriction\xi}$ is predense. Now we can apply the first statement.

If $\eta(s) = \lambda(s)$ is not $\Sigma_1$-projectible then the second statement follows by reflection to some $P^{<s\restriction\xi}$, where $\tilde{J}^{s\restriction\xi}_\xi$ is $\Sigma_1$-elementary in $\tilde{J}^{s\restriction\lambda(s)}_{\lambda(s)}$. Lastly, if $\eta(s) = \lambda(s)$ is $\Sigma_1$-projectible then the result follows by (b2) in the definition of condition. □

We are left with Extendibility, Distributivity and $\Sigma_f$-Density when $\kappa$ is a limit cardinal of $\tilde{J}_\beta$. For these properties we follow the approach of Section 4.3, making the necessary added verifications for strong coding.

**Lemma 7.20** (Distributivity). *Suppose that $\Sigma_f$-Density, Extendibility, Distributivity hold for $s_0 < s$ (i.e., for $s_0 \in S^\beta_\kappa$, $s_0$ a proper initial segment of s). Then Distributivity holds for s.*

*Proof.* First assume that $\kappa$ is regular in $\tilde{J}^{s\restriction\lambda(s)}_{\lambda(s)}$ and that $|s|$ is a limit of limit ordinals. Let $\langle D_i \mid \alpha \le i < \kappa\rangle$ belong to $\tilde{J}^s_{\lambda(s)}$, where $D_i$ is $i^+$-predense on $P^{<s}$ below $p \in P^{<s}$ for each $i$. We will show that there is $q \le_{\alpha^+} p$ meeting each $D_i$. We may assume that $\eta(s) < \lambda(s)$, as otherwise we may apply induction. By reflection choose $\eta < \eta(s)$ such that cofinality $(\eta) = \kappa$ in $\tilde{J}^s_{\lambda(s)}$, $p \in P^{<s\restriction\lambda^*}$ (where $\lambda^* = \min(\mathcal{O}(\kappa) - \eta)$), $D_i \cap P^{<s\restriction\lambda^*}$ is $i^+$-predense on $P^{<s\restriction\lambda^*}$ and $\langle D_i \cap P^{<s\restriction\lambda^*} \mid i < \kappa\rangle$ belongs to $\tilde{J}^{s\restriction\lambda^*}_{\lambda^*+k}$ for some $k < \omega$. We may assume that $k$ is small enough so that $\langle D_i \cap P^{<s\restriction\lambda^*} \mid i < \kappa\rangle$ is uniformly $\Delta_1\langle\tilde{J}_\eta, C_\eta\rangle$.

We shall inductively define $p^i$, $\eta_i$, $f_j$ for $j < i \le \kappa$ where $p = p^0 \ge p^1 \ge \cdots$ and set the desired $q$ to be $p^\kappa$. We may assume that $p$ meets $D^{s\restriction\eta}$ if $s \restriction \eta \in E$ and $D^{s\restriction\eta}$ is a subset of $P^{<s\restriction\lambda^*}$ which is $\alpha$-predense for all $\alpha \in \beta\text{-card}^+ \cap \kappa$. Let $x \in \tilde{J}_\eta$ be a parameter that defines $\langle D_i \cap P^{<s\restriction\lambda^*} \mid i < \kappa\rangle$ in a uniformly $\Delta_1$ way over $\langle\tilde{J}_\eta, C_\eta\rangle$.

Let $p^0 = p$ and let $\eta_0$ be least such that $p, x$ belong to $\tilde{J}_{\eta_0}$. Now suppose that $p^i$, $\eta_i$ have been defined. For any $\alpha \in \beta\text{-card} \cap \kappa$ let $H_i(\alpha) = \Sigma_1$ Hull of $\alpha \cup \{p, x, \kappa\}$ in $\langle\tilde{J}_{\eta_i}, C_{\eta_i}\rangle$ and define $f_i(\alpha) = H_i(\alpha) \cap \tilde{J}_{\alpha^{++}}$ if $i < \alpha \in H_i(\alpha)$, $f_i(\alpha) = \emptyset$ otherwise. Then $p^{i+1}$ = least $q \le p^i$ such that $q \restriction i^+ = p^i \restriction i^+$, $q \notin P^{<s\restriction\eta_i} = \cup\{P^{s\restriction\xi} \mid \xi < \eta_i, \xi \in \mathcal{O}(\kappa)\}$ and $q$ meets both $D_i$ and $\Sigma^{p^i}_{f_i}$. This is possible by $\Sigma_f$-Density and Extendibility for $P^{s\restriction\lambda^*}$. And $\eta_{i+1}$ is least so that $\eta_{i+1} \in \text{Lim}\, C_\eta$, $\langle\tilde{J}_{\eta_{i+1}}, C_{\eta_{i+1}}\rangle \vDash p^{i+1}$ meets $D_i$. For limit $\lambda$, $\eta_\lambda = \cup\{\eta_i \mid i < \lambda\}$ and $p^\lambda$ = greatest lower bound to $\langle p^i \mid i < \lambda\rangle$.

**Claim.** $p^\lambda$ is defined for limit $\lambda \le \kappa$.

*Proof of Claim.* Let $\bar{H}_\lambda(\alpha)$ equal $\langle\tilde{J}_{\bar{\eta}}, C_{\bar{\eta}}\rangle$, the transitive collapse of $H_\lambda(\alpha)$ and assume first that $\alpha$ belongs to $H_\lambda(\alpha)$. Let $\bar{P}$ be the image of $P^{<s\restriction\eta_\lambda}$ under collapse. If $D \in \tilde{J}_{\eta_\lambda}$ is predense on $P^{<s\restriction\eta_\lambda}$ then by distributivity for $P^{<s\restriction\xi}$, the fact that $\eta(s \restriction \xi) \le \eta_\lambda$ and the required predensity reduction, $p^i$ $\alpha'$-reduces $D$ for some $\alpha' \in \beta\text{-card}^+ \cap \kappa$ and some $i < \lambda$. Now suppose that in addition $D \in H_\lambda(\alpha)$. Then as $p^{i+1}$ meets $\Sigma^{p^i}_{f_i}$ we have that some $p^i, i < \lambda$, $\alpha'$-reduces $D$ for some $\alpha' \in \beta\text{-card}^+ \cap \alpha$. It follows that if $D \in H_\lambda(\alpha)$ and $D$ is predense on $P^{<s\restriction\eta_\lambda}_\alpha$ then some $p^i, i < \lambda$ meets $D$. Thus $\bar{G} = \{\bar{p} \in \bar{P} \mid \bar{p^i} \le \bar{p}$ for some $i < \lambda\}$ is $\bar{P}$-generic over $\bar{H}_\lambda(\alpha)$ (for predense $\bar{D} \in \bar{H}_\lambda(\alpha)$), where $\bar{p^i}$ = image of $p^i$ under collapse. Let $p^\lambda_\alpha$ be the function $t : [\alpha, \bar{\lambda}^*) \longrightarrow 2$ coded by $\cup\{p^i_\alpha \mid i < \lambda\} = t \restriction (\alpha^+$ of $\bar{H}_\lambda(\alpha))$ over $\tilde{J}_{\bar{\lambda}^*}$, where $\bar{H}_\lambda(\alpha) \simeq H_\lambda(\alpha)$ canonically extends to $\tilde{J}_{\bar{\lambda}^*} \overset{\sim}{\longrightarrow} \tilde{J}_{\lambda^*}$.

We argue that $p^\lambda_\alpha$ belongs to $S_\alpha$. As $\langle p^i_\alpha \mid i < \lambda'\rangle$ is $\Delta_1\langle J_{\bar{\eta}_{\lambda'}}, C_{\bar{\eta}_{\lambda'}}\rangle$ uniformly for limit $\lambda' \le \lambda$, it follows that $p^\lambda_\alpha \restriction \eta(p^\lambda_\alpha)$ is uniformly $\Delta_1\langle\tilde{J}_{\eta(p^\lambda_\alpha)}, C_{\eta(p^\lambda_\alpha)}\rangle$ (where $\eta(p^\lambda_\alpha) = \alpha^+$ of $\bar{H}_\lambda(\alpha)$). It remains to verify that $p^\lambda_\alpha$ is $\Sigma$-generic when $\bar{\eta} = \text{ORD}(\bar{H}_\lambda(\alpha))$ is recursively inaccessible and that $\bar{\eta}$ is $p^\lambda_\alpha$-inadmissible when $\bar{\eta}$ is a successor admissible. If $\lambda^*$ is not recursively inaccessible then we can guarantee that $\bar{\lambda}^*$ is not recursively inaccessible and that if $\lambda^*$ is a successor admissible then $\bar{\lambda}^*$ is inadmissible relative to $p^\lambda_\alpha$. If $\lambda^*$ is recursively inaccessible then as in the proof of Lemma 7.10, either $\lambda^*$ is not $\Sigma_1$-projectible to $\kappa$ and $\Sigma$-genericity follows by reflection, or $\lambda^*$ is $\Sigma_1$-projectible to $\kappa$ and we can modify our construction so as to guarantee that all $\Sigma_1$ sets have been "decided" by stage $\kappa$, the first instance of recursive inaccessibility for $\bar{\lambda}^*$. An inductive use of distributivity implies that $\Sigma$-genericity holds.

If $\alpha$ does not belong to $H_\lambda(\alpha)$ then let $\alpha^* = \min(H_\lambda(\alpha) - \alpha)$ and if $\alpha^* < \kappa$ then instead of $\bar{P} =$ collapse of $P_\alpha^{<s\restriction\lambda}$, use $\bar{P} =$ collapse of $P_{\alpha^*}^{<s\restriction\lambda}$. If $\alpha^* = \kappa$ then $p_\alpha^\lambda$ is the collapse of $s \restriction \lambda^*$ and hence belongs to $S_\alpha$.

We are ready to verify that $p^\lambda$ is a condition where $p_\alpha^\lambda$ is defined above and $(p_\alpha^\lambda)^* = H_1^{p_{\alpha+}^\lambda}(\alpha)$. (Note that $(p_\alpha^i)^* \subseteq (p_\alpha^\lambda)^*$ for $i < \lambda$.) It is clear for $\alpha \leq \kappa$, $\alpha \in \beta$-card$'$ that $p^\lambda \restriction \alpha$ codes $p_\alpha^\lambda$. Predensity Reduction (b1) holds by choice of $p^0$ and by Strong Relativized $\diamondsuit$(d), (e). Predensity Reduction (b2) does not apply unless $\alpha = \lambda = \kappa$, and then it holds due to our modification of the construction to guarantee $\Sigma$-genericity. Predensity (b3) does not apply. Restriction is clear by induction. Nonstationary Restraint holds when $\alpha \neq \lambda$ by induction and otherwise by diagonal intersection. Definability is clear by construction. □ (Claim)

Thus the lemma holds when $|s|$ is a limit of limits, $\kappa$ regular in $\tilde{J}_{\lambda(s)}^s$. Now suppose that $|s|$ is a limit of limits, $\kappa$ singular in $\tilde{J}_{\lambda(s)}^s$ and hence singular in $\tilde{J}_{\eta(s)}$. Let $\lambda < \kappa$ be regular in $\tilde{J}_{\eta(s)}$. It suffices to show: If $\langle D_i \mid i < \lambda \rangle \in \tilde{J}_{\lambda(s)}^s$, $D_i$ $i^+$-predense on $P^{<s}$, $p \in P^{<s}$ then there exists $q \leq p$, $q$ meets each $D_i$. For, if $\langle D_i' \mid i < \kappa \rangle$ are as in the hypothesis of Distributivity for $P^{<s}$ and $\langle \alpha_i \mid i < \lambda_0 \rangle \in \tilde{J}_{\eta(s)}$ is cofinal in $\kappa$, $\lambda_0 =$ cofinality of $\kappa$ in $\tilde{J}_{\eta(s)}$ then $D_i = \{q \mid q$ meets each $D_j'$, $\lambda_0 \leq j < \alpha_i^+\}$ is $i^+$-predense for each $i$ and hence there is $q_0 \leq p$, $q_0$ meets each $D_j'$, $\lambda_0 \leq j < \kappa$. Then apply our weakened form of distributivity once more (with $\lambda = \lambda_0$) to obtain the desired $q \leq q_0 \leq p$.

The proof that any $p \in P^{<s}$ can be extended to meet each $D_i, i < \lambda$ where $\langle D_i \mid i < \lambda \rangle \in \tilde{J}_{\lambda(s)}^s$, $D_i$ $i^+$-predense for each $i$, $\kappa > \lambda$ in $\tilde{J}_{\lambda(s)}^s$ is very similar to the case where $\kappa$ is regular in $\tilde{J}_{\lambda(s)}^s$. Again we may assume $\eta(s) < \lambda(s)$ and use reflection, this time to obtain $\eta < \eta(s)$ such that cofinality $(\eta) = \lambda$ in $\tilde{J}_{\eta(s)}$, $p \in P^{<s\restriction\lambda^*}$ ($\lambda^* = \min(\mathcal{O}(\kappa) - \eta)$), $D_i \cap P^{<s\restriction\lambda^*}$ $i^+$-predense on $P^{<s\restriction\lambda^*}$ for each $i$ and $\langle D_i \cap P^{<s\restriction\lambda^*} \mid i < \lambda \rangle$ uniformly $\Delta_1\langle \tilde{J}_\eta, C_\eta \rangle$. Then define $\langle p^i \mid i \leq \lambda \rangle$ as we defined $\langle p^i \mid i \leq \kappa \rangle$ before and $q = p^\lambda$ is as desired.

Finally suppose that $|s|$ is a limit but is not a limit of limit ordinals and write $s_0 = s \restriction (|s| - \omega)$. By Extendibility for $s_0$ we can assume that $p$ (as given in the hypothesis of Distributivity for $s$) belongs to $P^{s_0} - P^{<s_0}$. In fact we can assume that $p$ belongs to $P^{s_1} - P^{<s_1}$ where $s_1 = s \restriction |s_0| + k$ and $\langle D_i \mid i < \kappa \rangle$ is coded by $A \subseteq \kappa$, $A \in \Delta_k\langle \tilde{J}_{\eta(s_0)}, C_{\eta(s_0)} \rangle$. The fact that $p$ codes $s_1$ implies that there is $\gamma_0 < \kappa$ such that for each $i < \kappa$, $|p_\gamma| > |r_\gamma|$, $p_\gamma^* \supseteq r_\gamma^*$ for $r \in D_i$, $\gamma \geq \gamma_0 \cup i^+$. It follows that $p$ meets $D_i$ for $i \geq \gamma_0$ and reduces $D_i$ below $\gamma_0^+$ for all $i$. By induction $p$ can be extended to meet each $D_i$.

If $|s|$ is not a limit then the result follows by induction. □

**Remark.** The argument of the previous lemma also shows: If $\alpha < \kappa, \kappa$ $\Sigma_{n+1}(L_\kappa)$-regular, $\langle D_i \mid i < \alpha \rangle$ $\Sigma_n(L_\kappa)$, $D_i$ $i^+$-predense on $P^{<\kappa}$, $p \in P^{<\kappa}$ then there exists $q \in P^{<\kappa}$, $q \leq p$, $q$ meets each $D_i$: Inductively choose $p^{i+1}$ to extend $p^i$ and to meet $D^i$, $\Sigma_{f_i}^{p^i}$ where $f_i$ is defined using $H_i(\gamma) = \Sigma_1$ Hull of $\gamma \cup \{x\}$ in $L_{\kappa_{i+1}}$ where $L_{\kappa_{i+1}}$ is $\Sigma_n$-elementary in $L_\kappa$, $p^i \in P^{<\kappa_{i+1}}$ and $x$ defines $p$, $\langle D_i \mid i < \alpha \rangle$.

**Lemma 7.21.** *Suppose that Extendibility holds for $s_0 < s$ and $\Sigma_f$-Density, Distributivity hold for $s_0 \leq s$. The Extendibility holds for $s$.*

*Proof.* First assume that Lim $C_{\eta(s)}$ is unbounded in $\eta(s)$ and let $\langle \eta_i \mid i < \lambda \rangle$ be a final segment of $\eta(s) \cap \operatorname{Lim} C_{\eta(s)}$ such that the given $p \in P^{<s}$ belongs to $P^{<s \restriction \eta_0}$. (We may assume that $p$ belongs to $P^{<s}$.) Now as in the previous lemma, define $p^i$, $f_j$ for $j < i$ by induction on $i$: $p^0 = p$ and for $\gamma \in \beta$-card $\cap\, \kappa$, $H_i(\gamma) = \Sigma_1$ Hull of $\gamma \cup \{p, \kappa\}$ in $\langle \tilde{J}_{\eta_i}, C_{\eta_i} \rangle$, $f_i(\gamma) = H_i(\gamma) \cap \tilde{J}_{\gamma^{++}}$ if $\alpha < \gamma, i < \gamma \in H_i(\gamma)$ and $f_i(\gamma) = \emptyset$ otherwise. Then $p^{i+1}$ is the least $q \leq_\alpha p^i$ such that $q \restriction i^+ = p^i \restriction i^+$ and $q \in P^{s \restriction \eta_{i+1}} - P^{s \restriction \eta_i}$, $q$ meets $\Sigma_{f_i}^{p^i}$. For limit $\lambda' \leq \lambda$, $p^{\lambda'}$ = greatest lower bound to $\langle p^i \mid i < \lambda' \rangle$, if it exists.

The proof that $p^{\lambda'}$ exists for limit $\lambda' < \lambda$ is just as in the previous lemma. When $\lambda' = \lambda$ the proof is the same except one must use Distributivity at $s$ to conclude that $P^{<s}$ is $\kappa^+$-cc in $\tilde{J}^s_{\lambda(s)}$ (and $\kappa^+$ $\Sigma$-ccin $\tilde{J}^s_{\lambda(s)}$ if $\lambda(s)$ is recursively inaccessible).

Now suppose that Lim $C_{\eta(s)}$ is bounded in $\eta(s)$ and $\eta(s)$ is a limit of limit ordinals. Then we choose $\langle \eta_i \mid i < \omega \rangle$ to be a cofinal in $\eta(s)$ increasing $\Delta_1 \langle \tilde{J}_{\eta(s)}, C_{\eta(s)} \rangle$ sequence of limit ordinals, $\eta_0$ large enough so that $p \in P^{<s \restriction \eta_0}$ and $\tilde{J}_{\eta_0}$ contains a parameter $x$ defining $\langle \eta_i \mid i < \omega \rangle$. Using $H_i(\gamma) = \Sigma_1$ Hull of $\gamma \cup \{p, x, \kappa, C_{\eta(s)} \cap \eta_i\}$ in $\tilde{J}_{n_i}$, we can argue as in the previous case.

Suppose that $|s| = \eta(s)$ is a limit ordinal but not the limit of limit ordinals and let $s_0 = s \restriction |s| - \omega$. Now use $H_i(\gamma) = \Sigma_i$ Hull of $\gamma \cup \{p, \kappa, x\}$ in $\langle \tilde{J}_{\eta(s_0)}, C_{\eta(s_0)} \rangle$. Then defining $p^i, i < \omega$ as before we get $p^\omega = q$ which obeys all the requirements for being a condition in $P^s - P^{s_0}$.

Finally suppose that $|s|$ is a successor ordinal $|s_0| + k + 1$, where $s_0 = s \restriction |s_0|$, $|s_0|$ limit, $k < \omega$. We may assume by induction that $p \notin P^{<s \restriction |s_0| + k}$, $p \in P^{<s}$. Also assume that $s_0 \neq \emptyset$. Then $p$ is $\Delta_{k+2} \langle \tilde{J}_{\eta(s_0)}, C_{\eta(s_0)} \rangle$ and if $\kappa$ is $\Sigma_{k+2} \langle \tilde{J}_{\eta(s_0)}, C_{\eta(s_0)} \rangle$-regular then there is a CUB $C \subseteq \kappa$ such that $\alpha \in C \Longrightarrow p^*_\alpha = \emptyset$ and $C$ is $\Delta_{k+2} \langle \tilde{J}_{\eta(s_0)}, C_{\eta(s_0)} \rangle$. If $D = \{\alpha < \kappa \mid \alpha = \kappa \cap \Sigma_{k+2}$ Hull of $\alpha \cup \{x\}$ in $\langle \tilde{J}_{\eta(s_0)}, C_{\eta(s_0)} \rangle\}$ is unbounded in $\kappa$ (where $x$ is a parameter defining $p, C, \kappa$) then we can successively extend $p \restriction \alpha_i^+$, where $\langle \alpha_i \mid i < \lambda_0 \rangle$ is the increasing enumeration of $D$, so as to code $(s)_{\alpha_i}$. Note that $D \subseteq C$ so there is no concern that the restraints $p^*_{\alpha_i}$ will prevent this coding. (For $\alpha \notin D$, there is no conflict due to the different types of codings being used.) If $D$ is bounded in $\kappa$ but $\kappa$ is $\Sigma_{k+2} \langle \tilde{J}_{\eta(s_0)}, C_{\eta(s_0)} \rangle$-regular then $\kappa$ has $\Sigma_{k+3} \langle \tilde{J}_{\eta(s_0)}, C_{\eta(s_0)} \rangle$-cofinality $\omega$ and we can easily make the desired successive extensions of $p \restriction \alpha_1^+$, where $\alpha_0 < \alpha_1 < \cdots$ is a $\Sigma_{k+3} \langle \tilde{J}_{\eta(s_0)}, C_{\eta(s_0)} \rangle$ cofinal $\omega$-sequence. If $\kappa$ is $\Sigma_{k+2} \langle \tilde{J}_{\eta(s_0)}, C_{\eta(s_0)} \rangle$-singular then choose a $\Sigma_{k+2} \langle \tilde{J}_{\eta(s_0)}, C_{\eta(s_0)} \rangle$ sequence $\langle \alpha_i \mid i < \lambda_0 \rangle$ continuous and cofinal in $\kappa$ so that $\lambda_0 < \alpha_0$ and for limit $\lambda \leq \lambda_0$, $p \restriction \alpha_\lambda$ and $\langle \alpha_i \mid i < \lambda \rangle$ are $\Delta_{k+2}$ over the $\Sigma_{k+2}$ Hull of $\alpha_\lambda \cup \{\kappa, x\}$ in $\langle \tilde{J}_{\eta(s_0)}, C_{\eta(s_0)} \rangle$, uniformly in a parameter $x \in \tilde{J}_{\alpha_0}$. Now successively extend $p \restriction [\lambda_0^+, \alpha_i^+)$ to meet $\Sigma_{f_i}^{p^i \restriction \alpha_i^+}$ where $f_i$ is defined using $H_i(\gamma) = \Sigma_{k+3}$ Hull of $\gamma \cup \{\kappa, x\}$ in ($\Sigma_{k+2}$ Hull of $\alpha_i \cup \{\kappa, x\}$ in $\langle \tilde{J}_{\eta(s_0)}, C_{\eta(s_0)} \rangle$). At stage $\lambda_0$ one reaches a condition $q = p^{\lambda_0}$ which can be easily extended (by lengthening $q_\alpha$ with $s \upharpoonright [|s_0|, |s|)$) to code $s$. If $s_0 = \emptyset$ the argument is the same, with $\langle \tilde{J}_{\eta(s_0)}, C_{\eta(s_0)} \rangle$ replaced by $\tilde{J}_\kappa$. □

**Lemma 7.22.** *Suppose that Extendibility, $\Sigma_f$-Density hold for $s_0 < s$ and Distributivity holds for $s$. Then $\Sigma_f$-Density holds for $s$.*

*Proof.* We may assume that $|s| = |s_0| + \omega$, $s_0 = s \upharpoonright |s_0|$, $|s_0|$ limit as otherwise we can apply induction. And we may assume that $p \in P^{s \upharpoonright |s_0|+k} - P^{<s \upharpoonright |s_0|+k}$, $f$ is $\Delta_k \langle \tilde{J}_{\eta(s_0)}, C_{\eta(s_0)} \rangle$ in parameter $x$, where $p$ and $f$ are given as in the statement of $\Sigma_f$-Density (and where $\mathrm{Dom}(f)$ is witnessed to be thin in $\Delta_k \langle \tilde{J}_{\eta(s_0)}, C_{\eta(s_0)} \rangle$). Choose $\alpha_0 < \kappa$ large enough so that $(s \upharpoonright |s_0| + k)_\alpha$ lies on $p_\alpha$ for $\alpha_0 \le \alpha < \kappa$, $\alpha \in \beta$-card $\cap\, \kappa$. If $\kappa$ is $\Delta_{k+1} \langle \tilde{J}_{\eta(s_0)}, C_{\eta(s_0)} \rangle$-regular then we can choose a CUB $C \subseteq \kappa$ such that $\alpha \in C \Longrightarrow \alpha \ge \alpha_0$, $f(\alpha) = \emptyset$ where $C \cap \alpha$ is uniformly $\Delta_{k+1}$ over $\Sigma_{k+1}$ Hull of $\alpha \cup \{x, \kappa\}$ in $\langle \tilde{J}_{\eta(s_0)}, C_{\eta(s_0)} \rangle$; then we may successively extend $p$ on the intervals $[\alpha_i, \alpha_{i+1})$ to meet $\Sigma^p_{f \upharpoonright [\alpha_i, \alpha_{i+1})}$ without difficulty at limit stages, where $\langle \alpha_i \mid i < \lambda_0 \rangle$ is the increasing enumeration of $C$. If $\kappa$ is $\Delta_{k+1} \langle \tilde{J}_{\eta(s_0)}, C_{\eta(s_0)} \rangle$-singular then we can instead choose $\langle \alpha_i \mid i < \lambda_0 \rangle$ to be continuous, cofinal in $\kappa$ of ordertype $\lambda_0 < \alpha_0$ (increasing $\alpha_0$ if necessary) such that for limit $\lambda \le \lambda_0$, $f \upharpoonright \lambda$ and $\langle \alpha_i \mid i < \lambda \rangle$ are $\Delta_{k+1}$ over $\Sigma_{k+1}$ Hull of $\alpha_\lambda \cup \{x, \kappa\}$ in $\langle \tilde{J}_{\eta(s_0)}, C_{\eta(s_0)} \rangle$; then we can successively extend $p$ on $[\alpha_0, \alpha_i]$ to meet $\Sigma^{p^i}_{f \upharpoonright [\alpha_0, \alpha_i]}$ without difficulty at limit stages since $(s)_{\alpha_\lambda}$ lies on $p_{\alpha_\lambda}$ for all limit $\lambda$. In either case we obtain $q \le p$ meeting $\Sigma^p_{f \upharpoonright [\alpha_0, \kappa)}$ and then by induction we can further extend $q$ to meet $\Sigma^p_f$. □

This completes the verification of Extendibility, Distributivity, Chain Condition, Persistence and $\Sigma_f$-Density for the $P^s_\gamma$-forcings.

Now we are very close to proving Theorem 7.9.

**Definition.** Suppose $\beta$ is recursively inaccessible and $P \subseteq L_\beta$ is a $\Delta_1(L_\beta)$ partial-ordering. We say that *$\Delta_0$-forcing for $P$ is densely* $\Sigma_1$ if for some $\Sigma_1(L_\beta)$ relation $p \Vdash^* \varphi$ ($p \in P$, $\varphi$ $\Delta_0$ with parameters from $L_\beta$):

(1) $p \Vdash^* \varphi \Longrightarrow p \Vdash \varphi$

(2) $p \Vdash \varphi \Longrightarrow \exists q \le p (q \Vdash^* \varphi)$.

**Lemma 7.23.** *For $\beta \in$ RI, $\kappa \in \beta$-card, $\Delta_0$-forcing for $P^\beta_\kappa$ is densely $\Sigma_1$.*

*Proof.* By Distributivity and Chain Condition, if $\langle D_i \mid i < \gamma \rangle$, $\gamma < \beta$ is a $\Sigma_1(L_\beta)$ sequence of sets predense below $p \in P^\beta_\kappa$ then there exists $q \le p$, $\langle d_i \mid i < \gamma \rangle \in L_\beta$ such that $d_i \subseteq D_i$ for each $i < \gamma$ and each $d_i$ is predense below $q$. Now we show that given $p \in P^\beta_\kappa$ and $\Delta_0$ $\varphi$ we can effectively (i.e., in a $\Sigma_1(L_\beta)$ way) find $q \le p$ such that $q$ decides $\varphi$ (i.e., $q \Vdash \varphi$ or $q \Vdash\sim \varphi$ and we know effectively which of these occurs). Given this, the lemma follows as we may define $q \Vdash^* \varphi$ iff there exists $p$ such that $q \leq p$ arises effectively as above from $p$, $\varphi$ and the case of $q \Vdash \varphi$ occurs.

We describe how to find $q$, given $p$ and $\varphi$, by induction on $\varphi$: By Extendibility the base case (where $\varphi$ is atomic) is clear. If $\varphi$ is a negation $\sim \psi$, then the result is clear as deciding $\sim \psi$ is the same as deciding $\psi$. The interesting case is the quantifier case $\forall x \in a\psi(x)$. By induction we can effectively build $\Sigma_1(L_\beta)$ sets $D_x$ which are dense and such that $q \in D_x \Longrightarrow q$ decides $\psi(x)$ (and we effectively know whether $q \Vdash \psi(x)$ or $q \Vdash\sim \psi(x)$.) By the first statement of this proof we can find $q_0 \leq p$ which reduces the sequence $\langle D_x \mid x \in a\rangle$ to $\langle d_x \mid x \in a\rangle \in L_\beta$. (I.e., $d_x \subseteq D_x$ and $d_x$ is predense below $q_0$ for each $x \in a$.) Then either $r \in d_x \Longrightarrow r \Vdash \psi(x)$ for each $x \in a$ (and hence $q_0 \Vdash \forall x \in a\psi(x)$) or we can choose $q \leq q_0$ such that for some $x \in a$, $q \Vdash\sim \psi(x)$ (and hence $q \Vdash\sim \forall x \in a\psi(x)$). □

**Corollary 7.24.** *Suppose $\beta \in$ RI and $G \subseteq P^\beta$ is $P^\beta$ $\Sigma$-generic over $L_\beta$. Then $\langle L_\beta[G], G\rangle$ is admissible.*

*Proof.* First note that we may assume: $p \Vdash^* \varphi, q \leq p \Longrightarrow q \Vdash^* \varphi$. (If necessary redefine $\Vdash^*$ by: $p \Vdash^{**}\varphi$ iff $\exists q(p \leq q, q \Vdash^* \varphi)$.) We claim that for $\Delta_0$ $\varphi$ : $\langle L_\beta[G], G\rangle \models \varphi$ iff $\exists p \in G(p \Vdash^* \varphi)$. The proof is by induction on $\varphi$. If $\varphi$ is atomic the result of clear. If $\varphi$ is $\sim \psi$ then as the $\Sigma$-genericity of $G$ implies that $G$ intersects $\{p \mid p \Vdash^* \psi$ or $p \Vdash^*\sim \psi\}$ we get the result. Similarly if $\varphi$ is $\forall x \in a\psi(x)$ then we get the result as the $\Sigma$-genericity of $G$ implies that $G$ intersects $\{p \mid$ For each $x \in a, \forall q \leq p\exists r \leq q(r \Vdash^* \psi(x))\} \cup \{p \mid$ For some $x \in a, q \leq p \Longrightarrow\sim r \Vdash^* \psi(x)\}$, and hence $G$ contains $p$ such that either $p \Vdash^* \forall x \in a\psi(x)$ or $p \Vdash^*\sim \psi(x)$ for some $x \in a$.

Now suppose $\langle L_\beta[G], G\rangle \models \forall x \in a\exists y\varphi$ where $\varphi$ is $\Delta_0$. Then $T = \{(p, x) \mid p \Vdash^* \varphi(x, y)$, for some $y\}$ is persistent and $\Sigma_1$. Thus by $\Sigma$-genericity there is $p \in G$ such that either for all $x \in a$, $D_x = \{q \mid q \Vdash^* \varphi(x, y)$, some $y\}$ is dense below $p$ or for some $x \in a$, $q \leq p \Longrightarrow$ for all $y, \sim q \Vdash^* \varphi(x, y)$. Thus $\{q \mid$ For some $Y \in L_\beta[G], q \Vdash^* \forall x \in a\exists y \in Y\varphi(x, y)\}$ is dense below $p$ (the second possibility cannot occur as for each $x \in a$ there exists $r \in G$, $r \Vdash^* \varphi(x, y)$, for some $y$). By $\Sigma$-genericity there is $q \in G$, $q \Vdash^* \forall x \in a\exists y \in Y\varphi(x, y)$, for some $Y \in L_\beta[G]$ and hence $\langle L_\beta[G], G\rangle \models \forall x \in a\exists y \in Y\varphi(x, y)$. □

**Lemma 7.25.** *Suppose $G \subseteq P$ is $P$-generic over $L$. Then $L[G] = L[X]$ for some $X \subseteq \omega_1^L$ and:*

(a) *For all $\beta > \omega$, $\beta$ is $X \cap \omega_1^{L_\beta}$-admissible iff $\beta \in$ RI (where $\omega_1^{L_\beta} = \beta$ if $L_\beta \models \omega_1$ does not exist).*

(b) *$\langle L[G], G\rangle \models$ ZFC and $L[G]$, $L$ have the same cofinalities.*

*Proof.* By Persistence, $G \cap L_\beta$ is $P^\beta$ $\Sigma$-generic over $L_\beta$ for $\beta \in$ RI. By Extendibility, $G \cap L_\beta$ is $\Delta_1\langle L_\beta[X \cap \omega_1^{L_\beta}], X \cap \omega_1^{L_\beta}\rangle$. By Corollary 7.24 and the definition of $P^\beta$, we have (a). (b) follows by the remark after Lemma 7.20 and by Distributivity, Chain Condition. □

Finally we prove Theorem 7.9. Suppose $X \subseteq \omega_1^L$ codes a $P$-generic, as in Lemma 7.25. Now code $X$ by a real $R$, as in Jensen–Solovay [70]. Then $\beta > \omega$ is $R$-admissible iff $\beta$ is $X \cap \omega_1^{L_\beta}$-admissible. Thus $\Lambda(R) = \text{RI}$. As in the proof of Theorem 4.18 we can obtain such an $X$, and therefore such an $R$, in $L[0^\#]$. In fact $R <_L 0^\#$ as $L$, $L[R]$ have the same cofinalities.

## 7.3 Other Spectra

The proof of Section 7.2 yields:

**Theorem 7.26.** *Suppose $X \subseteq$ Admissible Ordinals satisfies:*

(a) *For some $\Sigma_2$ sentence $\varphi$, $\beta \in X$ iff either $\beta$ is regular in $\tilde{J}_{\beta+\omega}$ or $\langle \tilde{J}_\beta, C_\beta \rangle \models \varphi$.*

(b) *The same holds for some $\Pi_2$ sentence $\varphi$.*

*Then there exists $R <_L 0^\#$, $\Lambda(R) = X$.*

However there are some strict limitations upon the possibilities for $\Lambda(R)$:

**Examples.** (1) Suppose $X$ is the class of $L$-cardinals. Then $X \neq \Lambda(R)$ for any $R$. The reason is that $X$ fails to satisfy: $\alpha \in X \Longrightarrow \alpha$ is $X \cap \alpha$-admissible.

(2) Suppose $X = \{\alpha \mid L_\alpha \models \text{Power Set}\}$. Then $X \neq \Lambda(R)$ for any $R$. For, the $L[R]$-cardinal successor to $\omega_\omega^{L[R]}$ would then be greater than the $L$-cardinal successor to $\omega_\omega^{L[R]}$, and hence by Jensen covering, $0^\# \leq_L R$; but then $\Lambda(R) \subseteq$ the class of $L$-cardinals $\cup \alpha$ for some (countable) $\alpha$.

(3) Suppose $X = \{\alpha \mid \alpha$ a successor admissible, $L$-card$(\alpha)$ a successor $L$-cardinal$\} \cup \{\alpha \mid \alpha$ recursively inaccessible, $L$-card$(\alpha)$ a limit $L$-cardinal$\}$. Then $X \neq \Lambda(R)$ for any $R$. For if $X = \Lambda(R)$ then the class of $L$-cardinals is $\Delta_1(L[R])$ and hence by reflection $\kappa^+$ of $L[R] \neq \kappa^+$ of $L$ where $\kappa = \omega_\omega^{L[R]}$. So as in (b), $0^\# \leq_L R$ and we get a contradiction.

(4) Suppose $X = \{\alpha \mid \alpha$ is $\Sigma_1$ nonprojectible$\}$. Then $X \neq \Lambda(R)$ for any $R$, else the least $R$-admissible greater than $\omega_1$ would have cofinality $\omega$, which is impossible.

We also recall the following general limitative result:

**Theorem 7.27.** (a) *$R \in L \Longrightarrow \Lambda(R) \cup \alpha \supseteq \Lambda(0)$ for some $\alpha < \omega_1^L$.*

(b) *$R$ set-generic over $L \Longrightarrow \Lambda(R) \cup \alpha \supseteq \Lambda(0)$ for some $\alpha$.*

*Proof.* (a) Choose $\alpha$ so that $R \in L_\alpha$.
(b) Choose $\alpha$ so that $P \in L_\alpha$ where $R$ is $P$-generic over $L$. □

In fact there *cannot* be a simple characterization of admissibility spectra, by virtue of the following result.

**Theorem 7.28.** *Let* $X = \{A \subseteq \omega_1^L \mid A \in L$ *and for some real* $R$, $\omega_1^{L[R]} = \omega_1^L$ *and* $\Lambda(R) \cap \omega_1^L = A\}$. *Then* $X =_L 0^\#$.

*Proof.* We first show that $0^\#$ is constructible from $X$.

**Lemma 7.29.** *There is a* $\Delta_1(L_{\omega_2^L})$ *(in parameter* $\omega_1^L$*) sequence* $\langle T_n \mid n < \omega\rangle$ *of trees on* $\omega_2^L$ *such that for all* $n$, $n \in 0^\#$ *iff* $T_n$ *has an* $\omega_2$*-preserving branch (i.e.,* $T_n$ *has a branch* $b$ *of length* $\omega_2^L$ *such that* $L$, $L[b]$ *have the same* $\omega_2$*) iff* $T_n$ *has a* $\mathcal{P}(\omega_1^L)$*-preserving branch* $b$ *which in* $L[b]$ *has stationary intersection with* $(\operatorname{cof}\omega_1)^L = \{\alpha \mid L\text{-cof}(\alpha) = \omega_1^L\}$.

*Proof of Lemma* 7.29. First define $T_n^*$ to consist of all $s \in L$, $s : |s| \longrightarrow 2$, $|s| < \omega_2^L$ such that:

(∗) For all $\eta \leq |s|$ in $(\operatorname{cof}\omega_1)^L$, whenever $L_\beta[s \upharpoonright \eta] \models \mathrm{ZF}^- + \eta = \omega_2$ then $L_\beta[s \upharpoonright \eta] \models$ Every $n$-bad guess has been killed

where killing of guesses $(\alpha_1, \dots, \alpha_k)$ is defined as in Chapter 6 and where $(\alpha_1, \dots, \alpha_k)$ is *n-bad* if $I(\alpha_1, \dots, \alpha_k)$ disagrees with the Silver indiscernibles $I$ on the $n^{\text{th}}$ formula $\varphi_n$ (i.e., for $\beta_1 < \cdots < \beta_n$ in $I(\alpha_1, \dots, \alpha_k)$ and $j_1 < \cdots < j_n$ in $I$ we have $L_{\alpha_k} \models \varphi_n(\beta_1, \dots, \beta_n) \not\Leftrightarrow L \models \varphi_n(j_1, \dots, j_n)$). We may also require in (∗) that $s \upharpoonright \eta$ generically code over $L_\beta$ a generic that kills $n$-bad guesses in $L_\beta$. It follows by reflection that:

**Fact.** If $X$ is a $\mathcal{P}(\omega_1^L)$-preserving branch of $T_n^*$ then $X$ codes a generic for the forcing $P_n$ defined over $L$ whose generics kill all $n$-bad guesses.

Thus $n \in 0^\#$ iff $T_n^*$ has a $\mathcal{P}(\omega_1^L)$-preserving branch.

Now we define a constructible function $f_n : \omega_2^L \longrightarrow T_n^*$ such that if $T_n$ is the tree on $\omega_2^L$ defined by $\alpha <_{T_n} \beta$ iff $f_n(\alpha) <_{T_n^*} f_n(\beta)$ then $T_n$ is the desired tree. We define $f_n(\alpha)$ by induction on $\alpha$:

(1) If $f_n(\bar{\alpha})$ has length $\geq \alpha$ for some $\bar{\alpha} < \alpha$ or $L\text{-cof}(\alpha) = \omega$ then $f_n(\alpha)$ is the least element of $T_n^*$ of length $\geq \alpha$ not in $f_n[\alpha]$.

(2) If (1) fails then let $\beta$ be least such that $L_\beta \models \mathrm{ZF}^- + \alpha = \omega_2$ and for some least condition $p \in (P_n)^{L_\beta}$ and least name $\sigma$ in $L_\beta$ : $p \Vdash \sigma$ is a CUB subset of $\alpha$ disjoint from $f_n[b^G]$, $b^G$ denoting the generic branch through $T_n^* \upharpoonright \alpha$. If $\beta$ does not exist then define $f_n(\alpha)$ as in (1). Otherwise choose $f_n(\alpha)$ to be the least element of $T_n^*$ of length $\alpha$ coding a $(P_n)^{L_\beta}$-generic extending $p$.

Suppose that $b$ is a $\mathcal{P}(\omega_1^L)$-preserving branch through $T_n^*$. We claim that $f_n[b] \cap (\operatorname{cof}\omega_1)^L$ is a stationary branch through $T_n$. If not then (as $b$ generally codes a $P_n$-generic) there is $\beta > \omega_2^L$ satisfying $L_\beta \models \mathrm{ZF}^-$, $p \in (P_n)^{L_\beta}$ and $\sigma \in L_\beta$ such that $p \Vdash \sigma$ is a CUB subset of $\omega_2^L$ disjoint from $f_n[b^G]$, where $b^G$ denotes the generic branch through $T_n^*$. Choose $\beta, p, \sigma$ to be least. By reflection there is $\alpha \in (\operatorname{cof}\omega_1)^L \cap \omega_2^L$

so that $f_n(\alpha)$ is defined as in the last part of (2) above. Choose $c$ to be a branch through $T_n^*$ generically coding a $(P_n)^{L_\beta}$-generic extending $p$, where $c$ extends $f_n(\alpha)$; this is possible (using the countability of $\beta$ in $L[0^\#]$) provided $\alpha$ is chosen so that $\alpha = \omega_2^L \cap \text{Skolem Hull}(\alpha)$ in $L_\beta$ (so that the extension from $p_{\omega_1}$ to $f_n(\alpha)$ will obey the restraint imposed by $p^*_{\omega_1}$). But now we have a contradiction, as $\alpha$ is a limit of $\sigma^c$, a CUB subset of $\omega_2^L$, and $f_n(\alpha)$ lies on $c$.

If $n$ belongs to $0^\#$ then $T_n^*$ has a generic branch and hence a $\mathcal{P}(\omega_1^L)$-preserving branch; by the above, $T_n$ has a stationary such branch. If $n$ does not belong to $0^\#$ then $T_n^*$ has no $\mathcal{P}(\omega_1^L)$-preserving branch and hence $T_n$ has no $\mathcal{P}(\omega_1^L)$-preserving branch as such a branch $b$ would yield the $\mathcal{P}(\omega_1^L)$-preserving branch $f_n[b]$ through $T_n^*$.

□ (Lemma 7.29)

The next lemma is from Stanley [99].

**Lemma 7.30.** *There is a constructible function that to each $\Delta_1(L_{\omega_2^L})$ in parameter $\omega_1^L$ tree $T$ on $\omega_2^L$ associates a $\Delta_1(L_{\omega_2^L})$ in parameter $\omega_1^L$ set $X \subseteq \omega_2^L$ such that $T$ has a $\mathcal{P}(\omega_1^L)$-preserving branch $b$ whose intersection with* $(\text{cof}\,\omega_1)^L$ *is stationary in $L[b]$ iff $X$ has a $\mathcal{P}(\omega_1^L)$-preserving CUB subset. And, if $X$ has an $\omega_2$-preserving CUB subset then $T$ has an $\omega_2$-preserving branch.*

*Proof of Lemma* 7.30. Let $T$ be as in the lemma. Then there is a gap 1 morass at $\omega_1^L$ in $L$ such that:

(1) To each $\sigma \in \bigcup\{S_\alpha \mid \alpha \leq \omega_1^L\}$ is associated a tree $T_\sigma$ on $\sigma$.

(2) For $\sigma \in S_{\omega_1^L}$, $T_\sigma = T \restriction \sigma$.

(3) For $\sigma <_1 \tau$, $T_\sigma = \pi_{\sigma\tau}^{-1}[T_\tau]$ and for $\sigma <_0 \tau$, $T_\sigma = T_\tau \restriction \sigma$.

To obtain this morass, start with the morass constructed in Chapter 1, thinning out $S_{\omega_1^L}$ to consist of $\sigma$ such that $T \restriction \sigma$ is uniformly $\Delta_1(L_\sigma)$ in parameter $\omega_1^L$. Then define $T_\sigma$ for $\alpha(\sigma) < \omega_1^L$ to be the tree on $\sigma$ defined over $L_\sigma$ with parameter $\omega_1^{L_\sigma}$ in the same way as $T$ is defined over $L_{\omega_2^L}$ with parameter $\omega_1^L$. For $\alpha < \omega_1^L$ let $\sigma(\alpha)$ be the maximum of $S_\alpha$.

By induction on $\alpha < \omega_1^L$ we define $X_\alpha \subseteq \sigma(\alpha)$. For any $\sigma \in S_\alpha$ and $i < \alpha$, $\sigma(i)$ denotes the unique $\bar\sigma <_1 \sigma$ such that $\bar\sigma \in S_i$, if it exists. Now let $\langle \beta_i \mid i < \alpha\rangle$ be the least $\alpha$-sequence in $L_{\sigma(\alpha)}$ of pairwise $T_{\sigma(\alpha)}$-compatible elements of $\sigma(\alpha)$ such that if $\sigma_i = \beta_i{}^{\text{th}}$ element of $S_\alpha$ then $\{i < \alpha \mid \text{For all } j < i, \sigma_j(i) \in X_{\alpha(\sigma_j(i))}\} = \emptyset$. Then $X_\alpha$ consists of all $\sigma \in S_\alpha$ such that there is $\bar\sigma <_1 \sigma$, $\pi_{\bar\sigma\sigma}$ cofinal, $\bar\sigma \in X_{\alpha(\bar\sigma)}$ together with each $\sigma_i$. (Ignore the $\sigma_i$'s if $\langle \beta_i \mid i < \alpha\rangle$ does not exist.)

Let $X = (\text{cof}\,\omega_1)^L \cup \{\sigma \in S_{\omega_1^L} \mid \text{There exists } \bar\sigma <_1 \sigma, \pi_{\bar\sigma\sigma} \text{ cofinal}, \bar\sigma \in X_{\alpha(\bar\sigma)}\}$. If $X$ contains an $\omega_2$-preserving CUB subset $C$ then note that $\sigma, \tau \in \{\beta \mid \text{ot}(C \cap \beta) = \beta\} \cap (\text{cof}\,\omega_1)^L \Longrightarrow \sigma, \tau$ are $T$-compatible, as otherwise $\{i < \omega_1^L \mid \bigcup \text{Range}\,\pi_{\sigma(i)\sigma} \in X$ and $\bigcup \text{Range}\,\pi_{\tau(i)\tau} \in X\}$ is empty, by construction, in contradiction to the fact that $C \cap \sigma$,

$C \cap \tau$ are CUB in $\sigma, \tau$ of $L$-cofinality $\omega_1^L$. Therefore $T$ has an $\omega_2$-preserving branch. Conversely, suppose that $T$ has a $\mathcal{P}(\omega_1^L)$-preserving branch $b$, $b \cap (\mathrm{cof}\ \omega_1)^L$ stationary. For $\sigma \in S_{\omega_1^L}$, $\sigma$ of $L$-cofinality $\omega_1^L$ and $i < \omega_1^L$, let $\sigma^i$ denote $\bigcup \mathrm{Range}\ \pi_{\sigma(i)\sigma}$ (when $\sigma(i)$ is defined). By construction of $X$, if we let $X(\sigma)$ denote $\{i \mid \sigma^i \in X\}$ then $\{X(\sigma) \mid \sigma \in b^*\}$ generates a normal filter $\mathcal{F}$, where $b^* = \{\sigma \in b \mid \sigma = \mathrm{ot}(S_{\omega_1^L} \cap \sigma)\} \cap (\mathrm{cof}\ \omega_1)^L$. But now there is a $\mathcal{P}(\omega_1^L)$-preserving forcing (over $L[b]$) to add a CUB subset of $X$, using conditions which are closed, bounded subsets of $X$, making use of the fact that the $X(\sigma)$ for $\sigma \in b^*$ generate a normal filter to prove $\omega_2^L$-distributivity. □ (Lemma 7.30)

**Lemma 7.31.** *There is a constructible function that associates to each $\Delta_1(L_{\omega_2^L})$ in parameter $\omega_1^L$ subset $X$ of $\omega_2^L$, sets $Y \subseteq Z \subseteq \omega_1^L$ such that $X$ contains a $\mathcal{P}(\omega_1^L)$-preserving CUB subset $\Longrightarrow \Lambda(R) \cap Z = Y$ for some $\omega_2$-preserving real $R \Longrightarrow X$ contains an $\omega_2$-preserving CUB subset.*

*Proof of Lemma* 7.31. As in the proof of Lemma 7.30, we may choose a constructible morass such that to each $\sigma \in \bigcup\{S_\alpha \mid \alpha \leq \omega_1^L\}$ is assigned $X_\sigma \subseteq \sigma$ such that:

(1) For $\sigma \in S_{\omega_1^L}$, $X_\sigma = X \cap \sigma$.

(2) $\sigma <_0 \tau \Longrightarrow X_\sigma = X_\tau \cap \sigma$, $\sigma <_1 \tau \Longrightarrow X_\sigma = \pi_{\sigma\tau}^{-1}[X_\tau]$.

We set $Z = \bigcup\{S_\alpha \mid \alpha < \omega_1^L\}$ $Y = \{\sigma \in Z \mid \sigma$ is admissible and $\sigma$ belongs to $X_{\sigma^*} \cup \{\sigma^*\}$, where $\sigma^* = \max S_{\alpha(\sigma)}\}$. Now consider the forcing $P$ whose conditions are of the form $(s, s^*, t)$ where:

(1) $s : |s| \longrightarrow 2$, $|s| < \omega_1^L$ and $\sigma \leq |s| \Longrightarrow$ ($\sigma$ is $s \restriction \sigma$-admissible iff $\sigma$ belongs to $Y$).

(2) $t : |t| \longrightarrow 2$, $|t| < \omega_2^L$ and $\alpha(\sigma) = \omega_1^L$, $\sigma \leq |t| \Longrightarrow$ ($\sigma$ is $t \restriction \sigma$-admissible iff $\sigma$ is admissible and belongs to $X$).

(3) $s^*$ is restraint for coding $G_{\omega_1^L}$ by $G_\omega$.

$P$ factors as $P_{\omega_1^L} * Q$ where $Q = P_{\omega_1^L}^G$ and $G = G_{\omega_1^L}$ is the generic for $P_{\omega_1^L}$. And $P_{\omega_1^L} \Vdash Q$ is $\omega_1^L$-distributive and $\omega_2^L$-cc. Thus $P$ is $\omega_2$-preserving if $P_{\omega_1^L}$, consisting of $t$ as in 2) above, is $\omega_2^L$-distributive. If $X$ contains a $\mathcal{P}(\omega_1^L)$-preserving CUB subset then the $\omega_2^L$-distributivity of $P_{\omega_1^L}$ holds in such a $\mathcal{P}(\omega_1^L)$-preserving extension, and hence in $L$.

Finally the $P$-generic $G_\omega \subseteq \omega_1^L$ can be coded by a cardinal-preserving real preserving admissibles. The latter implication of the lemma is immediate since if $\Lambda(R) \cap Z = Y$ then $X$, $\Lambda(R)$ agree on the admissibles in $S_{\omega_1^L}$ and hence $X$ contains all $\alpha < \omega_2^L$ such that $L_\alpha[R]$ is $\Sigma_1$-elementary in $L_{\omega_2^L}[R]$. □ (Lemma 7.31)

Combining the previous three lemmas, we conclude that there is a constructible sequence $\langle (Y_n, Z_n) \mid n < \omega \rangle$ of pairs of subsets of $\omega_1^L$ such that $n \in 0^\#$ iff $\Lambda(R) \cap Z_n = Y_n$ for some $\omega_2$-preserving real $R$. It follows that $0^\#$ is constructible from $\{A \subseteq \omega_1^L \mid A \in L$ and for some real $R$, $\omega_2^{L[R]} = \omega_2^L$ and $\Lambda(R) \cap \omega_1^L = A\}$. The latter set is constructible from $0^\#$ as it equals $\{A \subseteq \omega_1^L \mid A \in L$ and for some real $R \in L[0^\#]$, $\omega_2^{L[R]} = \omega_2^L$ and $\Lambda(R) \cap \omega_1^L = A\}$ by the proof of Theorem 6.26.

Finally, we indicate how to replace the condition $\omega_2^{L[R]} = \omega_2^L$ by the weaker condition $\omega_1^{L[R]} = \omega_1^L$. Modify Lemma 7.29 as follows: Define trees $T_n^\kappa$ for each $L$-regular $\kappa > \omega$ such that $n \in 0^\#$ iff $T_n^\kappa$ has a $\kappa^+$-preserving branch iff $T_n^\kappa$ has a $\mathcal{P}(\kappa)$-preserving branch $b$ which in $L[b]$ has stationary intersection with $(\mathrm{cof}\,\kappa)^L$. Similarly modify Lemma 7.30, generalizing from $\omega_1^L$ to an arbitrary $L$-regular $\kappa > \omega$. Thus we have $\langle X_n^\kappa \mid n < \omega, \kappa\ L\text{-regular}\rangle$ with $X_n^\kappa \subseteq (\kappa^+)^L$ such that $n \in 0^\# \Longrightarrow X_n^\kappa$ has a $\mathcal{P}(\kappa)$-preserving CUB subset and $n \notin 0^\# \Longrightarrow X_n^\kappa$ has no $\kappa^+$-preserving CUB subset. For each ordinal $\gamma$ we consider $Z_n^\gamma = \{\sigma \mid L_\sigma \vDash \gamma$ is the largest cardinal and the $\Delta_1$ definition of $X_n^\kappa$ over $L_{\kappa^+}$ with parameter $\kappa$ yields a $\Delta_1$ definition of $X_n^\gamma \cap \sigma$ over $L_\sigma$ with parameter $\gamma\}$. By strong coding, if $n \in 0^\#$ then there is a cardinal-preserving real $R$ such that $\Lambda(R) \cap Z_n = Y_n$ where $Z_n = \cup\{Z_n^\gamma \mid \gamma < \omega_1^L\}$ and $Y_n = \cup\{X_n^\gamma \cap \sigma \mid \gamma < \sigma < \omega_1^L,\ \sigma \in Z_n^\gamma\}$. If $n \notin 0^\#$ then there is no $\omega_1^L$-preserving real $R$ such that $\Lambda(R) \cap Z_n = Y_n$, else $X_n^\kappa$ would contain a $\kappa^+$-preserving CUB subset, where $\kappa = \omega_2^{L[R]}$. So $0^\#$ is constructible from $\{A \subseteq \omega_1^L \mid A \in L$ and for some real $R$, $\Lambda(R) \cap \omega_1^L = A$ and $\omega_1^{L[R]} = \omega_1^L\}$. □

## Uncountable Admissibles

We close this section by mentioning without proof some generalizations to the uncountable of Sacks' Theorem (see Sacks [76]), that every countable admissible ordinal greater than $\omega$ is the least $R$-admissible for some real $R$.

**Definition.** Assume $V = L$ and let $\kappa$ be an uncountable cardinal. For $X \subseteq \kappa$, $\alpha(X)$ denotes the least ordinal $\alpha > \kappa$ such that $L_\alpha[X]$ is admissible.

**Theorem 7.32** (Friedman [81], Friedman [82]). *Assume $V = L$. Then $\alpha = \alpha(X)$ for some $X \subseteq \kappa$ iff:*

(a) ($\kappa$ regular) *$\alpha$ is admissible, $\kappa < \alpha < \kappa^+$, cofinality $(\alpha) = \kappa$ and $L_\alpha$ is closed under the function of $\beta \longmapsto \beta^{<\kappa}$.*

(b) ($\kappa$ singular of cofinality $\omega$) *$\alpha$ is admissible, $\kappa < \alpha < \kappa^+$, if $L_\alpha$ has a largest cardinal $\gamma$ then $\gamma$ has cofinality $\omega$ in $L$ and there is a 1-1 function $f : L_\alpha \longrightarrow \kappa$ such that $f^{-1}[\delta] \in L_\alpha$ for $\delta < \kappa$.*

(c) ($\kappa$ singular of uncountable cofinality) *$\alpha$ is the least admissible greater than some $\beta$, where $\kappa \leq \beta < \kappa^+$ and $L_\alpha \vDash$ cardinality $(\beta) = \kappa$.*

The following result combines Corollary 7.7 and Theorem 7.32:

**Theorem 7.33** (David–Friedman [85]). *Assume* $V = L$. *Let* $L_\alpha \models \mathrm{ZF}$, $\kappa < \alpha < \kappa^+$. *Then there exists* $X \subseteq \kappa$ *with* $\alpha$ *least such that* $L_\alpha[X] \models \mathrm{ZF}$ *iff:*

(a) ($\kappa$ regular) *There are* $\beta < \alpha$ *and* $\langle X_n \mid n < \omega \rangle$ *such that:*

(a1) $\forall \gamma < \kappa \, \forall f : \gamma \longrightarrow \beta \, (f \text{ bounded} \Longrightarrow f \in L_\alpha)$.

(a2) $X_n \in L_\alpha$, $L_\alpha \models \mathrm{Card}\, X_n < \beta$ *for each* $n$, $L_\alpha = \bigcup_n X_n$.

(a3) $L_\alpha \models \beta$ *regular.*

(b) ($\kappa$ singular, $L_\alpha \models \kappa$ regular) *There are* $\beta < \alpha$ *and* $\langle X_n \mid n < \omega \rangle$ *satisfying* (a2), (a3) *and:*

(b1) $\forall \lambda < \beta \, \exists f : L_\lambda \longrightarrow \kappa$, $f$ 1-1 *and* $\forall \gamma < \kappa$, $f^{-1}[\gamma] \in L_\alpha$.

(c) ($L_\alpha \models \kappa$ singular) *For some* $\beta$, $\kappa \leq \beta < \kappa^+$, $\alpha$ *is the least ordinal greater than* $\beta$ *such that* $L_\alpha \models \mathrm{ZF}$, *and* $L_\alpha \models$ *cardinality* $(\beta) = \kappa$.

## Chapter 8

# Further Applications of Class Forcing

In this final chapter we describe a number of results whose proofs take advantage of the method of class forcing. Three new class forcing techniques are introduced: First we discuss a *condensation condition* on a class $A$ which enables $A$ to be $\Delta_1$-*coded* by a real $R$ (in the sense that $A$ is $\Delta_1$-definable over $L[R]$). We apply this to prove a result about $L$-cofinality. Second we introduce *iterated class forcing*, which can be applied to establish the optimality of our result in Chapter 5 concerning generic saturation. Third we indicate how to *minimally code* a class $A$ by a real $R$ (in the sense that there is no inner model intermediate between $L[A]$ and $L[R]$). In addition, using techniques from earlier chapters, a number of assertions in descriptive set theory will be shown consistent with the axiom that $\aleph_1$ is inaccessible in $L[R]$ for every real $R$.

## 8.1 $\Delta_1$-Coding

The principal results of this section are taken from Friedman–Veličković [97]. A real $R$ $\Delta_1$-*codes* a class $A \subseteq \mathrm{ORD}$ iff $A$ is $\Delta_1$-definable over $L[R]$. Every $L$-amenable class $A$ is $\Delta_1$-coded by $0^\#$, as indicated in Chapter 3. The next result provides a converse to this result.

**Proposition 8.1.** *Suppose that* $L\text{-card} = \{\alpha \mid \alpha \text{ is a cardinal of } L\}$ *is* $\Sigma_1$ *over* $L[R]$, $R$ *a real. Then* $0^\# \leq_L R$.

*Proof.* Suppose that the $\Sigma_1$ definition has parameters less than $\kappa$, where $\kappa$ is a singular cardinal. As $\kappa^+$ is an $L$-cardinal, by reflection there must be unboundedly many $\alpha < \kappa^+$, $\alpha \in L$-card. But then $(\kappa^+)^L < \kappa^+$, which implies that $0^\#$ exists. As this argument can be carried out in $L[R]$, in fact $0^\# \leq_L R$. □

We introduce a sufficient condition for an $L$-amenable class to be $\Delta_1$-coded by a real which is class-generic over $L$. To motivate it we first indicate a necessary condition for $\Delta_1$-codability:

**Definition.** Suppose that $x$ is an extensional set (i.e., $\langle x, \in\rangle$ satisfies the axiom of extensionality). Let $\bar{x}$ denote the transitive collapse of $x$. For $A \subseteq \mathrm{ORD}$ we say that $x$ *preserves* $A$ if $\langle \bar{x}, \in, A \cap \bar{x}\rangle$ is isomorphic to $\langle x, \in, A \cap x\rangle$.

**Proposition 8.2.** *Suppose that* $A$ *is* $L$*-amenable and* $\Delta_1$*-coded by a real* $R$, $0^\# \not\leq_L R$. *Let* $\kappa$ *be an uncountable cardinal,* $\langle T, \in, \ldots\rangle$ *a transitive structure of cardinality at least* $\kappa$, *for a countable language. Then there is* $\langle x, \in, \ldots\rangle \prec \langle T, \in, \ldots\rangle$ *such that* $\mathrm{card}(x) = \kappa$, $x \in L$ *and* $x$ *preserves* $A$.

*Proof.* By strong covering (see Corollary 3.11) we can choose $y$ of cardinality $\kappa$ such that $\langle T, \in, \ldots\rangle \in y$, $\langle y, \in\rangle$ is $\Sigma_1$-elementary in $\langle L[R], \in\rangle$, $y$ contains the parameters in a $\Delta_1$ definition of $A$ over $L[R]$ and $x_0 = y \cap L$ belongs to $L$. Then $x = x_0 \cap T$ is as desired. □

**Definition.** For a set $x$ and ordinal $\delta$, $x[\delta]$ denotes $\{f(\gamma) \mid \gamma < \delta,\ f \in x,\ f$ a function whose domain includes $\gamma\}$. We say that $x$ *strongly preserves* $A \subseteq \mathrm{ORD}$ if $x[\delta]$ is extensional and preserves $A$ for each cardinal $\delta$. A sequence of extensional sets $t_0 \subseteq t_1 \subseteq \cdots$ is *tight* if it is continuous (i.e., $t_\lambda = \bigcup\{t_i \mid i < \lambda\}$ for limit $\lambda$) and for each $i$: $t_i \in t_{i+1} \cup \{t_{i+1}\}$ and $\langle \bar{t}_j \mid j < i\rangle$ belongs to the least $\mathrm{ZF}^-$ model containing $\bar{t}_i$ as an element which correctly computes $\mathrm{card}(\bar{t}_i)$.

## Condensation Condition

Suppose that $t$ is transitive, $\kappa$ is regular, $\kappa \in t$ and $x \in t$. Then:

(1) There is a tight $\kappa$-sequence $t_0 \prec t_1 \prec \cdots \prec t$ such that $x \in t_0$ and for each $i < \kappa$: $\mathrm{card}(t_i) = \kappa$ and $t_i$ strongly preserves $A$.

(2) If $\kappa$ is inaccessible then there exists $t_0 \prec t_1 \prec \cdots \prec t$ as above, but where $\mathrm{card}(t_i) = \aleph_i$.

**Theorem 8.3** ($\Delta_1$-Coding Theorem). *Suppose that $A$ is $L$-amenable and obeys the Condensation Condition in $L$. Then $A$ is $\Delta_1$-coded in a tame class-generic extension of $\langle L, A\rangle$ by a real $R$ such that $L$, $L[R]$ have the same cofinalities.*

**Remark.** $R$ need not exist in $L[0^\#]$ without further assumptions on $A$; see the corollary below.

*Proof.* As in the proof of Theorem 7.1, we use a modification of the forcing used to prove Theorem 4.1. The key change is in the definition of $S_\alpha$, the "strings" $s : [\alpha, |s|) \longrightarrow 2$, $\alpha \le |s| < \alpha^+$. For $s$ to belong to $S_\alpha$ we require:

(∗) If $L_\beta[A \cap \alpha, s \restriction \eta] \models \mathrm{ZF}^- + \eta = \alpha^+$ (or $\beta = \eta$ and $L_\beta[A \cap \alpha, s \restriction \eta] \models \mathrm{ZF}^- + \alpha$ is the largest cardinal) then $A \cap [\alpha, \beta) = \mathrm{Even\,Part}(X)$ where $X \subseteq [\alpha, \beta)$ is decoded in $L_\beta[A \cap \alpha, s \restriction \eta]$ from $A \cap \alpha$, $s \restriction \eta$ as in the Coding Theorem.

**Definition** (Coding Structures). For $s \in S_\alpha$ define $\mu^s$, $\mu^{<s}$ inductively by: $\mu^{<\emptyset_\alpha} = \alpha$ where $\emptyset_\alpha$ = empty string at $\alpha$, $\mu^{<s} = \bigcup\{\mu^{s\restriction\eta} \mid \eta < |s|\}$ for $s \ne \emptyset_\alpha$ and $\mu^s$ = least $\mu > \mu^{<s}$ such that $\mu'\mu = \mu$ for $0 < \mu' < \mu$, $s \in L_\mu$, $\langle L_\mu, A \cap \mu\rangle$ is amenable and $\langle L_\mu, A \cap \mu\rangle \models \mathrm{ZF}^- + \mathrm{card}(s) \le \alpha + A \cap \mu$ satisfies the Condensation Condition. And $\mathcal{A}^s = \langle L_{\mu^s}, A \cap \mu^s\rangle$.

The remainder of the definitions from the proof of Theorem 4.1 remain the same in the present context. As in the proof of Theorem 7.1, it remains to establish distributivity and extendibility.

**Lemma 8.4** (Distributivity for $R^s$). *Suppose $\alpha \in$ Card, $s \in S_{\alpha^+}$. Then $R^s$ is $\alpha^+$-distributive in $\mathcal{A}^s$.*

*Proof.* Let $\kappa$ be regular, $\kappa \leq \alpha$ and $\langle D_i \mid i < \kappa \rangle \in \mathcal{A}^s$ dense on $R^s$. Let $(u, u^*)$ belong to $R^s$ and choose a transitive $t \in \mathcal{A}^s$ such that $x = \{(u, u^*), A \cap \alpha^+, s, \langle D_i \mid i < \kappa \rangle\}$ belongs to $t$. Apply the Condensation Condition in $\mathcal{A}^s$ to obtain a tight $\kappa$-sequence $t_0 \prec t_1 < \cdots \prec t$ as in (1) of the condition. As in the proof of Lemma 4.3, build $(u, u^*) = (u_0, u_0^*) \geq (u_1, u_1^*) \geq \cdots$ so that $(u_{i+1}, u_{i+1}^*) \in t_{i+1}$ meets $D_i$ and for limit $\lambda \leq \kappa$, $A \cap \alpha$ and $u_\lambda$ code the images $\bar{A}, \bar{s}$ of $A, s$ under the transitive collapse of $t_\lambda$. Then $u_\lambda$ obeys the requirements for membership in $S_\alpha$, provided $A$ is replaced by $\bar{A}$ in the statement of those requirements. But $t_\lambda$ preserves $A$, so $\bar{A}$ is an initial segment of $A$ and $u_\lambda$ belongs to $S_\alpha$. The final condition $(u_\kappa, u_\kappa^*)$ extends $(u, u^*)$ and meets each $D_i$. □

**Lemma 8.5** (Extendibility for $P^s$). *Suppose $p \in P^s$, $s \in S_\alpha$, $X \subseteq \alpha$ and $X \in \mathcal{A}^s$. Then there exists $q \leq p$ such that $X \cap \beta \in \mathcal{A}^{q_\beta}$ for each $\beta \in$ Card $\cap\, \alpha$.*

*Proof.* Proceed as in the proof of Lemma 4.4. In the definition of $q$, the only new instances of $(*)$ to verify are for $s_\beta$ when Even$(Y \cap \beta)$ codes $s_\beta$ with domain $[\beta, |s_\beta|)$, $|s_\beta| < \beta^+$. Clearly $s_\beta$ obeys $(*)$ with $A$ replaced by $\bar{A}$ = inverse image of $A$ under the canonical embedding $\bar{\mathcal{A}}_\beta \longrightarrow \mathcal{A}$; so it suffices to guarantee that $\bar{A}$ is an initial segment of $A$. If $\alpha$ is not inaccessible in $\mathcal{A}^s$ then we choose $\mathcal{A}$ to be large enough so that $\alpha$ is singular in $\mathcal{A}$ and then there are only boundedly many $s_\beta$ as above; so we can apply induction on $\alpha$ to complete the proof. If $\alpha$ is inaccessible in $\mathcal{A}^s$ then we can choose $\mathcal{A}$ to be large enough so as to contain a witness to Condensation Condition (2), where $t = \tilde{\mathcal{A}}^s$, $\kappa = \alpha$ and $x = \{A \cap \alpha, s\}$. It follows then that the range of $\bar{\mathcal{A}}_\beta \longrightarrow \mathcal{A}$ preserves $A \cap \tilde{\mu}^s$ for all $\beta \in$ Card $\cap\, \alpha$ and hence $s_\beta$ belongs to $S_\beta$ for $s_\beta$ as above. □

**Lemma 8.6** (Distributivity for $P^s$). *Suppose $s \in S_{\beta^+}$, $\beta \in$ Card.*

(a) *If $\langle D_i \mid i < \beta \rangle \in \mathcal{A}^s$, $D_i$ $i^+$-dense on $P^s$ for each $i < \beta$ and $p \in P^s$ then there is $q \leq p$, $q$ meets each $D_i$.*

(b) *If $p \in P^s$, $f$ small in $\mathcal{A}^s$ then there is $q \leq p$, $q \in \Sigma_f^p$.*

*Proof.* Proceed as in the proof of Lemma 4.5. In the proof of the Claim we must verify that $p_\gamma^\lambda$ satisfies $(*)$. The fact that $p_\gamma^\lambda$ codes a generic over the collapse of $H_\lambda(\gamma) \cap L_\beta$ implies that $(*)$ is satisfied with $A$ replaced by $\bar{A}$ = inverse image of $A$ under the canonical embedding $H_\lambda(\gamma) \longrightarrow L_{\mu_\lambda}[A \cap \beta^+, s]$. The Condensation Condition guarantees that we can arrange for $\bar{A}$ to be an initial segment of $A$ as $H_\lambda(\gamma)$ can be made equal to $t_\lambda[\gamma]$ where $t_0 \prec t_1 < \cdots \prec t \in \mathcal{A}^s$ strongly preserve $A$. Part (2) of the Condensation Condition is used to treat inaccessibles. □

This completes the proof of Theorem 8.3. □

**Corollary 8.7.** *Suppose that $A$ is $L$-amenable, obeys the Condensation Condition in $L$ and preserves indiscernibles. Then $A$ is $\Delta_1$-definable over $L[R]$ for some indiscernible-preserving real $R$ such that $L$, $L[R]$ have the same cofinalities.*

*Proof.* Apply the proof of Theorem 8.3, and use the proof of Theorem 4.18 to argue that $R$ can be constructed inside $L[0^\#]$, preserving indiscernibles. □

We can apply the above to show that $L\text{-cof}\,\omega = \{\alpha \mid \alpha \text{ has } L\text{-cofinality } \omega\}$ is $\Delta_1$-definable in $L[R]$, where $R$ is a real not constructing $0^\#$.

**Lemma 8.8.** *There is a real $R_0$, class-generic over $L$, such that $R_0 <_L 0^\#$, $R_0$ preserves all $L$-cardinals with the exception of $\aleph_1^L$ and the Condensation Condition holds for $A = L\text{-cof}\,\omega$ in $L[R_0]$.*

*Proof.* Starting from $L$, Lévy collapse $\aleph_1^L$ to $\omega$ and then perform a Reverse Easton iteration where at regular stage $\kappa$, $C(\kappa)$ is added, where $C(\kappa)$ is a CUB subset of $\kappa$ disjoint from $L\text{-cof}\,\omega$. This preserves $L$-cardinals greater than $\aleph_1^L$ and we can choose a generic $\langle C(\kappa) \mid \kappa\ L\text{-regular}, \kappa \geq \aleph_2^L\rangle$ definable in $L[0^\#]$ which preserves indiscernibles. Choose $R_0 <_L 0^\#$ to code $\langle C(\kappa) \mid \kappa\ L\text{-regular}, \kappa \geq \aleph_2^L\rangle$, where $R_0$ is a real preserving all $L$-cardinals greater than $\aleph_1^L$, using the technique of Theorem 4.18.

Now we verify the Condensation Condition for $A = L\text{-cof}\,\omega$ in $L[R_0]$. Working in $L[R_0]$, let $t$ be transitive, $\kappa$ regular, $\kappa \in t$ and $x \in t$. We first verify (1) of the Condensation Condition. We may assume that $t$ has cardinality greater than $\kappa$.

Let $\alpha$ be regular, $\alpha > \kappa^+$ such that $t \in L_\alpha[R_0]$. Let $u_0 \prec u_1 \prec \cdots \prec L_\alpha[R_0]$ be a continuous tight sequence of length $\kappa^+$ where $\kappa, R_0, x, t \in u_0$ and for each $i < \kappa^+$: $\operatorname{card}(u_i) = \kappa$, $\langle u_j \mid j \leq i\rangle \in u_{i+1}$. Let $u = \bigcup\{u_i \mid i < \kappa^+\}$ and note that for regular $\lambda \in u$, $\sup(u \cap \lambda)$ has cofinality $\kappa^+$ and hence $L$-cofinality $\kappa^+$ (as only $\aleph_1^L$ is collapsed by $R_0$).

In fact, for any pair of cardinals $\delta < \lambda$, $\lambda$ regular if $\lambda \in u[\delta]$ then either $\delta = u[\delta] \cap \lambda$ or $\sup(u[\delta] \cap \lambda)$ has $L$-cofinality $\kappa^+$. For, if $\delta \neq u[\delta] \cap \lambda$ we may choose $i_0 < \kappa^+$ and $\gamma < \delta$ such that $\lambda \in u_{i_0}[\gamma]$, $u_{i_0}[\gamma] \cap [\delta, \lambda) \neq \emptyset$. Then for $i_0 \leq i < j < \kappa^+$, $\sup(u_i[\delta] \cap \lambda) < \sup(u_j[\delta] \cap \lambda)$. Thus $\sup(u[\delta] \cap \lambda)$ is the supremum of the increasing sequence $\langle \sup(u_i[\delta] \cap \lambda) \mid i_0 \leq i < \kappa^+\rangle$ and therefore has cofinality (and hence $L$-cofinality) $\kappa^+$.

Now build $v_0 \prec v_1 \prec \cdots \prec L_\alpha[R_0]$ just like the $u_i$'s but with the additional requirement that $u$ belongs to $v_0$. Note that if $\beta$ is regular, $\beta \in v_j$ then $v_j$ contains a constructible map from $\kappa^+$ cofinally into $\sup(u \cap \beta)$. Now we claim that for each $j < \kappa^+$, $v_j \cap u$ preserves $A$: Suppose $\beta \in v_j \cap u$, $L\text{-cof}(\beta) > \omega$ and let $\bar{\beta}$ denote the image of $\beta$ under the transitive collapse of $v_j \cap u$. Let $\gamma$ be the $L$-cofinality of $\beta$; clearly $\bar{\beta}$ and $\bar{\gamma}$ have the same $L$-cofinality so to show that $\bar{\beta}$ has $L$-cofinality $> \omega$ we may assume that $\beta = \gamma$ and hence that $\beta$ is regular (note that $\bar{\beta} = \beta$ when $\beta = \aleph_1^L$). Thus $\sup(u \cap \beta)$ has $L$-cofinality $\kappa^+$ and so $\bar{v}_j$ contains a constructible map from $\kappa^+ \cap v_j$ cofinally into ordertype $(u \cap \beta \cap v_j) = \bar{\beta}$. But $\kappa^+ \cap v_j$ belongs to $C(\kappa^+)$ and therefore has $L$-cofinality $> \omega$. So $\bar{\beta}$ has $L$-cofinality $> \omega$.

In fact $v_j \cap u$ strongly preserves $A$: If $\delta < \beta$, $\delta$ a cardinal, $\beta \in v_j \cap u$, $L\text{-cof}(\beta) > \omega$ then to show that $L\text{-cof}(\bar{\beta}) > \omega$ where $\bar{\beta}$ = image of $\beta$ under the transitive collapse of $(v_j \cap u)[\delta]$, we may again assume that $\beta$ is regular. Then either $\sup(u[\delta] \cap \beta)$ has cofinality $\kappa^+$ and belongs to $v_j[\delta]$, hence the above argument applies, or $u[\delta] \cap \beta = \delta$ and $(v_j \cap u)[\delta] \cap \beta = \delta \in C(\beta)$ and hence $\bar{\beta} = \delta$, which has $L$-cofinality $> \omega$.

Finally let $t_i = v_i \cap u \cap t$ for $i < \kappa$. Then each $t_i$ strongly preserves $A$ as $v_i \cap u$ does and $t$ is transitive. We must show that $\langle t_i \mid i < \kappa \rangle$ is tight. Note that $\bar{v}_i$ is of the form $L_{\bar{\alpha}}[R_0]$ and satisfies $\text{card}(\bar{t}_i) > \kappa$. So any $\text{ZF}^-$ model containing $\bar{t}_i$ and satisfying $\text{card}(\bar{t}_i) = \kappa$ must in addition contain $\bar{v}_i$ as an element and satisfy $\text{card}(\bar{v}_i) = \kappa$. Thus the tightness of $\langle v_i \mid i < \kappa \rangle$ implies that of $\langle t_i \mid i < \kappa \rangle$. This proves (1) of the Condensation Condition. To prove (2), perform the same argument but with $\text{card}(u_i) = \aleph_i$ and $u$ of cardinality $\kappa$. □

**Corollary 8.9.** *There exists a real $R <_L 0^\#$ such that $R$ is class-generic over $L$, $R$ preserves indiscernibles, $R$ preserves all $L$-cardinals greater than $\aleph_1^L$, and $L$-cof $\omega$ is $\Delta_1$ over $L[R]$.*

Recall that $\alpha$ is *quasi $R$-admissible* if the ordertype of each wellordering in $L_\alpha[R]$ is less than $\alpha$.

**Corollary 8.10.** *There is a real $R <_L 0^\#$ such that every quasi $R$-admissible has uncountable $L$-cofinality.*

*Proof.* Let $R_0$ be as in the previous corollary and for each Silver indiscernible $\kappa$ for $L[R_0]$ let $C(\kappa)$ be the $L[R_0]$-least CUB subset of $\kappa$ consisting of ordinals of uncountable $L$-cofinality, with the additional property that $\alpha \in C(\kappa) \implies (\alpha^+)^L \in C(\kappa)$. Then $C = \bigcup\{C(\kappa) \mid \kappa \text{ an indiscernible for } L[R_0]\}$ is CUB in ORD, consists of ordinals of $L$-cofinality $> \omega$ and $\alpha \in C \implies (\alpha^+)^L \in C$.

Now force over $\langle L[R_0], C\rangle$, with an Easton product of Lévy collapses, to get a model $M$ such that every uncountable cardinal of $M$ belongs to $C$. Let $R_1$ be a real coding $M$, $R_0 = \text{Even}(R_1)$ with $R_1 <_L 0^\#$. Thus every $L[R_1]$-cardinal $> \omega$ has uncountable $L$-cofinality and the latter is a $\Delta_1(L[R_1])$ property. By Theorem 7.6, we can force over $L[R_1]$ to obtain the desired real $R$. □

**Corollary 8.11.** *There is a real $R <_L 0^\#$ such that the function $f(\alpha) = [\alpha]^\omega \cap L$ is $\Delta_1$ over $L[R]$.*

*Proof.* As in the previous proof, produce a real $R_1$ such that every uncountable $L[R_1]$-cardinal has $L$-cofinality $> \omega$ and $\{\alpha \mid L\text{-cof}(\alpha) > \omega\}$ is $\Delta_1(L[R_1])$. By arranging for $R_1$ to code over $L_{\kappa^+}[R_1]$ a wellordering of ordertype the least element of $C$ (from the previous proof) greater than $\kappa^+$ for $\kappa \in C$, we also have that if $\alpha$ is quasi $R_1$-admissible then $L\text{-card}(\alpha)$ has uncountable $L$-cofinality.

Now the function $f(\alpha) = [\alpha]^\omega \cap L$ is $\Delta_1(L[R_1])$: It suffices to define $g$ such that $g$ is $\Delta_1(L[R_1])$ and $[\alpha]^\omega \cap L \subseteq L_{g(\alpha)}$. Define $g(\alpha)$ to be the least $\beta$ such that one of the following holds:

(1) $L$-cof$(\alpha) > \omega$ and $\beta \geq \cup\{g(\alpha') \mid \alpha' < \alpha\}$.

(2) $\beta \geq g(\alpha')$ for some $\alpha' < \alpha$ such that $L_\beta \models \text{card}(\alpha) \leq \alpha'$.

(3) $\beta > \alpha$, $\beta$ is quasi $R_1$-admissible.

Then either (1) or (2) holds for some $\beta$ or $\alpha$ is an $L$-cardinal of $L$-cofinality $\omega$, in which case any $\beta$ obeying (3) is at least $(\alpha^+)^L$. So $g$ is as desired. □

An *immune partition* is $F : \text{ORD} \longrightarrow 2$ such that neither $\{\alpha \mid F(\alpha) = 0\}$ nor $\{\alpha \mid F(\alpha) = 1\}$ contains an infinite constructible set.

**Corollary 8.12.** *There is a real $R <_L 0^\#$ such that some immune partition is $\Delta_1(L[R])$.*

*Proof.* First select $R_0$ as in the previous corollary and then by Friedman [85a] add $R_1$ over $L[R_0]$ so as to code an immune partition. Then in $L[R]$, where $R = (R_0, R_1)$, immunity is a $\Delta_1$ property and there is a $\Delta_1$ CUB class $C$ consisting of ordinals of uncountable $L$-cofinality. For each $\alpha \in C \cup \{0\}$ let $F_\alpha$ be the $L[R]$-least immune partition of $[\alpha, \alpha^*)$ where $\alpha^*$ = least element of $C$ greater than $\alpha$. Then $F = \cup\{F_\alpha \mid \alpha \in C \cup \{0\}\}$ is a $\Delta_1(L[R])$ immune partition and as before we can choose $R <_L 0^\#$. □

In analogy with Section 7.3, we consider the "characterization problem" for $\Delta_1$-definability in a real: Is there an exact constructible criterion for a subset of an $L$-cardinal $\kappa$ to be the intersection with $\kappa$ of a predicate which is $\Delta_1$-definable in $L[R]$ for some real $R$ that preserves $L$-cardinals? It can be shown that the answer is "Yes" when $\kappa$ is $\omega_2^L$. But the answer is "No" when $\kappa$ is $\omega_3^L$:

**Theorem 8.13.** *Let $S = \{X \subseteq \omega_3^L \mid X = \omega_3^L \cap A$ for some $A = \text{ORD}$, $A$ $\Delta_1$-definable in $L[R]$ for some real $R$ that preserves $L$-cardinals$\}$. Then $S =_L 0^\#$.*

*Proof.* The reduction $S \leq_L 0^\#$ follows as by absoluteness, $S$ is unchanged if we require "$R \in L[0^\#]$" in its definition (see the proof of Theorem 6.26).

Suppose that $X$ is a subset of the interval $[\omega_1^L, \omega_2^L)$. If $X$ contains a CUB subset of $\omega_2^L$ in a $\mathcal{P}(\omega_1^L)$-preserving (and cardinal-preserving) extension then $X \cup \{\omega_2^L\}$ obeys the Condensation Condition for $\kappa = \omega_1^L$ in this extension. Also in $L$ we may define $Y \subseteq \omega_1^L$ such that $Y \cup X \cup \{\omega_2^L\}$ obeys the Condensation Condition for $\kappa = \omega$ in $L$ (and hence in our $\mathcal{P}(\omega_1^L)$-preserving extension), by a straightforward $\diamondsuit$-like construction where $Y$ is built inductively by taking least counterexamples. Thus by the proof of Theorem 8.3, $Y \cup X \cup \{\omega_2^L\}$ belongs to $S$.

Conversely if $Y \cup X \cup \{\omega_2^L\}$ belongs to $S$ for some $Y \in L$, $Y \subseteq \omega_1^L$ then $X$ contains a CUB subset of $\omega_2^L$ in a cardinal-preserving extension. It now follows from Lemmas 7.29, 7.30 that $0^\#$ is constructible from $S$. □

Theorem 8.13 rules out any simple characterization of when an $L$-amenable predicate can be $\Delta_1$-definable in a real not constructing $0^\#$.

## 8.2 Iterated Class Forcing

The class forcings that we have considered in this book can be iterated, provided we are careful with the supports that are used. In this section we introduce the notions of *stratified* and $\Delta$-*stratified* forcing and show that these properties are preserved through iteration (with the appropriate support) and imply ZFC-preservation. The class forcing examples that we have considered are each either stratified or $\Delta$-stratified. We shall apply iterated class forcing to demonstrate the optimality of our result in Chapter 5 concerning Generic Saturation.

**Definition.** A pre-ordering $(P, \leq)$ is *stratified with witness A* iff $V = L[A]$, $\langle L[A], A\rangle$ is a ground model and:

(1) $(P, \leq)$ is $\langle L[A], A\rangle$-definable. A condition in $P$ is a function $p$: $\text{Card} \cap \alpha(p) \longrightarrow V$, $\alpha(p) \in \text{Card} = \{0\} \cup$ Infinite Cardinals. If $q$ extends $p$ as a function and $q(\gamma) = \emptyset$ for $\gamma \in \text{Dom}(q) - \text{Dom}(p)$ then we identify $p$ with $q$. We require that $p(\gamma) = \emptyset$ for singular $\gamma$ and that the conditions with constant value $\emptyset$ are the weakest in $P$. Lastly, $\{p \mid p \restriction \kappa^+ \in H_{\kappa^+}\}$ is dense for each $\kappa \in \text{Card}$ ($0^+$ denotes $\omega$).

(2) ($\kappa$-Density Reduction) Let $\kappa$ be infinite, regular and define $p \leq_\kappa q$ iff $p \leq q$ and $p \restriction \kappa^+ = q \restriction \kappa^+$. Then $p \leq q \Longrightarrow \exists r \leq_\kappa q \exists s \leq r$ ($s \restriction [\kappa^+, \infty) = r \restriction [\kappa^+, \infty)$ and $s \leq p$). Conversely, if $p \leq r, s$ and $r \restriction \kappa^+ = q \restriction \kappa^+$, $s \restriction [\kappa^+, \infty) = q \restriction [\kappa^+, \infty)$, then $p \leq q$. If $D$ is an $\langle L[A], A\rangle$-definable dense class, $p \in P$ then there exists $q \leq_\kappa p$ and $d \subseteq D$, $\text{card}(d) \leq \kappa$ such that every $r \leq q$ can be extended to $s$ meeting $d$, $r \restriction [\kappa^+, \infty) = s \restriction [\kappa^+, \infty)$. (We say that $q$ $\kappa^+$-*reduces D*.)

(3) ($\kappa$-Definable Closure) For $\kappa$ infinite, regular these are $\Pi_n\langle L[A], A\rangle$ operators $F_n(x, \kappa, p)$, $0 < n < \omega$ such that $F_n(x, \kappa, p) \leq_\kappa p$ for all $p$ and whenever $p_0 \geq_\kappa p_1 \geq_\kappa \cdots$ is a $\Pi_n\langle L[A], A\rangle$ (in parameters from $\kappa \cup \{x\}$) sequence of length $\lambda \leq \kappa$ such that for each $i < \lambda$, $p_{i+1} \leq_{\kappa_i} F_n(x, \kappa_i, p_i^*)$ for some $p_i^* \leq_{\kappa_i} p_i$ and regular $\kappa_i \geq \kappa$ then there is $p \leq_\kappa p_i$ for all $i < \lambda$.

**Theorem 8.14.** *Suppose that P is stratified with witness A. Then P preserves* ZFC *(relative to A) and cofinalities. If* $V \models$ GCH *then P preserves the* GCH.

*Proof.* Suppose that $\langle D_i \mid i < \kappa\rangle$ is an $\langle L[A], A\rangle$-definable sequence of dense classes, $\kappa$ infinite and regular, $p \in P$. Then using $\kappa$-Density Reduction and $\kappa$-Definable Closure there is a sequence $p = p_0 \geq_\kappa p_1 \geq_\kappa \cdots \geq_\kappa q$ of length $\kappa + 1$ where $p_{i+1}$ $\kappa^+$-reduces $D_i$ for each $i < \kappa$, and hence $q$ simultaneously $\kappa^+$-reduces each $D_i$, $i < \kappa$. Thus $P$ is pretame and cofinalities are preserved. The power set axiom is preserved as for regular $\kappa$, $\mathcal{P}(\kappa)$ exists in the $P$-generic extension by virtue of the last statement of (1) and $\kappa$-Density Reduction. For the same reason, the GCH is preserved at regular $\kappa$. When $\kappa$ is singular, choose $\langle \kappa_i \mid i < \text{cof}(\kappa)\rangle$ cofinally in $\kappa$, each $\kappa_i$ regular and apply both $\kappa_i$-Density Reduction and $\text{cof}(\kappa)$-Definable Closure. □

To preserve stratification under iteration we must discuss strong witnesses and diagonal supports.

**Definition.** $P$ is stratified with *strong witness* $A$ if in the definition of stratified with witness $A$ we have: $(P, \leq)$, Card are $\Delta_1\langle L[A], A\rangle$ and there is a $\Sigma_1\langle L[A], A\rangle$ function $f(x, \kappa, p) = (q, d)$ which given an index $x$ for a $\Sigma_1\langle L[A], A\rangle$ $D$ dense below a condition $p$ and an infinite, regular $\kappa$ produces $q \leq_\kappa p$ that $\kappa^+$-reduces $D$, as witnessed by $d$ (i.e., card$(d) \leq \kappa$, $d \subseteq D$ and every $r \leq q$ can be extended to $s$ such that $s \restriction [\kappa^+, \infty) = r \restriction [\kappa^+, \infty)$, $s$ meets $d$).

**Proposition 8.15.** *If $P$ is stratified then $P$ has a strong witness.*

*Proof.* If $P$ is stratified with witness $A$ then $A$ can be converted to a strong witness $A^*$ by taking $A^* = \Sigma_N$ satisfaction for $\langle L[A], A\rangle$, $N$ sufficiently large. □

Strong witnesses are used to control the level of definability of the forcing relation:

**Theorem 8.16.** *Suppose that $P$ is stratified with strong witness $A$. Then the forcing relation for $P$ restricted to $\Sigma_1^A$ sentences is "densely $\Sigma_1^A$": For some $\Sigma_1^A$ relation $\Vdash\!\vdash$ we have $p \Vdash\!\vdash \varphi \Longrightarrow p \Vdash \varphi$ and $p \Vdash \varphi \Longrightarrow$ for some $q \leq p$, $q \Vdash\!\vdash \varphi$.*

**Remark.** In the above statement we make the Assumption that for every $p \in P$ there is a $G$ $P$-generic over $\langle L[A], A\rangle$, $p \in G$. Alternatively the relation $\Vdash$ can be replaced by the syntactic forcing relation $\Vdash^*$ of Chapter 2.

*Proof.* It suffices to establish this for $\Delta_0$ sentences $\varphi$. We show by a $\Sigma_1^A$ induction on $\varphi$ that given $p$ we can "effectively" (i.e., via a $\Sigma_1^A$ procedure) find $q \leq p$ and $i \in \{0, 1\}$ such that either $q \Vdash \varphi, i = 1$ or $q \Vdash\sim \varphi, i = 0$. Then we can take $q \Vdash\!\vdash \varphi$ to mean that for some $p \geq q$, $(q, 1)$ arises from $p$ as above.

Note that by allowing bounded quantities we may assume that $\varphi$ contains no names other than the standard names $\hat{a}$ for elements $a$ of the ground model $L[A]$. We proceed by induction on $\varphi$, the only nontrivial case being the case of the bounded quantifier $\varphi \equiv \forall x \in a \psi$. By induction we can effectively extend $p$ to decide any instance $\psi(b)$, $b \in a$, via a condition $q_b \leq p$. If any $q_b$ forces $\psi(b)$ negatively we can take the desired $q \leq p$ to be $q_b$. Otherwise we can build a $\Sigma_1^A$ sequence $\langle D_b \mid b \in a\rangle$ of classes dense below $p$ such that $r \in D_b \Longrightarrow r \Vdash \psi(b)$. As $A$ is a strong witness we can effectively find $q \leq p$ and $d$ such that $d \in L[A]$ and $D_b \cap d$ is predense below $q$ for each $b \in a$; then $q \Vdash \forall x \in a \psi$. □

We are ready to discuss stratified iterations.

**Definition.** A *stratified iteration* is a sequence $\langle P(< i) \mid i < \alpha\rangle$, $\alpha \leq \infty$ (with $P(< 0) = \{\emptyset\}$, $P(< i+1) = P(\leq i) = P(< i) * P(i)$, $P(< \lambda) \subseteq$ Inverse limit$\langle P(< i) \mid i < \lambda\rangle$ for limit $\lambda$) such that for some class $A$, $A$ strongly witnesses that $P(0)$ is stratified, for each $i + 1 < \alpha$, $P(< i) \Vdash P(i)$ is stratified with strong witness $\langle A, G(< i)\rangle$ (where $G(< i)$ denotes the $P(< i)$-generic) via $F_n^i$, $f^i$, $P(i)$ and $f^i$ are $\Delta_1^{A,G_i}$ uniformly in $i + 1 < \alpha$ and $F_n^i$ is $\Pi_n^{A,G_i}$ uniformly in $i + 1 < \alpha$ for each $n > 0$.

Such an iteration has *short diagonal supports* if for $j < \alpha$, $p \in P(< j)$ and infinite, regular $\kappa$, $\{i \mid i < j \text{ and } p \restriction i \nVdash \forall\gamma < \kappa^+ p(i)(\gamma) = \emptyset\}$ is a subset of $\kappa^+$ of cardinality $< \kappa$ (and this is the only support restriction).

**Theorem 8.17** (Stratified Iterations). *Suppose that $\langle P(< i) \mid i \leq \alpha\rangle$ is a stratified iteration with short diagonal supports as witnessed by the class A and the* GCH *holds. Then $P(< \alpha)$ is isomorphic to a stratified forcing, via an isomorphism which is definable relative to A.*

*Proof.* We begin with two simplifying observations: First, note that by replacing $A$ by $\Sigma_1$ satisfaction for $\langle L[A], A\rangle$ in Theorem 8.16 we can in fact assume that for $i+1 < \alpha$, $P(< i) \Vdash$ the forcing relation for $P(i)$ is $\Sigma_1^A$ for $\Sigma_1^A$ sentences; it follows that the same holds with $\Sigma_1^A$ replaced by $\Sigma_n^A$ for arbitrary $n > 0$. Second, in the definition of stratified we may require that $\kappa_1 \leq \kappa_2$ infinite, regular and $p(\gamma) = \emptyset$ for all $\gamma < \kappa_2 \Longrightarrow F_n(x, \kappa_1, p) = F_n(x, \kappa_2, p)$. For, we may achieve this property by redefining $F_n$ to be $F_n^*(x, \kappa, p) = F_n(x, \kappa(p), p)$ where $\kappa(p) =$ least $\gamma$ such that $p(\gamma) \neq \emptyset$ if $p(\delta) = \emptyset$ for all $\delta < \kappa$, $\kappa(p) = \kappa$ otherwise.

The theorem is proved by induction on $\alpha$, maintaining the coherence property that the isomorphism of $P(< \alpha)$ with a stratified forcing $P^*(< \alpha)$ extend the isomorphisms of $P(< \beta)$ with $P^*(< \beta)$ for $\beta < \alpha$ (viewing $P(< \beta)$ as contained in $P(< \alpha)$ in the natural way). Also, $A$ will serve as a strong witness to the stratification of each $P^*(< \alpha)$ and the forcing relation for $P^*(< \alpha)$ restricted to $\Sigma_n^A$ sentences will be $\Sigma_n^A$ for each $n$ (uniformly in $\alpha$, for each $n$).

The result is trivial for $\alpha = 0$ or 1. Suppose that $\alpha = \beta + 1$ where $\beta > 0$. By induction $P(< \beta)$ is isomorphic to a stratified forcing $P^*(< \beta)$; let $\leq_\kappa^\beta$, $F_n^\beta(x, \kappa, p^\beta)$ come from the stratification of $P^*(< \beta)$ and let $\Vdash_\beta$ denote the forcing relation for $P^*(< \beta)$, $P^*(\beta)$ the $P^*(< \beta)$-name for $P(\beta)$. Then $P^*(< \beta) \Vdash_\beta P^*(\beta)$ is stratified and we let $\leq_\kappa$, $F_n(x, \kappa, q)$ denote (names for) the relations, functions arising from the stratification of $P^*(\beta)$.

Roughly speaking, we take $P^*(< \alpha)$ to consist of functions $f$ on an initial segment of Card such that for some $p^\beta$ in $P^*(< \beta)$ and some $q$, $p^\beta \Vdash_\beta q \in P^*(\beta)$ and for all $\kappa \in \text{Dom}(f)$, $f(\kappa) = (p^\beta(\kappa), q(\kappa))$ where $q(\kappa)$ is the canonical name obtained by applying the function named by $q$ to $\kappa$. However, we modify this in two ways: First, if $p^\beta(\kappa) = \emptyset$ and $p^\beta \Vdash_\beta q(\kappa) = \emptyset$ or undefined, then take $f(\kappa) = \emptyset$ instead of ($\emptyset$, name for $\emptyset$). Second, require that $\text{Dom}(f)$ contain $\text{Dom}(p^\beta)$ and $\text{rank}(q) < \cup\text{Dom}(f)$, so that $P^*(< \beta) \Vdash_\beta \text{Dom}(q) \subseteq \text{Dom}(f)$. It is straightforward to verify that $P(< \alpha)$ is isomorphic to $P^*(< \alpha)$ (ordered as $P^*(< \beta) * P^*(\beta)$) and that (a) and all but the last statement of (2), in the definition of stratified hold for $P^*(< \alpha)$.

Next we demonstrate that the last statement of (2), $\kappa$-Density Reduction, holds for $P^*(< \alpha)$. Suppose $D \subseteq P^*(< \alpha)$ is dense and $\langle L[A], A\rangle$-definable and $f \in P^*(< \alpha)$. View $f$ as a pair $(p^\beta, q) \in P^*(< \beta) * P^*(\beta)$. Consider $D^{G^*(<\beta)} = \{q_0 \in P^*(\beta) \mid (p_0^\beta, q_0) \text{ meets } D \text{ for some } p_0^\beta \in G^*(< \beta)\}$ where $G^*(< \beta)$ denotes the $P^*(< \beta)$-generic. Then $D^{G^*(<\beta)} \subseteq P^*(\beta)$ is forced by $P^*(< \beta)$ to be dense. By $\kappa$-Density Reduction for $P^*(\beta)$, $P^*(< \beta)$ also forces that $\{q_0 \in P^*(\beta) \mid q_0\ \kappa^+\text{-reduces } D^{G^*(<\beta)}\}$

is $\leq_\kappa$-dense on $P^*(\beta)$ (i.e., every condition in $P^*(\beta)$ can be $\leq_\kappa$-extended into this class). Thus $D^\beta = \{p_0^\beta \mid$ For some $q_0, d_0, p_0^\beta \Vdash_\beta$ Either $q \notin P^*(\beta)$ or $q_0 \leq_\kappa q, q_0\ \kappa^+$-reduces $D^{G^*(<\beta)}$ to $d_0 \subseteq D^{G^*(<\beta)}$, $\text{card}(d_0) \leq \kappa\}$ is dense on $P^*(<\beta)$. By $\kappa$-Density Reduction for $P^*(<\beta)$, let $p_0^\beta \leq_\kappa^\beta p^\beta\ \kappa^+$-reduce $D^\beta$.

Now we can form terms $q_0, d_0$ such that $p_0^\beta \Vdash_\beta q_0 \leq_\kappa q, q_0\ \kappa^+$-reduces $D^{G^*(<\beta)}$ to $d_0 \subseteq D^{G^*(<\beta)}$, $\text{card}(d_0) \leq \kappa$. For each $i < \kappa$ it is dense below $p_0^\beta$ to force some $p_{0,i}^\beta \in G^*(<\beta)$, $(p_{0,i}^\beta, d_0(i)) \in D$, where $d_0(i) = i^{\text{th}}$ element of $d_0$. Finally, by $\kappa$-Density Reduction and $\kappa$-Definable Closure for $P^*(<\beta)$, we can assume that $p_0^\beta$ $\kappa^+$-reduces all of these dense sets and hence $(p_0^\beta, q_0)\ \kappa^+$-reduces $D$.

To complete the successor case we need to define the operators $F_n^\alpha(x, \kappa, (p^\beta, q))$ and verify (c) in the definition of stratified. We set $F_n^\alpha(x, \kappa, (p^\beta, q)) =$ "least" $(p_0^\beta, q)$ such that $p_0^\beta \leq_\kappa^\beta F_n^\beta(x, \kappa, p^\beta)$ and $p_0^\beta \Vdash_\beta F_n(x, \kappa, q) = q_0$. We must explain the term "least": Note that this property of $(p_0^\beta, q_0)$ is $\Pi_n^A$ (in the parameters $x, \kappa, p^\beta, q$); we say that $(p_0^\beta, q_0)$ is "least" with this property if for some $\kappa \in \text{Dom}(p_0^\beta)$, $(p_0^\beta \upharpoonright \kappa, q_0)$ is least with this property in the canonical wellordering of $\langle L[A], A\rangle$ and all $(p_1^\beta, q_1)$ less than $(p_0^\beta \upharpoonright \kappa, q_0)$ in this ordering are witnessed to fail to have this property by stage $\lambda = \cup \text{Dom}(p_0^\beta)$, $\lambda$ least such stage. Thus $F_n^\alpha$ is $\Pi_n^A$. Also note that $F_n^\alpha(x, \kappa, (p^\beta, q))$ is defined as it is a dense property of $p_0^\beta$ to force a value for $F_n(x, \kappa, q)$ and hence by $\kappa$-Density Reduction for $P^*(<\beta)$ there is $p_0^\beta$ reducing this dense property to a set, with $p_0^\beta \leq_\kappa F_n^\beta(x, \kappa, p^\beta)$. Then we can form a name $q_0$ so that $p_0^\beta \Vdash_\beta F_n(x, \kappa, q) = q_0$.

The $\kappa$-Definable Closure of $P^*(<\alpha)$ follows from the $\kappa$-Definable Closure of $P^*(<\beta)$ and the $\kappa$-Definable Closure of $P^*(\beta)$. Also $\leq_\kappa^\alpha$ is $\Delta_1^A$, uniformly in $\kappa$, using the facts that $\leq_\kappa^\beta$ is uniformly $\Delta_1^A$, $\leq_\kappa$ is uniformly $\Delta_1^{A, G^*(<\beta)}$ and the fact that $\Vdash_\beta$ is $\Sigma_1^A$ for $\Sigma_1^A$ sentences.

Now we turn to the case where $\alpha$ is a limit ordinal. We take $P^*(<\alpha)$ to consist of all functions $f$ on an initial segment of Card such that for some $\langle f_\beta \mid \beta < \alpha\rangle$ in the short diagonal support limit of $\langle P^*(<\beta) \mid \beta < \alpha\rangle$, $f(\kappa) = \langle f_\beta(\kappa) \mid \beta < \alpha\rangle$ for all $\kappa$ in $\text{Dom}(f)$; we also require $\text{Dom}(f) \supseteq \text{Dom}(f_\beta)$ for each $\beta < \alpha$ and modify $f(\kappa)$ so as to be $\emptyset$ if $f_\beta(\kappa) = \emptyset$ for all $\beta < \alpha$. The $f$'s are ordered by ordering the corresponding $\langle f_\beta \mid \beta < \alpha\rangle$'s.

We must show that $\{f \in P^*(<\alpha) \mid f \upharpoonright \kappa^+ \in H_{\kappa^+}\}$ is dense for each $\kappa$. To perform an induction on $\kappa$ we prove a bit more: If $\gamma < \kappa$ belong to Card, $\gamma$ regular then $\{f \in P^*(<\alpha) \mid f \upharpoonright (\gamma, \kappa] \in H_{\kappa^+}\}$ is $\leq_\alpha$-dense (i.e., any $f$ can be extended into this set without altering $f \upharpoonright \gamma^+$). Now we assume that this holds for $P^*(<\beta)$, $\beta < \alpha$ and prove it for $P^*(<\alpha)$ by induction on $\kappa$. By virtue of short diagonal supports we may assume $\text{cof}(\alpha) < \kappa$ as otherwise our given $f$ has the property that for some $\beta_0 < \alpha$, $f_\beta$ is the $\emptyset$-function below $\kappa^+$ for all $\beta_0 \leq \beta < \alpha$ (where $f$ arises from $\langle f_\beta \mid \beta < \alpha\rangle$) and so we can apply induction at $\beta_0$. By induction on $\kappa$ we can first extend $f$ to guarantee that $f \upharpoonright (\gamma, \text{cof}(\alpha)]$ belongs to $H_{\text{cof}(\alpha)^+}$. So we may assume that $\gamma \geq \text{cof}(\alpha)$.

Choose a cofinal $\mathrm{cof}(\alpha)$-sequence $\alpha_0 < \alpha_1 < \cdots$ below $\alpha$ and successfully extend $f = f_0$ to $f_1, f_2, \ldots$ in $\mathrm{cof}(\alpha)$ steps so that $f_{i+1} \restriction \alpha_i$ on $(\gamma, \kappa]$ belongs to $H_{\kappa^+}$ and $f_{i+1} \restriction \alpha_j \geq_\gamma F_1^{\alpha_j}(x, \gamma, f_i \restriction \alpha_j)$ for all $j \leq i$, where $F_1^{\alpha_j}$ comes from Definable Closure for $P(< \alpha_j)$ and $x$ is a parameter encoding $f, \gamma, \kappa, \langle \alpha_i \mid i < \mathrm{cof}(\alpha) \rangle$. (We have slightly abused notation: $f_i \restriction \alpha_i$ should be the function $g_i(\delta) = f_i(\delta) \restriction \alpha_i$.) For limit $\lambda$, $f_\lambda$ is a condition and $f_{\mathrm{cof}(\alpha)}$ is as desired.

Next we define the $F_n^\alpha$'s, for the purpose of establishing Definable Closure for $P^*(< \alpha)$. If $\mathrm{cof}(\alpha) \leq \kappa$ we define $F_n^\alpha(x, \kappa, p)$ to be the "least" $q \leq_\kappa p$ such that $q \restriction \alpha_i \leq_\kappa F_n^{\alpha_i}(x, \kappa, p \restriction \alpha_i)$ for each $\alpha_i$ in a fixed $\mathrm{cof}(\alpha)$-sequence cofinal in $\alpha$, where "least" is defined as in the definition of $F_n^\alpha$ for successor $\alpha$. If $\kappa < \mathrm{cof}(\alpha) < \alpha$ and $\mathrm{cof}(\alpha)$ is not the successor of a regular cardinal then we obtain $F_n^\alpha(x, \kappa, p)$ by first choosing $q_0 \leq_{\mathrm{cof}(\alpha)} p$ so that $q_0 \restriction \alpha_i \leq_{\mathrm{cof}(\alpha)} F_n^{\alpha_i}(x, \mathrm{cof}(\alpha), p \restriction \alpha_i)$ for each $\alpha_i$ and then $q_1 \leq_\kappa q_0$ so that $q_1 \restriction \mathrm{cof}(\alpha) \leq_\kappa F_n^{\mathrm{cof}(\alpha)}(x, \kappa, q_0 \restriction \mathrm{cof}(\alpha))$. If $\kappa \leq \lambda < \lambda^+ = \mathrm{cof}(\alpha) < \alpha$ with $\lambda$ regular then we choose $q_0, q_1$ as above and then $q_2 \leq_\lambda q_1$ so that $q_2 \restriction \beta \leq_\lambda F_n^\beta(x, \lambda, q_1 \restriction \beta)$ for those $\beta$ such that $\mathrm{cof}(\alpha) \leq \beta$, $q_1 \restriction \beta \not\Vdash_\beta q_1(\beta)$ is the $\emptyset$-function below $\mathrm{cof}(\alpha)$. Finally if $\kappa < \alpha$, $\alpha$ regular then choose $q_0$ as above (setting $\alpha_i = i$) and then $q_1 \leq_\kappa q_0$ so that $q_1 \restriction \beta \leq_\kappa F_n^\beta(x, \kappa, q_0 \restriction \beta)$ where $\beta < \alpha$ is least so that $\beta \leq \beta' < \alpha \Longrightarrow q_0 \restriction \beta' \Vdash_{\beta'} q_0(\beta')$ is the $\emptyset$-function below $\alpha^+$. Our construction guarantees that if $q = F_n^\alpha(x, \kappa, p)$ and $\beta < \alpha$ then for some $\kappa' \geq \kappa$ and $q'$ we have $q \restriction \beta \leq_\kappa q' \leq_{\kappa'} F_n^\beta(x, \kappa', p')$ for some $p' \leq_\kappa p \restriction \beta$. Using the second observation made at the start of this proof it is straightforward to verify $\kappa$-Definable Closure for $P^*(< \alpha)$, making use of the previous sentence when $\alpha$ is regular.

It remains to establish the second part of $\kappa$-Density Reduction for $P^*(< \alpha)$. First suppose that $\alpha < \kappa^+$ and choose a $\mathrm{cof}(\alpha)$-sequence $\alpha_0 < \alpha_1 < \cdots$ cofinal in $\alpha$. Given $p \in P^*(< \alpha)$ and an $\langle L[A], A \rangle$-definable open dense $D$, use the $F_n^{\alpha_i}$ functions to successively extend $p \restriction \alpha_i$, producing $q \leq_\kappa p$ such that for each $i$, $q \restriction \alpha_i$ $\kappa^+$-reduces $D^{\alpha_i} = \{r \restriction \alpha_i \mid r \in D\}$. Then successively $\leq_\kappa$-extend $q = q_0$ to $q_1, q_2, \ldots$ so that for each $\gamma$ there is $s_{\gamma+1}$ defined on $\mathrm{Card} \cap \kappa^+$ so that $s_{\gamma+1} \cup q_{\gamma+1}$ belongs to $D$, yet is incompatible with each $s_{\gamma'+1} \cup q_{\gamma'+1}$ for $\gamma' < \gamma$. For each $i$ there is a stage $\gamma_i < \kappa^+$ such that for $\gamma \geq \gamma_i$, $(X_{\gamma+1} \cup q_{\gamma+1}) \restriction \alpha_i$ is compatible with some $(X_{\gamma'+1} \cup q_{\gamma'+1}) \restriction \alpha_i$ for some $\gamma' < \gamma_i$, since $q \restriction \alpha_i$ $\kappa^+$-reduces $D^{\alpha_i}$. Let $\gamma = \bigcup\{\gamma_i \mid i < \mathrm{cof}(\alpha)\} < \kappa^+$. Then $q_{\gamma+1}$ is undefined so for some $\gamma' < \kappa^+$, $q_{\gamma'}$ $\kappa^+$-reduces $D$.

Now suppose that $\alpha \geq \kappa^+$. We may assume that $\alpha = \kappa^+$. as short diagonal supports requires that conditions in $P^*(< \alpha)$ are trivial below $\kappa^+$ on all but fewer than $\kappa$ coordinates, all less than $\kappa^+$. We can assume that conditions in $D$ when restricted to $\mathrm{Card} \cap \kappa^+$ belong to $H_{\kappa^+}$ and therefore can choose $q \leq_\kappa p$ and $\alpha_0 < \kappa^+$ of cofinality $\kappa$ such that the conditions in $D$ which are trivial below $\kappa^+$ on coordinates $\geq \alpha_0$ form a set predense below $q$. If we extend $q$ to $q_0 \leq_\kappa q$ which $\kappa^+$-reduces $D^{\alpha_0}$ then $q_0$ in fact $\kappa^+$-reduces $D$. □

Long Easton forcing at Successors is not stratified (nor is Jensen coding). For this reason we introduce a modified notion, called $\Delta$-*stratification*.

**Definition.** $P$ is $\Delta$-*stratified* if it obeys the definition of stratified, but where (2), (3) are restricted to infinite successor cardinals $\kappa$ and where the following is added: If $0 < n < \omega$, $\kappa$ inaccessible and $p_0 \geq p_1 \geq \cdots$ is a $\Pi_n\langle L[A], A\rangle$ (in parameters from $\lambda \cup \{x\}$) sequence of length $\lambda \leq \kappa$ such that for each $i < \lambda$, $p_{i+1} \leq_{\kappa_i} F_n(x, \kappa_i, p'_i)$ for some $p'_i \leq_{\kappa_i} p_i$ and some regular $\kappa_i \geq \aleph_i$, then there is a condition $p$ such that $p \leq_{\aleph_i} p_i$ for each $i < \lambda$. $A$ is a *strong witness* to the $\Delta$-stratification of $P$ if it obeys the definition of strong witness to stratification, but with $\kappa$ restricted to infinite successor cardinals. A $\Delta$-*stratified iteration* is defined just like a stratified iteration, but with stratified replaced by $\Delta$-stratified everywhere. Such an iteration $\langle P(< i) \mid i < \alpha\rangle$ has *long diagonal supports* if for $j < \alpha$, $p \in P(< j)$ and infinite successor cardinals $\kappa$, $\{i < j \mid p \restriction i \nVdash \forall\gamma \leq \kappa,\ p(i)(\gamma) = \emptyset\}$ is a subset of $\kappa^+$ of cardinality $< \kappa$ and for inaccessible $\kappa \leq j$, $\{\bar\kappa < \kappa \mid \text{For some } j' \in [\bar\kappa, j), p \restriction j' \nVdash p(j')(\bar\kappa) = \emptyset\}$ is nonstationary in $\kappa$ (and this is the only support restriction.)

**Theorem 8.18.** *Suppose that $P$ is $\Delta$-stratified with witness $A$. Then $P$ preserves* ZFC *(relative to $A$) and cofinalities. If $V \models$* GCH *then $P$ preserves the* GCH.

*Proof.* As in the proof of Theorem 8.14, but using $\Delta$-stratification at $\kappa$ (and stratification at $\bar\kappa^+$, $\bar\kappa < \kappa$) to show that "cof $> \kappa$" is preserved for inaccessible $\kappa$. □

**Theorem 8.19** ($\Delta$-Stratified Iterations). *Suppose that $\langle P(< i) \mid i \leq \alpha\rangle$ is a $\Delta$-stratified iteration with long diagonal supports as witnessed by the class $A$ and the* GCH *holds. Then $P(< \alpha)$ is isomorphic to a $\Delta$-stratified forcing, via an isomorphism which is definable relative to $A$.*

*Proof.* We follow the proof of Theorem 8.17. Note that Theorem 8.16 still applies in the present context, as $P$ is stratified at cofinally many regular cardinals $\kappa$. Again proceed by induction on $\alpha$. For successor $\alpha$ we have (2), (3) at infinite successor cardinals by our earlier proof. For $\Delta$-stratification, given an inaccessible $\kappa$, $0 < n < \omega$ and the $\lambda$-sequence $p_0 \geq p_1 \geq \cdots$ use $\Delta$-stratification for $P^*(< \beta)$ and the fact $P^*(< \beta) \Vdash_\beta P^*(\beta)$ is $\Delta$-stratified (where $\alpha = \beta + 1$) to obtain $\bar p$ such that $\bar p \restriction \beta \leq_{\aleph_i} p_i \restriction \beta$ for each $i < \lambda$, $\bar p \restriction \beta \Vdash_\beta$ There is $\bar p(\beta) \leq_{\aleph_i} p_i(\beta)$ for each $i$; then $\leq_{\kappa^+}$-extend $\bar p \restriction \beta$ to $p \restriction \beta$ so that for some name $p(\beta)$, $p \restriction \beta \Vdash_\beta p(\beta) \leq_{\aleph_i} p_i(\beta)$ for each $i < \lambda$, using $\kappa^+$-Density Reduction. Thus $p \leq_{\aleph_i} p_i$ for each $i < \lambda$, as desired.

Now suppose that $\alpha$ is a limit ordinal, Define $P^*(< \alpha)$ as before. First we show that $\{f \in P^*(< \alpha) \mid f \restriction (\gamma, \kappa] \in H_{\kappa^+}\}$ is $\leq_\gamma$-dense for each $\gamma < \kappa$ in Card, $\gamma$ an infinite successor cardinal. This is proved by induction on $\kappa$, assuming the result for smaller $\alpha$. Using long diagonal supports we may assume that either $\alpha = \kappa$ is inaccessible or that $\text{cof}(\alpha) < \kappa$. If $\text{cof}(\alpha)$ is a successor cardinal or $\text{cof}(\alpha)^+ < \kappa$ then the previous argument can be applied, using $\text{cof}(\alpha)$-Definable Closure or $\text{cof}(\alpha)^+$-Definable Closure applied to $P^*(< \alpha_i)$, where $\alpha_0 < \alpha_1 < \cdots$ is a $\text{cof}(\alpha)$-sequence cofinal in $\alpha$. So we may assume that either $\alpha = \kappa$ is inaccessible or $\text{cof}(\alpha)^+ = \kappa$ where $\text{cof}(\alpha)$ is inaccessible. In the latter case successively $\leq_\gamma$-extend our given $f = f_0$ to $f_1, f_2, \ldots$ in $\text{cof}(\alpha)$ steps so that $(f_{i+1} \restriction \alpha_i)(\kappa) \in H_{\kappa^+}$ and $f_{i+1} \restriction \alpha_j \leq_{\aleph_j} F_1^{\alpha_j}(x, \gamma \cup \aleph_j, f_i \restriction \alpha_j)$ for all $j \leq i$, $x$ a parameter encoding $f, \gamma, \kappa, \langle \alpha_i \mid i < \text{cof}(\alpha)\rangle$. By induction we may $\leq_\gamma$-extend $f_{\text{cof}(\alpha)}$ to $g$ so that $g \restriction (\gamma, \text{cof}(\alpha)] \in H_\kappa$; then $g$ is as desired. If $\alpha = \kappa$

is inaccessible use Definable Closure to successively $\leq_\gamma$-extend $f = f_0$ to $f_1, f_2, \ldots$ in $\kappa$ steps so that $f_{i+1} \restriction \kappa_i = f_i \restriction \kappa_i$, $f_{i+1} \restriction (\kappa_i, \kappa_{i+1}) \in H_{\kappa_{i+1}^+}$ and $f_{i+1}(\kappa_i) = \emptyset$ for all $i$ where $\kappa_0 < \kappa_1 < \cdots$ is continuous and cofinal in $\kappa$, using the fact that $\{\gamma < \kappa \mid f(\gamma) \neq \emptyset\}$ is nonstationary in $\kappa$. Then $f_\kappa$ is as desired.

We define $F_n^\alpha(x, \kappa, p)$ as in the stratified case if $\operatorname{cof}(\alpha) \leq \kappa$ or $\alpha$ is a successor cardinal or $\operatorname{cof}(\alpha)$ is neither inaccessible nor the successor of an inaccessible. If $\operatorname{cof}(\alpha) > \kappa$ is inaccessible then let $\alpha_0 < \alpha_1 < \cdots$ be a $\operatorname{cof}(\alpha)$-sequence cofinal in $\alpha$ such that $\alpha_j \geq \aleph_{j+1} \cup \kappa$ for each $j < \operatorname{cof}(\alpha)$ and choose $F_n^\alpha(x, \kappa, p)$ to be a lower bound to $\langle p_j \mid j < \operatorname{cof}(\alpha)\rangle$ where $p_0 = p$, $p_{j+1}$ is "least" so that $p_{j+1} \restriction \alpha_{j'} \leq_{\kappa \cup \aleph_j} F_n^{\alpha_{j'}}(x, \kappa \cup \aleph_j, p_j \restriction \alpha_{j'})$ for all $j' \leq j$. If $\kappa \leq \lambda < \lambda^+ = \operatorname{cof}(\alpha) < \alpha$, $\lambda$ inaccessible then similarly modify the previous choice of $q_2$, enumerating the relevant $\beta$'s in $\lambda$ steps.

$\kappa$-Density Reduction for infinite successor cardinals $\kappa$ follows just as in the stratified case. If $\kappa$ is inaccessible then our construction of $P^*(<\alpha)$ implies that if $p_{i+1} \leq_{\aleph_i} F_n^\alpha(x, \aleph_i, p_i)$ for $i < \kappa$ then for cofinally many $\alpha' < \alpha$, $p_{i+1} \restriction \alpha' \leq_{\aleph_{i'}} F_n^{\alpha'}(x, \aleph_{i'}, p_i \restriction \alpha')$ for each $i$ (and some $i' \geq i$ depending on $\alpha', i$). Also if $\beta < \alpha$ and $p \leq_\kappa q$ in $P^*(<\beta+1)$, $p$ at $\beta = q$ at $\beta$ then $F_n^{\beta+1}(x, \kappa, p)$ at $\beta$ equals $F_n^{\beta+1}(x, \kappa, q)$ at $\beta$. So, given $\langle p_i \mid i < \lambda\rangle$ as in the hypothesis of $\Delta$-stratification at $\kappa$ for $P^*(<\alpha)$, we can obtain the desired lower bound $p$ by choosing $q \restriction \beta + 1$ to be a lower bound for $\langle p \restriction \beta \cup \{\langle \beta, p_i \text{ at } \beta\rangle\} \mid i < \lambda\rangle$ and taking $p(\beta) = q(\beta)$. □

**Examples.** (1) Easton forcing and Easton Forcing at Successors are equivalent to stratified forcings. We may take $F_n(x, \kappa, p)$ to simply be $p$.

(2) Long Easton at Successors, Thin Easton at Successors and Coherent Easton at Successors are each equivalent to $\Delta$-stratified forcings. Again we may take $F_n(x, \kappa, p)$ to be simply $p$.

(3) Reverse Easton Forcing $P$ is equivalent to a stratified forcing, using the decomposition (for infinite regular $\kappa$) of $P$ as $P \simeq P(\leq \kappa) * P(> \kappa)$, where $P(\leq \kappa)$ is $\kappa^+$-cc with a dense subset of cardinality $\leq \kappa^+$ and $P(\leq \kappa) \Vdash P(> \kappa)$ in $\kappa^+$-closed. The latter permits us to again take $F_n(x, \kappa, p) = p$.

(4) Amenable forcing is trivially equivalent to both a stratified and a $\Delta$-stratified forcing. To each condition $p$ associate $p^*$ (in the equivalent stratified or $\Delta$-stratified forcing) where $p^*(\kappa) = p \restriction \kappa$ for $\kappa$ infinite regular, $\kappa \leq \operatorname{Dom}(p)^+$ and $p^*(\kappa) = \emptyset$ otherwise.

(5) The forcing of Section 4.2, used to prove Jensen's Coding Theorem without $0^\#$, is equivalent to a $\Delta$-stratified forcing. To see this one must reindex: To each condition $p$ associate $p^*$ (in the equivalent $\Delta$-stratified version) where $p^*(\kappa^+) = p(\kappa)$ for all infinite cardinals $\kappa$ in $\operatorname{Dom}(p)$ and $p^*(\lambda) = \emptyset$ otherwise. We take $F_n(x, \kappa, p)$ to be the "least" $q \leq_\kappa p$ such that for $\kappa < \lambda \in \operatorname{Dom}(p)$, $\lambda \in \Sigma_n\langle L[A], A\rangle \operatorname{Hull}(\lambda \cup \{x, p\})$, $q \restriction [\lambda, \infty)$ meets all predense $D$ on $P \restriction [\lambda, \infty) = \{r \restriction [\lambda, \infty) \mid r \in P\}$ which belongs to $\Sigma_n\langle L[A], A\rangle \operatorname{Hull}(\lambda \cup \{x, p\})$. Then $\Delta$-stratification is verified in the way that the distributivity of $P$ was established, using the $\Sigma_f$'s.

(6) The forcing of $P$ of Section 4.3, used to prove Jensen's Coding Theorem in the general case, is equivalent to a forcing that is both stratified and $\Delta$-stratified. This equivalent forcing is defined as in (5) above and is $\Delta$-stratified. To obtain stratification, use the fact that $P$ factors as $P \upharpoonright [\kappa, \infty) * P^{G\upharpoonright[\kappa,\infty)}$ for all infinite cardinals $\kappa$, where $P \upharpoonright [\kappa, \infty) \Vdash P^{G\upharpoonright[\kappa,\infty)}$ is $\kappa^+$-cc. One change is required to verify $\kappa$-Density Reduction and $\kappa$-Definable Closure for inaccessible $\kappa$: Allow $p \upharpoonright \kappa$ to code only an initial segment of $p_\kappa$ (and to belong to the coding structure $\mathcal{A}^{p_\kappa}$); this gives us the freedom to extend at $\kappa$ (at $\kappa^+$ in the equivalent stratified version) without making extensions below $\kappa$.

Natural iterations of the above stratified ($\Delta$-stratified) forcings will form stratified ($\Delta$-stratified) iterations, provided short diagonal supports (long diagonal supports) are used.

## An Application

We show that in a sense our work in Section 5.3 cannot be improved. First recall some definitions.

**Definition.** An $L$-forcing $P$ (i.e., a tame forcing $P$ defined over a ground model $\langle L, A\rangle$) is $\lambda_0$, $\lambda$-*periodic* if there is a $P$-generic $G$ such that $I_{\lambda_0\lambda} = \{i_{\lambda_0+\lambda\alpha} \mid \alpha \in \mathrm{ORD}\}$ is a class of indiscernibles for $\langle L[G], A, G\rangle$, where $i_0 < i_1 < \cdots$ is the increasing enumeration of the Silver indiscernibles. $P$ is *almost* $\lambda_0$, $\lambda$-*periodic* if $P$ is $\lambda_0$, $\lambda$-periodic in a set-generic extension of $V$.

We showed in Section 5.3 (assuming that $\infty$ is $\omega + \omega$-Erdös) that any relevant $L$-forcing is almost $\lambda_0$, $\lambda$-periodic for some $\lambda_0$ and some countable $\lambda$.

**Theorem 8.20.** *Suppose $\alpha_0$, $\alpha$ are ordinals and $\alpha$ is countable in $L[0^\#]$. Then there is a relevant $L$-forcing $P$ which is* not *almost $\lambda_0$, $\lambda$-periodic for $\lambda_0 < \alpha_0$ or for $\lambda < \alpha$.*

*Proof.* By the solution to the $\Pi^1_2$-Singleton Problem there is a relevant $L$-definable forcing $P$ with a unique generic, which can be considered to be a real. This unique generic real $R$ can be made to satisfy $I^R = \text{Even } I$, where $I^R =$ Silver indiscernibles for $L[R]$ and Even $I = \{i_{2\gamma} \mid \gamma \in \mathrm{ORD}\}$.

Now suppose that $\alpha$ is an $L$-countable ordinal and define the (stratified) iteration $\langle P(< i) \mid i \leq \alpha\rangle$ as follows: $P(< i) \Vdash P(i)$ is the forcing for adding a unique generic real $R$ over the model $L[R(< i)]$, where the $P(< i)$-generic can be considered to be a real $R(< i)$, such that $I^{R,R(<i)} = \text{Even } I^{R(<i)}$. We let $R(< i+1)$ be the join of $R(< i)$ and the $P(i)$-generic real $R(i)$; for limit $\lambda \leq \alpha$, $R(< \lambda) = \text{Join } \langle R(< i) \mid i < \lambda\rangle$ using the $L$-least counting of $\lambda$. As $\alpha$ is countable in $L$ we may take full inverse limits at limit stages.

The forcing $P(< \alpha)$ adds a unique real $R(< \alpha)$ such that $I^{R(<\alpha)} = \{i_{2^\alpha\gamma} \mid \gamma \in \mathrm{ORD}\}$ where $2^\alpha$ is ordinal exponentiation, as the proof of relevance for Jensen coding applies to $P(< \alpha)$ as well.

If $\alpha$ is not $L$-countable define $P^*(<\alpha)$ by first Lévy collapsing $\alpha$ to $\omega$ and then over this extension, proceeding as above. As $\alpha$ is countable in $L[0^\#]$ there exists a $P^*(<\alpha)$-generic real in $L[0^\#]$ and any such real $R$ obeys: $I^R = \{i_{2^\alpha\gamma} \mid \gamma \in \text{ORD}\} - (\alpha+1)$.

Finally to prove the theorem, let $P = P^*(<\alpha) * Q$ where $Q$ adds a Cohen subset to $\alpha_0^+$ over $L[R]$, $R =$ the $P^*(<\alpha)$-generic real. Then $I_{\lambda_0\lambda}$ is not a class of indiscernibles for any $P$-generic for $\lambda_0 < \alpha_0$ or for $\lambda < \alpha$. □

## 8.3 Minimal Coding

In this section we establish the following strengthening of Theorem 4.1.

**Theorem 8.21.** *Suppose that* $A \subseteq \text{ORD}$ *and* $\langle L[A], A\rangle$ *is a model of* ZFC + GCH. *Then there is an* $\langle L[A], A\rangle$*-definable class forcing* $P$ *such that if* $G \subseteq P$ *is* $P$*-generic over* $\langle L[A], A\rangle$*:*

(a) $\langle L[A, G], A, G\rangle$ *is a model of* ZFC + GCH.

(b) $L[A, G] = L[R]$ *for some real* $R$ *and* $A, G$ *are definable over* $L[R]$ *from the parameter* $R$.

(c) $L[A]$ *and* $L[R]$ *have the same cofinalities.*

(d) $R$ *is minimal over* $L[A]$*: If* $x \in L[R]$ *then either* $x \in L[A]$ *or* $R \in L[A, x]$.

Thus a universe obeying GCH can be "coded minimally" by a real. Note that in clause (d) of the theorem, $x$ is any set constructible from $R$, not necessarily a real.

We first consider the following easier result, whose proof reveals the main ideas required for establishing Theorem 8.21.

**Theorem 8.22.** *There is a nonconstructible real* $R$ *such that:*

(a) $R$ *is* $L$*-minimal: If* $x \in L[R]$ *then either* $x \in L$ *or* $R \in L[x]$.

(b) $R$ *is not generic over* $L$ *for any forcing of* $L$*-cardinality* $\omega_1^L$.

Using the proof of Theorem 8.22, we will show how to minimally code any model $L[A]$ of ZFC + GCH by a real, when $A \subseteq \omega_2$ (of $L[A]$); Theorem 8.21 generalizes this to arbitrary $A \subseteq \text{ORD}$.

*Proof of Theorem* 8.22. Our goal is to produce a real $R$ that is $L$-minimal and in addition codes a predicate $B \subseteq \omega_2$ which cannot be coded by a forcing of cardinality $\omega_1$ (where $\omega_1, \omega_2$ denote $\omega_1^L, \omega_2^L$). The basic approach to $L$-minimality comes from Sacks [71], where an $L$-minimal real is obtained by forcing with perfect trees of height $\omega$. It is not difficult to generalize this to $\omega_1$, obtaining an $L$-minimal $x \subseteq \omega_1$, which can then be minimally coded by a real $R$. However the resulting $R$ will *not* be $L$-minimal as one will have $0 <_L x <_L R$.

Instead one must "mix" the forcing for producing $x$ with that for minimally coding $x$ by $R$, so as to obtain $0 <_L x =_L R$. Our conditions for doing this are of the form $(a, t, T)$, where $a$ is a finite string (of 0's and 1's), $t$ is a perfect subtree of $2^{<\omega}$ = set of all finite strings of 0's and 1's and $T$ is a "generalized ($\omega_1$-)tree". Roughly speaking, a generalized tree is a "perfect" collection of perfect subtrees of $2^{<\omega}$, all taken from a special collection of perfect subtrees of $2^{<\omega}$, coded the "canonical trees."

To facilitate an analogy between our treatment of perfect subtrees of $2^{<\omega}$ and that of generalized trees, we adopt a nonstandard definition of the former.

**Definition.** An *$\omega$-tree* is a function $t : 2^{<\omega} \longrightarrow 3$ such that for any $a \in 2^{<\omega}$ there is $b \in 2^{<\omega}$ such that $b$ extends $a$, $t(b) = 2$. A string *$a$ lies on $t$* if for $k < |a| = \text{length}(a)$, either $t(a \upharpoonright k) = 2$ or $t(a \upharpoonright k) = a(k)$. A real $R : \omega \longrightarrow 2$ *satisfies* $(a, t)$ if $a \subseteq R$ and for all $k$, $R \upharpoonright k$ lies on $t$. We let $[a, t]$ denote $\{R \mid R \text{ satisfies } (a, t)\}$ and $[t] = [\emptyset, t]$. We write $(a_0, t_0) \leq (a_1, t_1)$ ($(a_0, t_0)$ *extends* $(a_1, t_1)$) iff $[a_0, t_0] \subseteq [a_1, t_1]$ and $t_0 \leq t_1$ ($t_0$ *extends* $t_1$) iff $(\emptyset, t_0) \leq (\emptyset, t_1)$. Split$(t)$ denotes $\{a \mid a \text{ lies on } t, t(a) = 2\}$ and for $a \in \text{Split}(t)$, Split$(t)$-rank$(a)$ denotes the ordertype of $\{k < |a| \mid a \upharpoonright k \in \text{Split}(t)\}$. We set $\text{Split}(t, k) = \{a \in \text{Split}(t) \mid \text{Split}(t)\text{-rank}(a) = k\}$ and $\text{Split}^+(t, k) = \{a * j \mid a \in \text{Split}(t, k), j < 2\}$. We have $t_0 \leq_k t_1$ ($t_0$ *$k$-extends* $t_1$) if $t_0 \leq t_1$ and $\text{Split}(t_0, k) = \text{Split}(t_1, k)$. Suppose that $t$ is an $\omega$-tree and $t_a$, $a \in \text{Split}^+(t, k)$ are $\omega$-trees extending $t$. then $t(t_a \mid a \in \text{Split}^+(t, k))$ is defined to be the tree $t^* \leq_k t$ where: $t^*(b) = t_a(b)$ if $b$ extends some $a \in \text{Split}^+(t, k)$, $t^*(b) = t(b)$ otherwise. We say that $t^*$ *results from $t$, $t_a (a \in \text{Split}^+(t, k))$ via amalgamation at splitting level $k$.*

Now we provide the inductive definition of $R_\alpha$ = the canonical trees of rank $\alpha$, $\alpha < \omega_1$. If a real $R$ belongs to $[t]$ for some $t \in R_\alpha$ then $t$ is unique and is written $t^R_\alpha$; moreover $t^R_\beta$ is defined for all $\beta < \alpha$ as well.

**Case 1.** $\alpha = 0$
$R_0 = \{t\}$ where $t$ is the "full tree": $t(a) = 2$ for all $a \in 2^{<\omega}$.

**Case 2A.** $\alpha = \beta + 1$, $\beta$ even.
For each $t \in R_\beta$, $a$ lying on $t$ and $k \in \omega$ we select two $k$-extensions $t(a, k, 0)$, $t(a, k, 1)$ of $t$ upon which $a$ lies such that $(a, k, j) \neq (a^*, k^*, j^*) \Longrightarrow [t(a, k, j)] \cap [t(a^*, k^*, j^*)] = \emptyset$. Then $R_\alpha$ consists of all such $t(a, k, j)$.

**Case 2B.** $\alpha = \beta + 1$, $\beta$ odd.
For each $t \in R_{\leq\beta} = \bigcup\{R_\gamma \mid \gamma \leq \beta\}$, $k \in \omega$ and distinct $\langle t_a \mid a \in \text{Split}^+(t, k)\rangle = \vec{t}_k$ with each $t_a$ from $R_\beta$, we select a $k$-extension $t(\vec{t}_k)$ of $t(t_a \mid a \in \text{Split}^+(t, k))$ such that $(t, \vec{t}_k) \neq (t^*, \vec{t}^*_{k^*}) \Longrightarrow [t(\vec{t}_k)] \cap [t^*(\vec{t}^*_{k^*})] = \emptyset$. Then $R_\alpha$ consists of all such $t(\vec{t}_k)$.

**Case 3.** $\alpha$ limit.
Let $\langle \alpha_n \mid n \in \omega\rangle$ be the increasing $\omega$-sequence cofinal in $\alpha$ defined below. Suppose that $\langle t_n \mid n_0 \leq n\rangle$ is a sequence such that $t_n \in R_{\alpha_n+2}$ and $t_{n+1} \leq_n t_n$ for each $n \geq n_0$. Then we place $\text{Lim}\langle t_n \mid n_0 \leq n\rangle = t$ into $R_\alpha$, where $t$ is defined by: $t(a)$ = the eventual value of $t_n(a)$ if $a$ lies on each $t_n$, $n \geq n_0$; $t(a) = 0$ otherwise.

It remains to define $\langle \alpha_n \mid n \in \omega \rangle$. Let $\beta$ be the largest limit ordinal such that $\beta = \alpha$ or $\alpha$ is regular in $L_\beta$. If for some least $p \in L_\beta$ we have that $L_\beta = \Sigma_1 \operatorname{Hull}(\alpha \cup \{p\})$ in $L_\beta$, and in addition $S = \{\bar{\alpha} < \alpha \mid \bar{\alpha} = \alpha \cap \Sigma_1 \operatorname{Hull}(\bar{\alpha} \cup \{p\}) \text{ in } L_\beta\}$ has ordertype $\omega$, then let $\langle \alpha_n \mid n \in \omega \rangle$ be the increasing enumeration of $S$. Otherwise let $\langle \alpha_n \mid n \in \omega \rangle$ be the $L$-least increasing $\omega$-sequence cofinal in $\alpha$.

This completes the construction of the $R_\alpha$, $\alpha < \omega_1$. For $t$ a canonical tree (i.e., for $t \in \bigcup\{R_\alpha \mid \alpha < \omega_1\}$) we set $\alpha(t) =$ unique $\alpha$ such that $t \in R_\alpha$. Also let $R_{<\alpha} = \bigcup\{R_\beta \mid \beta < \alpha\}$ and (as in Case 2B) $R_{\leq\alpha} = R_{<\alpha+1}$.

If $R$ belongs to $[t]$ for some $t \in R_\alpha$, define $B^R \cap \alpha$ by: $\beta \in B^R \cap \alpha$ iff $\beta$ is even, $\beta < \alpha$ and $t^R_{\beta+1} = t^R_\beta(a, k, 1)$ for some $a, k$. We let Even denote the class of even ordinals.

An extension $t_0 \leq t_1$ of canonical trees is a *linear extension* if $R, S \in [t_0] \Longrightarrow t^R_\alpha = t^S_\alpha$ for $\alpha(t_1) \leq \alpha \leq \alpha(t_0)$. Extensions introduced in the construction of the $R_\alpha$'s with the exception of Case 2B are linear.

**Lemma 8.23** (Extendibility for Canonical Trees). *Suppose that $t$ is canonical, $k \in \omega$, $\alpha(t) < \alpha < \omega_1$ and $B \subseteq [\alpha(t), \alpha) \cap \text{Even}$. Then there is a linear extension $t^* \leq_k t$ such that $\alpha(t^*) = \alpha$, $R \in [t^*] \Longrightarrow B^R \cap [\alpha(t), \alpha) = B$.*

*Proof.* By induction on $\alpha$. The result is clear by induction when $\alpha$ is a successor ordinal, using Case 2A of the canonical trees construction. Suppose that $\alpha$ is a limit ordinal and let $\langle \alpha_n \mid n \in \omega \rangle$ be as defined in Case 3 of the canonical trees construction. Let $n_0$ be such that $\alpha(t) < \alpha_{n_0}$, $k \leq n_0$. By induction we may choose canonical $t \geq_k t_{n_0} \geq t_{n_0+1} \geq \cdots$ so that $t_n \in R_{\alpha_n+2}$, $t_{n+1} \leq_n t_n$, $R \in [t_{n_0}] \Longrightarrow B^R \cap [\alpha(t), \alpha_{n_0} + 2) = B \cap [\alpha(t), \alpha_{n_0}+2)$ and $R \in [t_{n+1}] \Longrightarrow B^R \cap [\alpha_n+2, \alpha_{n+1}+2) = B \cap [\alpha_n+2, \alpha_{n+1}+2)$. Then $t^* = \operatorname{Lim}\langle t_n \mid n_0 \leq n \rangle$ is the desired linear $k$-extension of $t$. □

## Generalized Trees

We develop the concept of generalized tree in analogy with our treatment of $\omega$-trees.

**Definition.** $R^{\text{Even}}$ denotes $\{t \mid t \text{ canonical}, \alpha(t) \text{ even}\}$. $X \subseteq R^{\text{Even}}$ is *unbounded* if for each canonical tree $t$ and each $k$, $t$ has a linear $k$-extension $t^* \in X$. $X \subseteq R^{\text{Even}}$ is *closed* if whenever $t_0 \geq t_1 \geq \cdots$ is a sequence, each $t_n \in X$ and $t = \operatorname{Lim}\langle t_n \mid n \in \omega \rangle$ is defined and belongs to $R^{\text{Even}}$ then $t$ belongs to $X$. $C \subseteq R^{\text{Even}}$ is *CUB* if $C$ is closed and unbounded.

**Definition.** A *generalized ($\omega_1$-)tree* is a function $T$ from $\{(t, a) \mid t \in R^{\text{Even}}, a \text{ lies on } t\}$ to 3 such that $\operatorname{Split}(T) = \{t \in R^{\text{Even}} \mid T(t, a) = 2 \text{ for all } a \text{ lying on } t\}$ is CUB.

**Definition.** For $t$ a canonical tree, the $\operatorname{Split}(T)$*-rank* of $t$ is defined as follows: If $\alpha(t) = 0$ then $\operatorname{Split}(T)\text{-rank}(t) = 0$. If $\alpha(t) = \beta + 1$, $\beta$ even and $t^*$ is the unique canonical tree such that $t \leq t^*$, $\alpha(t^*) = \beta$ then $\operatorname{Split}(T)\text{-rank}(t) = \operatorname{Split}(T)\text{-rank}(t^*) + 1$ if $t^* \in \operatorname{Split}(T)$, $\operatorname{Split}(T)\text{-rank}(t) = \operatorname{Split}(T)\text{-rank}(t^*)$ otherwise. If $\alpha(t) = \beta + 1$,

$\beta$ odd and $t = t^*(\langle t_a^* \mid a \in \text{Split}^+(t^*, k)\rangle)$ then Split$(T)$-rank$(t) = \max\{\text{Split}(T)\text{-rank}(t_a^*) \mid a \in \text{Split}^+(t^*, k)\}$. If $\alpha(t) = \alpha$ limit and $t = \text{Lim}\langle t_n \mid n_0 \leq n\rangle$ where $t_n \in R_{\alpha_n+2}$ then Split$(T)$-rank$(t) = \bigcup\{\text{Split}(T)\text{-rank}(t_n) \mid n_0 \leq n\}$. We set Split$(T, \alpha) = \{t \in \text{Split}(T) \mid \text{Split}(T)\text{-rank}(t) = \alpha\}$. For $\alpha < \omega_1$, $T_0$ *$\alpha$-extends* $T_1$ ($T_0 \leq_\alpha T_1$) iff $T_0 \leq T_1$ and Split$(T_0, \alpha) = $ Split$(T_1, \alpha)$.

**Remark.** An easy induction shows: If $t_0, t_1 \in \text{Split}(T, \alpha)$ are distinct then $[t_0] \cap [t_1] = \emptyset$.

**Definition.** A real $R$ *satisfies* $T$ if $\{\text{Split}(T)\text{-rank}(t_{2\alpha}^R) \mid \alpha < \omega_1\}$ is unbounded in $\omega_1$ and for all even $\alpha < \omega_1$, $t_\alpha^R$ is defined and $t_{\alpha+1}^R = t_\alpha^R(a, k, j)$ where $T(t_\alpha^R, a) = 2$ or $j$ *and* where $a \subseteq R$.

**Remark.** There is a "full" generalized tree defined by: $T(t, a) = 2$ for all $(t, a)$. Suppose that $R$ is a real and $t_\alpha^R$ is defined for all $\alpha$; it is *not* necessarily true that $R$ satisfies $T$. This is due to the requirement that $a \subseteq R$ when $t_{\alpha+1}^R = t_\alpha^R(a, k, j)$, for even $\alpha$.

**Definition.** A pair $(a, t)$ *lies on* $T$ iff $R \in [a, t] \Longrightarrow$ for all even $\alpha < \alpha(t)$, $t_{\alpha+1}^R = t_\alpha^R(b, k, j)$ where $T(t_\alpha^R, b) = 2$ or $j$ and $b \subseteq R$. A real $R$ *satisfies* $(a, t, T)$, where $(a, t)$ lies on $T$, if $R \in [a, t]$ and $R$ satisfies $T$. We let $[a, t, T]$ denote $\{R \mid R$ satisfies $(a, t, T)\}$ and $[T] = [\emptyset, 2^{<\omega}, T]$ (where $2^{<\omega}$ denotes the full $\omega$-tree $t(a) = 2$ for all $a$). We write $(a_0, t_0, T_0) \leq (a_1, t_1, T_1)$ ($(a_0, t_0, T_0)$ *extends* $(a_1, t_1, T_1)$) iff $[a_0, t_0, T_0] \subseteq [a_1, t_1, T_1]$ in all set-generic extensions, and $T_0 \leq T_1$ iff $(\emptyset, 2^{<\omega}, T_0) \leq (\emptyset, 2^{<\omega}, T_1)$.

**Definition** (Amalgamation). Suppose that $T$ is a generalized tree and $T_t, t \in \text{Split}^+(T, \alpha)$ are generalized trees extending $T$, where $\text{Split}^+(T, \alpha) = \{t_0(a, k, j) \mid t_0 \in \text{Split}(T, \alpha)\}$. Then $T(T_t \mid t \in \text{Split}^+(T, \alpha))$ is the generalized tree $T^*$ defined by: $T^*(t^*, a^*) = T_t(t^*, a^*)$ if $R \in [a^*, t^*] \Longrightarrow R \in [a, t]$ for some $t \in \text{Split}^+(T, \alpha)$ of the form $t = t_0(a, k, j)$; $T^*(t^*, a^*) = T(t^*, a^*)$ otherwise. We say that $T^*$ *results from* $T, T_t(t \in \text{Split}^+(T, \alpha))$ *via amalgamation at splitting level* $\alpha$.

**Lemma 8.24.** *Suppose that $T_0 \geq T_1 \geq \cdots$ forms an $\omega$-sequence of generalized trees. Then there is a generalized tree $T \leq T_n$ all $n$.*

*Proof.* If $t$ belongs to $R^{\text{Even}}$ then set $T(t, a) =$ eventual value of $T_n(t, a)$ (as $n \to \infty$) if exists; otherwise set $T(t, a) = 0$). Suppose that $R$ satisfies $T$; we claim that $t_{\alpha+1}^R = t_\alpha^R(a, k, j)$ where the eventual value of $T_n(t_\alpha^R, a) = 2$ or $j$, for each even $\alpha$. This is proved by induction on $\alpha$, using $T_0 \geq T_1 \geq \cdots$ and the fact that $R$ satisfies $T$. It follows that $T_n(t_\alpha^R, a) = 2$ or $j$ (where $t_{\alpha+1}^R = t_\alpha^R(a, k, j)$) for all $n$, all even $\alpha$ so $R$ satisfies each $T_n$.

We must show that Split$(T)$ is CUB. Note that $\text{Split}(T) = \bigcap\{\text{Split}(T_n) \mid n \in \omega\}$ so clearly Split$(T)$ is closed. To see that it is unbounded, given a canonical $t$ and $k \in \omega$ build an $\omega$-sequence of linear extension $t \geq_k t_k \geq t_{k+1} \geq \cdots$ such that $t_{n+1} \leq_n t_n$, $t_n$ belongs to Split$(T_n)$ and $\text{Lim}\langle t_n \mid k \leq n\rangle = t^*$ exists as a canonical tree. Then $t^*$ is a linear $k$-extension of $t$ and belongs to $\text{Split}(T) = \bigcap\{\text{Split}(T_n) \mid n \in \omega\}$ as each Split$(T_n)$ is closed. □

**Lemma 8.25.** *Suppose that $T_0 \geq T_1 \geq \cdots$ is an $\omega_1$-sequence of generalized trees such that $\alpha \leq \beta \Longrightarrow T_\beta \leq_\alpha T_\alpha$. Then there exists a generalized tree $T$ such that $T \leq_\alpha T_\alpha$ for all $\alpha < \omega_1$.*

*Proof.* Define $T$ as in the proof of the previous lemma. Again, $\mathrm{Split}(T)$ is clearly closed. To see that it is unbounded, build $t \geq_k t_k \geq t_{k+1} \geq \cdots$ as before but with $t_{n+1}$ of $\mathrm{Split}(T_{\alpha_n})$-rank at least $\alpha_n$ for even $n$. Then $\mathrm{Lim}\langle t_n \mid k \leq n\rangle = t^*$ is a linear $k$-extension of $t$ in $\mathrm{Split}(T_\alpha, \alpha)$ where $\alpha = \cup\{\alpha_n \mid n \in \omega\}$ and hence $t \in \mathrm{Split}(T)$. □

## The Forcing $P$

$P$ consists of all $(a, t, T)$ with $a$ lying on $t$, $(a, t)$ lying on $T$, ordered as specified earlier: $(a_0, t_0, T_0) \leq (a_1, t_1, T_1)$ iff $[a_0, t_0, T_0] \subseteq [a_1, t_1, T_1]$ where $[a, t, T] = \{R \mid a \subseteq R$, $R \in [t]$ and for even $\alpha$, $t^R_{\alpha+1} = t^R_\alpha(b, k, j)$ where $T(t^R_\alpha, b) = 2$ or $j$ and $b \subseteq R\}$. We shall show that forcing with $P$ produces an $L$-minimal real not generic over $L$ for any forcing of $L$-cardinality $\omega_1^L$.

**Lemma 8.26.** *Suppose $(a_1, t_1, T_1) \leq (a_2, t_2, T_2)$. Then there exists $(a_0, t_0, T_0) \leq (a_1, t_1, T_1)$ such that $a_0$ extends $a_2$, $t_0 \leq t_2$, $T_0 \leq T_2$.*

*Proof.* Clearly $a_2$ lies on $t_1$, as (in a set-generic extension) there is a real $R \in [a_1, t_1, T_1] \subseteq [a_2, t_2, T_2]$ and hence $a_2 \subseteq R \in [t_1]$. Also $a_1, a_2$ are compatible strings (for the same reason) so we take $a_0 = a_1 \cup a_2$. Next extend $(a_0, t_1)$ to $(a_0, t_0^*)$ lying on $T_1$ so that $\alpha(t_2) \leq \alpha(t_0^*)$, and choose $(a, t_2^*) \leq (a, t_2)$ with $(a, t_2^*)$ lying on $T_2$ and $\alpha(t_2^*) = \alpha(t_0^*)$. By fusion we can produce $(a_0, t_0)$ lying on $T_1$ such that $(a_0, t_0, T_1) \leq (a_0, t_1, T_1)$ and $t_0 \leq t_2$. Finally define $T_0(t, a) = T_2(t, a)$ if $[a, t] \subseteq [a^*, t^*]$ for some $a, t^*, \alpha(t^*) = \alpha(t_0)$, $[a^*, t^*] \cap [a_0, t_0] = \emptyset$ and $T_0(t, a) = T_1(t, a)$ otherwise. Then $(a_0, t_0, T_0) \leq (a_0, t_0, T_1)$ and $T_0 \leq T_2$. □

**Lemma 8.27** (Density Reduction at $\omega_1$). *Suppose that $(a, t, T) \in P$ and $\langle D_\alpha \mid \alpha < \omega_1\rangle$ are open dense on $P$. Then there exists $(a, t, T^*) \leq (a, t, T)$ such that for each $\alpha < \omega_1$, $\{(a^*, t^*, T^*) \mid (a^*, t^*, T^*) \in D_\alpha\}$ is predense below $(a, t, T^*)$.*

*Proof.* Fix $\alpha < \omega_1$. For each $t^* \in \mathrm{Split}^+(T, \alpha)$ and each $a^*$ lying on $t^*$ choose $T_{a^*,t^*} \leq T$ such that if $(a^*, t^*, T') \leq (a, t, T)$, $(a^*, t^*, T') \in D_\alpha$ for some $T'$ then $(a^*, t^*, T_{a^*,t^*})$ has these properties. This is possible by Lemma 8.26. Now by enumerating the possible $a^*$ lying on $t^*$, we may choose $T_{t^*}$ to satisfy the previous for all $a^*$ lying on $t^*$. Form $T^*_\alpha = T(T_{t^*} \mid t^* \in \mathrm{Split}^+(T, \alpha))$. Now by enumerating the $\alpha < \omega_1$, we may choose $T^* \leq T$ to satisfy the above for $D_\alpha$, for all $a^*$ lying on $t^*$, $t^* \in \mathrm{Split}^+(T^*, \alpha)$, simultaneously for all $\alpha$.

We claim that for each $\alpha$, $D_\alpha^{T^*} = \{(a^*, t^*, T^*) \mid (a^*, t^*, T^*) \in D_\alpha\}$ is predense below $(a, t, T^*)$: Suppose $(a_0, t_0, T_0) \leq (a, t, T^*)$. We may extend $(a_0, t_0, T_0)$ to guarantees that $(a_0, t_0, T_0) \in D_\alpha$ and $t_0 \in \mathrm{Split}^+(T^*, \alpha)$ for some $\alpha$. By construction, $(a_0, t_0, T^*)$ belongs to $D_\alpha$, hence to $D_\alpha^{T^*}$ and is compatible with $(a_0, t_0, T_0)$. □

**Lemma 8.28** (Density Reduction at $\omega$). *Suppose that $(a, t, T) \in P$ and $\langle D_n \mid n \in \omega\rangle$ are open dense on $P$. Then there exists $(a, t^*, T^*) \leq (a, t, T)$ such that for each $n$, $D_n^{t^*,T^*} = \{(a^*, t^*, T^*) \mid (a^*, t^*, T^*) \in D_n\}$ is predense below $(a, t^*, T^*)$.*

*Proof.* By the previous lemma we may choose $(a, t, T^*) \leq (a, t, T)$ such that each $D_n^{T^*}$ is predense below $(a, t, T^*)$. Now let $\alpha_0 < \alpha_1 < \cdots$ be the first $\omega$-many $\bar{\alpha}$ such that $\bar{\alpha} = \omega_1 \cap \Sigma_1 \operatorname{Hull}(\bar{\alpha} \cup \{(a, t, T^*), \langle D_n^{T^*} \mid n \in \omega\rangle\})$ in $L_{\omega_2}$ and let $\alpha = \bigcup\{\alpha_n \mid n \in \omega\}$. Choose $n_0$ so that $a$ can be extended into $\operatorname{Split}^+(t, n_0)$ and build $t = t_{n_0} \geq t_{n_0+1} \geq \cdots$ so that $t_n \in R_{\alpha_n+2}$, $t_{n+1} \leq_n t_n$ and for $i < n$ there is $b \in \operatorname{Split}^+(t_{n+1}, n)$ such that $(b, t_{n+1}, T^*) \in D_i$. By the canonical trees construction, $t^* = \operatorname{Lim}\langle t_n \mid n_0 \leq n\rangle$ belongs to $R_\alpha$ and hence $(a, t^*, T^*)$ is as desired. □

**Corollary 8.29.** *$P$ preserves cofinalities.*

Now we prove:

**Lemma 8.30.** *Suppose that $G$ is $P$-generic and let $R = \bigcup\{a \mid (a, t, T) \in G\}$. Then $R$ is $L$-minimal: $x \in L[R]$, $x \notin L \Longrightarrow R \in L[x]$.*

*Proof.* We may assume that $x$ is a set of ordinals. Let $(a, t, T) \in P$, $(a, t, T) \Vdash x \subseteq \mathrm{ORD}$, $x \notin L$. Then for some $\alpha$ there are $(a_0, t_0, T_0), (a_1, t_1, T_1) \leq (a, t, T)$ such that $(a_0, t_0, T_0) \Vdash \alpha \in x$, $(a_1, t_1, T_1) \Vdash \alpha \notin x$. We may assume that $t_0 \neq t_1$ belong to $\operatorname{Split}^+(T, i)$ for some $i$ and that $T_0, T_1 \leq T$. So we may form $T^* = T(T_u \mid u \in \operatorname{Split}^+(T, i))$ where $T_u = T$ if $u \neq t_0, t_1$ and $T_{t_0} = T_0$, $T_{t_1} = T_1$. Then $(a_0, t_0, T^*) \Vdash \alpha \in x$, $(a_1, t_1, T^*) \Vdash \alpha \notin x$.

Now consider $D_n = \{(a^*, t^*, T^*) \mid (a^*, t^*, T^*) \leq (a, t, T)$ and whenever $b_0 \neq b_1$ belong to $\operatorname{Split}^+(t^*, n)$ and extend $a$, there is $\alpha$ such that $(b_0, t^*, T^*) \Vdash \alpha \in x$, $(b_1, t^*, T^*) \Vdash \alpha \notin x$ or vice-versa$\}$. Then $D_n$ is open dense below $(a, t, T)$ for all $n$. By Density Reduction at $\omega$ we may choose $(a, t^*, T^*) \leq (a, t, T)$ to meet each $D_n$. But then $(a, t^*, T^*) \Vdash R \in L[x]$. □

Finally:

**Lemma 8.31.** *Suppose $R$ is as in Lemma* 8.30. *The $R$ does not belong to any $P_0$-generic extension of $L$ for $P_0$ of $L$-cardinality $\omega_1^L$.*

*Proof.* Recall that $\alpha \in B^R$ iff $\alpha$ is even, $t^R_{\alpha+1} = t^R_\alpha(a, k, 1)$ for come $a, k$. For any condition $(a, t, T)$ and $\beta < \omega_2^L$ there is $(a, t, T^*) \leq (a, t, T)$ forcing that some ordinal $\gamma \in (\beta, \omega_2^L)$ is admissible but inadmissible relative to $R$, $(a, t, T)$ (where $R$ denotes the generic real), as we may force $B^R$, $T$ to code a wellordering of $\omega_1$ of length $\gamma$. It follows that there are unboundedly many admissible $\gamma < \omega_2^L$ which are $R$-inadmissible. This effect cannot be achieved via a constructible forcing of $L$-cardinality $\omega_1^L$. □

This completes the proof of Theorem 8.22. □

We next embellish the above construction to show that a subset of $\omega_2$ can be minimally coded by a real.

**Theorem 8.32.** *Suppose that* $A \subseteq \omega_2^{L[A]}$ *and* $L[A] \models$ GCH. *Then there is a forcing* $P \in L[A]$ *such that if* $G$ *is* $P$*-generic over* $L[A]$*:*

(a) $L[A, G] = L[R]$ *for some real* $R$.

(b) $L[A], L[R]$ *have the same cofinalities.*

(c) $R$ *is minimal over* $L[A]$*: If* $x \in L[R]$ *then either* $x \in L[A]$ *or* $R \in L[A, x]$.

*Proof.* We must modify the forcing used to prove Theorem 8.22 so as to code $A$ by the generic real $R$. Conditions are now of the form $p = (a, (t, p^*_\omega), (T, p^*_{\omega_1}))$ where $p^*_\omega, p^*_{\omega_1}$ "restrain" $t, T$, respectively so as to guarantee that $R$ codes $A$. In addition, we must make major changes to the notion of canonical tree (due to the fact that $A \cap \omega_1$ may fail to be "reshaped") and introduce the notion of canonical generalized tree.

## Canonical Trees

Again we inductively define $R_\alpha, \alpha < \omega_1$ but we modify Cases 2B, 3. In Case 2B, we add two $k$-extensions $t(\vec{t}_k, 0), t(\vec{t}_k, 1)$ of $t(t_a \mid a \in \text{Split}^+(t, k))$, so that $(t, \vec{t}_k, j) \neq (t^*, \vec{t}^*_{k^*}, j^*) \Longrightarrow [t(\vec{t}_k, j)] \cap [t^*(\vec{t}^*_{k^*}, j^*)] = \emptyset$. This effects a change in the definition of $B^R \cap \alpha$ for $R \in [t]$, some $t \in R_\alpha : \beta \in B^R \cap \alpha$ iff either $\beta$ is even, $t^R_{\beta+1} = t^R_\beta(a, k, 1)$ for some $a, k$ or $\beta$ is odd, $t^R_{\beta+1} = t(\vec{t}_k, 1)$ for some $t, \vec{t}_k$. Now we consider the new version of Case 3.

**Case 3.** $\alpha$ limit.
We include two types of trees in $R_\alpha$. First, suppose that $t_0 \geq t_1 \geq t_2 \geq \cdots$ is a sequence of linear extensions such that $\alpha(t_0) < \alpha(t_1) < \cdots$ is cofinal in $\alpha$ and $\text{Lim}\langle t_n \mid n \in \omega\rangle = t$ exists; then we place $t$ in $R_\alpha$. To describe the second type of tree to be placed into $R_\alpha$, we need a definition. We say that a real $R$ is *special at* $\alpha$ if it meets the following conditions:

(1) $\alpha > \omega$, $L_\alpha[A]$ is locally countable (i.e, $L_\alpha[A] \models \omega$ is the largest cardinal).

(2) For some least limit ordinal $\gamma$, $R$ codes some $\bar{A} \subseteq \beta = \omega_2$ of $L_\gamma[\bar{A}]$ such that $\alpha$ is countable in $L_{\gamma+\omega}[\bar{A}]$, $\alpha = \omega_1$ of $L_\gamma[\bar{A}]$.

(3) For some least $p \in L_\gamma[\bar{A}]$ we have that $L_\gamma[\bar{A}] = \Sigma_1 \text{ Hull}(\alpha \cup \{p\})$ in $L_\gamma[\bar{A}]$ and $S = \{\bar{\alpha} < \alpha \mid \bar{\alpha} = \alpha \cap \Sigma_1 \text{ Hull}(\bar{\alpha} \cup \{p\}) \text{ in } L_\gamma[\bar{A}]\}$ has ordertype $\omega$.

(4) Let $t^R_n = t^R_{\alpha_n}$ where $\alpha_0 < \alpha_1 < \cdots$ is the increasing enumeration of $S$. Then $t^R_n$ is defined and $\text{Lim}\langle t^R_n \mid n_0 \leq n\rangle = t^R$ exists for some $n_0$, where $t^R_{n_0} \geq t^R_{n_0+1} \geq \cdots$ forms a sequence of nonlinear extensions.

In (2), when we say that "$R$ codes $\bar{A}$," we intend this in the same sense that our desired $P$-generic real $R$ will code $A$ over $L_{\omega_2}[A]$; we use the Recursion Theorem to provide an index for this decoding procedure.

Now if $t^R = \text{Lim}\langle t^R_n \mid n_0 \leq n\rangle = t$ is independent of $R \in [t]$ then we put $t$ into $R_\alpha$.

This completes the construction of the $R_\alpha$'s. As before we have:

**Lemma 8.33** (Extendibility for Canonical Trees). *Suppose that $t$ is canonical, $k \in \omega$, $\alpha(t) < \alpha < \omega_1$ and $B \subseteq [\alpha(t), \alpha)$. Then there is a linear extension $t^* \leq_k t$ such that $\alpha(t^*) = \alpha$, $R \in [t^*] \Longrightarrow B^R \cap [\alpha(t), \alpha) = B$.*

## Generalized Trees

Generalized trees are defined as before. However now we need to define the notion of *canonical* generalized tree in order to code $A \subseteq \omega_2$. Thus we shall define $R_\alpha^{\omega_1}$, $\omega_1 \leq \alpha < \omega_2$ in analogy to $R_\alpha$, $\alpha < \omega_1$. If a real $R$ belongs to $[T]$ for some $T \in R_\alpha^{\omega_1}$ then $T$ is unique and is written $T_\alpha^R$; moreover $T_\beta^R$ is defined for all $\omega_1 \leq \beta < \alpha$ as well.

**Case 1.** $\alpha = \omega_1$.
$R_\alpha^{\omega_1} = \{T\}$ where $T$ is the "full generalized tree" $T(t, a) = 2$ for all $t \in R^{\text{Even}} = \bigcup\{R_\alpha \mid \alpha \text{ even}\}$, $a$ lying on $t$.

**Case 2A.** $\alpha = \beta + 1$, $\beta$ even.
For each $T \in R_\beta^{\omega_1}$, $(a, t)$ lying on $T$ and $\gamma < \omega_1$ we select a $\gamma$-extension $T(a, t, \gamma)$ of $T$ upon which $(a, t)$ lies such that $(a_0, t_0, \gamma_0) \neq (a_1, t_1, \gamma_1) \Longrightarrow [T(a_0, t_0, \gamma_0)] \cap [T(a_1, t_1, \gamma_1)] = \emptyset$ in all set-generic extensions. Then $R_\alpha^{\omega_1}$ consists of all such $T(a, t, \gamma)$.

**Case 2B.** $\alpha = \beta + 1$, $\beta$ odd.
For each $T \in R_{\leq\beta}^{\omega_1} = \bigcup\{R_\gamma^{\omega_1} \mid \gamma \leq \beta\}$, $\gamma < \omega_1$ and distinct $\langle T_t \mid t \in \text{Split}^+(T, \gamma)\rangle = \vec{T}_\gamma$ with each $T_t$ from $R_\beta^{\omega_1}$, we select two $\gamma$-extensions $T(\vec{T}_\gamma, 0)$, $T(\vec{T}_\gamma, 1)$ of $T(T_t \mid t \in \text{Split}^+(T, \gamma))$ such that $(T, \vec{T}_\gamma, j) \neq (T^*, \vec{T}^*_{\gamma^*}, j^*) \Longrightarrow [T(\vec{T}_\gamma, j)] \cap [T^*(\vec{T}^*_{\gamma^*}, j^*)] = \emptyset$ in all set-generic extensions. Then $R_\alpha^{\omega_1}$ consists of all such $T(\vec{T}_\gamma, j)$.

**Case 3.** $\alpha$ limit, $\alpha > \omega_1$.
We include two types of generalized trees in $R_\alpha^{\omega_1}$. First, suppose that $T_0 \geq T_1 \geq \cdots$ is a $\lambda$-sequence of linear extensions such that $\alpha(T_0) < \alpha(T_1) < \cdots$ is cofinal in $\alpha$ and $\text{Lim}\langle T_\beta \mid \beta < \lambda\rangle = T$ (defined as in the proof of Lemma 8.24) exists; then we put $T$ into $R_\alpha^{\omega_1}$. (The extension $T_0 \geq T_1$ is *linear* if $R, S \in [T_1] \Longrightarrow T_\alpha^R = T_\alpha^S$ for all $\alpha \in [\alpha(T_0), \alpha(T_1))$, in all set-generic extensions.) To describe the second type of generalized tree to be added to $R_\alpha^{\omega_1}$ we need a definition. We say that $\alpha$ is *special* if the following conditions are met:

(1) $L_\alpha[A] \models \omega_1$ is the largest cardinal and $\alpha$ is not regular in $L[A \cap \alpha]$.

(2) Let $\beta$ be the largest limit ordinal such that $\beta = \alpha$ or $\alpha$ is regular in $L_\beta[A \cap \alpha]$. Then in fact $\beta > \alpha$, $L_\beta[A \cap \alpha] \models \alpha = \omega_2$ and for some least $p \in L_\beta[A \cap \alpha]$, $L_\beta[A \cap \alpha] = \Sigma_1 \text{ Hull}(\alpha \cup \{p\})$ in $L_\beta[A \cap \alpha]$. In addition, $S = \{\bar{\alpha} < \alpha \mid \bar{\alpha} = \alpha \cap \Sigma_1 \text{ Hull}(\bar{\alpha} \cup \{p\}) \text{ in } L_\beta[A \cap \alpha]\}$ has limit ordertype $\lambda \leq \omega_1^{L[A]}$.

In this case let $\langle \alpha_i \mid i < \lambda \rangle$ be the increasing enumeration of $S$. If $T = \text{Lim}\langle T_i \mid i_0 \leq i < \lambda \rangle$ where $T_{i_0} \geq T_{i_0+1} \geq \cdots$, $T_i$ belongs to $R^{\omega_1}_{\alpha_i}$ for each $i \geq i_0$, the extensions $T_i \geq T_{i+1}$ are nonlinear and $A \cap \alpha = \{\gamma < \alpha \mid 2\langle \gamma, \delta \rangle + 1 \in B^R$ for unboundedly many $\delta < \alpha\}$ for all $R \in [T]$ in a set-generic extension, then we put $T$ into $R^{\omega_1}_{\alpha}$.

This completes the construction of the $R^{\omega_1}_{\alpha}$, $\omega_1 \leq \alpha < \omega_2$. If a real $R$ belongs to $[T]$ for some $T \in R^{\omega_1}_{\alpha}$ then we define $B^R \cap [\omega_1, \alpha)$ by: $\beta \in B^R \cap [\omega_1, \alpha)$ iff $\beta$ is odd, $\beta \in [\omega_1, \alpha)$ and $T^R_{\beta+1} = T(\vec{T}_\gamma, 1)$ for some $T, \vec{T}_\gamma$.

The natural analog of Lemma 8.33 holds for generalized trees:

**Lemma 8.34** (Extendibility for Canonical Generalized Trees). *Assume that $T$ is canonical, $\gamma < \omega_1$, $\alpha(T) < \alpha < \omega_2$ and $B \subseteq [\alpha(T), \alpha) \cap$ Odd, where Odd denotes the class of odd ordinals. Then there is a linear extension $T^* \leq_\gamma T$ such that $\alpha(T^*) = \alpha$, $R \in [T^*] \Longrightarrow B^R \cap [\alpha(T), \alpha) = B$.*

## The $\omega_2$-Coding Forcing $P$

A condition in $P$ is $p = (a, (t, p^*_\omega), (T, p^*_{\omega_1}))$ where $a$ lies on $t$, $(a, t)$ lies on $T$, $t$ and $T$ are canonical and $p^*_\omega$ is a countable subset of $\omega_1$, $p^*_{\omega_1}$ is a cardinality $\leq \omega_1$ subset of $\omega_2$. We write $a_p, t_p, T_p$ for $a, t, T$. Extension is defined by $p_0 \leq p_1$ iff in all set-generic extensions, $[a_{p_0}, t_{p_0}, T_{p_0}] \subseteq [a_{p_1}, t_{p_1}, T_{p_1}]$, $p^*_{1_\omega} \subseteq p^*_{0_\omega}$, $p^*_{1_{\omega_1}} \subseteq p^*_{0_{\omega_1}}$ and:

(1) $R \in [a_{p_0}, t_{p_0}, T_{p_0}]$, $\alpha \in p^*_\omega$, $\alpha(t_{p_1}) \leq 2\langle \alpha, \beta \rangle + 1 < \alpha(t_{p_0}) \Longrightarrow (\alpha \in A$ iff $2\langle \alpha, \beta \rangle + 1 \in B^R \cap \alpha(t_{p_0}))$.

(2) $R \in [a_{p_0}, t_{p_0}, T_{p_0}]$, $\alpha \in p^*_{\omega_1}$, $\alpha(T_{p_1}) \leq 2\langle \alpha, \beta \rangle + 1 < \alpha(T_{p_o}) \Longrightarrow (\alpha \in A$ iff $2\langle \alpha, \beta \rangle + 1 \in B^R \cap \alpha(T_{p_0}))$.

In the above, $\langle \cdot, \cdot \rangle$ denotes the Gödel pairing function. Thus generically $P$ produces a real $R$ such that for $\alpha < \omega_1$, $\alpha \in A$ iff $2\langle \alpha, \beta \rangle + 1 \in B^R \cap \omega_1$ for sufficiently large $\beta < \omega_1$ and for $\alpha < \omega_2$, $\alpha \in A$ iff $2\langle \alpha, \beta \rangle + 1 \in B^R \cap \omega_2$ for sufficiently large $\beta < \omega_2$. We must show that $R$ preserves cofinalities and is minimal over $L[A]$.

First note that using amalgamation for canonical generalized trees can establish the natural analogue of Lemma 8.26:

**Lemma 8.35.** *Suppose $q \leq p$. Then there exists $r \leq q$ such that $a_r$ extends $a_p$, $t_r \leq t_p$, $T_r \leq T_p$.*

Now we establish Density Reduction.

**Lemma 8.36** (Density Reduction at $\omega_1$). *Suppose that $p \in P$ and $\langle D_\alpha \mid \alpha < \omega_1 \rangle$ are open dense on $P$. Then there exists $q \leq p$ such that $(a_q, (t_q, q^*_\omega)) = (a_p, (t_p, p^*_\omega))$ and for each $\alpha < \omega_1$, $D^{q,1}_\alpha = \{r \in D_\alpha \mid (T_r, r^*_{\omega_1}) = (T_q, q^*_{\omega_1})\}$ is predense below $q$.*

*Proof.* As in the first part of the proof of Lemma 8.27 we may use amalgamation to show that the present lemma holds for a single $D_\alpha$: For all $p$ and $\alpha$ there is $q \leq p$,

$(a_q, (t_q, q^*_\omega)) = (a_p, (t_p, p^*_\omega))$, $T_q \leq_\alpha T_p$ such that $D^{q,1}_\alpha$ is predense below $q$. Now let $\alpha_0 < \alpha_1 < \cdots$ be the first $\omega_1$-many $\bar\alpha$ such that $\bar\alpha = \omega_2 \cap \Sigma_1$ Hull$(\bar\alpha \cup \{p, \langle D_\alpha \mid \alpha < \omega_1\rangle\})$ in $L_{\omega_3}[A]$ and let $\alpha = \cup\{\alpha_\gamma \mid \gamma < \omega_1\}$. Note that $\alpha_\lambda$ is special for sufficiently large limit $\lambda < \omega_1$ and hence by Case 3 of the construction of the canonical generalized trees, we may build $p \geq p_{i_0} \geq p_{i_0+1} \geq \cdots$ of length $\omega_1 + 1$ such that for all $i < \omega_1$, $(a_{p_i}, (t_{p_i}, p^*_{i_\omega})) = (a_p, (t_p, p^*_\omega))$, $T_{p_i} \in R^{\omega_1}_{\alpha_i}$, $T_{p_{i+1}} \leq_{\alpha_i} T_{p_i}$ is a nonlinear extension with $D^{p_{i+1},1}_{i-i_0}$ predense below $p_{i+1}$. Finally let $q$ be $p_{\omega_i+1}$. □

**Lemma 8.37** (Density Reduction at $\omega$). *Suppose that $p \in P$ and $\langle D_n \mid n \in \omega\rangle$ are open dense on $P$. Then there exists $q \leq p$ such that $a_q = a_p$ and for each $n$, $D^{q,0}_n = \{r \in D_n \mid ((t_r, r^*_\omega), (T_r, r^*_{\omega_1})) = ((t_q, q^*_\omega), (T_q, q^*_{\omega_1}))\}$ is predense below $q$.*

*Proof.* By Lemma 8.36 we may assume that each $D^{p,1}_n = \{r \in D_n \mid (T_r, r^*_{\omega_1}) = (T_p, p^*_{\omega_1})\}$ is predense below $p$, $p^*_{\omega_1} = \alpha(T_p)$ and $T_p$ is the type of generalized tree added in the second part of Case 3. In particular each $R \in [T]$ codes the structure $L_\beta[A \cap \alpha(T)]$ where $\beta$ is the largest limit ordinal such that $\beta = \alpha$ or $\alpha$ is regular in $L_\beta[A \cap \alpha(T)]$, and $T$ is the limit of $\langle T_i \mid i_0 \leq i < \lambda\rangle$ where $T_i = T^R_{\alpha_i}$ and $\langle \alpha_i \mid i < \lambda\rangle$ is defined canonically over $L_\beta[A \cap \alpha(T)]$. Thus each $R \in [T]$ codes the structure needed to produce $T$ from $\langle T^R_\alpha \mid \alpha < \alpha(T)\rangle$.

Set $T = T_p$ and consider the forcing $R^T$, whose conditions are of the form $p = (a_p, t_p, p^*)$ where $(a_p, (t_p, p^*)), (T, \emptyset)$ belongs to $P$. It suffices to show that Density Reduction at $\omega$ holds for $R^T$ in $\mathcal{A}^T = L_{\beta+\omega}[A \cap \alpha(T)]$: If $\langle D_n \mid n \in \omega\rangle \in \mathcal{A}^T$ are open dense on $R^T$ and $(a, t_0, p^*_0) \in R^T$ then there is $(a, t_1, p^*_1) \leq (a, t_0, p^*_0)$ such that for each $n$, $D^{a,t_1,p^*_1}_n = \{(a, t, p^*) \in D_n \mid (t, p^*) = (t_1, p^*_1)\}$ is predense below $(a, t_1, p^*_1)$. Let $\alpha_0 < \alpha_1 < \cdots$ be the first $\omega$-many ordinals $\bar\alpha$ such that $\bar\alpha = \omega_1 \cap \Sigma_1$ Hull$(\bar\alpha \cup \{(a, t_0, p^*_0), \langle D_n \mid n \in \omega\rangle\})$ in $\mathcal{A}^T$. We may build $(a, t_0, p^*_0) = p_0 \geq p_1 \geq \cdots$ in $R^T$ so that for all $n < \omega$, $a_{p_n} = a$, $D^{p_{n+1}}_n$ is predense below $p_{n+1}$, $t_{p_{n+1}} \leq_n t_{p_n}$ is nonlinear, $t_{p_n} \in R_{\alpha_n}$ and setting $t = \text{Lim}\langle t_n \mid n \in \omega\rangle$ we have that $R \in [t] \Longrightarrow R$ codes $\bar A =$ the image of $A \cap \alpha(T)$ under the transitive collapse of $\Sigma_1$ Hull$(\alpha \cup \{p_0, \langle D_n \mid n \in \omega\rangle\})$ in $\mathcal{A}^T$, where $\alpha = \cup\{\alpha_n \mid n \in \omega\}$. Note that if the latter transitive collapse is written as $L_\gamma[\bar A]$ and $\bar\gamma$ limit, $\alpha < \bar\gamma < \gamma$, $R$ codes $\bar{\bar{A}}$ in $L_{\bar\gamma}[R]$ then either $\bar{\bar{A}} = \bar A$ or $\bar{\bar{A}} \in L_\gamma[\bar A]$; in either case $\alpha$ is regular in $L_{\gamma+\omega}[\bar{\bar{A}}]$. It follows that $\gamma$ is least such that over $L_\gamma[R]$, $R$ codes $\bar A$ and $\alpha$ is collapsed in $L_{\gamma+\omega}[\bar A]$. But then by Case 3 of the construction of canonical trees, $t$ belongs to $R_\alpha$ and hence $(a, t, p^*)$ is the desired extension of $(a, t_0, p^*_0)$, where $p^* = \cup\{p^*_n \mid n \in \omega\}$. □

**Corollary 8.38.** *$P$ preserves cofinalities.*

**Lemma 8.39.** *Suppose that $G$ is $P$-generic over $L[A]$ and $R = \cup\{a_p \mid p \in G\}$. The $R$ is minimal over $L[A]$.*

*Proof.* As in the proof of Lemma 8.30, using amalgamation for canonical generalized trees, amalgamation for canonical trees and Density Reduction at $\omega$. □

This completes the proof of Theorem 8.32. □

## The General Case

We now attack Theorem 8.21. Of course we assume, as in the proof of the usual Coding Theorem, that $L_\alpha[A] = H_\alpha$ for infinite cardinals $\alpha$. As usual, Card denotes the class of infinite cardinals (of our ground model $L[A]$), $\text{Card}^+ = \{\alpha^+ \mid \alpha \in \text{Card}\}$ and $\text{Card}' = \{\alpha > \omega \mid \alpha \text{ a limit cardinal}\}$.

A key definition is that of *canonical* (generalized) *$\alpha$-tree* for $\alpha \in$ Card. The collection $R^\alpha$ of canonical $\alpha$-trees is defined by induction on $\alpha \in$ Card.

**Definition** (of $R^\omega$). $R^\omega = \bigcup\{R^\omega_\alpha \mid \omega \leq \alpha < \omega_1\}$. We define $R^\omega_\alpha$ by induction on $\alpha$.

**Case 1.** $\alpha = \omega$.
$R^\omega_\omega = \{t\}$ where $t$ is the "full tree" $t(a) = 2$ for all $a \in 2^{<\omega}$.

**Case 2A.** $\alpha = \beta + 1$, $\beta$ even.
For each $t \in R^\omega_\beta$, $a$ lying on $t$ and $k \in \omega$ we select two $k$-extensions $t(a, k, 0)$, $t(a, k, 1)$ of $t$ upon which $a$ lies such that $(a, k, j) \neq (a^*, k^*, j^*) \Longrightarrow [t(a, k, j)] \cap [t(a^*, k^*, j^*)] = \emptyset$. Then $R^\omega_\alpha$ consist of all such $t(a, k, j)$.

**Case 2B.** $\alpha = \beta + 1$, $\beta$ odd.
For each $t \in R^\omega_{\leq\beta} = \bigcup\{R^\omega_\gamma \mid \gamma \leq \beta\}$, $k \in \omega$ and $\langle t_a \mid a \in \text{Split}^+(t, k)\rangle = \vec{t}_k$ with each $t_a$ from $R^\omega_\beta$, we select two $k$-extensions $t(\vec{t}_k, 0)$, $t(\vec{t}_k, 1)$ of $t(t_a \mid a \in \text{Split}^+(t, k))$ such that $(t, \vec{t}_k, j) \neq (t^*, \vec{t}^*_{k^*}, j^*) \Longrightarrow [t(\vec{t}_k, j)] \cap [t^*(\vec{t}^*_{k^*}, j^*)] = \emptyset$. Then $R^\omega_\alpha$ consists of all such $t(\vec{t}_k, j)$.

**Case 3.** $\alpha$ limit, $\alpha > \omega$.
We include two types of trees in $R^\omega_\alpha$. An extension of canonical trees $t_0 \leq t_1$ from $R^\omega_{<\alpha} = \bigcup\{R^\omega_\beta \mid \beta < \alpha\}$ is *linear* if $R, S \in [t_0]$, $\alpha(t_1) \leq \beta \leq \alpha(t_0) \Longrightarrow t^R_\beta = t^S_\beta$ where $t^R_\beta =$ unique $t \in R^\omega_\beta$ such that $R \in [t]$. If $t_0 \geq t_1 \geq \cdots$ is a sequence of linear extensions such that $\alpha(t_0) < \alpha(t_1) < \cdots$ is cofinal in $\alpha$ (where $\alpha(t) =$ unique $\beta$ such that $t \in R^\omega_\beta$) and $t = \text{Lim}\langle t_n \mid n \in \omega\rangle$ exists then we place $t$ in $R^\omega_\alpha$. We refer to such $t$ as the *linear* elements of $R^\omega_\alpha$.

To describe the nonlinear elements of $R^\omega_\alpha$ we need a definition. We say that a real $R$ is *special at $\alpha$* if it meets the following conditions:

(1) $L_\alpha[A]$ is locally countable.

(2) $R$ canonically codes a structure $\mathcal{A}^R$ arising in the proof of Extendibility, Distributivity or $\Sigma_f$-Density for some $R^T$, $T$ a $\gamma$-tree for some $\gamma \in$ Card (in the sense of $\mathcal{A}^R$) where $T$ is canonically determined by $\mathcal{A}^R$ and $\alpha = \omega_1$ of $\mathcal{A}^R$.

(3) $\mathcal{A}^R$ gives rise to an $\omega$-sequence of nonlinear extensions $t_0 \geq t_1 \geq \cdots$ from $R^\omega_{<\alpha}$ with well-defined limit $t^R$ and $\alpha = \bigcup\{\alpha(t_n) \mid n \in \omega\}$.

(4) $S \in [t^R] \Longrightarrow t^S$ is defined and $\mathcal{A}^R = \mathcal{A}^S$, $t^R = t^S$.

Then we put $t^R$ into $R^\omega_\alpha$.

This completes the construction of the $R_\alpha^\omega$'s. The meaning of clauses (2), (3) in Case 3 will clarify after we have given the proofs of Extendibility, Distributivity and $\Sigma_f$-Density for the $R^T$ forcings. At this point we only mention that the Recursion Theorem must be used to provide a good definition of "$R$ canonically codes $\mathcal{A}^R$," as this coding makes use of the procedure for decoding $A$ from our intended $P$-generic real $R$.

**Definition** (of $R^\gamma$, $\gamma \in \text{Card}^+$). Write $\gamma = \delta^+$, $\delta \in \text{Card}$. We first define the concept of *$\gamma$-tree*. Let $R^\delta_{\text{Even}}$ denote $\{t \in R^\delta \mid \alpha(t) \text{ is even}\}$. Then $D^\gamma = \{d \mid \text{Dom}(d) = \{0\} \cup (\text{Card} \cap \gamma), d(0) \in 2^{<\omega}, d(\beta) \in R^\beta \text{ for } \beta \in \text{Card} \cap \gamma, d \restriction \beta \text{ lies on } d(\beta) \text{ for } \beta \in \text{Card}^+ \cap \gamma, d(\beta) = d \restriction \beta \text{ for } \beta \in \text{Card}' \cap \gamma\}$ and $D^\gamma_{\text{Even}} = \{d \in D^\gamma \mid d(\delta) \in R^\delta_{\text{Even}}\}$.

**Definition.** $X \subseteq R^\delta_{\text{Even}}$ is *unbounded* if for each $\alpha < \delta$, each $t \in R^\delta$ has a linear $\alpha$-extension in $X$. $X$ is *closed* if whenever $t_0 \geq t_1 \geq \cdots$ is a sequence of elements of $X$ with limit equal to a well-defined $t \in R^\delta_{\text{Even}}$ then $t$ belongs to $X$. We say that $X$ is CUB if it is both closed and unbounded.

**Definition.** A *$\gamma$-tree* is a function $T\colon D^\gamma_{\text{Even}} \longrightarrow 3$ such that $\text{Split}(T) = \{t \in R^\delta_{\text{Even}} \mid T(d) = 2 \text{ for all } d \text{ such that } d(\delta) = t\}$ is CUB. A real $R$ *satisfies* $T$, if for all even $\alpha \in [\delta, \gamma)$, $t^R_\alpha$ is defined, $\{\text{Split}(T)\text{-rank}(t^R_{2\alpha}) \mid \alpha < \gamma\}$ is cofinal in $\gamma$ and $t^R_{\alpha+1} = t^R_\alpha(d, \gamma, j)$ where $T(d * t^R_\alpha) = 2$ or $j$ and $R \in [d]$. (Split($T$)-rank is defined below.)

**Definition.** $d \in D^\gamma$ *lies on* $T$ if $R \in [d] \Longrightarrow$ for all even $\alpha \in [\delta, \alpha(d, (\delta)))$, $t^R_{\alpha+1} = t^R_\alpha(e, \gamma, j)$ where $T(e * t^R_\alpha) = 2$ or $j$ and $R \in [e]$. A real $R$ *satisfies* $d * T$, where $d$ lies on $T$, if $R$ satisfies $T$ and $R \in [d]$. We let $[d * T]$ denote $\{R \mid R \text{ satisfies } d * T\}$ and $[T] = [\emptyset * T]$ where $\emptyset$ denotes the sequence $\emptyset(\beta) = \text{Full } \beta\text{-Tree}$. We write $d_0 * T_0 \leq d_1 * T_1$ iff $[d_0 * T_0] \subseteq [d_1 * T_1]$ in all set-generic extensions and $T_0 \leq T_1$ iff $\emptyset * T_0 \leq \emptyset * T_1$.

**Definition** (Split($T$)-rank). For $T$ a $\gamma$-tree and $t \in R^\delta$ we define the Split($T$)-rank of $t$ as follows: If $\alpha(t) = \delta$ then $\text{Split}(T)\text{-rank}(t) = 0$. If $\alpha(t) = \beta + 1$, $\beta$ even and $t^*$ is the unique canonical $\delta$-tree such that $t \leq t^*$, $\alpha(t^*) = \beta$ then $\text{Split}(T)\text{-rank}(t) = \text{Split}(T)\text{-rank}(t^*) + 1$ if $t^* \in \text{Split}(T)$, $\text{Split}(T)\text{-rank}(t) = \text{Split}(T)\text{-rank}(t^*)$ otherwise. If $\alpha(t) = \beta + 1$, $\beta$ odd and $t = t^*(\langle t^*_u \mid u \in \text{Split}^+(t^*, \gamma)\rangle)$ then $\text{Split}(T)\text{-rank}(t) = \bigcup\{\text{Split}(T)\text{-rank}(t^*_u) \mid u \in \text{Split}^+(t^*, \gamma)\}$. If $\alpha(t) = \alpha$ limit and $t = \text{Lim}\langle t_i \mid i < \lambda\rangle$ with $t_i \in R^\delta_{<\alpha} = \bigcup\{R^\delta_\beta \mid \beta < \alpha\}$ then $\text{Split}(T)\text{-rank}(t) = \bigcup\{\text{Split}(T)\text{-rank}(t_i) \mid i < \lambda\}$. We set $\text{Split}(T, \alpha) = \{t \in \text{Split}(T) \mid \text{Split}(T)\text{-rank}(t) = \alpha\}$. For $\alpha < \gamma$, $T_0$ *$\alpha$-extends* $T_1$ ($T_0 \leq_\alpha T_1$) iff $T_0 \leq T_1$ and $\text{Split}(T_0, \alpha) = \text{Split}(T_1, \alpha)$.

**Definition** (Amalgamation). Suppose that $T$ is a $\gamma$-tree and $T_t$, $t \in \text{Split}^+(T, \alpha)$ are $\gamma$-trees extending $T$, where $\text{Split}^+(T, \alpha) = \{t_0(d, \gamma', j) \mid t_0 \in \text{Split}(T, \alpha)\}$. Then $T(T_t \mid t \in \text{Split}^+(T, \alpha))$ is the $\gamma$-tree $T^*$ defined by: $T^*(d^* * t^*) = T_t(d^* * t^*)$ if $[d^* * t^*] \subseteq [d * t]$ for some $t \in \text{Split}^+(T, \alpha)$ of the form $t = t_0(d, \gamma', j)$; $T^*(d^* * t^*) = T(d^* * t^*)$ otherwise. We say that $T^*$ *results from* $T$, $T_t (t \in \text{Split}^+(T, \alpha))$ *via amalgamation at splitting level* $\alpha$.

As before we have:

**Lemma 8.40.** *Suppose that $T_0 \geq T_1 \geq \cdots$ forms a $\delta$-sequence of $\gamma$-trees. Then there is a $\gamma$-tree $T \leq T_i$ for each $i$.*

**Lemma 8.41.** *Suppose that $T_0 \geq T_1 \geq \cdots$ forms a $\gamma$-sequence of $\gamma$-trees such that $\alpha \leq \beta \Longrightarrow T_\beta \leq_\alpha T_\alpha$. Then there exists a $\gamma$-tree $T \leq_\alpha T_\alpha$ for each $\alpha$.*

We shall also need a more general form of amalgamation:

**Definition** (Generalized Amalgamation). Suppose that $T$ is a $\gamma$-tree, $t$ is a $\bar{\gamma}$-tree for some $\bar{\gamma} \in \mathrm{Card}^+$, $\bar{\gamma} \leq \gamma$ and $T_0$, $T_u$, $u \in \mathrm{Split}^+(t, \eta)$ are $\gamma$-trees extending $T$. Then $T(T_0, T_u \mid u \in \mathrm{Split}^+(t, \eta))$ is the $\gamma$-tree $T^*$ defined by: $T^*(d^* * u^*) = T_u(d^* * u^*)$ if $[d^* * u^*] \subseteq [d * u]$ for some $u \in \mathrm{Split}^+(t, \eta)$ of the form $u = u_0(d, \gamma', j)$; $T^*(d^* * u^*) = T_0(d^* * u^*)$ otherwise.

Now we define $R^\gamma_\alpha$, $\delta \leq \alpha < \gamma$ by induction on $\alpha$.

**Case 1.** $\alpha = \gamma$.
$R^\gamma_\gamma = \{T\}$ where $T$ is the "full $\gamma$-tree" defined by $T(d) = 2$ for all $d \in D^\gamma_{\mathrm{Even}}$.

**Case 2A.** $\alpha = \beta + 1$, $\beta$ even.
For each $T \in R^\gamma_\beta$, $d \in D^\gamma$ lying on $T$ and $\eta < \gamma$ we select two $\eta$-extensions $T(d, \eta, 0)$, $T(d, \eta, 1)$ of $T$ upon which $d$ lies such that $(d_0, \eta_0, j_0) \neq (d_1, \eta_1, j_1) \Longrightarrow [T(d_0, \eta_0, j_0)] \cap [T(d_1, \eta_1, j_1)] = \emptyset$ in all set-generic extensions. We place all such $T(d, \eta, j)$ into $R^\gamma_\alpha$.

**Case 2B.** $\alpha = \beta + 1$, $\beta$ odd.
For each $T \in R^\gamma_{\leq\beta} = \cup\{R^\gamma_\eta \mid \eta \leq \beta\}$, $\eta < \gamma$ and $\langle T_0, \langle T_u \mid u \in \mathrm{Split}^+(t, \eta)\rangle\rangle = \vec{T}_\eta$ with $T_0$ and each $T_u$ extensions of $T$ from $R^\gamma_\beta$, we select two $\eta$-extensions $T(\vec{T}_\eta, 0)$, $T(\vec{T}_\eta, 1)$ of $T(T_0, T_u \mid u \in \mathrm{Split}^+(t, \eta))$ such that $(T, \vec{T}_\eta, j) \neq (T^*, \vec{T}^*_{\eta^*}, j^*) \Longrightarrow [T(\vec{T}_\eta, j)] \cap [T^*(\vec{T}^*_{\eta^*}, j^*)] = \emptyset$ in all set-generic extensions. We place all such $T(\vec{T}_\eta, j)$ into $R^\gamma_\alpha$.

**Case 3.** $\alpha$ limit, $\alpha > \gamma$.
We include two types of $\gamma$-trees in $R^\gamma_\alpha$. An extension $T_0 \leq T_1$ of canonical $\gamma$-trees from $R^\gamma_{<\alpha} = \cup\{R^\gamma_\beta \mid \beta < \alpha\}$ is *linear* if $R, S \in [T_0]$, $\alpha(T_1) \leq \beta \leq \alpha(T_0) \Longrightarrow T^R_\beta = T^S_\beta$ (in all set-generic extensions) where $T^R_\beta =$ the unique $T \in R^\gamma_\beta$ such that $R \in [T]$. If $T_0 \geq T_1 \geq \cdots$ is a sequence of length $\lambda$ of linear extensions in $R^\gamma_{<\alpha}$ such that $\cup\{\alpha(T_i) \mid i < \lambda\} = \alpha$ and $\mathrm{Lim}\langle T_i \mid i < \lambda\rangle = T$ is well-defined then we place $T$ into $R^\gamma_\alpha$. We refer to such $T$ as the *linear* elements of $R^\gamma_\alpha$.

To describe the nonlinear elements of $R^\gamma_\alpha$ we need a definition. We say that a real $R$ (in a set-generic extension) is *special at* $\alpha$ if it meets the following conditions:

(1) $L_\alpha[A] \models \gamma$ is the largest cardinal.

(2) $R$ canonically codes a structure $\mathcal{A}^R$ arising in the proof of Extendibility, Distributivity or $\Sigma_f$-Density for some $R^T$, $T$ a $\mu$-tree for some $\mu \in$ Card (in the sense of $\mathcal{A}^R$) where $T$ is canonically determined by $\mathcal{A}^R$ and $\alpha = \gamma^+$ of $\mathcal{A}^R$.

(3) $\mathcal{A}^R$ gives rise to a $\lambda$-sequence of nonlinear extensions $T_0 \geq T_1 \geq \cdots$ from $R^\gamma_{<\alpha}$ with well-defined limit $T^R$ and $\alpha = \cup\{\alpha(T_i) \mid i < \lambda\}$.

(4) $S \in [T^R] \Longrightarrow \mathcal{A}^S, T^S$ are defined and $\mathcal{A}^S = \mathcal{A}^R$, $T^S = T^R$ (in all set-generic extensions).

Then we put $T^R$ into $R^\gamma_\alpha$.

This completes the construction of the $R^\gamma_\alpha, \gamma \leq \alpha < \gamma^+$. As before, the meaning of clauses (2), (3) in Case 3 will clarify after we have examined the proof of Extendibility, Distributivity and $\Sigma_f$-Density for the $R^T$ forcings. For now we state:

**Lemma 8.42** (Distributivity for $R^T$, $T \in R^{\gamma^+}$, $\gamma \in \text{Card}^+$). *If $\langle D_i \mid i < \gamma\rangle \in \mathcal{A}^T$, each $D_i$ $\gamma$-predense on $R^T$ (i.e., $p \in R^T \Longrightarrow \exists q (q \restriction \gamma = p \restriction \gamma$, $q$ extends an element of $D_i$) and $p \in R^T$ then there exists $q \leq p$, $q$ meets each $D_i$, $i < \gamma$ (i.e., $q$ extends an element of $D_i$, each $i < \gamma$).*

*Proof.* Follow the proof of Lemma 4.6. The point now is that we interpret the nonlinear extensions added via Case 3 of the $R^\gamma$ construction so as to guarantee that a $\gamma$-tree arising at limit stages does indeed exist as a condition in $R^\gamma$. □

**Definition** (of $R^\gamma$, $\gamma \in \text{Card}'$). We base this heavily on the ideas of Section 4.3, where the Coding Theorem is proved in the general case.

**Definition.** A *$\gamma$-tree* is a sequence $T = \langle p_\delta \mid \delta \in \{0\} \cup \text{Card} \cap \gamma\rangle$ where $p_0 \in 2^{<\omega}$, $p_\delta \in R^\delta$ for $\delta \in \text{Card} \cap \gamma$, $T \restriction \delta$ lies on $p_\delta$ for $\delta \in \text{Card}^+ \cap \gamma$ and $p_\delta = T \restriction \delta$ for $\delta \in \text{Card}' \cap \gamma$. A real $R$ *satisfies* this $\gamma$-tree if $R$ satisfies $p_\delta$ for each $\delta \in \{0\} \cup \text{Card} \cap \gamma$. We write $[T]$ for the set of reals which satisfy $T$ and $T_0 \leq T_1$ ($T_0$ *extends* $T_1$) iff $[T_0] \subseteq [T_1]$ in all set-generic extensions. For $\delta \in \text{Card}^+ \cap \gamma$, $T_0 \leq_\delta T_1$ iff $T_0 \leq T_1$ and $T_0 \restriction \delta = T_1 \restriction \delta$.

We come now to the construction of $R^\gamma_\alpha$, $\gamma \leq \alpha < \gamma^+$. Associated with each $T \in R^\gamma = \cup\{R^\gamma_\alpha \mid \gamma \leq \alpha < \gamma^+\}$ will be a forcing $R^T$ as well as a structure $\mathcal{A}^T$, defined with $T$ by induction on $\alpha(T) =$ the unique $\alpha$ such that $T \in R^\gamma_\alpha$.

**Case 1.** $\alpha = \gamma$.
$R^\gamma_\gamma$ contains only the $\gamma$-tree $T = \langle p_\delta \mid \delta \in \{0\} \cup \text{Card} \cap \gamma\rangle$ where $p_0 = \emptyset$ and $p_\delta =$ the full $\delta$-tree for $\delta \in \text{Card}^+ \cap \gamma$. We define $\mathcal{A}^T = L_{\mu^T}[A \cap \gamma]$ where $\mu^T =$ least $\mu > \gamma$ such that $\mu' \mu = \mu$ for $0 < \mu' < \mu$.

Now we define $R^T$. A *quasicondition* in $R^T$ is a sequence $p = \langle (p_\delta, p^*_\delta) \mid \delta \in \{0\} \cup \text{Card} \cap \gamma\rangle \in \mathcal{A}^T$ where $\langle p_\delta \mid \delta \in \{0\} \cup \text{Card} \cap \gamma\rangle$ is a $\gamma$-tree and for each $\delta$, $p^*_\delta$ is a bounded subset of $\delta^+$ (taking $0^+ = \omega$). Quasiconditions are ordered by: $p \leq q$ iff for each $\delta$, $p_\delta \le q_\delta$, $p^*_\delta$ contains $q^*_\delta$ and $R \in [p_\delta]$, $\alpha(q_\delta) \le 2\beta+1 < \alpha(p_\delta)$, $\beta \in q^*_\delta$, $\beta = \langle \beta_0, \beta_1 \rangle \Longrightarrow 2\beta+1 \in B^R$ iff $\beta_0 \in A$ (in set-generic extensions). A *condition* in $R^T$ is a quasicondition $p$ such that:

(1) If $\gamma$ is inaccessible in $\mathcal{A}^T$ then there is a CUB $C \subseteq \gamma$, $C \in \mathcal{A}^T$ such that $\bar{\gamma} \in C \Longrightarrow p^*_{\bar{\gamma}} = \emptyset$.

(2) There is a ("good approximation to $\mathcal{A}^T$") $\mathcal{A} = \langle \tilde{J}_\mu[A \cap \gamma, T], A \cap \gamma, T \rangle$ where $\gamma \le \mu < \mu^T$, $\mu$ limit such that for some $k$: $p$ is $\Delta_1\langle L_\gamma[A], g^{\mathcal{A}}_k \rangle$ and $g^{\mathcal{A}}_k \restriction \bar{\gamma} \in \mathcal{A}^{p_{\bar{\gamma}}}$ for sufficiently large $\bar{\gamma} \in \mathrm{Card}' \cap \gamma$ (where as in Section 4.3: $g^{\mathcal{A}}_k(\delta) = \Sigma_k\,\mathrm{Hull}(\delta \cup \{x\})$ in $\mathcal{A}$, where $x$ = least parameter such that $\mathcal{A} = \Sigma_k\,\mathrm{Hull}(\gamma \cup \{x\})$ in $\mathcal{A}$).

(3) A third requirement, to be described at the end of the $R^\gamma$ construction.

Before turning to Case 2 we consider $\Sigma_f$-Density for $R^T$, $T$ = the unique element of $R^\gamma_\gamma$.

**Definition.** Suppose that $t$ is a canonical $\delta$-tree for some $\delta \in \{\omega\} \cup \mathrm{Card}^+ \cap \gamma$. Then $R^t = \{p \restriction \delta \mid p \text{ is a quasicondition}, p_\delta = t\}$. And we define $\mathcal{A}^t = L_{\mu^t}[A \cap \delta, t]$ as follows: If $t$ arises as in Case 1 of the definition of $R^\delta$ (i.e., $t$ is the full $\delta$-tree) then $\mu^t$ = least $\mu > \delta$ such that $\mu'\mu = \mu$ for $0 < \mu' < \mu$. If $t$ is the immediate successor to $t^* \in R^\delta_\beta$ (i.e., $t \in R^\delta_{\beta+1}$ and $t \le t^*$) then $\mu^t$ = least $\mu > \mu^{t^*}$ such that $\mu'\mu = \mu$ for $0 < \mu' < \mu$. If $\alpha(t)$ limit, $t$ linear then $\mu^t$ = least $\mu > \cup\{\mu^{t^*} \mid t \le t^* \text{ a linear extension}, t^* \in R^\delta_{<\alpha(t)}\}$ such that $0 < \mu' < \mu \Longrightarrow \mu'\mu = \mu$. Finally if $\alpha(t)$ limit, $t$ nonlinear and $R \in [t] \Longrightarrow R$ canonically codes $\mathcal{A}$, then $\mu^t$ = least $\mu > \mathrm{ORD}(\mathcal{A})$ such that $0 < \mu' < \mu \Longrightarrow \mu'\mu = \mu$.

**Definition.** $X \subseteq \gamma$ is *thin* in $\mathcal{A}^T$ if $X \in \mathcal{A}^T$ and for each $\mathcal{A}^T$-inaccessible $\bar{\gamma} \le \gamma$, $\mathcal{A}^T \models X \cap \bar{\gamma}$ is not stationary in $\bar{\gamma}$. A function $f : \mathrm{Card} \cap \gamma \longrightarrow \mathcal{A}^T$ in $A^T$ is *small in* $\mathcal{A}^T$ if for $\bar{\gamma} \in \mathrm{Card} \cap \gamma$, $\mathrm{Card}(f(\bar{\gamma})) \le \bar{\gamma}$ in $\mathcal{A}^T$ and $\mathrm{Support}(f) = \{\bar{\gamma} \in \mathrm{Card} \cap \gamma \mid f(\bar{\gamma}) \ne \emptyset\}$ is thin in $\mathcal{A}^T$. For $p \in R^T$, $f$ small in $\mathcal{A}^T$ we define $\Sigma^p_f$ = all $q \le p$ in $R^T$ such that whenever $\bar{\gamma} \in \mathrm{Card} \cap \gamma$, $D \in f(\bar{\gamma})$, $D$ predense on $R^{p_{\bar{\gamma}^+}}$, $D \in \mathcal{A}^{p_{\bar{\gamma}^+}}$ we have that $q$ *reduces* $D$ below $\bar{\gamma}$, in the sense that for some $\delta \in \mathrm{Card}^+$, $\delta \le \bar{\gamma}$, $\forall r \le q \exists s$ ($s$ meets $D$, $s \le r$ and $s \restriction [\delta, \gamma) = r \restriction [\delta, \gamma)$).

**Lemma 8.43.** *For $p \in R^T$, $f$ small in $\mathcal{A}^T$ there exists $q \le p$, $q \in \Sigma^p_f$.*

*Proof.* We assume by induction that this result holds for $f$ such that $\mathrm{Support}(f)$ is bounded in $\gamma$. Now follow the proof of Lemma 4.12 for the general case. We interpret the nonlinear extensions added via Case 3 of the $R^\delta$ constructions, $\delta \in \mathrm{Card} \cap \gamma$, so as to guarantee that conditions arising at limit stages in the proof of Lemma 4.12 do indeed exist. □

**Case 2A.** $\alpha = \beta + 1$, $\beta$ even.
For each $T = \langle p_\delta \mid \delta \in \{0\} \cup \mathrm{Card} \cap \gamma \rangle$ in $R^\gamma_\beta$, each $p^* = \langle p^*_\delta \mid \delta \in \{0\} \cup \mathrm{Card} \cap \gamma \rangle \in \mathcal{A}^T$

such that $p = \langle (p_\delta, p_\delta^*) \mid \delta \in \{0\} \cup \text{Card} \cap \gamma \rangle \in R^T$ and each $\delta \in \text{Card}^+ \cap \gamma$ we shall introduce two $\gamma$-trees $\hat{T}(p^*, \delta, 0)$, $\hat{T}(p^*, \delta, 1)$ in $R^\gamma_{\beta+1}$, $\delta$-extending $T$, where $T^*$ *$\delta$-extends* $T\,(T^* \leq_\delta T)$ iff $T^*$ extends $T$ and $T^* \restriction \delta = T \restriction \delta$. To carry this out we first note:

**Lemma 8.44** (Distributivity for $R^T$). *If $\langle D_i \mid i < \gamma \rangle \in \mathcal{A}^T$, each $D_i$ is $i^+$-predense on $R^T$ (i.e., $\forall q \exists r \leq q (r \restriction i^+ = q \restriction i^+$, $q$ extends an element of $D_i$)) and $p \in R^T$ then there exists $q \leq p$, $q$ meets each $D_i$, $i < \gamma$, (i.e., $q$ extends an element of $D_i$, for each $i < \gamma$).*

*Proof.* Follow the end of the proof of Lemma 4.10, again interpreting the nonlinear extensions added in Case 3 of the $R^\delta$ constructions to guarantee that conditions arising at limit stages do indeed exist. □

Of course we inductively assume that $\Sigma_f$-Density (in the sense of Lemma 8.43) holds for $R^T$, for each $T \in R^\gamma_\beta$. Now for each $T$, $p^*$, $\delta$, $j = 0$ or $1$ build a $\delta$-extension $T(p^*, \delta, j)$ as in the last part of the proof of Lemma 4.11 (taking $C^s$ from that proof to be $\{\mu_0^n \mid n \in \omega\}$ where $\mu_0$ is least so that $\mu^T = \cup\{\mu_0^n \mid n \in \omega\} = \mu_0^\omega$); this is possible using $\Sigma_f$-Density and Distributivity for $R^T$. To know that the result produces a $\gamma$-tree, again interpret Case 3 of the $R^\delta$ constructions to ensure this, when $\delta \neq \gamma \cap H_\omega(\delta)$, and otherwise use reflection from $\gamma$ to $\delta$ (where $H_\omega$ is defined as in the proof of Lemma 4.11). We can also arrange that $(T_0, p_0^*, \delta_0, j_0) \neq (T_1, p_1^*, \delta_1, j_1) \implies [\hat{T}_0(p_0^*, \delta_0, j_0)] \cap [\hat{T}_1(p_1^*, \delta_1, j_1)] = \emptyset$ (in all set-generic extensions), where $\hat{T}(p^*, \delta, j)$ is the $\gamma$-tree resulting from the "quasicondition" $T(p^*, \delta, j)$. We place all such $\hat{T}(p^*, \delta, j)$ into $R^\gamma_\alpha$.

Now let $T^* = \hat{T}(p^*, \delta, j)$ belong to $R^\gamma_\alpha$; we shall define $R^{T^*} \subseteq \mathcal{A}^{T^*} = L_{\mu^{T^*}}[A \cap \gamma, T^*]$, where $\mu^{T^*} =$ least $\mu > \mu^T$ such that $0 < \mu' < \mu \implies \mu'\mu = \mu$. A *quasicondition* in $R^{T^*} - R^T$ is a sequence $q = \langle (q_\delta, q_\delta^*) \mid \delta \in \{0\} \cup \text{Card} \cap \gamma \rangle \in \mathcal{A}^{T^*}$ where $\langle q_\delta \mid \delta \in \{0\} \cup \text{Card} \cap \gamma \rangle$ is a $\gamma$-tree extending $T^*$, for each $\delta$, $q_\delta^*$ is a bounded subset of $\delta^+$ (taking $0^+ = \omega$) and where $q$ extends $T(p^*, \delta, j)$. Quasiconditions are ordered as before. A *condition* in $R^{T^*}$ is either an element of $R^T$ or is a quasicondition $q$ in $R^{T^*} - R^T$ such that:

(1) If $\gamma$ is inaccessible in $\mathcal{A}^{T^*}$ then there is a CUB $C \subseteq \gamma$, $C \in \mathcal{A}^{T^*}$ such that $\bar{\gamma} \in C \implies q_{\bar{\gamma}}^* = \emptyset$.

(2) There is a ("good approximation" to $\mathcal{A}^{T^*}$) $\mathcal{A} = \langle \tilde{J}_\mu[A \cap \gamma, T^*], A \cap \gamma, T^* \rangle$ where $\mu^T \leq \mu < \mu^{T^*}$, $\mu$ limit such that for some $k$ : $q$ is $\Delta_1 \langle L_\gamma[A], g_k^{\mathcal{A}} \rangle$ and $g_k^{\mathcal{A}} \restriction \bar{\gamma} \in \mathcal{A}^{q_{\bar{\gamma}}}$ for sufficiently large $\bar{\gamma} \in \text{Card}' \cap \gamma$ (where $g_k^{\mathcal{A}}$ is defined as before).

As in Case 1, $\Sigma_f$-Density (Lemma 8.43) holds for $R^{T^*}$, by interpreting Case 3 of the $R^\delta$ constructions, $\delta \in \text{Card} \cap \gamma$, so as to guarantee that conditions arising at limit stages do indeed exist.

**Case 2B.** $\alpha = \beta + 1$, $\beta$ odd.
For each $T \in R^{\gamma}_{\leq\beta} = \cup\{R^{\delta}_{\eta} \mid \eta \leq \beta\}$, each $\eta < \gamma$ and $\langle T_0, \langle T_u \mid u \in \text{Split}^{+}(T_{\eta^+}, \eta)\rangle\rangle = \vec{T}_\eta$ consisting of extensions of $T$ from $R^{\gamma}_{\beta}$ we define the $\gamma$-tree $T(\vec{T}_\eta) = T^* = \langle T^*_{\bar{\gamma}} \mid \bar{\gamma} \in \{0\} \cup \text{Card} \cap \gamma\rangle$ by: $T^*_{\bar{\gamma}} = T_{\bar{\gamma}}(\langle T_{0_{\bar{\gamma}}}, \langle T_{u_{\bar{\gamma}}} \mid u \in \text{Split}^{+}(T_{\eta^+}, \eta)\rangle\rangle)$ if $\bar{\gamma} \in \text{Card}^{+} \cap \gamma$, $\bar{\gamma} \geq \eta^+$ and $T^*_{\bar{\gamma}} = T_{0_{\bar{\gamma}}}$ otherwise. For this to be well-defined we additionally assume that $\langle \alpha(T_{u_{\bar{\gamma}}}) \mid \bar{\gamma} \in \text{Card}^{+} \cap \gamma\rangle$ is independent of $u$ and that $\alpha(T_{u_{\eta^+}}) \geq \alpha(T_{\eta^+})$. The proceed as in Case 2A for each such $T^* = T(\vec{T}_\eta)$, building two $\delta$-extensions $\hat{T}^*(p^*, \delta, 0)$, $\hat{T}^*(p^*, \delta, 1)$ for each choice of restraint $p^\delta$, $\delta \in \text{Card}^{+} \cap \gamma$ and $j < 2$. We assume that $\mathcal{A}^{T_u}$ is constant for $u \in \{0\} \cup \text{Split}^{+}(T_{\eta^+}, \eta)$ and that $T^*, p^* \in \mathcal{A}^T$ for such $u$. The set $R^{T^*}$ is defined just like $R^T$ for $T \in R^{\gamma}_{\beta}$.

**Case 3.** $\alpha$ limit, $\alpha > \gamma$.
As before, we add both linear and nonlinear $\gamma$-trees to $R^{\gamma}_{\alpha}$. First we discuss the linear case, which is patterned closely after the treatment in Section 4.3.

For each $T \in R^{\gamma}_{<\alpha} = \cup\{R^{\gamma}_{\beta} \mid \beta < \alpha\}$ and restraint $p^*$ with $(T, p^*) = \langle (T_{\bar{\gamma}}, p^*_{\bar{\gamma}}) \mid \bar{\gamma} \in \{0\} \cup \text{Card} \cap \gamma\rangle \in R^T$ we aim to introduce canonical extensions $T^*(T, p^*, b)$ for each $b \subseteq [\alpha(T), \alpha)$ belonging to $S(T, p^*) =$ "the $(T, p^*)$-reshaped strings." The latter is defined to consist of all $b : [\alpha(T), |b|) \longrightarrow 2$ such that for $\eta \leq |b|$, $L[A \cap \gamma, T, p^*, b \restriction \eta] \models \text{card}\,\eta \leq \gamma$. We note that the $\emptyset$ string belongs to $S(T, p^*)$ as inductively it may be verified that $L[A \cap \gamma, T] \models \text{card}\,\alpha(T) = \gamma$.

For each $b \in S(T, p^*)$ we define $\mathcal{A}^{<b}$, $\mathcal{A}^b$ and $\hat{\mathcal{A}}^b$ as in Section 4.3, but always using $A \cap \gamma$, $T$, $p^*$ instead of just $A \cap \gamma$. Also, we take $\mathcal{A}^{<\emptyset}$ to be $\hat{\mathcal{A}}^T$ (defined inductively).

Now we have the following versions of Relativized $\Box$ and $\diamondsuit$:

## Relativized $\Box(T, p^*)$

There exists $\langle C^b \mid b \in S(T, p^*)\rangle$ such that:

(a) $b \neq \emptyset \Longrightarrow C^b$ is CUB in $\mu^{<b}$, ordertype $C^b \leq \gamma$, $C^b \in \mathcal{A}^b$.

(b) $\mu \in \text{Lim}\, C^b \Longrightarrow \mu = \mu^{b \restriction \eta}$ for some $\eta \leq |b|$ and $C^{b\restriction\eta} = C^b \cap \mu$.

(c) $\langle \mathcal{A}^{<b}, C^b\rangle$ is collapsible.

(d) $b \neq \emptyset$, $D \subseteq \mathcal{A}^{<b}$, $D \in (\mathcal{A}^{<b})^{+} \Longrightarrow D$ is $\Delta_1\langle \mathcal{A}^{<b}, C^b\rangle$.

## Relativized $\diamondsuit(T, p^*)$

There exists $\langle D^b \mid b \in S(T, p^*)\rangle$ such that:

(a) $D^b \subseteq \mathcal{A}^{<b}$, $\langle D^t \mid t \text{ an initial segment of } b\rangle \in \mathcal{A}^b$.

(b) If $D \subseteq \mathcal{A}^{<b}$, $D \in \hat{\mathcal{A}}^b \neq \mathcal{A}^{<b}$ then $\{\eta < |b| \mid D^{b\restriction\eta} = D \cap \mathcal{A}^{<b\restriction\eta}\}$ is stationary in $\hat{\mathcal{A}}^b$.

(c) If $\mu^{<b\restriction\eta} \in \operatorname{Lim} C^b$, $\eta < |b|$ then $D^{b\restriction\eta} = \emptyset$. If $\hat{\mathcal{A}}^b \models |b|^{++}$ exists then $D^b = \emptyset$. And if $\pi : \langle \mathcal{A}^{<\bar{b}}, \bar{C}\rangle \longrightarrow \langle \mathcal{A}^{<b}, C^b\rangle$ is $\Sigma_1$-elementary, $\pi(\bar{\gamma}) = \gamma$ where $\bar{b} \in S(T \restriction \bar{\gamma}, p^* \restriction \bar{\gamma})$ then $D^{\bar{b}} = \pi^{-1}[D^b]$.

Now suppose that $b \in S(T, p^*)$, $|b| = \alpha$. We take $R^{<b}$ to consist of all $(T_0, p_0^*) \leq (T, p^*)$ with $p_0^* \in R^{T_0}$, $T_0$ a linear extension of $T$, $T_0 \in R^\gamma_{<\alpha}$ such that $R \in [T_0] \Longrightarrow b^R$ agrees with $b$ on $[\alpha(T), \alpha(T_0))$ in set-generic extensions where $b^R = \{\beta \mid T_R^{\beta+1} = \hat{T}(p^*, \delta, 1) \text{ or } \hat{T}^*(p^*, \delta, 1) \text{ for some } T, p^*, \delta\}$. We have the following version of Lemma 4.10:

**Lemma 8.45.** *If $\langle D_i \mid i < \gamma\rangle \in \hat{\mathcal{A}}^b$, each $D_i$ is $i^+$-predense on $R^{<b}$ and $(T_0, p_0^*) \in R^{<b}$ then there exists $(T_1, p_1^*) \leq (T_0, p_0^*)$ meeting each $D_i$, $i < \gamma$.*

Now using Lemma 8.45, we can use the proof of Lemma 4.11 (Extendibility) to construct a canonical extension $(T^*, p^{**})$ of $(T, p^*)$ as the (linear) limit of conditions in $R^{<b}$. We set $T^*(T, p^*, b)$ equal to $T^*$ and write $p^{**}$ as $p^{**}(T, p^*, b)$. The linear $\gamma$-trees in $R^\gamma_\alpha$ are the $T^*(T, p^*, b)$ for some $T, p^*, b$ as above.

For $T^* = T^*(T, p^*, b)$ a linear element of $R^\gamma_\alpha$ we take $\mathcal{A}^{T^*}$ to be $\mathcal{A}^b$ and define $R^{T^*}$ to consist of $\cup\{R^{T_0} \mid T^* \leq T_0 \leq T, T_0 \neq T^*\}$ together with all quasiconditions $q \in \mathcal{A}^{T^*} - \mathcal{A}^{<b}$ such that:

(1) If $\gamma$ is inaccessible in $\mathcal{A}^{T^*}$ then there is a CUB $C \subseteq \gamma$, $C \in \mathcal{A}^{T^*}$ such that $\bar{\gamma} \in C \Longrightarrow q^*_{\bar{\gamma}} = \emptyset$.

(2) There is a *good approximation* $\langle \mathcal{A}, C\rangle$ to $\mathcal{A}^{T^*}$ such that for some $k$: $q$ is $\Delta_1\langle L_\gamma[A], g_k^{\langle \mathcal{A},C\rangle}\rangle$ and $g_k^{\langle \mathcal{A},C\rangle} \restriction \bar{\gamma} \in \mathcal{A}^q$ for sufficiently large $\bar{\gamma} \in \text{Card}' \cap \gamma$ (where a good approximation to $\mathcal{A}^{T^*} = \mathcal{A}^b$ is defined just like in Section 4.3, using $A \cap \gamma, T, p^*$ instead of just $A \cap \gamma$).

(3) If $D^b$ is contained in $R^{<b}$ and is $\beta$-predense for all $\beta \in \text{Card}^+ \cap \gamma$ then $q$ meets $D^b$.

This completes the definition of the set of linear $T^*$ in $R^\gamma_\alpha$ and of $\mathcal{A}^{T^*}$, $R^{T^*}$ for such $T^*$. We also define $\Sigma^p_f$ for $p \in R^{T^*}$, $f$ small in $\mathcal{A}^{T^*}$ as before and then the proof of Lemma 4.12 can be adapted to show:

**Lemma 8.46.** *For $T^*$ linear, $T^* \in R^\gamma_\alpha$ and $p \in R^{T^*}$, $f$ small in $\mathcal{A}^{T^*}$ there exists $q \leq p$, $q \in \Sigma^p_f$.*

The nonlinear elements of $R^\gamma_\alpha$ are described as follows: A real $R$ (in a set-generic extension) is *special at* $\alpha$ if it meets the following conditions:

(1) $L_\alpha[A] \models \gamma$ is the largest cardinal.

(2) $R$ canonically codes a structure $\mathcal{A}^R$ arising in the proof of Extendibility, Distributivity or $\Sigma_f$-Density for some $R^T$, $T$ a $\mu$-tree for some $\mu \in$ Card (in the sense of $\mathcal{A}^R$) where $T$ is canonically determined by $\mathcal{A}^R$ and $\alpha = \gamma^+$ of $\mathcal{A}^R$.

(3) $\mathcal{A}^R$ gives rise to a $\lambda$-sequence of nonlinear extensions $T_0 \geq T_1 \geq \cdots$ from $R^\gamma_{<\alpha}$ with well-defined limit $T^R$, and $\alpha = \bigcup\{\alpha(T_i) \mid i < \lambda\}$.

(4) $S \in [T^R] \Longrightarrow \mathcal{A}^S, T^S$ are defined and $\mathcal{A}^S = \mathcal{A}^R$, $T^S = T^R$ (in all set-generic extensions).

Then we put $T^* = T^R$ into $R^\gamma_\alpha$. We take $\mathcal{A}^{T^*} = L_\mu[A \cap \gamma, T^*]$ where $\mu =$ the least $\mu > \alpha$ such that $0 < \mu' < \mu \Longrightarrow \mu'\mu = \mu$ and $L_\mu[A \cap \gamma, T^*] \models \text{Card}(\alpha) = \gamma$. And $R^{T^*}$ consists of all quasiconditions in $A^{T^*}$ extending $(T^*, \emptyset)$ and obeying (1), (2) from the linear case. The proof of $\Sigma_f$-density is as before, so Lemma 8.46 holds without the assumption that $T^*$ is linear.

This completes the construction of $R^\gamma_\alpha$, $\gamma \leq \alpha < \gamma^+$, for all $\gamma \in$ Card. We only must specify the meaning of (3) in the definition of $R^T$, $T$ the full $\gamma$-tree ($\gamma \in$ Card′):

(3) Suppose that $D \subseteq P^\gamma = \{r \restriction \delta \mid r$ a quasicondition, $\delta \in \text{Card}^+ \cap \gamma\}$ is dense and $\Sigma_n$-definable over $\langle L_\gamma[A], A \cap \gamma\rangle$, where $\gamma$ is $\Sigma_{n+5}$-regular over $\langle L_\gamma[A], A \cap \gamma\rangle$. Then $q$ reduces $D$ below $\bar{\gamma}$ for some $\bar{\gamma} \in \text{Card} \cap \gamma$.

This clause does not interfere with the construction of conditions in $R^T$, as we may as before establish Distributivity for $P^\gamma$, for $\Sigma_n\langle L_\gamma[A], A\cap\gamma\rangle$ sequences of dense sets, using Case 3 of our constructions to ensure the existence of lower bounds at limit stages. Note that by reflection we therefore have Distributivity for $P = \bigcup\{P^\gamma \mid \gamma \in \text{Card}'\}$ and therefore a $P$-generic does code $A$ by a real, preserving cofinalities. Finally, we have:

**Lemma 8.47.** *If $R$ is a $P$-generic real then $R$ is minimal over $L[A]$.*

*Proof.* It suffices to observe:

**Claim.** Suppose $p \Vdash \sigma \subseteq \text{ORD}$, $\sigma \notin L[A]$. Then there exist $p_0, p_1 \leq p$ and $\alpha$ such that $p_0 \Vdash \alpha \notin \sigma$, $p_1 \Vdash \alpha \in \sigma$ and $p_0, p_1$ agree everywhere except at 0.

Given the Claim, we can use amalgamation and Density Reduction to build $p_0 \geq p_1 \geq \cdots$ with lower bound $p$ such that $s, t$ incompatible strings in $2^{<\omega}$ lying on $p_\omega \Longrightarrow (s, p \restriction \text{Card})$ and $(t, p \restriction \text{Card})$ force incompatible facts about $\sigma$. It follows that $p \Vdash R \in L[A, \sigma]$, where $R$ is the generic real.

The Claim follows from the amalgamations built into the construction of $P$. □

## 8.4 Further Applications to Descriptive Set Theory

Solovay [70] established the consistency of a number of regularity properties for projective sets of reals, using a natural model in which $\omega_1$ *is inaccessible to reals*, (i.e., $\omega_1$ is an inaccessible cardinal in $L[R]$ for each real $R$). In this section we construct other models with this property, which can be applied to the study of regularity properties for projective sets and projective prewellorderings.

A set of reals is $\mathbf{\Sigma}^1_1$ if it is the continuous image of a Borel set and is $\mathbf{\Pi}^1_1$ if its complement is $\mathbf{\Sigma}^1_1$. It is $\mathbf{\Sigma}^1_{n+1}$ if it is the continuous image of a $\mathbf{\Pi}^1_n$ set and is $\mathbf{\Pi}^1_{n+1}$ if its complement is $\mathbf{\Sigma}^1_{n+1}$. A set of reals is $\mathbf{\Delta}^1_n$ if both it and its complement are $\mathbf{\Sigma}^1_n$. Similar definitions apply to $k$-ary relations on the reals. It a set of reals (or $k$-ary relation in reals) is $\mathbf{\Sigma}^1_n$ for some $n$ then we say that it is *projective*.

### Regularity Properties

**Definition.** Measure ($\mathbf{\Sigma}^1_n$) is the assertion that every $\mathbf{\Sigma}^1_n$ set of reals is Lebesgue Measurable. Category ($\mathbf{\Sigma}^1_n$) is the assertion that every $\mathbf{\Sigma}^1_n$ set of reals has the Baire Property, i.e., has meager symmetric difference with some Borel set. Perfect ($\mathbf{\Sigma}^1_n$) is the assertion that any uncountable $\mathbf{\Sigma}^1_n$ set of reals contains a perfect closed subset. Similar definitions apply to $\mathbf{\Pi}^1_n$, $\mathbf{\Delta}^1_n$.

In ZFC one may prove Measure ($\mathbf{\Sigma}^1_1$), Category ($\mathbf{\Sigma}^1_1$), Perfect ($\mathbf{\Sigma}^1_1$). In Gödel's model $L$ one has ~Measure ($\mathbf{\Delta}^1_2$), ~Category ($\mathbf{\Delta}^1_2$), ~Perfect ($\mathbf{\Pi}^1_1$) using the fact that in $L$ there is a $\mathbf{\Delta}^1_2$ wellordering of the reals (and the Kondo–Addison Uniformization Theorem for $\mathbf{\Pi}^1_1$). By extending ZFC slightly we get:

**Theorem 8.48** (Solovay [69]). *Assume that $\omega_1$ is inaccessible to reals. Then the following hold: Measure ($\mathbf{\Sigma}^1_2$), Category ($\mathbf{\Sigma}^1_2$), Perfect ($\mathbf{\Sigma}^1_2$).*

Our next result implies that the previous theorem is optimal. The proof is based entirely on ideas from David [83].

**Theorem 8.49.** *Assume the consistency of an inaccessible cardinal. Then there is a model in which:*

(a) *$\omega_1$ is inaccessible to reals.*

(b) *There is a $\Delta^1_3$ wellordering of the reals, and hence ~Measure ($\Delta^1_3$), ~Category ($\Delta^1_3$).*

(c) *~Perfect ($\Pi^1_2$).*

**Remark.** We use $\Sigma^1_n$, $\Pi^1_n$, $\Delta^1_n$ to denote the "effective" versions of $\mathbf{\Sigma}^1_n$, $\mathbf{\Pi}^1_n$, $\mathbf{\Delta}^1_n$; see Moschovakis [80] for details.

*Proof of Theorem* 8.48. Let $\kappa$ be the least $L$-inaccessible; our model is obtained by iterated forcing over $L$.

First we define the $\alpha^+$-Suslin tree $T_\alpha$, where $\alpha$ is a successor cardinal. $T_\alpha$ has a unique node on level 0 and exactly 2 immediate successors on level $\beta + 1$ to each node on level $\beta$, for $\beta < \alpha^+$. If $\beta < \alpha^+$ is a limit ordinal of cofinality $< \alpha$ then level $\beta$ assigns a top to each branch through $T_\alpha \mid \beta = T_\alpha$ below level $\beta$. If $\beta < \alpha^+$ is a limit ordinal of cofinality $\alpha$ then let $\beta^*$ be the largest limit ordinal such that $\beta^* = \beta$ or $L_{\beta^*} \models \beta$ is a cardinal and obtain level $\beta$ by assigning a top to each branch $b$ through $T_\alpha \mid \beta$ such that $b \in L_{\beta^*+\omega}$ and $b$ is generic for $T_\alpha \mid \beta$ over $L_{\beta^*}$.

Now choose $F : \kappa \times \omega \longrightarrow \text{Card}^+$ to be the $L$-least injection, where $\text{Card}^+$ now denotes all successor cardinals greater than $\kappa$. The forcing $P(\gamma, n), \gamma < \omega_1$ and $n < \omega$, is designed to produce a "good" real $R$ coding a branch through $T_{F(\gamma,n)}$: First add a generic branch through $T_{F(\gamma,n)}$ and then code this branch by a real $R$ using David's Trick (see Chapter 6). Thus in defining the strings $s : [\alpha, |s|) \longrightarrow 2$ in $S_\alpha$, require that for $\xi \le |s|$ and $\eta > \xi$, if $L_\eta(s \restriction \xi) \models \xi = \alpha^+ + \text{ZF}^- + F(\gamma, n)$ is defined then $L_\eta(s \restriction \xi) \models \text{Even}(s \restriction \xi)$ codes a branch through $T_{F(\gamma,n)}$. The definition of $T_{F(\gamma,n)}$ ensures that this forcing is $\le F(\gamma, n)$-distributive when restricted to $[F(\gamma, n), \infty)$. Moreover if $R$ is $P(\gamma, n)$-generic then $T_\alpha$ remains $\alpha^+$-Suslin in $L[R]$ unless $\alpha = F(\gamma, n)$ and for every $\eta$, if $L_\eta(R) \models \text{ZF}^- + F(\gamma, n)$ is defined then $L_\eta(R) \models T_{F(\gamma,n)}$ is not $F(\gamma, n)^+$-Suslin.

The forcing $P(\gamma)$, $\gamma < \omega_1$ is designed to produce a "good" real $R$ such that for all $n$, $n \in R$ iff $R$ codes a branch through $T_{F(\gamma,n)}$. A condition in $P(\gamma)$ is an element $p$ of $\prod_n P(\gamma, n)$, where $p(n)(0)$ is $(\emptyset, \emptyset)$ for all but finitely many $n$. Extension is defined by: $q \le p$ iff $q(n) \le p(n)$ whenever $n = \langle n_0, n_1 \rangle$ and either $q(n_0)_0(n_1)$ is undefined or takes the value 1. The generic real is $R = \{\langle n_0, n_1 \rangle \mid \text{for all } p \in G,\ p(n_0)_0(n_1) = 1 \text{ or is undefined}\}$. This forcing preserves cofinalities and if $R$ is $P(\gamma)$-generic then $T_\alpha$ remains $\alpha^+$-Suslin in $L[R]$ unless $\alpha = F(\gamma, n)$ for some $n \in R$, in which case whenever $L_\eta(R) \models \text{ZF}^- + F(\gamma, n)$ is defined, then $L_\eta(R) \models T_{F(\gamma,n)}$ is not $F(\gamma, n)^+$-Suslin.

More generally, suppose that $R$ is a real and $\gamma$ is countable in $L[R]$. Then we may similarly define the forcing $P(R, \gamma)$ for adding a real $R^*$ preserving cofinalities over $L[R]$ such that $R = \text{Even}(R^*)$ and, provided $T_{F(\gamma,n)}$ remains $F(\gamma, n)^+$-Suslin in $L[R]$, $T_{F(\gamma,n)}$ remains $F(\gamma, n)^+$-Suslin in $L[R^*]$ unless $n \in R^*$, in which case whenever $L_\eta(R) \models \text{ZF}^- + F(\gamma, n)$ is defined, we have $L_\eta(R) \models T_{F(\gamma,n)}$ is not $F(\gamma, n)^+$-Suslin.

Lastly we describe the desired forcing iteration $P = \text{"Limit"} \langle P(< \gamma) \mid \gamma < \kappa \rangle$: Of course $P(< 0)$ is the trivial forcing. At limit stages $\gamma \le \kappa$, we take $P(< \gamma)$ to be the Inverse Limit of the $P(< \delta), \delta < \gamma$ with the added restriction that $\langle p_0(\delta), p_1(\delta)(0) \rangle$ is nontrivial for only finitely many $\delta < \gamma$ where $p(\delta) = \langle p_0(\delta), p_1(\delta) \rangle$. For $\delta < \kappa$ we take $P(\le \gamma) = P(< \gamma) * Q(\gamma)$ where $Q(\gamma)$ is obtained as follows: First collapse $\omega_1$ of $L[G(< \gamma)]$ to $\omega$ via the Lévy collapse forcing $Q_0(\gamma)$, and using the generic real $R^0_\gamma$ choose a real $R^1_\gamma$ coding the ordinal $\gamma$ as well as a wellordering of the reals in $L[G(< \gamma)]$ extending the wellorderings coded by $R^1_\delta$ for $\delta < \gamma$. Then $Q(\gamma) = Q_0(\gamma) * P(R^1_\gamma, \gamma)$, producing a real $R^*_\gamma$.

By the method of iterated class forcing (see Section 8.2) $P$ preserves cofinalities at or above $\kappa$. Thus in a $P$-generic extension, $\kappa = \omega_1$ and the reals are well-ordered by the union of the wellorderings coded by the $R^1_\gamma$, where $R^1_\gamma = \text{Even}(R^*_\gamma)$.

Let $X = \{R^*_\gamma \mid \gamma < \kappa\}$. Then $X$ is $\Pi^1_2$ as $R \in X$ iff for every $\eta$, if $L_\eta(R) \models \text{ZF}^-$ then $L_\eta(R) \models$ For some $\gamma, n \in R$ iff $T_{F(\gamma,n)}$ is not $F(\gamma,n)^+$-Suslin for all $n$. Thus the reals in our $P$-generic extension carry a $\Delta^1_3$ wellordering: $R_0 \leq R_1$ iff for some $R \in X$, $\text{Even}(R)$ codes a wellordering in which $R_0 \leq R_1$. If $X$ contained a perfect closed subset then the reals would have a $\boldsymbol{\Pi}^1_2$ (and hence $\boldsymbol{\Delta}^1_2$) wellordering, as $X$ carries a $\Pi^1_2$ wellordering. But this contradicts the fact that $\omega_1$ is inaccessible to reals in our model. □

Another axiom with consequences for regularity properties of projective sets is Martin's Axiom (MA). (We take MA to include the hypothesis $\sim$ CH.)

**Theorem 8.50.** *MA implies Measure* ($\boldsymbol{\Sigma}^1_2$), *Category* ($\boldsymbol{\Sigma}^1_2$).

Again this is optimal.

**Theorem 8.51.** *This is a model of MA in which:*

(a) $\omega_1 = \omega_1^L$.

(b) *There is a* $\Delta^1_3$ *wellordering of the reals.*

*Proof.* Perform the $\omega_2$-iteration for obtaining a model of MA, using finite support, but at odd stages $\alpha + 1$ add a real $R_\alpha$ as follows: $R_\alpha$ codes a branch through $T_{F(\alpha,n)}$ iff $n \in R_\alpha$ where $F : \omega_2 \times \omega \longrightarrow \text{Card}^+$ is as in the previous proof, with $\kappa$ replaced by $\omega_2$. Also require that $\text{Even}(R_\alpha) = (R^0_\alpha, R^1_\alpha)$ where $R^0_\alpha, R^1_\alpha$ belong to $L[G(<\alpha)]$ and $R^0_\alpha \leq R^1_\alpha$ in the canonical wellordering of the reals of that model. Use countable support on the components of the forcing that add the $R_\alpha$'s, except that once again $p(\alpha)(0)$ should be nontrivial for only finitely many $\alpha < \omega_2$.

Cofinalities are preserved by "factoring" into (Even Stages × Odd Stages at 0) × (Odd Stages at or above $\omega$). Finally, $R_0 \leq R_1$ iff there exists $R_\alpha$ such that $\text{Even}(R_\alpha) = (R_0, R_1)$ and thus as $\{R_\alpha \mid \alpha < \omega_2\}$ is $\Pi^1_2$, we get a $\Delta^1_3$ wellordering. □

**Remark.** Perfect ($\Pi^1_1$) fails in the above model, as this property implies that $\omega_1^L$ is countable. (See Jech [78].) It is not known if (a) can be replaced by "$\omega_1$ is inaccessible to reals" in the previous theorem (assuming the consistency of a weakly compact cardinal; this is a necessary assumption for the consistency of MA $+\omega_1$ inaccessible to reals).

Theorem 8.49 generalizes to higher levels of the projective hierarchy. Recall that $\kappa$ is *Mahlo* if $\kappa$ is inaccessible and $\{\bar{\kappa} < \kappa \mid \bar{\kappa} \text{ regular}\}$ is stationary.

**Theorem 8.52.** *Assume the consistency of a Mahlo cardinal. Then there is a model in which:*

(a) *Measure* ($\boldsymbol{\Sigma}^1_3$), *Category* ($\boldsymbol{\Sigma}^1_3$). *Perfect* ($\boldsymbol{\Sigma}^1_3$).

(b) *There is a* $\Delta^1_4$ *wellordering of the reals.*

(c) $\sim$*Perfect* $(\Pi^1_3)$.

*Proof.* We modify the forcing $P(R, \gamma)$ used in the proof of Theorem 8.49. Choose $F : \kappa \times \kappa \times \omega \longrightarrow \text{Card}^+$ where $\text{Card}^+$ now denotes the set of infinite successor cardinals *less* than $\kappa$. We assume that $F(\gamma, \delta, n) > \gamma$ all $(\gamma, \delta, n)$. The forcing $P(\gamma, n)$, $\gamma < \omega_1$ and $n < \omega$, produces a real $R$ coding a branch through $T_{F(\gamma,\delta,n)}$ for all $\delta < \kappa$. (There is no use here of David's trick.) The forcing $P(\gamma)$ produces a real $R$ such that for all $n$, $n \in R$ iff $R$ codes a branch through $T_{F(\gamma,\delta,n)}$ for all $\delta < \kappa$. More generally, suppose that $R$ is a real, $\gamma$ is countable in $L[R]$ and $T_{F(\gamma,\delta,n)}$ is $F(\gamma, \delta, n)^+$-Suslin in $L[R]$ for each $\delta, n$. Then we define $P(R, \gamma)$ to add a real $R^*$ such that $n \in R^*$ iff $R^*$ codes a branch through $T_{F(\gamma,\delta,n)}$ for all $\delta$. *However*, we impose the key requirement that in all of the above forcings, conditions have domain bounded in $\kappa$, so that each of these forcings is essentially of cardinality $\kappa$.

The iteration $P$ is defined as before, where at stage $\gamma < \kappa$, $Q(\gamma)$ first adds a Lévy collapse of $\gamma$, chooses a real $R^1_\gamma$ coding a wellordering of the reals in $L[G(<\gamma)]$ (and canonically specifies the ordinal $\gamma$) and applies $P(R^1_\gamma, \gamma)$. However, once again we restrict the iteration so that we can view $P$ as a subset of $L_\kappa$, by taking a direct limit at stage $\kappa$.

Again let $X = \{R^*_\gamma \mid \gamma < \kappa\}$. Then in the generic extension, $X$ is $\Pi^1_3$ as $R \in X$ iff for all $\delta$, $R$ codes a branch through $T_{F(\gamma,\delta,n)}$ exactly if $n \in R$, where $\gamma$ is canonically specified by $R$. It follows that the reals carry a $\Delta^1_4$ wellordering.

Now suppose that $P \Vdash \varphi$ where $\varphi$ is $\Sigma^1_3$. Let $M$ be a sufficiently elementary submodel of $L$ containing $\kappa$ as an element such that $M \cap \kappa = \bar{\kappa} < \kappa$. The $\bar{P} = P \cap L_{\bar{\kappa}}$ forces $\varphi$ in $\bar{M}$ = transitive collapse of $M$. If $\bar{P}$ is $\bar{\kappa}$-preserving in $L$ then by persistence $\bar{P}$ forces $\varphi$ in $L$ and we have shown that any $\Sigma^1_3$ sentence forced by $P$ is forced by some $\bar{P} \in L_\kappa$. To obtain $\bar{\kappa}$-preservation, note that by Mahloness we can choose $\bar{\kappa}$ to be regular and then apply the usual $\Delta$-stratification arguments to the forcing $\bar{P} = P \cap L_{\bar{\kappa}}$.

Also note that if $L[G]$ is a $P$-generic extension then for every real $R \in L[G]$, $\kappa$ is Mahlo in $L[R]$, again using the $\Delta$-stratification of $P$. It follows then by relativization of the above argument that if $R \in L[G]$, $S$ is generic over $L[R]$ for a forcing $P_0 \in L_\kappa[R]$ and $L[G] \models \varphi(R, S)$ where $\varphi$ is $\Sigma^1_3$ then $\varphi(R, S)$ holds in an extension of $L[R, S]$ via a forcing of cardinality $< \kappa$ (which we may assume to be the Lévy collapse of some $\alpha < \kappa$; note that as $\kappa$ is inaccessible to reals and equals $\omega_1^{L[G]}$, such generic extensions can be found inside $L[G]$).

It therefore follows that in $L[G]$, if $Y$ is a $\mathbf{\Sigma}^1_3$ set of reals then for every real $R$ and forcing $P \in L_\kappa[R]$ for adding a real, there is a Borel set $B$ such that $Y$ and $B$ agree on reals which are $P$-generic over $L_\kappa[R]$. If $P$ is Random forcing, then we get the Lebesgue measurability of $Y$; if $P$ is Cohen forcing then we get the Baire property for $Y$. Finally note that if $Y$ has an element not constructible from its defining parameter $R$ then it has such an element in a generic extension of $L[R]$ via a forcing in $L_\kappa[R]$, and therefore a perfect set of such elements.

As before $X$ witnesses the failure of Perfect $(\Pi^1_3)$. □

**Remark.** To generalize the previous argument, we must replace $L$ by a sufficiently $\Sigma^1_3$ correct model. Thus, assuming the consistency of "every set has a sharp" together with a Mahlo cardinal, one obtains a model of Measure ($\boldsymbol{\Sigma}^1_4$), Category ($\boldsymbol{\Sigma}^1_4$), Perfect ($\boldsymbol{\Sigma}^1_4$), ~Perfect ($\Pi^1_4$) with a $\Delta^1_5$ wellordering of the reals. However the author does not know if this use of #'s is necessary.

## Prewellorderings

A *prewellordering* is a reflexive, transitive well-founded relation. A wellordering is obtained by identifying two elements $a, b$ when $a \leq b$, $b \leq a$; the *length* of the prewellordering is the ordertype of its associated wellordering. $\boldsymbol{\delta}^1_n$ denotes the supremum of the lengths of all $\boldsymbol{\Delta}^1_n$ prewellorderings of the reals.

**Theorem 8.53** (Classical, see Jech [78]). $\boldsymbol{\delta}^1_1 = \omega_1$.

Kunen and Martin showed that $\boldsymbol{\delta}^1_2$ is at most $\omega_2$ (see Martin [77]). The next result shows that this result is the best possible.

**Theorem 8.54.** *It is consistent with* ZFC *that* $\boldsymbol{\delta}^1_2 = \omega_2$.

*Proof.* Recall the almost disjoint coding method of Jensen–Solovay [70]: If $V = L$ and $A \subseteq \omega_1$ then by a ccc forcing one may add a real $R$ such that $A \in L[R]$. In fact one obtains that $A$ is $\Delta_1(L_{\omega_1}[R])$, using conditions $(s, s^*)$ where $s \in 2^{<\omega}$ and $s^*$ is a finite subset of $\{b_\xi \mid \xi \in A\}$ where $\langle b_\xi \mid \xi < \omega_1\rangle$ is a $\Delta_1(L_{\omega_1})$ sequence of almost disjoint subsets of $\omega$. Let $P^A$ denote this forcing.

Define $P = \sum\{P^A \mid A \subseteq \omega_1\}$ where finite support is used. Then by the usual $\Delta$-system argument (used to prove the ccc for Cohen's forcing to make $2^\omega = \omega_2$) we have that $P$ is ccc. But in a $P$-generic extension of $L$, every constructible subset of $\omega_1$ is $\Delta_1(L_{\omega_1}[R])$ for some real $R$ and therefore by considering subsets of $\omega_1$ which code wellorderings of arbitrary length $< \omega_2$, we see that for every $\alpha < \omega_2$ there is a real $R$ and a $\Sigma^1_2$ in $R$ wellordering of the reals of length $\alpha$. This easily gives us a $\boldsymbol{\Delta}^1_2$ prewellordering of all the reals of length $\alpha$. □

Using the Condensation Condition of Section 8.1, we can simultaneously have $\omega_1$ inaccessible to reals:

**Theorem 8.55** (Friedman–Woodin [96]). *Assuming the consistency of an inaccessible, there is a model in which* $\boldsymbol{\delta}^1_2 = \omega_2$ *and* $\omega_1$ *is inaccessible to reals.*

*Proof.* Suppose that $\kappa$ is the least inaccessible and $V = L$. Let $\langle \alpha_i \mid i < \kappa^+\rangle$ be the increasing list of all $\alpha \in (\kappa, \kappa^+)$ such that $L_\alpha =$ Skolem Hull of $\kappa$ in $L_\alpha$. For each $i < \kappa^+$ define $f_i : \kappa \longrightarrow \kappa$ by $f_i(\gamma) =$ ordertype(ORD $\cap$ Skolem Hull of $\gamma$ in $L_{\alpha_i}$). We may identify $f_i$ with a subset of $\kappa$. Using the absoluteness of the definition of the $f_i$'s we have the following:

**Lemma 8.56.** *The sequence $\langle f_i \mid i < \kappa^+ \rangle$ obeys the "joint Condensation Condition": Suppose $t$ is transitive, $\delta$ regular, $\delta \in t$, $x \in t$. Then there is a tight $\delta$-sequence $x_0 \prec x_1 \prec \cdots \prec t$ such that* $\mathrm{Card}(x_i) = \delta$, *$x \in x_0$ and each $x_i$ strongly preserves the $f_j$ for $j \in x_i \cap \kappa^+$ (and if $\delta = \kappa$ we can alternatively require* $\mathrm{Card}(x_i) = \aleph_i$*).*

Now we use a "diagonally supported" product of $\Delta_1$-codings: For each $i < \kappa^+$ let $P(i)$ be the forcing from Section 8.1 to $\Delta_1$-code $f_i$ by a real. Then $P$ consists of all $p \in \prod\{P(i) \mid i < \kappa^+\}$ such that for infinite cardinals $\gamma$, $\{i \mid p(i)(\gamma) \neq (\emptyset, \emptyset)\}$ has cardinality at most $\gamma$ and in addition $\{i \mid p(i)(0) \neq (\emptyset, \emptyset)\}$ is finite.

$P$ preserves cofinalities: For successor cardinals $\gamma < \kappa$ the forcing $P$ factors as $P_\gamma * P^{G_\gamma}$ where $P_\gamma$ forces that $P^{G_\gamma}$ has the $\gamma^+$-cc. Also the "joint Condensation Condition" implies that the proof of $\Delta_1$-coding can be applied to show that $P_\gamma$ is $\gamma^+$-distributive and that $P$ is $\Delta$-distributive at $\kappa$.

Thus in a cofinality-preserving extension of $L$ we have produced $\kappa^+$ reals $\langle R_i \mid i < \kappa^+ \rangle$ where $R_i$ $\Delta_1$-codes $f_i$ and hence there are wellorderings of $\kappa$ of any length $< \kappa^+$ which are $\Sigma_1$ in a real. Finally Lévy collapse to make $\kappa$ equal to $\omega_1$ and we get $\delta^1_2 = \kappa^+ = \omega_2$ with $\omega_1$ inaccessible to reals. □

There is no explicit bound on $\delta^1_3$ provable in ZFC, even with the added hypothesis that $\omega_1$ is inaccessible to reals.

**Theorem 8.57** (Friedman [94a]). *Assuming the consistency of an inaccessible, there is a model in which $\omega_1$ is inaccessible to reals and there is a $\Pi^1_2$ wellordering of some set of reals of length $\kappa$, for any pre-chosen $L$-definable cardinal $\kappa$ (and hence $\delta^1_3 \geq \kappa$).*

*Proof.* First we show how to force a set of reals $X$ over $L$ so that cofinalities are preserved, $X$ has cardinality $\kappa$ in $L(X)$ and $X$ has a $\Pi^1_2$ definition which is uniform over all set-generic extensions of $L(X)$. To accomplish this we use the $\gamma^+$-Suslin trees $T_\gamma$ as in the proof of Theorem 8.49 and a diagonally supported product as in the proof of the previous theorem.

Fix an $L$-definable 1-1 function $F : \kappa \times \omega \times \mathrm{ORD} \longrightarrow$ Successor $L$-cardinals greater than $\kappa$. The forcing $P(\gamma, n)$, $\gamma < \kappa$ and $n \in \omega$, produces a real coding branches through each $T_{F(\gamma,n,\delta)}$, using David's trick as in the proof of Theorem 8.49. Then $P(\gamma)$, $\gamma < \kappa$ produces a real $R(\gamma)$ such that $n \in R(\gamma)$ iff $R(\gamma)$ codes a branch through each $T_{F(\gamma,n,\delta)}$, also as in the proof of Theorem 8.49.

$P$ is the diagonally-supported product of the $P(\gamma)$, $\gamma < \kappa$, as defined in the previous proof. Then $P$ preserves cofinalities and if we let $X = \{R(\gamma) \mid \gamma < \kappa\}$ where $R(\gamma)$ denotes the $P(\gamma)$-generic real added by $P$, we have: $R \in X$ iff for every $\mathrm{ZF}^-$-model $M$ containing $R$, $M \models$ For some $\gamma < \kappa^M$, $n \in R$ iff $R$ codes a branch through each $T^M_{F^M(\gamma,n,\delta)}$, where $\kappa^M$, $T^M_\alpha$, $F^M$ are defined in $M$ as $\kappa$, $T_\alpha$, $F$ are defined in $L$. This gives $\Pi^1_2$ definition of $X$ which is uniform over all set-generic extensions of $L(X)$.

Now we can guarantee that $Y = \{(R(0), R(\gamma_1), R(\gamma_2)) \mid 0 < \gamma_1 \leq \gamma_2 < \kappa\}$ also has such a uniform $\Pi^1_2$ definition: Design $R(0)$ so that $n \in R(0)$ iff $\mathrm{Even}(R(0))$ codes a branch through each $T_{F(0,n,\delta)}$ and also so that $\mathrm{Odd}(R(0))$ codes $\{(R(\gamma_1), R(\gamma_2)) \mid$

$0 < \gamma_1 \le \gamma_2 < \kappa\}$. This requires only a small modification of the forcing $P$. Now $R$ belongs to $Y$ iff $R = (R_0, R_1, R_2)$ where $R_0 = R(0)$, $R_1$ and $R_2$ belong to $X$ and $(R_1, R_2)$ belongs to the set coded by $R_0$. Since $R(0)$ is a $\Pi^1_2$-singleton uniformly over all set-generic extensions of $L(X)$, this is the desired uniform $\Pi^1_2$ definition of $Y$. Finally collapse to make the least $L$-inaccessible equal to $\omega_1$ to obtain a model with $\omega_1$ inaccessible to reals and a $\Pi^1_2$ wellordering of length $\kappa$. □

# Some Open Problems

1. Is there an $L$-definable forcing $P$ for which $P$-forcing is *not* definable?

2. Can one code a class by a real preserving $\Pi^n_m$-indescribability for $n > 1$?

3. Suppose that $\alpha$ and $\beta$ are countable ordinals, $\beta$ nonzero. Then is there a real $R$ such that $I^R = I_{\alpha,\beta}$ and $R$ preserves $L$-cofinalities?

4. Define *$n$-generic over $L$* as follows: $R$ is *0-generic over $L$* iff $R$ is generic over $L$. $R$ is *$n+1$-generic over $L$* iff $R$ is generic over an inner model of $L[S]$, where $S$ is $n$-generic over $L$. Does $n+1$-genericity imply $n$-genericity for some $n$? Is there a real $R <_L 0^\#$ which is *not* $n$-generic over $L$ for any $n$?

5. Is $0^\#$ generic over some proper inner model of $L[0^\#]$?

6. Can one prove that $L[0^\#]$ is generically saturated over $L$ in the theory ZFC $+0^\#$ exists?

7. Is $L[0^\#]$ the *least* inner model which is generically saturated over $L$?

8. Is there a reasonable notion of "forcing" with the property that every real either constructs $0^\#$ or can be obtained by "forcing" over $L$?

9. Is there a real $R$ such that $0 <_L R <_L 0^\#$ which is the unique solution to a $\Pi^1_2$ formula $\varphi$ which *provably in* ZFC has at most one solution?

10. Is there a simple characterization of the reals which belong to a countable $\Pi^1_2$ set?

11. Assuming only the consistency of an inaccessible cardinal, is it consistent for each $n$ that all $\Sigma^1_n$ sets of reals be Lebesgue Measurable and have the Baire and Perfect Set properties, while there is a $\Delta^1_{n+1}$ wellordering of the reals?

12. Assuming only the consistency of a weakly compact cardinal, is it consistent to have Martin's Axiom, $\omega_1$ inaccessible to reals, and a $\Delta^1_3$ wellordering of the reals?

13. Is it consistent for $\Delta^1_3$-reducibility and $L$-reducibility to coincide?

14. Assuming only the consistency of an inaccessible cardinal, is it consistent for Post's Problem to fail in HC $=$ the hereditarily countable sets?

15. Is there a *remarkable real*; i.e., a real $R <_L 0^\#$ such that $R$ is not generic over $L$, $R$ is a $\Pi^1_2$-singleton, $\Lambda(R) =$ the recursively inaccessible ordinals and $R$ has minimal $L$-degree? It has not yet been shown that there is a real $R <_L 0^\#$ which has more than one of these properties simultaneously.

16. Is it consistent that any parameter-free $\Sigma^1_3$ sentence true in a class forcing extension of $V$, be already true in $V$?

# References


Barwise [75] Barwise, J., Admissible Sets and Structures, Perspectives in Mathematical Logic, Springer-Verlag, Berlin–Heidelberg–New York 1975.

Beller–Jensen–Welch [82] Beller, A., Jensen, R. B., Welch, P., Coding the Universe, London Math. Soc. Lecture Note Ser. 47, Cambridge University Press, Cambridge 1982.

Cohen [66] Cohen, P. J., Set Theory and the Continuum Hypothesis, W. A. Benjamin Inc., New York–Amsterdam 1966.

David [82] David, R., Some Applications of Jensen's Coding Theorem, Ann. of Math. Logic 22 (1982), 177–196.

David [82a] David, R., A Very Absolute $\Pi^1_2$-Singleton, Ann. of Math. Logic 23 (1982), 101–120.

David [83] David, R., $\Delta^1_3$ Reals, Ann. of Pure Appl. Logic 23 (1983), 121–125.

David [89] David, R., A Functorial $\Pi^1_2$-Singleton, Adv. Math. 74 (1989), 258–268.

David–Friedman [85] David, R., Friedman, S. D., Uncountable ZF Ordinals, in: Recursion Theorey, Proc. Sympos. Pure Math. 42, Amer. Math. Soc., Providence, RI 1985, 217–222.

Devlin–Jensen [75] Devlin, K. J., Jensen, R. B., Marginalia to a Theorem of Silver, in: Logic Conference Kiel 1974, Lecture Notes in Math. 499, Springer-Verlag, Berlin–Heidelberg–New York 1975, 115–142.

Donder [85] Donder, H.-D., Another Look at Gap-1 Morasses, in: Proc. Sympos. Pure Math. 42, Amer. Math. Soc., Providence, RI 1985, 223–236.

Donder–Jensen–Stanley [85] Donder, H.-D., Jensen, R. B., Stanley, L. J., Condensation-Coherent Global Square Systems, in: Proc. Sympos. Pure Math. 42, Amer. Math. Soc., Providence, RI 1985, 237–258.

Easton [70] Easton, W. B., Powers of Regular Cardinals, Ann. of Math. Logic 1 (1970), 139–178.

Friedman [81] Friedman, S. D., Uncountable Admissibles II: Compactness, Israel J. Math. 40 (1981), 129–149.

Friedman [82] Friedman, S. D., Uncountable Admissibles I: Forcing, Trans. Amer. Math. Soc. 270 (1982), 61–73.

Friedman [83] Friedman, S. D., Some Recent Developments in Higher Recursion Theory, J. Symbolic Logic 48 (1983), 629–642.

Friedman [84] Friedman, S. D., Infinitary Logic and $0^{\#}$, Proc. Amer. Math. Soc. Summer Inst. at Boulder, Contemp. Math. 31, Amer. Math. Soc., Providence, RI 1984, 99–107.

Friedman [85a] Friedman, S. D., An Immune Partition of the Ordinals, in: Lecture Notes in Math, 1141, Springer-Verlag, Berlin–Heidelberg–New York 1985, 141–147.

Friedman [85b] Friedman, S. D., An Introduction to the Admissibility Spectrum, in: Stud. Logic Found. Math. 114, North Holland, Amsterdam 1985, 129–139.

Friedman [85c] Friedman, S. D., Fine Structure Theory and its Applications, Proc. Sympos. Pure Math. 42, Amer. Math. Soc., Providence, RI 1985, 259–269.

Friedman [87] Friedman, S. D., Strong Coding, Ann. Pure Appl. Logic 35 (1987), 1–98.

Friedman [90] Friedman, S. D., The $\Pi^1_2$-Singleton Conjecture, J. Amer. Math. Soc. 3 (1990), 771–791.

Friedman [94a] Friedman, S. D., A Large $\Pi^1_2$ Set, Absolute for Set Forcings, Proc. Amer. Math. Soc. 122 (1994), 253–256.

Friedman [94b] Friedman, S. D., Iterated Class Forcing, Math. Res. Letters 1 (1994), 427–436.

Friedman [94c] Friedman, S. D., The Genericity Conjecture, J. Symbolic Logic 59 (1994), 606–614.

Friedman [94d] Friedman, S. D., Minimal Universes, Adv. Math. 104 (1994), 59–65.

Friedman [95] Friedman, S. D., Provable $\Pi^1_2$-Singletons, Proc. Amer. Math. Soc. 123 (1995), 2873–2874.

Friedman [97a] Friedman, S. D., Coding without Fine Structure, J. Symbolic Logic 62 (1997), 808–815.

Friedman [97b] Friedman, S. D., The $\Sigma^*$ Approach to the Fine Structure of $L$, Fundamenta Math. 154 (1997), 133–158.

Friedman [98] Friedman, S. D., Generic Saturation, J. Symbolic Logic 63 (1998), 158–162.

Friedman [99a] Friedman, S. D., Strict Genericity, Lecture Notes in Pure and Appl. Math. 203, Marcel Dekker, New York 1999, 129–139.

Friedman [99b] Friedman, S. D., David's Trick, in: Sets and Proofs, London Math. Soc. Lecture Note Ser. 258, Cambridge University Press, Cambridge 1999, 67–71.

Friedman [99c] Friedman, S. D., New $\Sigma^1_3$ Facts, Proc. Amer. Math. Soc. 127 (1999), 3707–3709.

Friedman [99d] Friedman, S. D., Class Forcing, to appear in: Handbook of Set Theory, Kluwer Academic Publishers.

Friedman–Veličković [97] Friedman, S. D., Veličković, B., $\Delta_1$-Definability, Ann. Pure Appl. Logic 89 (1997), 93–99.

Friedman–Woodin [96] Friedman, S. D., Woodin, W. H., $\delta^1_2$ without Sharps, Proc. Amer. Math. Soc. 124 (1996), 2211–2213.

Harrington [77] Harrington, L., Long Projective Wellorderings, Ann. Math. Logic 12 (1977), 1–24.

Harrington–Kechris [77] Harrington, L., Kechris A. S., $\Pi^1_2$-Singletons and $0^\#$, Fundamenta Math. 95 (1977), 167–171.

Jech [78] Jech, T., Set Theory, Pure Appl. Math., Academic Press, New York–San Francisco–London 1978.

Jensen [72] Jensen, R. B., The Fine Structure of the Constructible Hierarchy, Ann. Math. Logic 4 (1972), 229–308.

Jensen–Solovay [70] Jensen, R. B., Solovay, R. M., Some Applications of Almost Disjoint Sets, in: Mathematical Logic and Foundations of Set Theory, North Holland, Amsterdam–New York–Oxford 1970. 84–104.

Kunen [80] Kunen, K., Set Theory: An Introduction to Independence Proofs, Stud. Logic Found. Math. 102, North Holland, Amsterdam–New York–Oxford 1980.

Magidor [90] Magidor, M., Representing Sets of Ordinals as Countable Unions of Sets in the Core Model, Trans. Amer. Math. Soc. 317 (1990), 91–126.

Mansfield [70] Mansfield, R., Perfect Subsets of Definable Sets of Real Numbers, Pacific J. Math. 35 (1970), 451–457.

Martin [77] Martin, D. A., Descriptive Set Theory: Projective Sets, in; Handbook of Mathematical Logic, Stud. Logic Found. Math. 90, North Holland, Amsterdam–New York–Oxford 1977, 783–815.

Martin–Solovay [69] Martin, D. A., Solovay, R. M., A Basis Theorem for $\Sigma^1_3$ Sets of Reals, Ann. of Math. 89 (1969), 138–159.

Moschovakis [80] Moschovakis, Y. M., *Descriptive Set Theory*, Stud. Logic Found. Math. 100, North Holland, Amsterdam–New York–Oxford 1980.

Paris [74] Paris, J. B., Patterns of Indiscernibles, Bull. London Math. Soc. 6 (1974), 183–188.

Sacks [71] Sacks, G. E., Forcing with Perfect Closed Sets, in: Proc. Sympos. Pure Math. 13, Amer. Math. Soc., Providence, RI 1971, 331–355.

Sacks [76] Sacks, G. E., Countable Admissible Ordinals and Hyperdegrees, Adv. Math. 19 (1976), 213–262.

Sacks [90] Sacks, G. E., Higher Recursion Theory, Perspect. Math. Logic, Springer-Verlag, Berlin–Heidelberg–New York 1990.

Shelah [82] Shelah, S., Proper Forcing, Lecture Notes in Math. 940, Springer-Verlag, Berlin–Heidelberg–New York 1982.

Shoenfield [61] Shoenfield, J. R., The Problem of Predicativity, in: Essays on the Foundations of Mathematics, Magnes Press, 1961, 132–142.

Shoenfield [71] Shoenfield, J. R., Unramified Forcing, in: Proc. Sympos. Pure Math. 13, Amer. Math. Soc., Providence, RI 1971, 357–382.

Silver [71] Silver, J. H., Some Applications of Model Theory in Set Theory, Ann. Math. Logic 3 (1971), 45–110.

Solovay [65] Solovay, R. M., $2^{\aleph_0}$ Can Be Anything It Ought To Be, in: The Theory of Models: Proceedings of the 1963 International Symposium at Berkeley, North Holland, Amsterdam–New York–Oxford 1965, 435.

Solovay [67] Solovay, R. M., A Nonconstructible $\Delta^1_3$ Set of Integers, Trans. Amer. Math. Soc. 127 (1967), 50–75.

Solovay [69] Solovay, R. M., On the Cardinality of $\Sigma^1_2$ Sets of Reals, in: Symposium Commemoratory Kurt Gödel, Springer-Verlag, Berlin–Heidelberg–New York 1969, 58–73.

Solovay [70] Solovay, R. M., A Model of Set Theory in which Every Set of Reals is Lebesgue Measurable, Ann. of Math. 92 (1970), 1–56.

Stanley [88] Stanley, M. C., Backwards Easton Forcing and $0^\#$, J. Symbolic Logic 53 (1988), 809–833.

Stanley [94] Stanley, M. C., A $\Pi^1_2$-Singleton Incompatible with $0^\#$, Ann. Pure Appl. Logic 66 (1994), 27–88.

Stanley [97] Stanley, M. C., A Non-Generic Real Incompatible with $0^\#$, Ann. Pure Appl. Logic 85 (1997), 157–192.

Stanley [99] Stanley, M. C., Forcing Closed Unbounded Subsets of $\omega_2$, to appear.

# Index

Extension of conditions in $P^{\infty}$, 94
Hyperclass forcing, 93
Indiscernible preservation, 98
$\infty$, 93
$\mu(S)$, 93
$P_L(\infty)$, 99
$R$ codes the Sat operator, 93
Sat operator, 93
Seq $X$, 94

Homogeneous set, 86

$I$, 52
$I_{\alpha,\beta}$, 91, 100, 101, 113
$I$ enumerated as $\langle i_\alpha \mid \alpha \in \mathrm{ORD}\rangle$, 91
Indiscernible preservation, 62, 90
Iterated class forcing, 175
Almost $\lambda_0$, $\lambda$-periodic, 182
$\Delta$-stratified, 179
$\Delta$-stratified iteration, 180
Densely $\Sigma_1^A$, 176
Examples, 181
$\lambda_0$, $\lambda$-periodic, 182
Long diagonal supports, 180
Short diagonal supports, 177
Stratified forcing, 175
Stratified iteration, 176
Strong witness to stratification, 176
Strong witness to $\Delta$-stratification, 180
Witness to stratification, 175

$J$-hierarchy, 1–5
$J_\alpha$, 2
$J_{\alpha,n}$, 2
$\widetilde{J}_\alpha$, 3
$J$-model, 12
Jensen's Covering Theorem, 55

$\kappa$-chain condition ($\kappa$-cc), 29
$\kappa$-distributive, 29, 37
Killing admissibles, 140

$L$-indiscernibles, 51
$\Lambda(R)$, 66
$\Lambda^*(R)$, 143
$\Lambda_T(R)$, 66
Large cardinal preservation, 86
Lévy hierarchy, 1
Local $\Pi_2^1$-singleton, 129
Long Easton forcing, 41
Long Easton forcing at Successors, 61

Magidor covering, 57
Mahlo cardinal, 42, 86, 204
Minimal coding, 183–201
$[T]$, 186, 194, 196
$[a, t, T]$, 186
$[a, t]$, 184
$[d * T]$, 194
$[t]$, 184
$(a_0, t_0)$ extends $(a_1, t_1)$, $(a_0, t_0) \leq (a_1, t_1)$, 184
$(a_0, t_0, T_0)$ extends $(a_1, t_1, T_1)$, $(a_0, t_0, T_0) \leq (a_1, t_1, T_1)$, 186
$a$ lies on $t$, 184
$a_p, t_p, T_p$, 191
$\mathcal{A}^T$, 196
$\mathcal{A}^t$, 197
$(a, t)$ lies on $T$, 186
$\alpha$ is special, 190
$\alpha(T)$, 196
$\alpha(t)$, 185
Amalgamation, 184, 194
Amalgamation $t(t_a \mid a \in \mathrm{Split}^+(t, k))$, 184
Amalgamation $T(T_t \mid t \in \mathrm{Split}^+(T, \alpha))$, 186
Amalgamation for generalized trees, 186
$B^R \cap \alpha$, 185
$B^R \cap [\omega_1, \alpha)$, 191
Canonical generalized tree, 190
Canonical trees, 184, 189
Canonical $\alpha$-tree, 193
Closed subset of $R^{\delta}_{\mathrm{Even}}$, 194
Closed subset of $R^{\mathrm{Even}}$, 185
Condition in $R^T$, 197